Constructive Approaches to Mathematical Models

Contributors

W. N. Anderson, Jr.
I. Babuška
A. Ben-Israel
A. Ben-Tal
Garrett Birkhoff
C. E. Blair
Raoul Bott
D. G. Bourgin
P. L. Chow
C. V. Coffman
Bernard D. Coleman
R. J. Duffin
B. C. Eaves
G. J. Fix
David Gale
Jerome A. Goldstein
P. R. Gribik
M. D. Gunzburger
William W. Hager
Dov Hazony
R. G. Jeroslow
L. A. Karlovitz
M. Lachance
C. E. Lemke
T. D. Morley
R. W. Newcomb
R. A. Nicolaides
J. Osborn
Jong-Shi Pang
L. E. Payne
R. N. Pederson
E. E. Rosinger
E. B. Saff
E. A. Saibel
James T. Sandefur, Jr.
D. H. Shaffer
G. E. Trapp
R. S. Varga
H. F. Weinberger
Douglass J. Wilde
A. H. Zemanian
Clarence Zener

Constructive Approaches to Mathematical Models

Proceedings of a Conference
in Honor of R. J. Duffin
Pittsburgh, Pennsylvania
10–14 July 1978

edited by
C. V. COFFMAN and G. J. FIX
DEPARTMENT OF MATHEMATICS
CARNEGIE-MELLON UNIVERSITY
PITTSBURGH, PENNSYLVANIA

1979

ACADEMIC PRESS
A Subsidiary of Harcourt Brace Jovanovich, Publishers
New York London Toronto Sydney San Francisco

ACADEMIC PRESS, INC.
111 Fifth Avenue, New York, New York 10003

United Kingdom Edition published by
ACADEMIC PRESS, INC. (LONDON) LTD.
24/28 Oval Road, London NW1 7DX

Library of Congress Cataloging in Publication Data
Main entry under title:

Constructive approaches to mathematical models.

Includes bibliographies.
1. Mathematical models--Congresses.
I. Duffin, Richard James. II. Coffman,
Charles Vernon. III. Fix, George J.
QA401.C66 001.4'24 79-51673
ISBN 0-12-178150-X

PRINTED IN THE UNITED STATES OF AMERICA

79 80 81 82 9 8 7 6 5 4 3 2 1

Contents

Part I General Talks

Some Problems Arising from Mathematical Models

R. J. DUFFIN

Some Recollections from 30 Years Ago

RAOUL BOTT

Transforms

D. G. BOURGIN

Part II Graphs and Networks

Matrix Operations Induced by Electrical Network Connections —A Survey

W. N. ANDERSON, JR., AND G. E. TRAPP

Pulse Control in Dynamic Systems

DOV HAZONY

MOS Neuristor Lines

R. W. NEWCOMB

Some Models in the Social and Behavioral Sciences Based on Proportioning Networks

A. H. ZEMANIAN

Part III Mathematical Programming

A Helly-Type Theorem and Semiinfinite Programming

A. BEN-TAL, E. E. ROSINGER, AND A. BEN-ISRAEL

Lagrange Dual Programs with Linear Constraints on the Multipliers

C. E. BLAIR AND R. G. JEROSLOW

A View of Complementary Pivot Theory (or Solving Equations with Homotopies)

B. C. EAVES

Selected Applications of Semiinfinite Programming

P. R. GRIBIK

Part IV Differential Equations

Analysis of Finite Element Methods Using Mesh Dependent Norms

I. BABUŠKA AND J. OSBORN

The Fundamental Mode of Vibration of a Clamped Annular Plate Is Not of One Sign

C. V. COFFMAN, R. J. DUFFIN, AND D. H. SHAFFER

On Explicit Norm Inequalities and Their Applications in Partial Differential Equations

L. E. PAYNE

Genetic Wave Propagation, Convex Sets, and Semiinfinite Programming

H. F. WEINBERGER

Part V Mathematical Models

Analysis of a Stochastic Reynolds Equation and Related Problems

P. L. CHOW AND E. A. SAIBEL

Optimization Methods for Solving Nonoptimization Problems

DAVID GALE

CLARENCE ZENER

Part VI Related Areas

The Algebra of Multivariate Interpolation

GARRETT BIRKHOFF

On the Optimal Shape of a Flexible Cable

BERNARD D. COLEMAN

Newton's Method for Polynomials with Real Roots and Their Divided Differences

R. N. PEDERSON

List of Contributors

Numbers in parentheses refer to the pages on which the authors' contributions begin.

W. N.Anderson, Jr. (53), Department of Mathematics, East Tennessee State University, Johnson City, Tennessee 37601

I. Babuška (251), Institute for Physical Science and Technology and Department of Mathematics, University of Maryland, College Park, Maryland 20742

A. Ben-Israel (127), Department of Mathematics, University of Delaware, Newark, Delaware 19711

A. Ben-Tal (127), Department of Computer Science, TECHNION-Israel Institute of Technology, Haifa, Israel

Garrett Birkhoff (345), Department of Mathematics, Harvard University, Cambridge, Massachusetts 02138

C. E. Blair (137), Department of Business Administration, University of Illinois, Urbana, Illinois 61801

Raoul Bott (33), Department of Mathematics, Harvard University, Cambridge, Massachusetts 02138

D. G. Bourgin (41), Department of Mathematics, University of Houston, Houston, Texas 77004

P. L. Chow (321), Department of Mathematics, Wayne State University, Detroit, Michigan 48202

C. V. Coffman (267), Department of Mathematics, Carnegie-Mellon University, Pittsburgh, Pennsylvania 15213

Bernard D. Coleman (365), Department of Mathematics, Carnegie-Mellon University, Pittsburgh, Pennsylvania 15213

R. J. Duffin (3, 267), Department of Mathematics, Carnegie-Mellon University, Pittsburgh, Pennsylvania 15213

B. C. Eaves (153), Department of Operations Research, Stanford University, Stanford, California 94305

G. J. Fix (375), Department of Mathematics, Carnegie-Mellon University, Pittsburgh, Pennsylvania 15213

David Gale (327), Department of Industrial Engineering and Operations Research, University of California, Berkeley, Berkeley, California 94729

Jerome A. Goldstein (395), Department of Mathematics, Tulane University, New Orleans, Louisiana 70118

P. R. Gribik (171), Computer System and Services Department, Pacific Gas and Electric Company, San Francisco, California 94105

M. D. Gunzburger (375), Department of Mathematics,University of Tennessee, Knoxville, Tennessee 37916

William W. Hager (189), Department of Mathematics, Carnegie-Mellon University, Pittsburgh, Pennsylvania 15213

Dov Hazony (75), Department of Electrical Engineering and Applied Physics, Case Western Reserve University, Cleveland, Ohio 44106

R. G. Jeroslow (137), Department of Mathematics, Carnegie-Mellon University, Pittsburgh, Pennsylvania 15213, and College of Industrial Management, Georgia Institute of Technology, Atlanta, Georgia 30332

L. A. Karlovitz (413), Georgia Institute of Technology, School of Mathematics, Atlanta, Georgia 30332

M. Lachance (421), Department of Mathematics, University of South Florida, Tampa, Florida 33620

C. E. Lemke (203), Department of Mathematical Sciences, Rensselaer Polytechnic Institute, Troy, New York 12181

T. D. Morley (439), Department of Mathematics, University of Illinois, Urbana, Illinois 61801

R. W. Newcomb (87), Electrical Engineering Department, Applied Mathematics Program, University of Maryland, College Park, Maryland 20742

R. A. Nicolaides (375), Department of Mathematics, Carnegie-Mellon University, Pittsburgh, Pennsylvania 15213

J. Osborn (251), Department of Mathematics, University of Maryland, College Park, Maryland 20742

Jong-Shi Pang (231), Graduate School of Industrial Administration, Carnegie-Mellon University, Pittsburgh, Pennsylvania 15213

L. E. Payne (279), Department of Mathematics, Cornell University, Ithaca, New York 14853

R. N. Pederson (447), Department of Mathematics, Carnegie-Mellon University, Pittsburgh, Pennsylvania 15213

E. E. Rosinger (127), Department of Computer Science, TECHNION-Israel Institute of Technology, Haifa, Israel

E. B. Saff (421), Department of Mathematics, University of South Florida, Tampa, Florida 33620

E. A. Saibel (321), Engineering Sciences Division, U.S. Army Research Office, Research Triangle Park, North Carolina 27709

James T. Sandefur, Jr. (395), Department of Mathematics, Georgetown University, Washington, D.C. 20057

D. H. Shaffer (267), Westinghouse Research Laboratories, Pittsburgh, Pennsylvania 15235

G. E. Trapp (53), Department of Computer Sciences, West Virginia University, Morgantown, West Virginia 26506

R. S. Varga (421), Department of Mathematics, Kent State University, Kent, Ohio 44242

H. F. Weinberger (293), School of Mathematics, University of Minnesota, Minneapolis, Minnesota 55455

Douglass J. Wilde (243), Mechanical Engineering Department, Stanford University, Stanford, California 94305

A. H. Zemanian (113), Department of Electrical Sciences, State University of New York at Stony Brook, Stony Brook, New York 11794

Clarence Zener (335), Carnegie-Mellon University, Pittsburgh, Pennsylvania 15213

Preface

The idea for a conference in honor of Dick Duffin occurred spontaneously among several of his close associates and former students. There was no objective reason for the conference, i.e., it was not given to celebrate a particular birthday, nor is Dick Duffin retiring. On the contrary, there was simply a general desire to honor this profound mathematician, inspiring teacher, and noble human being. The conference was organized around the four areas in which Dick Duffin has made substantial contributions plus a special session that covered closely related areas. The response to the call for papers was overwhelming. There were 108 talks during the week, and 111 registered participants. The editors profoundly regret the inevitable space limitations that prevented including several of the excellent contributed papers. We also wish to express our gratitude to R. Bott and G. Dantzig for their stimulating evening sessions.

One amusing sidelight that arose during the special evening session chaired by Dick Duffin was a series of "Duffinisms" with at least one for every letter in the alphabet. A selected list of these follows.

DUFFINISMS

A anagraph, almost definite operators, automatic scrutation, asymmetric norm, anti-entropy
B bipolynomials, brothers Markov
C connection boxes, collipses, canonical programs, confluent, confluences, conjoined networks, connected transformations, constrained inverse
D duality gap, degree of difficulty, discounted least squares, democracy principle
E extrapolator, extremal width, even network, elliptic flow systems
F frames, four current conjecture, flattened Bessel functions, flattened exponentials
G geometric programming, gyration operator
H homogenized programs
I infinite programs, impedance potential
J junctors, just vectors, just subspaces
K Kirchhoffian subspaces
L lower network, length–width inequality
M multigroups, most action principle
N network discriminant, "Dear Sirs: Nash is a mathematical genius." (letter of recommendation)
O odd network, overconsistent, over-damping, operator lattice
P posynomial, pre-Columbian Banach spaces, perfect duality, pseudo-powers, powerful numbers, positively connected, panharmonic, plus independence, plus-up independence, potentially definite, pseudo-Hamiltonians, parallel addition
Q quasi-circuits, quasi-linear replacement
R routers, reversed geometric programs, regular programs, reliable networks
S signomials, scrutators, superconsistent, subconsistent, superfeasible, Sylvester unisignants, synthetic networks, structure operator, Schur polynomials
T tapered fins, tapered walls
U ultraconsistent, upper network, up independence, uncover the degree (of a rational matrix function)
V vectrix
W Wang algebra
X exponomial
Y Yakawan potential
Z Duffin, Peterson, Zener, *Geometric Programming* (book)

Part I

GENERAL TALKS

Some Problems Arising from Mathematical Models†

R. J. DUFFIN

0. INTRODUCTION

The development of mathematics has often been aided by the use of models from science and technology. There are three main reasons why models help: (i) attention is focused on significant problems; (ii) the intuition is aided in perceiving complex relations; (iii) new concepts are suggested. This paper describes problems arising from models which have interested me. The models come from physics, chemistry, engineering, and economics.

1. THE DIRICHLET PROBLEM FOR THE WAVE EQUATION‡

We are given two photographs of a vibrating string, one at time $t = 0$ and another at a later time $t = \alpha$, as illustrated in Fig. 1. Is it possible to determine the state at intermediate times? This puzzle led David Bourgin and me to study the Dirichlet problem for the wave equation. The wave equation is

$$\frac{\partial^2 y}{\partial x^2} = \frac{\partial^2 y}{\partial t^2},$$

and the region of concern is the rectangle R, $0 \leqq x \leqq 1$, $0 \leqq t \leqq \alpha$. There are solutions of the form $y = \sin(\pi n x)\sin(\pi n t)$ for any integer n. Then if α is a rational number, say $\alpha = m/n$, it is seen that given y at $t = 0$ and $t = \alpha$ does not determine y at intermediate times. On the other hand, we found that if α was irrational and y was of class C^2 in the rectangle R, then y was uniquely determined.

† Sections 0–15 reprinted with permission of American Mathematical Society from Duffin, Some problems of mathematics and science, *Bull. Amer. Math. Soc.* **80** (1974), 1053–1070, © 1974 by the American Mathematical Society.

‡ See [1].

ISBN 0-12-178150-X

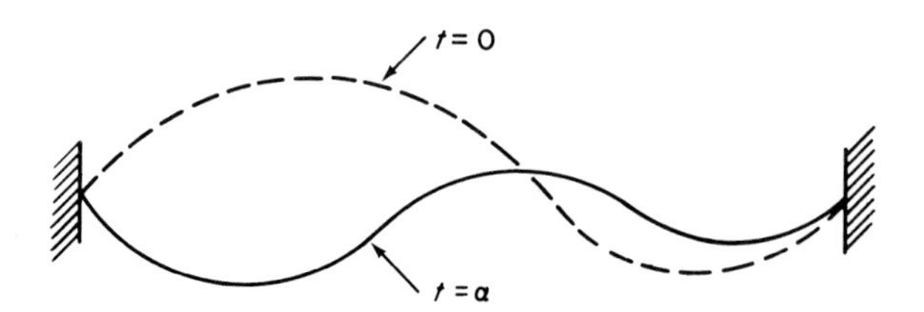

Fig. 1. The vibrating string.

Thus the irrationality of α established the uniqueness of the Dirichlet problem. To establish existence we had to assume that α could not be approximated too rapidly by rationals. For example, suppose that for suitable positive constant A and integer k, $|\alpha - m/n| > A/n^k$ for all integers m and n. Thus if the values of y at times $t = 0$, $t = \alpha$ are functions of class C^{3+k}, then there exists a solution of the wave equation at intermediate times. It is of interest to note that the above inequality holds if α is an algebraic number of degree k.

This mathematical problem which we solved was inspired by the physical model of a vibrating string. However it is not clear what the mathematical solution really implies for the physical situation. Thus what is the physical significance of the statement that a time interval may be an irrational number but not a rational number?

2. HEAVY PHOTON†

Maxwell's equations give the wave equation for the photon. The photon has zero rest mass. Proca proposed the following modification of Maxwell's equations for a particle of rest mass m,

$$\frac{\partial \varphi_j}{\partial x_i} - \frac{\partial \varphi_i}{\partial x_j} = \mu f_{ij}, \qquad \sum_1^4 \frac{\partial f_{ij}}{\partial x_i} = \mu \varphi_j. \tag{1}$$

Here $\mu^2 = -m^2c^2$ and $x_4^2 = -c^2t^2$. These equations are somewhat unwieldy to manipulate. Taking a hint from Dirac, I found that the system of linear equations (1) may be expressed as a single vector wave equation

$$\sum_1^4 \beta_i \frac{\partial \psi}{\partial x_i} = \mu \psi, \tag{2}$$

Here ψ is the column vector with ten components

$$\psi = (\varphi_1, \varphi_2, \varphi_3, \varphi_4, f_{14}, f_{24}, f_{34}, f_{23}, f_{31}, f_{12}), \tag{3}$$

† See [2, 3].

and β_1, β_2, β_3, and β_4 denote 10 by 10 constant matrices. They were found to satisfy the following identities:

$$\beta_i^3 = \beta_i, \qquad \beta_i\beta_j^2 + \beta_j^2\beta_i = \beta_i, \qquad \beta_i\beta_j\beta_k + \beta_k\beta_j\beta_i = 0. \tag{4}$$

These have since been termed the Duffin, Petiau, Kemmer commutation relations. Let

$$L_{ij} = \left(x_i\frac{\partial}{\partial x_j} - x_j\frac{\partial}{\partial x_i}\right) - (-1)^{1/2}\hbar S_{ij}, \tag{5}$$

where $S_{ij} = \beta_i\beta_j - \beta_j\beta_i$. Then the angular momentum operator L_{ij} is seen to commute with the operator of Eq. (2). Moreover, the matrix $(-1)^{1/2}S_{ij}$ has eigenvalues 1, 0, and -1. The physical interpretation of this is that the Proca particle has spin 1.

There are various other properties of the β matrices: If the β are transformed as components of a four vector, then the commutation relations are invariant under a Lorentz transformation. Again the β generate an interesting semisimple algebra.

The conclusion to be drawn from this study is that the matrices implicit in a system of partial differential equations may give rise to an algebraic formalism both elegant and significant.

3. BERNSTEIN INEQUALITY FOR NONANALYTIC FUNCTIONS†

The following theorem of S. Bernstein plays a central role in approximation theory.

A. *Let $p(x)$ be a trigonometric polynomial of degree n,*

$$p(x) = \sum_0^n (a_\nu \cos \nu x + b_\nu \sin \nu x).$$

If $|p(x)| \leqq 1$ for all x, then $|dp/dx| \leqq n$.

It seemed desirable to extend this theorem to nonanalytic functions. To do this Albert Schaeffer and I proved the following theorem.

B. *Let f be a function such that for some integer m and for all x,*

$$[f(x)]^2 \leqq 1, \qquad [f^{(m)}(x)]^2 + [f^{(m-1)}(x)]^2 \leqq 1.$$

Then for $k = 1, 2, \ldots, m$,

$$[f^{(k)}(x)]^2 + [f^{(k-1)}(x)]^2 \leqq 1.$$

† See [4].

Proof. The proof begins by considering the case $m = 2$. Consider the function

$$\varphi(x) = [f'(x)]^2 + [f(x)]^2.$$

If $\varphi(x) \leqq 1$ does not hold, then φ has a local maximum at some point x_0 and $\varphi(x_0) > 1$. Thus $\varphi'(x_0) = 2f'(x_0)[f(x_0) + f''(x_0)] = 0$. If the first factor is zero, then $\varphi(x_0) = [f(x_0)]^2 \leqq 1$. If the second factor is zero, then $\varphi(x_0) = [-f''(x_0)]^2 + [f'(x_0)]^2 \leqq 1$. This contradiction proves the theorem in the case $m = 2$. The proof is completed by induction.

Theorem B proves Bernstein's inequality. To see this let $f(x) = p(x/\lambda)$ where λ is some constant greater than n. It follows that as $m \to \infty$, $f^{(m)}(x) \to 0$ uniformly. Thus for m sufficiently large, the conditions of Theorem B are satisfied so

$$1 \geqq |f'(x)| = |p'(x/\lambda)/\lambda|.$$

Allowing $\lambda \to n$ shows that $|p'| \leqq n$. ■

4. A REFINEMENT OF MARKOFF'S INEQUALITY†

The following theorem of A. Markoff plays an important role in approximation theory.

Let $p(x)$ *be a polynomial of degree* n,

$$p(x) = \sum_0^n a_i x^i.$$

If

(a) $$|p(x)| \leqq 1 \quad \text{for} \quad -1 \leqq x \leqq 1,$$

then $|p'(x)| \leqq n^2$ *in the same interval.*

Albert Schaeffer and I found a refinement of this theorem to the effect that the conclusion holds when (a) is replaced by the weaker condition

(a*) $$|p(\cos k\pi/n)| \leqq 1 \quad \text{for} \quad k = 0, 1, \ldots, n.$$

The proof employed rather involved complex variable techniques.

In recent years a more general approach to such problems has been developed by use of functional analysis. Is it possible, by such methods, to extend the refined Markoff inequality to a wider class of functions?

† See [5].

5. REPRESENTATION OF FOURIER INTEGRALS AS SUMS†

Fourier integrals are an indispensible tool in almost all branches of applied mathematics. Since these integrals are difficult to evaluate numerically, I sought to replace the integrals by series.

Given an arbitrary function $\varphi(x)$, let functions $f(x)$ and $g(x)$ be defined by the sums:

$$f(x) = \varphi(x) - (1/3)\varphi(x/3) + (1/5)\varphi(x/5) - \cdots, \tag{1a}$$

$$g(x) = (1/x)\varphi(1/x) - (1/x)\varphi(3/x) + (1/x)\varphi(5/x) - \cdots. \tag{1b}$$

Then $f(x)$ and $g(x)$ are Fourier sine transforms; that is,

$$g(x) = \int_0^\infty \sin\left(\frac{\pi x t}{2}\right) f(t)\, dt, \tag{2a}$$

$$f(x) = \int_0^\infty \sin\left(\frac{\pi x t}{2}\right) g(t)\, dt, \tag{2b}$$

under mild restriction on φ.

An intuitive proof of this can be based on the well-known Poisson summation formula for the sine transform,

$$f(x) - f(3x) + f(5x) - \cdots = (1/x)g(1/x) - (1/x)g(3/x) + (1/x)g(5/x)\cdots. \tag{3}$$

This holds for sine transform pairs (f, g). If (1a) and (1b) are substituted in (3), it is seen that both sides reduce to

$$\sum_1^\infty \sum_1^\infty \frac{\alpha_n \alpha_m}{n} \varphi\left(\frac{mx}{n}\right), \tag{4}$$

where $\alpha_1 = 1, \alpha_2 = 0, \alpha_3 = -1$, etc. Since x is arbitrary, this is an indication of the validity of (1) and (2).

The left side of (3) is a Möbius series $M(x)$,

$$M(x) = f(x) - f(3x) + f(5x) - \cdots. \tag{5}$$

The inversion of this series is

$$f(x) = M(x) - \mu_3 M(3x) + \mu_5 M(5x) - \cdots, \tag{6}$$

where μ_n is the well-known Möbius symbol. Applying this inversion to the right side of (3) gives

$$f(x) = \sum_{n=1}^\infty \sum_{m=1}^\infty \frac{\mu_n \alpha_n \alpha_m}{nx} g\left(\frac{m}{nx}\right). \tag{7}$$

This is a direct representation of the Fourier sine transform as a double sum.

† See [6–9].

6. HADAMARD'S CONJECTURE ON THE CLAMPED PLATE†

The following conjecture was made by Hadamard in his 1908 prize memoir on the elastic plate.

A. *If a perpendicular force is applied at some point of a thin, flat, elastic plate which is rigidly clamped on its boundary, then the displacement of the plate is of one sign at all points.*

Suppose that the plate is the (x, y) plane; then a displacement $w(x, y)$ of the plate satisfies the biharmonic equation

$$\left(\frac{\partial^2}{\partial x^2} + \frac{\partial^2}{\partial y^2}\right)^2 w = 0 \tag{1}$$

in a region R where no forces are applied. By an argument appealing to the reciprocity principle (conservation of energy), it may be shown that Hadamard's conjecture is equivalent to the following assertion.

B. *Suppose a function w is biharmonic in a region R, and that on the boundary $w \geqq 0$ and $-\partial w/\partial n \geqq 0$, where n is the exterior normal. Then $w \geqq 0$ throughout R.* (This statement could be interpreted as a maximum principle for biharmonic functions.)

The assertion (B) is false, as the following counterexample shows. First it is observed that if q is any complex constant, then $e^{iqx} \sinh qy$ and $e^{iqx} \cosh qy$ are both harmonic functions. Moreover if h is a harmonic function, yh is biharmonic. Thus

$$f = e^{iqx}(-\cosh qy + y \coth q \sinh qy)$$

is a biharmonic function. If $q = 1.12 + i2.10$, then $q + \cosh q \sinh q = 0$, and it is easy to check that $f = 0$ and $\partial f/\partial y = 0$ on the lines $y = \pm 1$. Let $w = \mathrm{Re}(f)$. Then $w(x, 0) = -e^{-2.1x} \cos 1.12x$, so f has nodal lines which intersect the x axis at $1.12x = \pi/2 + n\pi$. Let the region R be bounded by

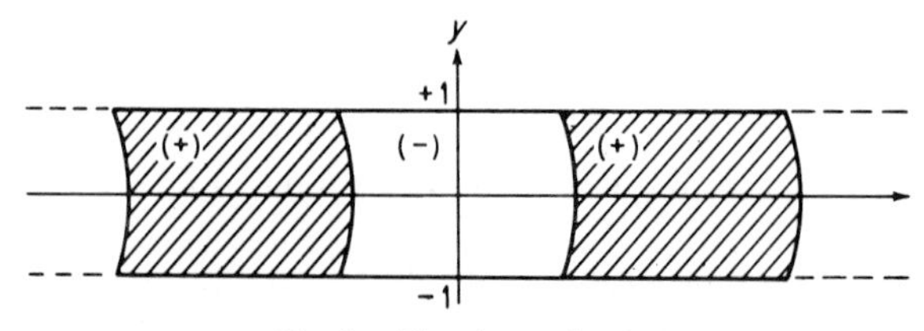

Fig. 2. The clamped strip.

† See [10, 11].

the lines $y = \pm 1$ and the nodal lines which intersect the x axis at $1.12x = \pm 3\pi/2$. Then R is divided into three congruent sections as is shown in Fig. 2. Then $w \leqq 0$ in the central section and $w \geqq 0$ in the others. Thus $w = 0$ and $-\partial w/\partial n \geqq 0$ on the boundary of R. However $w(0, 0) < 0$, and this contradicts statement (B). Hence statements (A) and (B) are both false.

The question still remained of whether or not there is any correct analog of the maximum principle for the biharmonic equation. The following is one such principle.

C. *Let w be biharmonic in a region R, let (a, b) be a point in R, and let r be the vector with components $x - a$, $y - b$. Then*

$$w(a, b) \leqq \max_{\partial R}[w - r \cdot \nabla w + r^2 \Delta w/4],$$

where ∂R denotes the boundary of R.

Proof. A short calculation shows that if w is biharmonic, then $w - r \cdot \nabla w + r^2 \Delta w/4$ is harmonic. Thus the inequality follows from the classical maximum principle. ∎

Presumably such maximum principles for biharmonic functions could have applications in elasticity.

7. SZEGÖ'S CONJECTURE ON THE CLAMPED PLATE†

The normal modes of vibration of a clamped plate satisfy the biharmonic wave equation

$$\Delta^2 w = \lambda^4 w. \tag{1}$$

The boundary conditions are, of course,

$$w = 0, \qquad \frac{\partial w}{\partial n} = 0. \tag{2}$$

Szegö assumed that the gravest mode of vibration is free of nodal lines. Under this hypothesis he proved that of all plates of a given area, the circular plate has the gravest tone.

Studies made in collaboration with Alfred Schild and Douglas Shaffer indicate that Szegö's conjecture is not universally valid. In particular we

† See [12–16].

solved Eq. (1) for a ring-shaped plate. In this geometry the wave equation is separable, and solutions have the form

$$w = f_n(r) \cos n\theta. \tag{3}$$

The radial part $f_n(r)$ is of the form

$$f_n(r) = aJ_n(\lambda r) + bY_n(\lambda r) + cI_n(\lambda r) + dK_n(\lambda r),$$

where J_n, Y_n, I_n, and K_n are standard notation for the Bessel functions. The constants a, b, c, d are chosen to satisfy the boundary condition (2) on both the inner and outer circles.

We found that if the outer circle had a diameter of over 715 times the diameter of the inner circle, then the gravest mode of vibration has a diametral nodal line. This corresponds to $n = 1$ in Eq. (3). We also found the conjecture to be false in a simply connected domain.

8. THE WANG ALGEBRA OF NETWORKS†

Consider a network of electrical conductors such as shown in the figure. To determine the joint conductance of the network, one could set up Kirchhoff's equations and solve for the current flow through the battery.

K. T. Wang managed an electrical power plant in China, and in his spare time sought simple rules for solving the network equations. Wang's rules were published in [21]. Wang could not write in English so his paper was actually written by his son, then a college student. Raoul Bott and I recognized that Wang's rules actually define an algebra. We restated the rules as three postulates for an algebra:

$$xy = yx, \qquad x + x = 0, \qquad xx = 0.$$

To apply the Wang algebra to the network shown in Fig. 3, let the conductances of the various branches be a, b, c, d, and e. Also regard these symbols as independent generators of a Wang algebra. A *star element* of

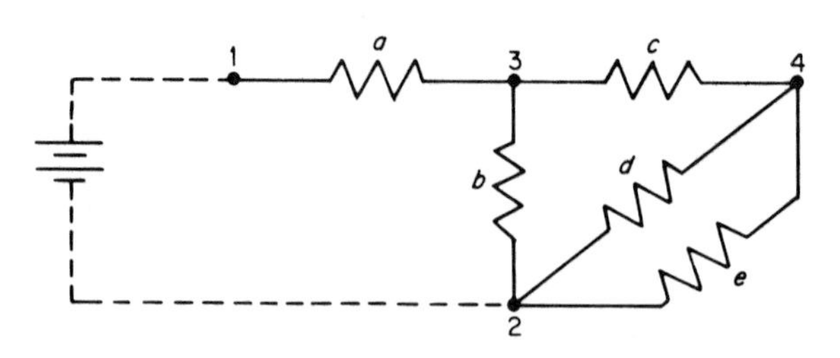

Fig. 3. A simple network.

† See [17–21].

the algebra is defined as the sum of the branches meeting at a node. Thus the star element at node 3 is $a + b + c$.

An algorithm for finding the joint conductance between nodes 1 and 2 follows. First form P, the Wang product of all star elements except those at nodes 1 and 2. Thus

$$P = (a + b + c)(c + d + e).$$

Using the postulates gives

$$P = ac + ad + ae + bc + bd + be + cd + ce.$$

Next form the Wang product T of all stars except one. Then

$$T = aP = abc + abd + abe + acd + ace. \qquad \text{(All trees!)}$$

Then the joint conductance K between nodes 1 and 2 is given as the ratio

$$K = \frac{T}{P} = \frac{abc + abd + abe + acd + ace}{ac + ad + ae + bc + bd + be + cd + ce}.$$

The Wang algebra has interesting and important connections with matroid theory, totally unimodular matrices, and Grassmann algebra. In fact Wang algebra is Grassmann algebra over the mod 2 field. The trees of any graph are given by the above algorithm for T.

9. SAMPLING OF PARTICLE SIZE BY PLANAR SECTIONING†

In some aluminum–silicon alloys, most of the silicon is distributed as small particles throughout an aluminum matrix. It is of importance in metallurgy to know the distribution of particle size. However it is very difficult to observe directly the size distribution. Instead, linear or planar samples are observed from a cross section of the alloy as shown in Fig. 4. The problem thereby posed is the determination of the true size distribution of the particle from the observation of linear or planar samples.

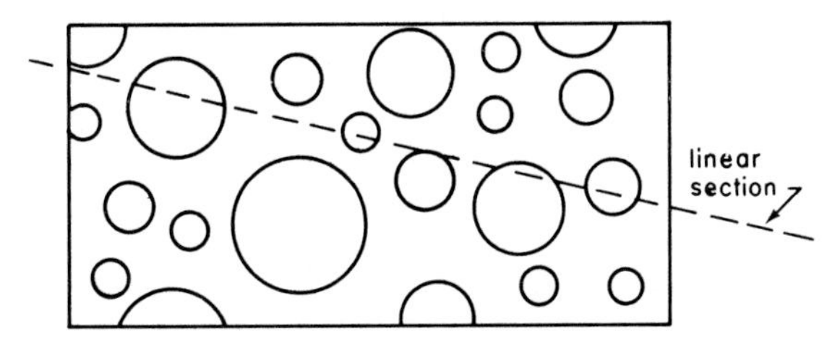

Fig. 4. A planar section of particles.

† See [22–24].

In a study made in collaboration with Russell Meussner and Frederick Rhines, it was found that if the particles are assumed to be spherical, then the size distribution of linear or planar sections is related to the true size distributions by certain integral equations. These integral equations are analogous to Abel's equation and the fractional integration of Herman Weyl [22], and it proved possible to obtain the resolvent kernels.

The following distribution functions were related:

(a) The sphere distribution function $G_3(s)$ is the average number of spheres per cubic centimeter having diameter greater than s.

(b) The circle distribution function $G_2(s)$ is the average number of circles per square centimeter having diameter greater than s. The circles are the intersection of a plane with the spherical particles.

(c) The segment distribution function $G_1(s)$ is the average number of segments per centimeter having length greater than s. The segments are the intersection of a line with the spherical particles.

Theorem 1. *The cumulative distribution functions are related by the following Stieltjes integrals*:

$$G_2(s) = -\int_s^\infty (n^2 - s^2)^{1/2}\, dG_3(n), \tag{1a}$$

$$G_3(s) = -\frac{2}{\pi}\int_s^\infty (n^2 - s^2)^{-1/2}\, dG_2(n), \tag{1b}$$

$$G_1(s) = -\frac{\pi}{4}\int_s^\infty (n^2 - s^2)\, dG_3(n), \tag{2a}$$

$$G_3(s) = -(2/\pi s)\, dG_1/dS, \tag{2b}$$

$$G_1(s) = -\int_s^\infty (n^2 - s^2)^{1/2}\, dG_2(n), \tag{3a}$$

$$G_2(s) = -\frac{2}{\pi}\int_S^\infty (n^2 - s^2)^{-1/2}\, dG_1(n). \tag{3b}$$

A distribution function may be termed Gaussian if it has the form $G(s) = A\exp(-ks^2)$, where A and k are constants. (This is not the same as the Gaussian density function.)

Theorem 2. *If any two of the distribution functions G_1, G_2, and G_3 are proportional, then they are all Gaussian. If any one of them is Gaussian, all are Gaussian.*

The proof of Theorem 2 follows from the relations of Theorem 1.

10. RAYLEIGH QUOTIENT FOR DISSIPATIVE SYSTEMS†

Consider a mechanical system of n degrees of freedom vibrating about a position of static equilibrium. If the vibrations are small then Newton's equation can be written in the linear form

$$A\frac{d^2Q}{dt^2} + CQ = 0,$$

where A and C are n by n symmetric positive definite matrices of constants. A normal mode of motion is of the form $Q = qe^{iwt}$, where the vector q is independent of the time, and w is the frequency of vibration. Thus $w^2Aq - Cq = 0$. The Rayleigh quotient is

$$R = [c(v)/a(v)]^{1/2},$$

where a and c are the quadratic forms (Av, v) and (Cv, v). The minimum frequency satisfies the relation

$$w = \min_v [a(v)/c(v)]^{1/2}$$

for arbitrary real vectors v. The maximum frequency satisfies a similar relation.

When frictional forces are introduced, Newton's equations take the form

$$A\frac{d^2Q}{dt^2} + B\frac{dQ}{dt} + CQ = 0,$$

where B is also a symmetric positive definite matrix. Again let $a(v)$, $b(v)$, and $c(v)$ denote the corresponding quadratic forms. A solution of the form qe^{-kt} satisfies the relation $k^2a(q) - kb(q) + c(q) = 0$. To make sure that k was real, I defined an *overdamped system* by the relation

$$b^2(v) - 4a(v)c(v) > 0$$

for arbitrary real vectors v. The nonlinear eigenvalue k corresponds to a motion of exponential decay (like radioactivity). It was found that the correspondent of the Rayleigh quotient is the functional

$$R(v) = [b \pm (b^2 - 4ac)^{1/2}]/2a.$$

Then the largest decay constant is given by

$$k = \max_v \left[\frac{b + (b^2 - 4ac)^{1/2}}{2a}\right].$$

† See [25, 26].

The smallest decay constant satisfies

$$k = \min_{v} \left[\frac{b - (b^2 - 4ac)^{1/2}}{2a} \right].$$

A similar theory holds when the frictional forces are replaced by gyroscopic forces. Then the matrix B is skew symmetric.

11. OPTIMIZATION OF COOLING FINS†

A common problem of heat transfer is the design of machinery so that the structure can dissipate excess heat. For example cooling fins are used on cylinders of air-cooled engines as shown in Fig. 5. Suppose for such an example that the fin is not permitted to exceed a given weight. It is not difficult to see that the fin should taper, narrowing in the direction of heat flow. The optimum design problem arising may be phrased in this way—find the taper, a thickness function $p(x)$, which gives the maximum dissipation of heat for a given weight of fin.

If the fin is of constant width, the weight of the fin is proportional to $\int_0^L p(x)\,dx$, where L is the length of the fin. The heat dissipated to the air is proportional to $\int_0^L u(x)\,dx$, where $u(x)$ is the temperature of the fin relative to the air. Of course, u has a prescribed value u_0 on the cylinder, $x = 0$. If the fin is thin the equation of heat conduction has the form

$$\frac{d}{dx}\left[p(x)\frac{du}{dx}\right] = cu,$$

where c is constant.

Conceivably this problem could be treated by solving the differential equation for various choices of p and L and retaining the best solution.

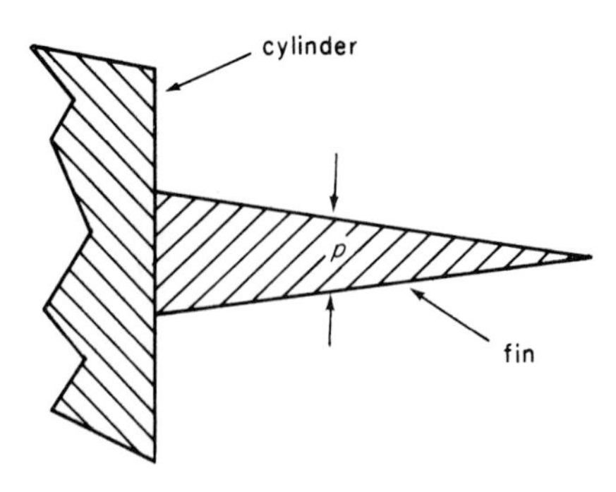

Fig. 5. Cooling fin cross section.

† See [27–31].

Such an approach would be very difficult to implement. Instead the differential equation was regarded as the Euler equation of a variational problem. Then a grand variational problem was set up in which both the temperature function u and the taper function p were variables. The Euler equation for this grand variational problem proved to be very simple. The condition on the temperature was simply *that the temperature gradient should have constant magnitude at all points of the fin.* This relation greatly simplifies the equation of heat conduction. It then results that the taper p is obtained by quadrature and the design problem is completely solved.

The above condition on the gradient was first obtained for the above special geometry. Later more general models were used and the same criterion was found to hold. The gradient condition might be stated as a "democracy" criterion thus: *To obtain the maximum dissipation of heat from a cooling fin of a given weight, the fin should be so proportional that each part of the fin carries the same heat current.*

12. PREDICTION ALGORITHMS BASED ON DISCOUNTED LEAST SQUARES†

In applied science it is often required to make extrapolations into the future based on observations obtained at regular time intervals in the past. A standard approach to such a problem is to extrapolate with a polynomial fitted to the data by least squares.

Theodore Schmidt and I were presented with such a situation where the observations of the distant past should be given less weight than those more recent. This discounting can be achieved by use of an exponential weight factor. Thus suppose that $y_1, y_2, y_3, \ldots$ is a sequence of real numbers giving the observation at previous times $x = 1, 2, 3, \ldots$. The central problem is to predict y_0, the value at $x = 0$. Let $p(x)$ be a polynomial, say,

$$p(x) = a + bx + cx^2.$$

Then the coefficients are chosen to minimize

$$E = \sum_{n=1}^{\infty} \theta^n [y_n - p(n)]^2.$$

Here the constant $\theta < 1$ is the discount factor. Then the predicted value of y_0 is defined as $y_0{}^* = p(0)$.

When the minimization is carried out, a "long formula" of the following type results:

$$y_0^* = \sum_{n=1}^{\infty} Q_n y_n.$$

† See [32–36].

Here the constants Q_n can be obtained explicitly. However there is a much better algorithm termed "the short formula" given as

$$y_0^* = 3(y_1 + \theta\delta_1) - 3(y_2 + \theta^2\delta_2) + (y_3 + \theta^3\delta_3).$$

Here $\delta_k = y_k^* - y_k$, where y_k is the predicted value based on the previous values $y_{k+1}, y_{k+2}, \ldots$. Thus the short formula gives the predicted value of y in terms of the last three observed values and last three predicted values.

More generally if $p(x)$ is a polynomial of degree $m - 1$, then the short formula is

$$y_k^* = \sum_{j=1}^{m} (-1)^{j+1} \begin{bmatrix} m \\ j \end{bmatrix} (y_{j+k} + \theta^j \delta_{j+k}).$$

Here $[{}^m_j]$ is the binomial coefficient.

For some applications it seemed desirable to use trigonometric polynomials or exponential polynomials instead of algebraic polynomials. Then it again proves possible to have a short formula

$$y_k^* = -\sum_{j=1}^{m} (g_j y_{j+k} + f_j \theta^j \delta_{j+k}).$$

Here the coefficients g_j and f_j are given by explicit formulas involving the polynomial terms.

13. CONVOLUTION PRODUCTS†

The points of the complex plane with integer coordinates form a lattice which breaks up the plane into unit squares. A function f is said to be discrete analytic on one of these squares if the difference quotient across one diagonal is equal to the difference quotient across the other diagonal

$$\frac{f(z + 1 + i) - f(z)}{i + 1} = \frac{f(z + i) - f(z + 1)}{i - 1}.$$

This definition was introduced by Rufus Isaacs and Jacqueline Ferrand (Lelong). This definition leads directly to difference equation analogs of the Cauchy–Riemann equations and the Laplace equation. Many of the theorems of continuous function theory can be extended to this discrete function theory.

If $f(z)$ and $g(z)$ are two analytic functions, then their product $f(z)g(z)$ is also an analytic function. This important property does not seem to carry over to discrete function theory in any simple way. However there

† See [37, 38].

is another type of product in the classical theory termed the convolution and defined as

$$f(z) * h(z) = \int_0^z f(z - w)g(w)\,dw.$$

The integration is performed along any contour connecting 0 and z. Charles Duris and I found that the same formula gives a product of two discrete analytic functions if the integral is interpreted in the following way. The contours are restricted to the lattice lines. Thus a contour is made up of unit line segments. The value of a discrete function at a point of the line segment is defined to be the average of the values at the end points. Thus the convolution product becomes the sum

$$f * g = \sum_{n=1}^{m} \frac{f(z - z_n) + f(z - z_{n-1})}{2} \frac{g(z_n) + g(z_{n-1})}{2} (z_n - z_{n-1}),$$

where $0 = z_0, z_1, z_2, \ldots, z_m = z$ is a chain of lattice points.

The convolution product of discrete analytic functions is again a discrete analytic function. The product is both commutative and associative. Thus the convolution product yields an algebra for discrete function theory.

These algorithms can be extended to arbitrary partial difference equations with constant coefficients in the plane. With each such partial differential equation, Joan Rohrer and I associated a convolution product. Given any two solutions of the partial difference equation, the convolution product is again a solution.

14. CONVEXITY AND CHEMICAL EQUILIBRIUM†

Consider a mixture of gases in a reaction chamber and suppose that the temperature T and pressure P are maintained constant. Then according to Gibbs, the mixture is in equilibrium if the free energy G is a minimum. For example consider the chemical reaction

$$2H_2 + O_2 = 2H_2O \qquad \text{(steam)}. \tag{1}$$

Let x_1 be the number of molecules of H_2, x_2 be the number of molecules of O_2, and x_3 be the number of molecules of H_2O. If it is assumed that there is only a negligible number of other types of molecules present, then $G = G(x_1, x_2, x_3)$. This function is to be minimized subject to the mass balance constraints

$$2x_1 + 2x_3 = e_1, \qquad 2x_2 + x_3 = e_2. \tag{2}$$

† See [39–42].

Here e_1 is the total number of hydrogen atoms in the chamber, and e_2 is the total number of atoms of oxygen in the chamber. Thus the equilibrium state is obtained by minimizing G subject to the mass balance constraints.

To obtain the form of the free energy function G, chemists assume the perfect gas laws of Boyle, Charles, and Dalton. Then G is given as a definite logarithmic formula involving constants which chemists determine experimentally. Moreover, it is found that $G(x)$ so evaluated is a convex function. Thus, in modern terminology, the Gibbs procedure is termed a *convex program.*

As is well known, each convex program has a dual program concerning the maximization of a concave function G^*. Then the theory of such programs gives the duality inequality

$$G(x_1, x_2, x_3) \geqq M \geqq G^*(y_1, y_2). \tag{3}$$

Here M is the minimum of the primal program and also the maximum of the dual program. The dual variable y_1 is interpreted as the chemical potential of atomic hydrogen, i.e., $y_1 = \partial G/\partial e_1$. Likewise y_2 is the chemical potential of atomic oxygen.

The same mathematical theory often applies to entirely different physical situations. Nevertheless it was somewhat surprising when Avriel, Passy, and Wilde pointed out that the above described duality theory of chemical equilibrium of perfect gases is essentially identical with the duality theory of geometric programming. Geometric programming concerns the economics of engineering design. The basic problem is to minimize the cost of construction and operation of a device or system.

Of course the assumption that the various compounds in a chemical reaction obey perfect gas laws is at best an approximation. Zener and I wondered whether or not the duality inequality is just an approximation. By making some mild assumptions of a physical nature, we were able to show the true Gibbs function is convex, and that there exists a concave function G^* which satisfies the duality inequality (3). We termed G^* the anti-Gibbs function.

15. ASSOCIATIVE NETWORK OPERATIONS†

Kirchhoff defined a network as an interconnection of resistors at nodal points. For the present purpose a network may be regarded as a black box. The nodes of the network on the surface of the box are called terminals and are ordered $1, 2, \ldots, n$. The network may or may not have other nodes

† See [43–46].

inside the box. William Anderson, George Trapp, and I have studied the problem of the interconnection of such network boxes to form larger networks. Thereby we have been led to some interesting algebraic questions.

Consider two network boxes R and S, each having n terminals. How can R and S be conjoined to form a larger network X also having n terminals? As a first approach to this question we defined a device termed a *junctor*. A junctor is a $3n$-terminal network box. The terminals are separated into three equal banks A, B, and J. A and B are termed input banks, and J is termed the output bank. Inside the junctor box the terminals are interconnected by wires of zero resistance. Such connections are subject to certain mild restrictions to ensure desirable physical properties. Thus unconnected terminals or short circuits are not allowed.

The two network boxes are R and S, and are "plugged" into the input banks A and B of the junctor. Then the output bank J forms the terminals of a new n-terminal network X. Thus interconnections can be symbolized as

$$X = J(R, S).$$

A junctor can be diagrammed as a triangular box. Thus Fig. 6 shows a junctor for interconnecting 4 terminal networks termed the hybrid junctor. The dotted lines denote the internal connections.

Let T be another n-terminal network box and suppose that there are two identical junctors J. Then a new n-terminal box Y can be constructed as

$$Y = J(X, T) = J(J(R, S), T).$$

This raises the question of the associativity of the junctor operations. Thus, is

$$(*) \qquad J(J(R, S), T) = J(R, J(S, T))?$$

The equality here is interpreted to mean that the interconnection between R, S, T on the left and right are electrically equivalent. It is not difficult to show graphically that the hybrid junctor has the associativity property.

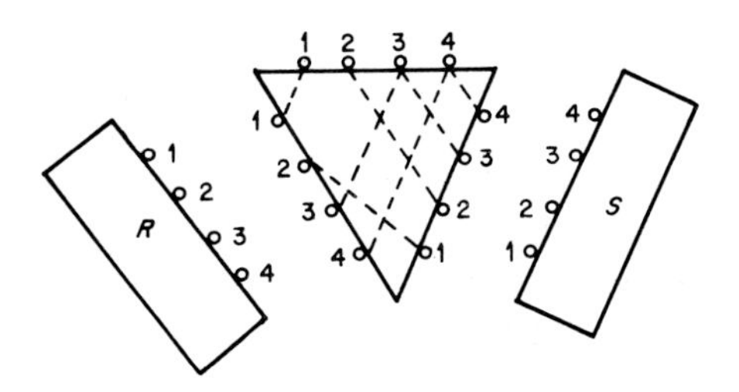

Fig. 6. The hybrid junctor.

To study the general associativity problem we introduced $3n$ by $3n$ adjacency matrices for a junctor. The matrix elements are 0 or 1. Thus 1 means an internal connection and 0 means no connection. Our study led to simple necessary and sufficient conditions to ensure the associativity condition.

16. TRANSMISSION OF SIGNALS WITHOUT DISTORTION

The simplest case of distortionless signaling occurs for systems satisfying the wave equation

$$\frac{\partial^2 V}{\partial x^2} = \omega^2 \frac{\partial^2 V}{\partial t^2}.$$

This is, of course, the equation of vibrating string and has the general solution

$$V(t, x) = \varphi(t - \omega x) + \psi(t + \omega x).$$

Here φ and ψ are arbitrary functions. Thus $\varphi(t - \omega x)$ is a wave traveling to the right with velocity ω^{-1}. This solution is of great practical importance because this wave is transmitted to a distant point x without distortion. There is, of course, a delay time $\tau = \omega x$.

The equation for spherical waves in n dimensions is

$$\frac{\partial^2 V}{\partial r^2} + \frac{n - 1}{r}\frac{\partial V}{\partial r} = \omega^2 \frac{\partial^2 V}{\partial t^2}.$$

If $n = 3$, the general solution of this equation is

$$V(t, r) = \frac{\varphi(t - \omega r)}{r} + \frac{\psi(t + \omega r)}{r}.$$

Again it is seen that a wave can be transmitted without distortion, but there is an attenuation inversely proportional to distance. It is a very happy circumstance that we live in a three-dimensional world because signals can be broadcast without distortion only if $n = 1$ or $n = 3$. This has been discussed by Courant [47].

We wish to introduce a model which is sufficiently general to include the two previous wave equations as special cases. This model is the electrical transmission line. The line can be inhomogeneous and dissipative. Thus suppose that a unit length of the line shown in the schematic diagram of Fig. 7 has series resistance r, series inductance l, shunt conductance g, and

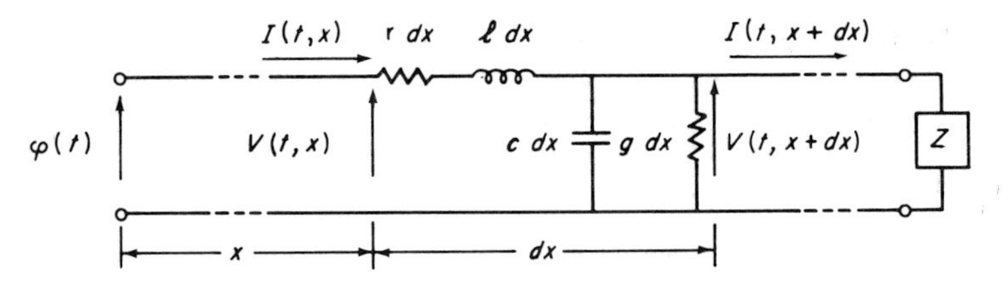

Fig. 7. Differential segment of a transmission line.

shunt capacitance c. Then applying Kirchhoff's laws gives the transmission line equations:

$$-\frac{\partial V}{\partial x} = r(x)I + l(x)\frac{\partial I}{\partial t},$$

$$-\frac{\partial I}{\partial x} = g(x)V + c(x)\frac{\partial V}{\partial t}.$$

Here V is the voltage and I is the current. The current can be eliminated, and it is found that the voltage satisfies

$$\left[\left(r + l\frac{\partial}{\partial t}\right)\frac{\partial^2}{\partial x^2} - \left(r' + l'\frac{\partial}{\partial t}\right) - \left(r + l\frac{\partial}{\partial t}\right)^2\left(g + c\frac{\partial}{\partial t}\right)\right]V = 0.$$

This may be termed the general equation of telephony.

Let us term the line right distortionless if, given an arbitrary function $\varphi(t)$, there is a solution of the form

$$V(t, x) = f(x)\varphi[t - \tau(x)], \qquad I(t, x) = h(x)\varphi[t - \tau(x)].$$

Here τ is the delay time and $\tau(0) = 0$.

Heaviside [48] investigated the possibility of distortionless transmission in a dissipative telephone line. He proved that the condition for no distortion is

$$r = gl/c$$

provided that r, c, g, and l are constant along the line.

In many practical realizations of the transmission line model r, c, g, and l are not constant but vary along the line. This question was studied in [49] and it was shown that:

Theorem. *A transmission line permits right distortionless signaling if and only if the nonnegative functions $r(x)$, $g(x)$, $l(x)$, and $c(x)$ satisfy the equation*

$$r = g\zeta^2 - \frac{d}{dx}\zeta,$$

where $\zeta = (l/c)^{1/2}$.

The delay time is given as

$$\tau = \int_0^x (lc)^{1/2}\,dx.$$

If both right and left distortionless signaling is required then, in addition, ζ must be constant.

17. WAVE MOTION AND THE PALEY–WIENER THEOREM†

The wave equation in three space is

$$\frac{\partial^2 u}{\partial x^2} + \frac{\partial^2 u}{\partial y^2} + \frac{\partial^2 u}{\partial z^2} = \frac{\partial^2 u}{\partial t^2}.$$

The Fourier transform of u is denoted by $U = \Psi u$ and is defined as

$$U(\xi, \eta, \zeta, t) = (2\pi)^{-3/2} \iiint_{-\infty}^{\infty} e^{i(\xi x + \eta y + \zeta z)} u(x, y, z, t)\,dx\,dy\,dz.$$

We shall assume that u satisfies the initial conditions

$$u(x, y, z, 0) = 0, \qquad \text{for all} \quad x, y, z,$$

and

$$u_t(x, y, z, 0) = 0, \qquad \text{for} \quad x^2 + y^2 + z^2 > b^2.$$

Then the Fourier transform of the wave equation is

$$-\rho^2 U(\xi, \eta, \zeta, t) = \frac{\partial^2 U}{\partial t^2}(\xi, \eta, \zeta, t),$$

where $\rho^2 = \xi^2 + \eta^2 + \zeta^2$. Solving this equation gives

$$U = F(\xi, \eta, \zeta)\sin(\rho t)/\rho.$$

Here $F = \Psi f$, where

$$f(x, y, z) = u_t(x, y, z, 0).$$

Introduce spherical coordinates ρ, θ, φ so

$$\xi = \rho \sin\theta \cos\varphi, \qquad \eta = \rho \sin\theta \sin\varphi, \qquad \zeta = \rho\cos\theta.$$

Thus, if $r^2 = x^2 + y^2 + z^2$, then

$$F(\rho, \theta, \varphi) = (2\pi)^{-3/2} \iiint e^{i\rho r \cos(\rho, r)} f(x, y, z)\,dx\,dy\,dz.$$

† See [50, 51].

Since f has compact support we may assume that $r \leqq b$. Thus if θ and φ are fixed the integral is seen to be an entire function of ρ. Moreover, even if ρ is complex valued,

$$|F| \leqq (2\pi)^{-3/2} \iiint |f| \, dx \, dy \, dz \, e^{b|\rho|} = Ae^{b|\rho|}.$$

Thus F is in an entire function of ρ of exponential type b. It is this fact which suggests application of the Paley–Wiener theorem.

The kinetic energy of a solution u at time t is defined to be

$$K = \iiint u_t^2 \, dx \, dy \, dz \doteq \|u_t\|^2.$$

The total initial energy is therefore given as

$$E = \iiint f^2 \, dx \, dy \, dz \doteq \|f\|^2.$$

By the Parseval theorem the Fourier transform does not change the norm so $\|u_t\|^2 = \|U_t\|^2$ or

$$K = \|F(\xi, \eta, \zeta) \cos \rho t\|^2.$$

Using the identity $2 \cos^2 \theta = 1 + \cos 2\theta$ gives

$$2K = \iiint |F|^2 \, d\xi \, d\eta \, d\zeta + \iiint |F|^2 \cos 2\rho t \, d\xi \, d\eta \, d\zeta.$$

Transforming the second integral to polar coordinates and carrying out the integrations with respect to the angular variables gives

$$(*) \qquad 2K = E + \int_0^\infty \cos 2\rho t H(\rho) \, d\rho.$$

Jerome Goldstein [50] noted that the Riemann–Lebesgue lemma requires that Fourier transforms vanish at infinity and consequently (*) gives the asymptotic relation

$$2K \to E \quad \text{as} \quad t \to \infty.$$

In other words the kinetic energy tends to one-half the initial energy.

Now observe that

$$H(\rho) = \rho^2 \int_0^{2\pi} d\varphi \int_0^\pi FF^* \sin \theta \, d\theta.$$

Thus $|H| \leqq 4\pi A^2\rho^2 e^{2b|\rho|}$, and H is seen to be an entire function of exponential type $2b$. In addition, symmetry considerations show that H is an even function of ρ. Thus

$$(**) \qquad 2\int_0^\infty (\cos \rho s) H(\rho)\, d\rho = \int_{-\infty}^\infty e^{i\rho s} H(\rho)\, d\rho.$$

Since $H(\rho)$ is an entire function of exponential type $2b$ and is absolutely integrable on the real axis it follows from the Paley–Wiener theorem that its Fourier transform vanishes outside the interval $[-2b, 2b]$. Applying this result to relation $(*)$ gives

$$2K = E \qquad \text{for} \quad t > b.$$

Thus after a time b the kinetic energy becomes constant and equals half the initial energy. Minor modification of the proof shows that this result remains true for arbitrary initial conditions of compact support in a sphere of radius b.

This theorem also holds for one-dimensional wave motion. However the statement of the theorem is false for two-dimensional wave motion because the corresponding function $H(\rho)$ is not even and relation $(**)$ does not hold.

18. SYLVESTER UNISIGNANT OPERATORS

Sylvester [52] proposed a special format for writing a system of n linear equations in n variables. For instance if $n = 2$ the Sylvester format is:

$$(\mathrm{S}) \qquad \begin{aligned} (A + E)x_1 - Ex_2 &= y_1, \\ -Fx_1 + (B + F)x_2 &= y_2. \end{aligned}$$

The determinant of the system is

$$D = \det\begin{vmatrix} A + E & -E \\ -F & B + F \end{vmatrix} = AB + AF + BE.$$

The inverse matrix of system (S) is seen to be

$$R = \begin{bmatrix} \dfrac{B + F}{D} & \dfrac{E}{D} \\[2ex] \dfrac{F}{D} & \dfrac{A + E}{D} \end{bmatrix}.$$

Thus the determinant of the system (S) and its cofactors are free of negative signs. For this reason Sylvester termed them "unisignants." The concepts extend readily to systems of order n.

Morley and I [53] studied the Sylvester system (S) when the coefficients A, B, E, F are linear operators rather than scalars and where x_i and y_i are vectors. The difficulty with a direct generalization is that the determinant concept does not extend to operators. To obtain a proper generalization we modified Sylvester's idea. For us a *unisignant is any rational function of A, B, E, F which can be obtained by the operations of addition, multiplication, and inversion.* Subtraction is not allowed.

Let A^+ be a short notation for the inverse of the operator A. Then we found the following unusual form for the inverse matrix R:

$$R = \begin{bmatrix} [A + E(B + F)^+ B]^+ & [B + (B + F)E^+ A]^+ \\ [A + (A + E)F^+ B]^+ & [B + F(A + E)^+ A]^+ \end{bmatrix}.$$

Thus the matrix elements of R are unisignant operators.

The general Sylvester system has the form

$$y_i = \sum_{j=0}^{n} G_{ij}(x_i - x_j), \qquad i = 1, \ldots, n,$$

and $x_0 = 0$. If the G_{ij} are operators then the inverse matrix has matrix elements R_{ij} and they are again found to be unisignant expressions in the G_{ij}.

If the coefficients G_{ij} are positive scalars then the R_{ij} satisfy certain inequalities. To obtain analogs of these inequalities for operators a special partial order is introduced as follows. The generating operators G_{ij} are declared to be "positive." Then given two rational functions P and Q we say $P \geqq Q$ if $P - Q$ is a unisignant form. By this convention rational functions become a *partially ordered system.* Relative to this partial order the following inequalities are shown.

$$R_{jj} \geqq R_{ij} \geqq 0.$$

Next suppose that the Sylvester system (S) satisfies the symmetry property

$$G_{ij} = G_{ji}. \tag{$*$}$$

Then the system can be interpreted as the equations defining a steady flow of current in an electrical network with $n + 1$ nodes. The x_i are the voltages of the nodes relative to the ground node ($x_0 = 0$). The y_i are the currents entering the nodes from outside. The G_{ij} are the conductances of the $n(n - 1)/2$ branches of the network. The R_{ij} are interpreted as transfer resistances of the network. In particular R_{jj} is interpreted as the joint resistance between node j and the ground node 0.

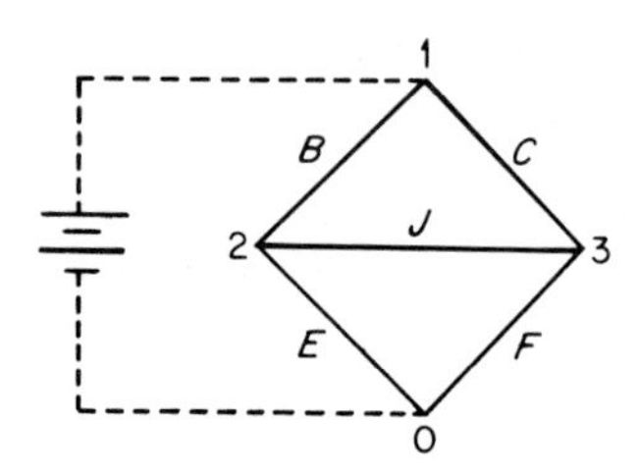

Fig. 8. The Wheatstone bridge.

The symmetry property (*) induces other relations. Thus let ρ_{ij} be the joint resistance between any two nodes i and j when other nodes are insulated. Thus $\rho_{ij} \geqq 0$. Moreover, the following "triangle inequality" holds:

$$\rho_{ij} \leqq \rho_{ik} + \rho_{kj}.$$

In the scalar case this was proved in [54].

An interesting special network of five operators B, C, E, F, J is the Wheatstone bridge connection shown in Fig. 8. Then the joint conductance between terminals 1 and 0 is found to be

$$K = F\gamma C + (E + F\gamma J)[B + E + J\gamma(C + F)]^{+}(B + J\gamma C),$$

where $\gamma = (C + F + J)^{+}$.

All the considerations to this point are formal, proceeding under the assumption that no singular operators are developed. To avoid this assumption suppose that B, C, E, F, J are positive semidefinite operators. Then by a limit analysis it can be shown that the expression for K is always well defined. However A^{+} now denotes a generalized inverse.

19. A DETERMINANT ALGORITHM FOR LINEAR INEQUALITIES

Fourier treated linear inequality systems by eliminating variables one at a time [55]. By use of determinants it proves possible to eliminate variables in blocks rather than one by one [56]. An interesting special case is the elimination of variables two at a time. This algorithm will be used to solve the following example problem.

Problem. *Consider the following system* (S) *of inequalities on affine functions* F_i:

(S)

$$\begin{aligned}
F_1 &= -6x - y + \lambda_1 \geqq 0,\\
F_2 &= -x \qquad + \lambda_2 \geqq 0,\\
F_3 &= x + y + \lambda_3 \geqq 0,\\
F_4 &= 4x + 5y + \lambda_4 \geqq 0,\\
F_5 &= x - y + \lambda_5 \geqq 0.
\end{aligned}$$

For what values of the parameters $(\lambda_1, \lambda_2, \ldots, \lambda_5)$ can this system be satisfied for some values of the variables x and y?

Solution. The method begins by forming a nonnegative summation of three of the affine functions F_i so that in the sum F' the variables x and y do not appear. For instance

$$F' = 2F_1 + 7F_3 + 5F_5 = 2\lambda_2 + 7\lambda_3 + 5\lambda_5 \geqq 0$$

is a restriction on λ_i. Then it is said that F_1, F_3, F_5 form a *permissible triad*. Thus a new inequality system S' is formed out of all permissible triads.

A systematic way is needed to form the triad sums. If $i < j$ let Δ_{ij} be the determinant formed from the ith and jth rows of the first and second columns of the S matrix. If $i > j$ let $\Delta_{ij} = -\Delta_{ji}$. Then define F_{ijk} as the linear combination

$$F_{ijk} = F_i\Delta_{jk} + F_j\Delta_{ki} + F_k\Delta_{ij}.$$

Then it results that the variables x and y do not appear in F_{ijk}. The coefficients Δ are now called *cofactors*.

It is clear that if the cofactors Δ_{jk}, Δ_{kj}, and Δ_{ij} are nonnegative, then F_{ijk} is a permissible triad sum. For example, if $i = 1$, $j = 5$, and $k = 3$, we see that $\Delta_{15} = 7$, $\Delta_{53} = 2$, and $\Delta_{31} = 5$, so F_{153} is the same as F' in the previous example.

To facilitate the selection of integers (i, j, k) giving permissible F_{ijk}, a directed graph is formed as shown in Fig. 9. The vertices of the graph correspond to the affine functions F_i and are denoted by integers in circles. The edges of the graph correspond to the determinants Δ_{ij}. Thus if $\Delta_{ij} > 0$ then the edge connecting vertex i and vertex j is directed from i to j. The edge is labeled with the value Δ_{ij}. Thus the graph is termed a *cofactor graph*.

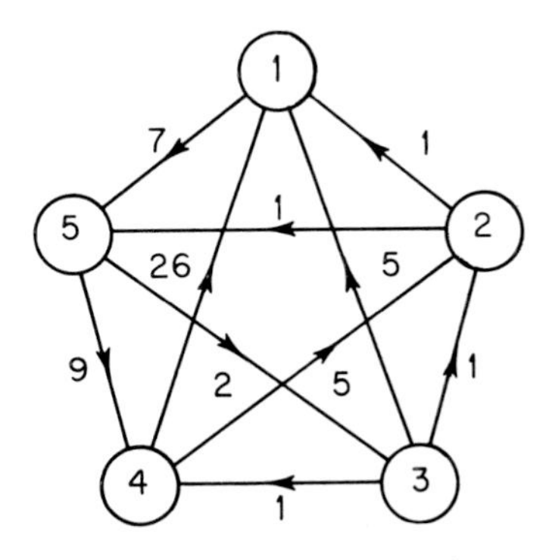

Fig. 9. The cofactor graph.

It is now observed that having Δ_{ij}, Δ_{jk}, Δ_{ki} all nonnegative is equivalent to having (i, j, k) form a directed triangular circuit of the cofactor graph. Thus permissible triads correspond to directed triangles. Inspection of the cofactor graph shows that there are four directed triangular circuits: (1, 5, 3), (2, 5, 3), (2, 5, 4), (4, 1, 5). Thus the eliminant system (S′) may now be written down by selection of these directed triangles:

$$
\text{(S}'\text{)} \qquad
\begin{aligned}
2\lambda_1 + \quad 7\lambda_3 \quad + \ 5\lambda_5 &\geqq 0, \\
2\lambda_2 + \lambda_3 \quad + \ \lambda_5 &\geqq 0, \\
9\lambda_2 \quad + \lambda_4 + 5\lambda_5 &\geqq 0, \\
9\lambda_1 \quad + 7\lambda_4 + 26\lambda_5 &\geqq 0.
\end{aligned}
$$

Thus if system (S) has a solution (x, y), then (S′) is true. Conversely it can be shown that if (S′) is true, then values of x and y exist to satisfy (S).

An example with four variables (x_1, x_2, x_3, x_4) and six inequalities is shown in tableau form in Fig. 10. First x_1 and x_2 are eliminated giving system (S′). Then x_3 and x_5 are eliminated from (S′) giving the final eliminant system (S″). The relations marked with a $*$ are redundant and are dropped. A relation is redundant if it contains the same or more parameters λ_i appearing in

	Row	x_1	x_2	x_3	x_4	λ_1	λ_2	λ_3	λ_4	λ_5	λ_6	Triangles			
(S)	1	−1	1	−5	−3	1									
	2	−1	1	2	−1		1								
	3	2	−2	−2	4			1							
	4	1	−2	3	2				1						
	5	1	1	−6	−5					1					
	6			1	−1						1				
(S′)	7			−12	−2	2		1	0			1	4	3	
	8			−15	0	3			2	1		1	4	5	
	9			2	2		2	1	0			2	4	3	
	10			6	6		3		2	1		2	4	5	
	11			4	4		4	2		0		3	5	2	*
	12			−24	−4	4		2		0		5	1	3	*
	13			1	−1						1	6			
(S″)	14					8	28	18			20	7	13	9	
	15					24	42	12	28	14	60	7	13	10	*
	16					12	30	15	8	4	30	8	13	9	*
	17					36	45		54	27	90	8	13	10	

Fig. 10. Tableau of eliminant systems.

another relation of the eliminant. Thus system S has a solution (x_1, x_2, x_3, x_4) if and only if

$$
(S'') \qquad \begin{aligned}
8\lambda_1 + 28\lambda_2 + 18\lambda_3 \qquad\qquad\qquad\qquad + 20\lambda_6 &\geqq 0, \\
36\lambda_1 + 45\lambda_2 \qquad\qquad + 54\lambda_4 + 27\lambda_5 + 90\lambda_6 &\geqq 0.
\end{aligned}
$$

20. CLARK'S THEOREM EXTENDED TO CONVEX PROGRAMS

Let $g(x), g_1(x), \ldots, g_p(x)$ be convex functions on a space X. Then the standard convex program can be stated as follows.

Program A. *Seek the value* $MA = \inf g(x)$ *subject to* x *being in the set* α *defined as*:

$$\alpha = \{x \mid g_k(x) \leqq 0, k = 1, \ldots, p\}.$$

This minimization problem has been extensively studied by introducing the Lagrange function

$$L(x, \lambda) = g(x) + \sum_{1}^{p} \lambda_k g_k(x).$$

For given values of the Lagrange multipliers λ_k let a function $\varphi(\lambda)$ be defined as

$$\varphi(\lambda) = \inf_{x} L(x, \lambda).$$

Then an associated maximization problem, termed the dual program, is defined as follows.

Program B. *Seek the value* $MB = \sup \varphi(\lambda)$ *subject to* λ *being in the set* β *defined as*:

$$\varphi(\lambda) > -\infty \qquad \textit{and} \qquad \lambda_k \geqq 0, \quad k = 1, \ldots, p.$$

The sets α and β are termed the feasible sets. The functions $g(x)$ and $\varphi(\lambda)$ are termed the objective functions.

In the linear case

$$g_k(x) = \sum_{j=1}^{n} c_{kj} x_j + b_k$$

then Program A becomes the standard linear program A_L.

Program A_L. *Seek* $MA = \inf \sum_{j=1}^{n} c_j x_j$ *subject to* x *being in the set* α *defined as*:

$$\alpha = \left\{x \,\middle|\, \sum_{j=1}^{n} c_{kj} x_j + b_k \leqq 0, k = 1, \ldots, p\right\}.$$

The dual objective function $\varphi(x)$ can be calculated explicitly and the dual program in the linear case becomes the following.

Program $\mathbf{B_L}$. *Seek* $MB = \sup \sum_{k=1}^{p} \lambda_k b_k$ *subject to* x *being in the set* β *defined as*: λ_k such that

$$\sum_{k=1}^{p} \lambda_k c_{kj} + c_j = 0, \quad j = 1, \ldots, n \qquad \textit{and} \qquad \lambda_k \geqq 0, \quad k = 1, \ldots, p.$$

Clark [57] established the following theorem for a pair of dual linear programs in the standard form: *If the set of feasible points of one program is bounded, then the set of feasible points of the other program is unbounded.* The convention is adopted that the empty set is neither bounded nor unbounded. It was shown in [58] that Clark's theorem applies unchanged for the pair of dual convex programs A and B. The proof used the Beltrami–Courant penalty function to reduce the constrained program A to an unconstrained problem.

For linear programs the values of dual programs are equal, that is $MA = MB$. This important relation does not always hold for dual convex programs. There can be a "duality gap." However it was also shown in [58] that *there is no duality gap if one of the feasible sets is bounded.*

REFERENCES

1. D. G. Bourgin and R. J. Duffin, The Dirichlet problem for the vibrating string equation, *Bull. Amer. Math. Soc.* **45** (1939), 851–858. MR **1**, 120.
2. R. J. Duffin, On the characteristic matrices of covariant systems, *Phys. Rev.* **54** (1938), 1114.
3. R. J. Duffin, On wave equation vector-matrices and their spurs, *Phys. Rev.* **77** (1950), 683–685. MR **11**, 543.
4. R. J. Duffin and A. C. Schaeffer, On the extension of a functional inequality of S. Bernstein to non-analytic functions, *Bull. Amer. Math. Soc.* **46** (1940), 356–363. MR **1**, 205.
5. R. J. Duffin and A. C. Schaeffer, A refinement of an inequality of the Brothers Markoff, *Trans. Amer. Math. Soc.* **50** (1941), 517–528. MR **3**, 235.
6. R. J. Duffin, Representation of Fourier integrals as sums. I, *Bull. Amer. Math. Soc.* **51**, (1945), 447–455. MR **6**, 266.
7. R. J. Duffin, Representation of Fourier integrals as sums, II, *Proc. Amer. Math. Soc.* **1** 1950), 250–255. MR **11**, 592; MR **12**, 1002.
8. R. J. Duffin, Representation of Fourier integrals as sums, III, *Proc. Amer. Math. Soc.* **8** (1957), 272–277. MR **18**, 893.
9. H. F. Weinberger, Fourier Transforms of Mobius Series, Ph.D. Thesis, Carnegie-Mellon University, Pittsburgh, Pennsylvania, 1950.
10. R. J. Duffin, On a question of Hadamard concerning super-biharmonic functions, *J. Math. Phys.* **27** (1949), 253–258. MR **10**, 534.
11. R. J. Duffin, The maximum principle and biharmonic functions, *J. Math. Anal. Appl.* **3** (1961), 399–405. MR **26**, 1617.
12. G. Szegö, On membranes and plates, *Proc. Nat. Acad. Sci. U.S.A.* **36** (1950), 210–216. MR **11**, 757.
13. R. J. Duffin and D. H. Schaffer, On the modes of vibration of a ring-shaped plate, *Bull. Amer. Math. Soc.* **58** (1952), 652.
14. R. J. Duffin, Nodal lines of a vibrating plate, *J. Math. Phys.* **31** (1953), 294–299. MR **14**, 601.
15. R. J. Duffin and A. Schild, The effect of small constraints on natural vibrations, *Proc. Symp. Appl. Math.* **5** (1954), 155–163.

16. R. J. DUFFIN AND A. SCHILD, On the change of natural frequencies induced by small constraints, *J. Math. Mech.* **6** (1957), 731–758. MR **19**, 1101.
17. R. BOTT and R. J. DUFFIN, On the Wang algebra of networks, *Bull. Amer. Math. Soc.* **57** (1951), 136.
18. R. BOTT AND R. J. DUFFIN, On the algebra of networks, *Trans. Amer. Math. Soc.* **74** (1953), 99–109. MR **15**, 95.
19. R. J. DUFFIN, An analysis of the Wang algebra of networks, *Trans. Amer. Math. Soc.* **93** (1959), 114–131. MR **22**, 49.
20. R. J. DUFFIN, Network models, *Math. Aspects Electr. Network Theory, SIAM-AMS Proc.*, Vol. 3, pp. 65–91, Amer. Math. Soc., Providence, Rhode Island, 1971.
21. K. T. WANG, On a New Method of Analysis of Electrical Networks, Mem. No. 2, Nat. Res. Inst. Engrg., Academia Sinica, 1934.
22. G. H. HARDY, E. E. LITTLEWOOD, AND G. POLYA, "Inequalities," Cambridge Univ. Press, London, 1934.
23. R. A. MEUSSNER, The Growth of Silicon Particles in an Aluminum Matrix during Isothermal Heat Treatment, Ph.D. Thesis, Carnegie-Mellon Univ., Pittsburgh, Pennsylvania, 1952.
24. R. J. DUFFIN, R. A. MEUSSNER, AND F. N. RHINES, Statistics of Particle Measurement and of Particle Growth, Tech. Rep. No. 32, Carnegie–Mellon Univ., Pittsburgh, Pennsylvania, 1953.
25. R. J. DUFFIN, A minimax theory for overdamped networks, *J. Rational Mech. Anal.* **4** (1955), 221–233. MR **16**, 979.
26. R. J. DUFFIN, The Rayleigh–Ritz method for dissipative or gyroscopic systems, *Quart. Appl. Math.* **18** (1960–1961), 215–221. MR **22**, 12775.
27. R. J. DUFFIN, A variational problem relating to cooling fins, *J. Math. Mech.* **8** (1959), 47–56. MR **21**, 2477.
28. R. J. DUFFIN AND D. K. MCLAIN, Optimum shape of a cooling fin on a convex cylinder, *J. Math. Mech.* **17** (1968), 769–784.
29. R. J. DUFFIN, Optimum heat transfer and network programming, *J. Math. Mech.* **17** (1968), 759–768.
30. S. BHARGAVA AND R. J. DUFFIN, Network models for maximization of heat transfer under weight constraints, *Networks* **2** (1972), 355–365.
31. S. BHARGAVA AND R. J. DUFFIN, Dual extremum principles relating to cooling fins, *Quart. J. Appl. Math.* **31** (1973), 27–41.
32. R. J. DUFFIN AND T. W. SCHMIDT, Simple formula for prediction and automatic scrutation, *J. Amer. Rocket Soc.* **30** (1960), 364–365.
33. R. J. DUFFIN, Discounted least squares, *Proc. 4th Ordnance Confer. Oper. Res., Army Res. Office, Durham, N.C., 1960.*
34. R. J. DUFFIN AND T. W. SCHMIDT, An extrapolator and scrutator, *J. Math. Anal. Appl.* **1** (1960), 215–227. MR **22**, 11510.
35. R. J. DUFFIN AND P. WHIDDEN, An exponomial extrapolator, *J. Math. Anal. Appl.* **3** (1961), 526–536. MR **36**, 2301.
36. R. J. DUFFIN, Extrapolating time series by discounted least squares, *J. Math. Anal. Appl.* **20** (1967), 325–341. MR **36**, 1078.
37. R. J. DUFFIN AND C. S. DURIS, A convolution product for discrete function theory, *Duke Math. J.* **31** (1964), 199–220. MR **29**, 429.
38. R. J. DUFFIN and J. ROHRER, A convolution product for the solutions of partial difference equations, *Duke Math. J.* **35** (1968), 683–698. MR **39**, 1831.
39. R. J. DUFFIN, E. L. PETERSON, AND C. ZENER, "Geometric Programming: Theory and Application," Wiley, New York, 1967. MR **35**, 5225.
40. R. J. DUFFIN AND C. ZENER, Geometric programming, chemical equilibrium and the anti-entropy function, *Proc. Nat. Acad. Sci. U.S.A.* **63** (1969), 629–636.

41. R. J. DUFFIN AND C. ZENER, Geometric programming and the Darwin–Fowler method in statistical mechanics, *J. Chem. Phys.* **74** (1970), 2419–2423.
42. R. J. DUFFIN, Duality inequalities of mathematics and science, *in* "Nonlinear Programming" (J. B. Rosen, O. L. Mangasarian, and K. Ritter, eds.), Academic Press, New York, 1970, pp. 401–423.
43. W. N. ANDERSON, JR. AND R. J. DUFFIN, Series and parallel addition of matrices, *J. Math. Anal. Appl.* **26** (1969), 576–594. MR **39**, 3904.
44. R. J. DUFFIN, Network models, *Math. Aspects Electr. Network Theory, SIAM-AMS Proc.*, Vol. 3, Rhode Island, 1971. Amer. Math. Soc., Providence, pp. 65–91.
45. R. J. DUFFIN AND G. E. TRAPP, Hybrid Addition of Matrices—A Network Theory Concept, Math. Rep. 70–44, Carnegie-Mellon Univ., Pittsburgh, Pennsylvania, 1970; *Applicable Anal.* **2** (1972), 241–254.
46. W. N. ANDERSON, JR., R. J. DUFFIN, AND G. E. TRAPP, Tripartite graphs to analyze the interconnection of networks, *in* "Graph Theory and Applications," (Y. Alavi, D. R. Lick, and A. T. White, eds.), Springer-Verlag, Berlin and New York, 1972, pp. 7–12.
47. R. COURANT, Hyperbolic partial differential equations and applications, *in* "Modern Mathematics for the Engineer," (E. F. Beckenbach, ed.), McGraw–Hill, New York, 1956, pp. 92–109.
48. O. HEAVISIDE, Electrical papers, *Electrician* **2** (1887), 125.
49. V. BURKE, R. J. DUFFIN, AND D. HAZONY, Distortionless wave propagation in inhomogeneous media and transmission lines, *Quart. Appl. Math.* **34** (1976), 183–194.
50. J. A. GOLDSTEIN, An asymptotic property of solutions of wave equations, *Proc. Amer. Math. Soc.* **23** (1969), 359–363.
51. R. J. DUFFIN, Equipartition of energy in wave motion, *J. Math. Anal. Appl.* **32** (1970), 386–391.
52. J. J. Sylvester, On the change of systems of independent variables, *Quart. J. Math.* **1** (1855), 42–56.
53. T. D. MORLEY AND R. J. DUFFIN, Operator Networks Treated by Sylvester Unisignants, Carnegie-Mellon Univ., Pittsburgh, Pennsylvania, 1978.
54. R. J. DUFFIN, Distributed and lumped networks, *J. Math. Mech.* **8** (1959), 793–826.
55. R. J. DUFFIN, "On Fourier's Analysis of Linear Inequality Systems," Mathematical Programming Study, No. 1, North–Holland Publ., Amsterdam, 1974, pp. 71–95.
56. J. F. BUCKWALTER AND R. J. DUFFIN, Linear Inequality Systems Treated by Determinants, Carnegie-Mellon Univ., Pittsburgh, Pennsylvania, 1978.
57. F. E. CLARK, Remark on the constraint sets in linear programming, *Amer. Math. Monthly* **68** (1967), 351–352.
58. R. J. DUFFIN, Clark's theorem on linear programs holds for convex programs, *Proc. Nat. Acad. Sci. U.S.A.* **75** (1978), 1624–1626.

Sections 0–15 of this paper were prepared under Research Grant DA-ARO-D-31-124-71-G17. Sections 0–15 were part of an address delivered before the annual meeting of the American Mathematical Society in San Francisco on January 16, 1974.

AMS Primary 35L05, 81A06, 42A08, 26A82, 42A68, 73K10, 94A20, 82A65, 70J05, 80A20, 62M10, 30A95, 80A30; Secondary 41A10

DEPARTMENT OF MATHEMATICS
CARNEGIE-MELLON UNIVERSITY
PITTSBURGH, PENNSYLVANIA

Some Recollections from 30 Years Ago

RAOUL BOTT

These evening sessions are to be informal and I see—now at the eleventh hour—that I was put down on the program as a coordinator. So forgive me if instead of conducting a panel I simply address you in two different roles. First I would like to recall some moments from my own past which were pivotal in my life, and in which it was very great good blessing to have John L. Synge, Dick Duffin, and Carnegie Tech as my guardian angels. And then if time permits I would like to give you a little bit of the flavor of the mathematician which I encountered in Princeton after I left here in 1949.

As some of you know I graduated in electrical engineering from McGill in 1946 with a masters degree, and then immediately, in the capricious ways of youth, decided to switch to mathematics. McGill was willing to take me on in this new role, but only if I first spent three years or so, to obtain a B.A. in arts, and as this plan did not appeal to me at all I cast about for better terms at some other school. One of my friends and benefactors at McGill was Professor Loyd Williams, who suggested Carnegie Tech, where John L. Synge was just forming a brand new graduate school in mathematics and presumably needed some graduate students.

Accordingly I hitchhiked to Pittsburgh in April of 1947 to put my case to him in person. Those of you who know John L. Synge also know his impressive presence. At that time he wore an eye patch over one eye and, I believe, a filter in his nose because of the Pittsburgh smog. The total effect was unforgettable! Synge was very kind and immediately proposed that I enter as a graduate student heading for a masters degree. We then looked up the newly formulated rules of the department and determined that it would again take three years until I had completed the requirements for a masters.

Synge sensed my deep disappointment and without saying anything gazed out the window into the gray mist outside for a long time. Then he suddenly turned to me and said the magic words, "Let's look at the requirements for a doctors degree." Well, these turned out to be completely flexible:

ISBN 0-12-178150-X

two years residency, qualifying exams, and a thesis. I was therefore delighted with my good fortune when Synge accepted me into this program, and indeed, two years later I was able to walk out of Carnegie with my doctors degree.

At that time I of course did not know just how fortunate this whole turn of events had been. I did not know that I would be working and collaborating with Dick Duffin, or that I would have in Professor Rosenbach, the Assistant Chairman at the time, a boss who would take me apartment hunting, help me buy furniture, and, God bless him, bring the whole administration of Carnegie Tech into the fray when the Immigration Department started to deport me.

But to properly acknowledge these and all other great kindnesses, which my young wife Phyllis and I experienced during our two years here, would be too long a story to relate. All I can say really, is thank you Jean and Dick and all of you who made Carnegie such a supportive environment. Over the years I had put this down in part to your inexperience at that time in putting hurdles in front of graduate students. But at this meeting I was glad to hear that—at least among Dick's students—that situation still prevails.

But let me conclude these reminiscences with an anecdote concerning the theorem which Dick and I were lucky enough to discover here 30 years ago and which now bears our name. This theorem gives an algorithm for synthesizing a two-point network without the use of ideal transformers, and so resolved a question which had been open for some 20 years since the pioneering work of Brune and Foster in network synthesis.

I had brought this problem with me from my electrical engineering days, although I have forgotten now just how it came to fascinate me so. In any case, Dick immediately became interested and we spent many hours puzzling over it. Then one day I came across the beautiful paper of I. R. Richards which contained a quite new transformation of "positive real functions" which we right away sensed to have some bearing on our problem. In particular I recall an afternoon when Dick and I tried desperately to fit this theorem to our problem and just could not manage to do it. Finally after six we packed up for home, exhausted and slightly dejected. I walked home along Forbes Street, and suddenly it became clear to me that actually the solution had been staring us in the face all that afternoon. I rushed home and picked up the phone to call Dick. The line was busy. Dick, it turned out a few minutes later, had been trying to reach me at precisely the same time with the same observation!

But I cannot let you get away without making you work a little, so let me now take up my second role and tell you a little about the subject which captivated me soon after I left here to go to the Institute for Advanced Study in Princeton in the fall of 1949. Partly I want to show off a beautiful piece of pure mathematics but partly I also feel that this branch of analysis–topology, might be useful in more applied problems and is at the same time nearly unknown in that domain. I am referring to "Morse theory" whose founder Marston Morse, also was a dear friend and benefactor, and who after a long and fruitful career just died last year at the age of 84.

Let me start by explaining one of Morse's celebrated theorems. Consider a smooth closed surface S in $\mathbb{R}^3$, say, and assume that S has the topological type of the usual sphere:

$$x^2 + y^2 + z^2 = 1.$$

If you wish, imagine a slightly distorted version of this sphere. Now take two points P and Q on S, and consider the problem of finding all the *geodesics* on S joining P to Q. Recall here that such a geodesic is a smooth parametrical curve $\mathbf{x}(t)$, $0 \leqq t \leqq 1$, on S with $\mathbf{x}(0) = P$, $\mathbf{x}(1) = Q$, and with the property that the length of the curve $L\{\mathbf{x}(t)\}$ is *stationary* with respect to variations of the path keeping the endpoints fixed. Thus $\mathbf{x}(t)$ is stationary if and only if

$$\frac{d}{d\alpha} L\{\mathbf{x}(t)_\alpha\}\bigg|_{\alpha=0} = 0$$

for any family of paths $\mathbf{x}_\alpha(t)$ varying smoothly with α subject to $\mathbf{x}_\alpha(0) = P$, $\mathbf{x}_\alpha(t) = Q$.

One's instinct is of course that there should be at least one such geodesic, namely, the one along which the length becomes a minimum. Afid indeed when S is convex one can visualize this minimal geodesic:

Take a string and pin it to S at P. Then pull the string tight and lay it along S till it passes through Q. If the string segment between P and Q maintains its pressure even when you pluck it slightly between P and Q, then it will have traced out the minimal geodesic on S. Of course if S is not convex you must imagine the string confined to S before you pull it tight.

In any case it was known long before Morse that this problem, or really any similar problem in the calculus of variations, had at least one solution, i.e., that a minimal geodesic existed. Wht Morse showed was that, whatever the shape of S, there must always be an *infinite number of geodesics joining P to Q, and that these geodesics could not all be of bounded length.*

I know that the concern of many applied questions is the development of algorithms for minimizing or maximizing a function, and that in these questions the other stationary points are of no interest. Still, there are famous questions which involve all stationary points of a function. For instance the spectral theorem for the operator

$$L = \frac{\partial^2}{\partial x^2} + \frac{\partial^2}{\partial y^2} + \frac{\partial^2}{\partial z^2}$$

on functions f, defined on $x^2 + y^2 + z^2 \leqq 1$ and subject to the boundary condition

$$f \equiv 0 \quad \text{on} \quad x^2 + y^2 + z^2 = 1,$$

is again a theorem concerning the existence of an infinite number of critical points. Indeed properly interpreted the solutions of

$$Lf = \lambda f$$

are simply the critical points of the function

$$E(f) = (Lf, f)$$

on the unit sphere $(f, f) = 1$ of a suitable Hilbert space.

And indeed, as I hope to explain to you, in the Morse theory both the results on the number of geodesics and on the number of eigenvalues make sense from the same point of view, although of course the technicalities are quite different.

Now the main idea of the Morse theory is simply that the *topological complexity* of a space M, on which a *suitably restricted function* f is defined, forces f to have a certain number of critical points.

The simplest instance of this relation is of course the maximum principle: On a compact space S a continuous function f assumes its minimum.

Now clearly if S consists of several connected pieces, then this argument applies to each piece, and hence we have the inequality:

(*) *The number of relative minima of f on $S \geqq$ the number of connected pieces into which S falls.*

In topology the number $R_0(S)$, of connected pieces of a space S is called the 0th Betti number of S and the appropriate *higher dimensional analog* of this concept leads to the higher *Betti numbers* $R_i(S)$, $i = 1, 2, \ldots$, of the space. The "counting series" of these numbers is called the *Poincaré series* of S and is denoted by $P_t(S)$:

$$P_t(S) = \sum_{i \geqq 0} R_i(S) t^i.$$

There are many, many quite different approaches to the definition of the $R_i(S)$, and it is in fact one of the joys of the subject that however one extends the notion of connectedness the same invariants $R_i(S)$ seem to make their appearance.

The most geometric definition of these is very roughly the following one. Consider two points P and Q in S. The points are (arcwise) connected in S if and only if we can put a curve on S with P and Q as endpoints. Analogously if C_1 and C_2 are two closed curves on S, we say that C_1 and C_2 are homo-

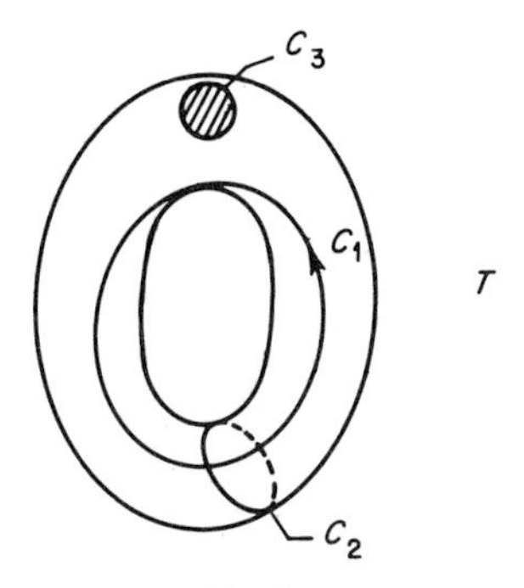

Fig. 1

logous if and only if we can span a surface between them in S whose boundary is precisely C_1 and C_2.

The number of distinct "homology classes" of closed curves now gives rise to the first Betti number $R_1(S)$. And so on! For instance for the "torus" T (Fig. 1), the Poincaré series is given by

$$P_t(T) = 1 + 2t + t^2$$

with the curves C_1 and C_2 representing the two distinct types of "interesting" closed curve on T.

A curve of the type C_3 which is itself the boundary of a little disk—or the complement of the disk!—is considered uninteresting and is said to be homologous to zero.

Now the beautiful observation of Morse was that under certain very general conditions on a space S *and a function f on it, the higher Betti numbers also force f to have extrema.*

More precisely, if $M_i(f)$ denotes the extrema of f *with precisely i linearly independent directions of steepest descent*, then the unequality $(*)$ is generalized by him to

$$(**) \qquad M_i(f) \geqq R_i(S).$$

Let us test this theorem on our torus T, and let f be the function on T which assigns to a point on T its z-coordinate. The extrema of f are then pretty clearly the intersection of the z-axis with T, and we clearly have that A is a minimum, B and C are saddle points with one direction of steepest descent, and D, the maximum· has precisely two (see Fig. 2).

Thus if we in general define the Morse series of f by

$$\mathcal{M}_t(f) = \sum M_i(f)t^i,$$

then, in this instance, we obtain the formula

$$\mathcal{M}_t(f) = 1 + 2t + t^2.$$

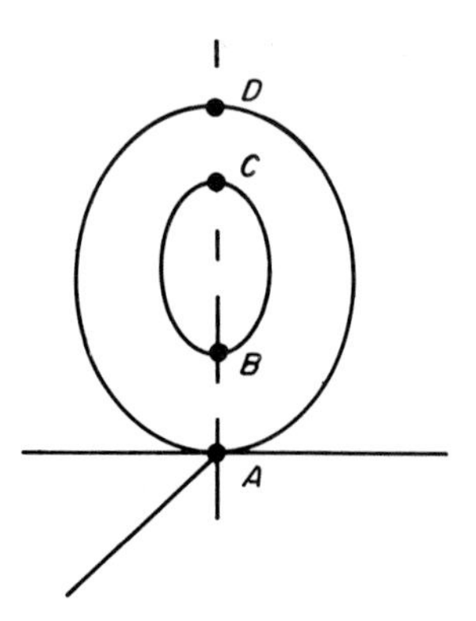

Fig. 2

Thus for this function

$$\mathcal{M}_t(f) = P_t(T),$$

and one correspondingly calls this f a *perfect Morse function on T.*

In short, a *perfect Morse function* has the minimum number of extrema forced on f by the topology of S. Actually the famous Morse inequalities are a little better than (**), and I might as well write them down for you: In the notation introduced already they assert that

$$\mathcal{M}_t(f) - P_t(f) = (1 + t)Q_t,$$

where $Q_t = q_0 + q_1 t + \cdots$ is a series with *nonnegative integer coefficients.*

Well let us return now to the problems at hand. Clearly these ideas give a clue on how to proceed.

In each case we must construct a space S appropriate to the problem, study its topology, and then verify that the Morse theory is valid.

For the geodesic problem the appropriate space can be taken to be the space Ω of piecewise smooth curves on S joining P to Q, topologized appropriately. Its Poincaré series can then be computed and is found to be

$$P_t(\Omega) = 1 + t + t^2 + t^3 + \cdots.$$

Next one finds that the compactness conditions on f necessary for the inequalities to hold are verified and hence concludes that there are an infinite number of geodesics joining P to Q. In fact the above formula leads one to expect saddle points of every *index of steepest descent.*

To apply this theory to the spectral theorem, we must take note that the function

$$(Lf, f)$$

is invariant under the transformation $f \mapsto \lambda f$, with λ a complex number of absolute value 1. Thus, properly speaking, this function is defined on the space of rays $R(H)$ of an appropriate Hilbert space H. Finally for this space one computes that

$$P_t\{R(H)\} = 1 + t^2 + t^4 + \cdots.$$

So once again, provided certain compactness conditions are satisfied—and these arise from the ellipticity of the equation—we expect an infinite number of eigenvalues.

But this is enough exposition for an evening session of a hard day. I hope this very rough account will however sufficiently raise the interest of some of you in this subject for you to pursue it further. Not just in its own right, for I am quite certain that some quite marvelous applications of this theory are yet to be found. Morse theory was my first love after network theory, and it was my good fortune some 20 years ago to make some quite unexpected applications of it to homotopy theory. Since then after many love affiairs—for as I learned already from Dick—fidelity is not necessarily a virtue in mathematics, I find myself, in collaboration with M. Atiyah, again occupied with Morse theory. And this time in an exciting new setting related to the theory of the Yang–Mills equation in field theory on the one hand and certain questions in algebraic geometry on the other.

DEPARTMENT OF MATHEMATICS
HARVARD UNIVERSITY
CAMBRIDGE, MASSACHUSETTS

Transforms

D. G. BOURGIN

Perhaps since my acquaintanceship with Dick Duffin dates back to the period when he was a mere fledgling, I may have some tenuous claim on inaugurating these revels. I interpret the temporal order relation in the program as a suggestion to relate to his early concerns in mathematics. Dick had enrolled in a class of mine on something to do with transforms. However, when at the end of the course he materialized in my office with a notebook stuffed with contributions to various open problems I had suggested both formally and informally in my lectures, he was transformed from a neophyte who asked disconcerting, discerning, direct questions into a scientific collaborator and then into a dear friend for over 40 years. We wrote three papers together, even though we never got around to publishing a variety of other results we had obtained, so it is natural for me to concern myself in the body of my address with two (dissimilar) topics, both related to transforms. The first is somewhat abstract and is near my recent interests.

1. FIRST TOPIC

Assume then for X and Y, topological spaces, that

$$f: X \to Y \tag{1.1}$$

is upper semicontinuous (usc), with both $\operatorname{Im} f$ and $f(x)$ compact in the sense that the graph

$$\Gamma(f) = \{x, y \mid y \in f(x)\}$$

is closed in $X \times Y$. Such transformations enter naturally in great profusion. For instance, the inverse of a continuous map is usc. Say f *dominates* g and write $g \prec f$ if f and g are usc and $g(x) \subset f(x)$ for all x. We shall draw on a reservoir of spaces that we loosely refer to as *nice spaces*. These are connected

ISBN 0-12-178150-X

and locally connected and, depending on context, include polyhedra, absolute and neighborhood retracts, Banach spaces, etc.

The Kakutani theorem asserts

If X is a closed disk in R^n and f is an usc self-map where $f(x)$ is a closed convex set, there is a fixed point $\bar{x}$ in the sense that

$$\bar{x} \in f(\bar{x}). \tag{1.2}$$

This theorem and the many results based on it have almost invariably required point images to be convex, or at worst acyclic, that is to say, to be *pointlike.*

Lefschetz [1] and O'Neill [2] have suggested formal, hardly realizable methods of dealing with set-valued maps by using a multiplicity of homomorphisms (whose existence is not assured) to obtain a *set* of Lefschetz numbers on the one hand, or a limit of smoothing homomorphisms on chains on the other, and almost restricted to the case that R_1, the first Betti number, differs from 0.

My advance consisted in replacing the heavy requirements of existence of homology group homomorphisms in all dimensions when Lefschetz numbers are involved, by a method requiring only that such a homomorphism exist in a single prescribed dimension, or above a certain dimension [3–5].

The difficulties inherent in the subject are occasioned by the fact that various procedures natural for single-valued maps must be discarded for set-valued maps. Thus if $f(x)$ is acyclic for each x, $f^2(x)$ need not be acyclic for some x or even for any x. For instance with X the unit circle S^1 let $f(x)$ be the semicircle symmetrically placed opposite x for each x. Then $f^2(x)$ is S^1, a nonacyclic set. The Brouwer fixed point theorem, and indeed homotopy theory, leans on the fact that S^n and in particular S^2 is not deformable over itself to a point. However, set-valued usc deformations are easy to define. Thus for each t in the unit interval $0 \leqq t \leqq 1$, let h map the north pole into the horizontal circle at height $2 - 2t$. Let h move all other points proportionately to t along the great circles joining them with the south pole so that $h(\ ,1)$ maps S^2 into this point. Finally Strother [6] published a counterexample to the set-valued extension of the Brouwer theorem when $f(x)$ can be nonacyclic. A somewhat simpler example is afforded by projecting all points in the n-disk D^n, not including the origin, radially onto the bounding sphere S^{n-1}. The origin maps into all of S^{n-1}. Combine this map with the antipodal reflection through the center. The resulting transformation is usc, but has no fixed point.

The points with nonacyclic images under f are referred to as *singular* and are partitioned according to dimension by

$$\sigma_r = \{x \mid H^r(f(x)) \neq 0\}, \tag{1.3}$$

where H^r refers to cohomology over the rationals and is tacitly assumed throughout this lecture to be finitely generated. We refer to this as *condition F*. The desired advance is possible because of recent generalizations of the Vietoris–Begle theorem. These generally are formulated and demonstrated in completely different ways by me [3] and by Skljarenko [7] and neither of our results includes the other. However, for the applications we have in mind Skljarenko's results are more general and establish that if there are relatively few singular points, and $\sigma_r = \varnothing$ for $r > p$, then $H^r(X)$ is isomorphic to $H^r(Y)$ for $r > p$. (However, if the dimensions of the singular sets are nonzero, then the value of the bound, p, must be increased to some well-defined value $\bar{p}$.) Call maps satisfying these conditions *admissible*.

The specific technical device, suggested originally by Eilenberg and Montgomery [8], is the factorization of f in (1.1) described in the diagram

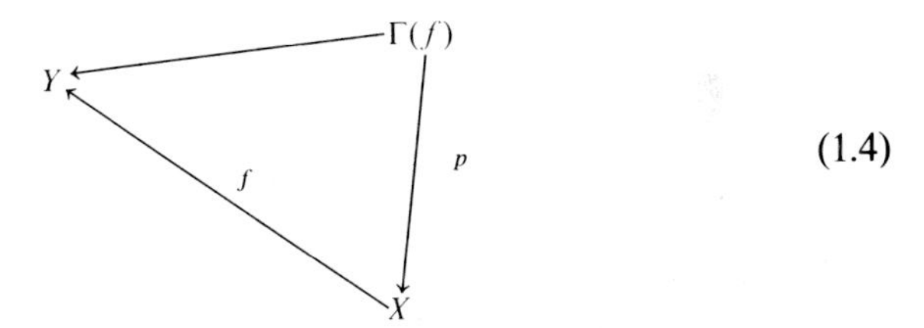

(1.4)

that is to say, with p and q projections onto X and Y from $X \times Y$,

$$f = qp^{-1}. \tag{1.5}$$

We may remark that any admissible g with $g \prec f$ can replace f. The generalized Vietoris–Begle theorem [3], [7] is applicable to define homomorphisms $f^*(n)$: $H^n(Y) \to H^n(X)$ in a restricted range of dimensions [even though the conditions that $f^*(n)$ exist in all dimensions, as would be required for the definition of the Lefschetz number are not met].

My extension of the Kakutani–Brouwer fixed point theorem can be stated in the form:

Theorem A. *If f is an usc transformation of the disk D^{n+1} to D^{n+1} where $\bar{p} \leqq n - 2$, then f admits a fixed point.*

Write $\dot{D}$ for the boundary of D^{n+1} so $\dot{D}$ is homeomorphic to S^n under k. The demonstration is effected by modifying a classical lemma, viz.,

Theorem B. *There is no usc map f of D^{n+1} to the annulus $D^{n+1} \times I$ with f restricted by $kf(x) \cap -kx = \varnothing$ for $x \in \dot{D} = \dot{D} \times 0$.*

The introduction of the annulus is necessitated by the fact that the proof proceeds by reflecting $f(x)$ through X onto $\dot{D}$ to give a well-defined map F when there is no fixed point. However, to assure $F(x)$ is a homeomorph of

$f(x)$ and is therefore acyclic when $f(x)$ is, one need modify the definition of F and map into an annulus rather than into $\dot{D}$.

Theorem A has possible application to min–max theorems like those underlying game theory. Thus for X and Y convex compacta in R^k and R^l, respectively, and f a continuous real function on $X \times Y$ define the maps $M: Y \to X$, $N: X \to Y$, and $g: X \times Y \to X \times Y$ by

$$M(y) = \{x \mid f(x, y) \leqq \max_X f(x, y)\} \subset X,$$

$$N(x) = \{y \mid f(x, y) \geqq \min_Y f(x, y)\} \subset Y, \tag{1.6}$$

$$g(x, y) = M(y) \times N(x).$$

Then application of Theorem A yields

Theorem C. *If M and N are usc and if there are only a finite number of singular points for N and M with none in dimensions above p and q, respectively, where $k \geqq p + 2$ and $l \geqq 1 + 2$, then there is a saddle point $\bar{x}$, $\bar{y}$ for which*

$$\min f(\bar{x}, y) \leqq f(\bar{x}, \bar{y}) \leqq \max f(x, \bar{y}).$$

An even stronger theorem is stated in [4], but it must be owned that condition F [even in the case of acyclicity of $M(x)$ and $N(x)$] may present difficulties in application, that are avoided in the conventional bilinear format for $f(x, y)$, because of obvious convexity.

Applications of Theorem A to problems of analysis would be of interest, say, in the field of differential or integral equations. What are analytical correspondences to the conditions in Theorem C? Perhaps a generalized game could be defined involving the strategies and payoffs and constraints in some implicit relation.

Our ideas prevail even for nonfinite dimensions and serve to define an extension of the Leray–Schauder index or degree. I shall first summarize my original presentation.

Suppose then D is a convex body in the Banach space E with boundary $\dot{D}$. For simplicity assume D is symmetric about the origin θ and let the subscript N indicate intersection with an N-dimensional linear subspace E_N. We require that $E_N \subset E_{N+1}$. Thus $\dot{D}_N = \dot{D} \cap E_N$, etc.

Let F be a compact usc mapping of D to E taking points into closed sets and let p be the projection $\Gamma(F \mid \dot{D}) \to \dot{D}$. Let

$$f = I - F, \tag{1.7}$$

where I is the identity map. It is easily established that there are projections Q_N on $\overline{F(\dot{D}_N)}$ to E_N with 0 disjunct from the image of $\dot{D}_N$ under the map

$$(I - F)_N \dot{D}_N = (p_N - Q_N q_N) p_N^{-1} \dot{D}_N. \tag{1.8}$$

For F *admissible* then, provided F admits no fixed point [cf. (1.1)] on $\dot{D}$, $I - F$ maps D into an annulus $A_N = S^{N-1} \times \mathrm{I}$ or shell surrounding 0. Hence a degree d_N can be defined in terms of the induced homomorphisms by

$$f_N^* = p_N^{*-1}(p_N^* - Q_N^* q_N^*): H^{N-1}(A_N) \approx H^{N-1}(S^{N-1}) \to H^{N-1}(\dot{D}_N). \quad (1.9)$$

In spite of appearance, d_N for large N is independent of N (as well as choice of Q_N and E_N). It is therefore designated as $d(f)$. It is homotopy invariant when no fixed points are introduced on the boundary, and the fact that it is nonvanishing guarantees F has a fixed point [5].

The restriction to a solid sphere D can, of course, be weakened by admissible deformation retractions, but indeed nothing in the formal developments of [5] is affected when D is the closure of a bounded open subset of E, though of course in general $\dot{D}_N$ and $\dot{D}_N$ are no longer single components. In fact, we assume now that $\dot{D}$ and $\dot{D}_N$ have a *finite number* of components. $H^{N-1}(\dot{D}_N)$ has at least as many generators as there are components of $\dot{D}_N$, and the degree is to be defined by $\sum m_i(N)$ where for γ_1 a generator of $H^{N-1}(S^{N-1})$ and $\{\rho_i(N) | i = 1, \ldots, k\}$ the generators of $H^{N-1}(\dot{D}_N)$,

$$f_N^*(\gamma_1) = \sum m_i(N)\rho_i(N). \quad (1.10)$$

Each component of $\dot{D}_N$ is $N - 1$ spherelike. The demonstration (cf. [5]), that the degree is independent of N for N large is circumvented by interpreting the groups as Cech groups and taking direct limits. Thus since the image of D is an annulus around 0, (1.9) becomes

$$f^*: H^{\infty-1}(S) = \underset{\rightarrow}{L} H^{N-1}(S^{N-1}) \to H^{\infty-1}(\dot{D}) = \underset{\rightarrow}{L} H^{N-1}(\dot{D}_N), \quad (1.11)$$

where S is the sphere in E. Write m_i for $m_i(\infty)$, whence our degree definition is

$$d(F) = \sum m_i. \quad (1.12)$$

The introduction of the infinite cohomology is more than a formal device since by [9] for N finite the Alexander–Pontrjagin duality isomorphisms are available making $\dot{D}_N$ and $E_N - \dot{D}_N$ correspond, whence

$$H^{\infty-1}(S) \approx H_0^{\sim}(S), \quad (1.13a)$$

$$H^{\infty-1}(\dot{D}) \approx H_0^{\sim}(E - \dot{D}), \quad (1.13b)$$

where singular homology is understood. This provides a clearer interpretation of m_i than as the limit of $m_i(N)$. (The possibility of a generalization of the degree as developed here with use of [9] was suggested by me in [10]. My student M. Colvin has since elaborated a somewhat more detailed (unpublished) exposition along similar lines.) The presentation above indicates that if D is connected, then

$$d(f) = m_1$$

since only ρ_1 enters. Moreover, it makes transparent that, for a finite number of fixed points, additivity maintains. More precisely, let $\{D_i | i = 1, \ldots, r\}$ be open neighborhoods of the r fixed points of F with disjunct closures, and let D be the closure of a connected open set containing $\cap \overline{D_i}$. Then from (1.11) and (1.13b), it is evident that

$$d = \sum d_i$$

where $d_i = d(f \,|\, \dot{D}_i)$.

I cite, in closing, an unpublished extension to coincidences or fuzzy coincidences of two compact maps F and G on $X \to Y$ subject to a certain relation R in $Y \times Y$, that is to say, x is a coincidence if $(F(x), G(x)) \subset R$. We require of R that it satisfy a restriction exploited in work by A. D. Wallace (cf. [1, p192]).

2. SECOND TOPIC

My second topic is more practical. It is possible to discern an evolutionary trend starting with the papers we wrote for the *American Journal* and for the *Bulletin of the American Mathematical Society* on Heaviside Laplace transforms and their generalizations. The questions I will be concerned with stem from an early article of mine [11].

Consider the transformation

$$Tf = \int_0^x h(x, t) f(t)\, dt = h^* f \tag{2.1}$$

for the case that $h(x, y)$ is nonzero only in the wedge

$$0 \leqq y < x < \infty. \tag{2.2}$$

(Sufficient regularity conditions are discussed in [11].) Such transformations enter in the classical theory of the Volterra integral equation

$$\begin{aligned} u(x) &= f(x) + \lambda \int_0^x h(x, t) u(t)\, dt \\ &= f(x) + \lambda T u(x) \end{aligned} \tag{2.3}$$

for which one of the earliest representations of a solution is

$$u(x) = f(x) + \sum \lambda^n \int_0^x h^n(x, y) f(y)\, dy, \tag{2.4}$$

where the nth power on h refers to the nth convolution. Thus

$$T(h) = h^2(x, y) = \int_y^x h(x, t) h(t, y)\, dt$$

and $h^n(x, y)$ is defined inductively.

Volterra was especially concerned with functions of the closed cycle, that is to say, with $h(x, y) = 1$, or more generally $h(x, y)$ a function of $x - y$. These functions are intimately connected with the Heaviside symbolic calculus. It has long been recognized that these functions and the resulting operation T are associated with a kernel e^{-xp} and involve the Laplace transform. This association is now course book material, but how was one to guess that $x - y$ and e^{-xp} go together [cf. (2.8)]?

If $h = 1$ or $x - y$, subject to (2.2), then $h(x, y)$ is easy to compute, namely $(x - y)^n/n!$ or $(x - y)^{2n-1}/(2n - 1!)$. However, under the innocuous appearing change of $x - y$ and $x + y$, already

$$h^2(x, y) = \tfrac{1}{6}5x^3 + 9(x^2y + xy^2) + 5y^3$$

and there seems no obvious directive for h^n. Indeed the associate kernel to $x + y$ is $e^{-pt^2/2}D_{1/2}(2pt)$ where $D_{1/2}$ is the parabolic cylinder function of order $\frac{1}{2}$.

The key concept of my paper [11] involved the recognition that the associate kernel $H(p, x)$ to $h(x, y)$ was determined by the prescription

$$H(p, x) = p \int_x^\infty H(p, t)h(t, x)\, dt. \tag{2.5}$$

This yields at once

$$\frac{H(p, x)}{p^m} = \int_x^\infty H(p, t)h^m(t, x)\, dt \tag{2.6}$$

and more generally even for nonanalytic ψ

$$\psi\left(\frac{1}{p}\right)H(p, x) = \int_x^\infty H(p, t)\psi(h(t, x))\, dt.$$

Indeed m need not be an integer nor even real in (2.6) and one is led to the analog of fractional integration and differentiation.

One defines an associated transformation to T by

$$(Su)p = \int_0^\infty H(p, t)u(t)\, dt. \tag{2.7}$$

If x is viewed as a parameter, then by reference to the domain (or support) of h, namely (2.2), evidently $u(t) = h(t, x)$ implies that (2.5) follows from (2.7).

Suppose S admits a left inverse which I write S^{-1}. Then the solution of (2.3) is

$$u = S^{-1}\left(\left(\frac{p}{p - \lambda}\right)Sf\right).$$

A particular representation is of interest, namely, in view of the relation of (2.1) and (2.6),

$$u = S^{-1} \sum \left(\frac{\lambda}{p}\right)^n Sf$$

$$= \sum \lambda^n (h^n f),$$

which is recognized as (2.4).

When can one invert (2.1) practically? I showed that if $h(x, y)$ satisfies an mth order differential equation in y with possibly variable coefficients, then $H(p, x)$ satisfies a sort of adjoint mth order differential equation.

Thus suppose $h(x, y) = 1$. Then

$$\frac{dh}{dy} = 0,$$

and a straightforward argument using (2.1) yields the adjoint equation

$$\frac{\partial}{\partial x} H(p, x) + pH(p, x) = 0,$$

the solution of which is

$$H(p, x) = e^{-px} \tag{2.8}$$

so that e^{-px} appears as an ineluctable consequence rather than as a mysterious concomitant.

For

$$h(x, y) = x + y,$$

$$\frac{d^2h}{dy^2} = 0,$$

and the ensuing differential equation for H yields the kernel mentioned earlier. Application to various polynomial kernels for $h(x, y)$ have been made in University of Illinois theses by Ray Langebartel in 1949 and by his student Robert Curry in 1972. However, the full potential of the method has not been realized as yet.

It is not necessarily true that $H(p, x)$ is uniquely determined. Moreover, the inverse problem of starting with $H(p, x)$ and determining $h(x, y)$ as the solution of (2.5) has not been considered. While my students and I have used the condition that h satisfy a differential equation in order to determine H, other restrictions on $h(x, y)$ ought to be exploited. For $h(x, y) = x + y$, my treatment of S^{-1} [11] is vaguely reminiscent of the introduction of the Mellin transform for the inversion of the Laplace transform, but this is only one possible type of inversion. A central open problem is, then, the representation or representations of S^{-1}, possibly even making use of the spectrum of some correlated expressions for S to bring everything into L_2 and so achieve a self-adjoint operator. As expected, this spectrum is continuous when $h = x + y$. In another direction, considerations analogous to the treatment of counterparts of the Dirichlet series intervention in Bourgin–Duffin [12] would very likely be fruitful.

We have now come full cycle to a range of questions that were central in Dick Duffin's coming of age. Possibly some of you may be persuaded to turn your attention to them, and in any case I trust that you, too, may have the experience of welcoming a student who enters your office with a notebook crammed with developments for these or other extensions.

REFERENCES

1. S. Lefschetz, On coincidences of transformations, *Bol. Sci. Mat.* (1957), 16–25.
2. B. O'Neill, Induced homology homomorphisms for set-valued maps, *Pacific J. Math.* **7** (1957), 1179–1184.
3. D. G. Bourgin, Cones and Vietoris–Begle type theorems, *Trans. Amer. Math. Soc.* **174** (1972), 155–183.
4. D. G. Bourgin, Fixed point and min–max theorems, *Pacific J. Math.* **45** (1973), 403–412.
5. D. G. Bourgin, A generalization of the mapping degree, *Canad. J. Math.* **26** (1974), 1109–1117.
6. W. L. Strother, On an open question concerning fixed points, *Proc. Amer. Math. Soc.* **4** (1953), 988–993.
7. E. G. Skljarenko, Some applications of the theory of sheaves in general topology, *Uspekhi Mat. Nauk* **19**, No. 6 (1964), (120, 47–70; *Russian Math. Surveys* **19** (1964), 41–62 (MR 30 No. 1490).
8. S. Eilenberg and D. Montgomery, Fixed point theorems for multi-valued transformations, *Amer. J. Math.* **68** (1946), 214–222.
9. K. Geba and A. Granas, Infinite dimensional cohomology theories, *J. Math. Pures Appl.* **52** (1973), 145–270.
10. D. G. Bourgin, Fixed point theory for non-acyclic set-valued maps, *Southwest Confer., Lafayette, La.*, 1974.
11. D. G. Bourgin, Associated transforms and convolutions, *Rev. Cienc.* **469** (1949), 5–46.
12. D. G. Bourgin and R. J. Duffin, The Heaviside Operational Calculus, *Amer. J. Math.* **59** (1937), 489–505.

DEPARTMENT OF MATHEMATICS
UNIVERSITY OF HOUSTON
HOUSTON, TEXAS

Part II

GRAPHS AND NETWORKS

Matrix Operations Induced by Electrical Network Connections—A Survey

W. N. ANDERSON, JR.

G. E. TRAPP

Motivated by Kirchhoff's current laws, a linear subspace—termed a confluence—is constructed. The natural duality of current and voltage is preserved in this model. It is shown that every confluence induces a mapping between two sets of positive operators. As an example, the confluence generated by a resistive network induces a mapping from the branch impedance matrix of the network to the port impedance matrix. The shorted operator, one particular induced map, is considered in detail because it is fundamental in the theory of confluences. Various other examples and special cases are also discussed. Generalizations to infinite and frequency dependent cases are also considered. Finally, a survey of related work is given.

1. INTRODUCTION

The authors of this paper are both former students of R. J. Duffin, and he is responsible for stimulating the work presented here.

R. J. Duffin has greatly influenced the development of mathematical models derived from electrical network theory. A survey of his work reveals that he uses elementary physical principles to motivate elegant mathematical theories; an overview of Duffin's contributions in this area may be found in [38]. In this paper, we consider one aspect of Duffin's work—algebraic theories derived from electrical networks. We model our approach after that presented in the fundamental paper of Bott and Duffin [23], where by specifying precisely the physical laws governing an electrical network, they develop a rich algebraic theory.

Our starting point is a special type of a subspace of a vector space; we term this type of a subspace a confluence. An example of a confluence is provided by considering the set of all the allowable current flows in an electrical network. We do not develop the network theory applications of confluences until later in the paper, instead in the first two sections we present the basic properties of abstract confluences and prove a fundamental existence theorem. This theorem states that every confluence induces a mapping on the set of positive operators. Many interesting properties of this

ISBN 0-12-178150-X

mapping—termed a Φ-operator—are presented, and in particular, a special case of a Φ-operator, the shorted operator, is thoroughly investigated.

After developing the theory of confluences and Φ-operators, we then describe two different electrical network situations in which confluences and Φ-operators arise. One particular subclass of Φ-operators is considered in more detail; the operators in this subclass are called routers. A router is an order preserving semigroup operation defined on the set of positive operators. Special cases of routers have been thoroughly investigated and we present a brief summary of these cases.

We discuss the extension of confluences and Φ-operators to a Hilbert space setting and to an operator valued function setting. Finally, we present after generalizations of this work.

To begin, we will need the following notation and definitions. We assume that we have a complex finite dimensional inner product space. We denote the inner product by $\langle , \rangle$. A matrix A is Hermitian positive semidefinite (HSD) if and only if $\langle Ax, x\rangle$ is real, positive, and equals 0 only when Ax equals 0. If A and B are HSD, then by the partial ordering, denoted $A \geqq B$, we mean that $A - B$ is HSD.

2. CONFLUENCES AND Φ-OPERATORS

Let E and P denote respectively finite dimensional complex vector spaces. Let $V = E \oplus P$ with inner product

$$\langle (a_1, c_1), (a_2, c_2)\rangle_V = \langle a_1, a_2\rangle_E + \langle c_1, c_2\rangle_P.$$

We assume that fixed orthonormal bases are given for E and P, then their union is a basis for V. All matrices will be written with respect to these fixed bases.

Definition. A *confluence* is a subspace Γ of V such that

(i) for all $c \in P$, there is an $a \in E$ so that $(a, c) \in \Gamma$;
(ii) if $(0, c) \in \Gamma$, then $c = 0$.

We will see in later sections that confluences arise naturally in electrical network theory. Bott and Duffin [23, 35] exhibit many interesting duality results for electrical networks. To obtain a theory of duality in our setting, we need the following definition and theorem.

Definition. If Γ is a confluence, then the *dual confluence* is defined by

$$\Gamma' = \{(\alpha, \gamma) \in V \mid \langle \alpha, a\rangle_E = \langle \gamma, c\rangle_P \quad \text{for all} \quad (a, c) \in \Gamma\}.$$

Theorem 1. *If Γ is a confluence, then*

(a) *Γ' is a confluence,*
(b) $\dim \Gamma + \dim \Gamma' = \dim V$,
(c) $(\Gamma')' = \Gamma$.

The proof is an easy exercise in linear algebra and is omitted.

Since a confluence is a subspace, it may be represented by a basis matrix. The following lemma exhibits the basis representation for a confluence and the dual confluence, and using the basis matrices, we obtain an alternate definition of a confluence.

Lemma 1. *There exist matrices K, L, M, and N such that the columns of J_Γ and $J_{\Gamma'}$ form bases of Γ and Γ', respectively, where J_Γ and $J_{\Gamma'}$ are given by*

$$J_\Gamma = \begin{bmatrix} K^* & L^* \\ I & 0 \end{bmatrix}, \qquad J_{\Gamma'} = \begin{bmatrix} M^* & N^* \\ I & 0 \end{bmatrix}.$$

Moreover, the vector (a, c) is in Γ if and only if the following equivalent conditions hold:

(i) *$Ma = c$ and $Na = 0$;*
(ii) *there is a vector d such that $K^*c + L^*d = a$.*

Similarly (α, γ) is in Γ' if and only if

(iii) *$K\alpha = \gamma$ and $L\alpha = 0$;*
(iv) *there is a vector δ so that $M^*\gamma + N^*\delta = \alpha$.*

Proof. By condition (i) of the definition of a confluence, we may choose K^* and M^* as shown, the remaining columns (L^* and N^*) are chosen to be a basis for all vectors of the form $(a, 0)$ in Γ, and $(\alpha, 0)$ in Γ', respectively. Part 2 of this lemma follows directly from the definitions of J_Γ and $J_{\Gamma'}$. ■

To this point, we have considered a confluence as a subspace of a linear space. We may also think of a confluence as a partial linear function from E to P. The adjective partial is used because the domain of the function may not be all of E. The following definition and lemma are a natural consequence of the function concept.

Definition. Let Γ be a confluence:

$$\mathrm{dom}(\Gamma) = \{a \in E \mid (a, c) \in \Gamma \quad \text{for some } c \in P\};$$

$$\ker(\Gamma) = \{a \in E \mid (a, 0) \in \Gamma\}.$$

Lemma 2. $\ker(\Gamma') = (\mathrm{dom}(\Gamma))^\perp$.

Proof. Let $e \in (\mathrm{dom}(\Gamma))^\perp$; then $\langle e, a \rangle = 0$ for all $a \in \mathrm{dom}(\Gamma)$. This is equivalent to saying $\langle e, a \rangle = \langle 0, c \rangle$ for all $(a, c) \in \Gamma$, thus from the definition of Γ', $(e, 0) \in \Gamma'$. Therefore, $e \in \ker(\Gamma')$. Since each of these steps is reversible we are done. ■

We have now completed our discussion of confluences. Our next goal is to show that this seemingly elementary concept of a confluence yields a surprisingly powerful result in operator theory. We begin by showing that a confluence induces a correspondence between vectors in P and vectors in E.

Theorem 2. *Let Γ be a confluence and let A be an HSD operator on E. Then given a vector $c \in P$, there exists a vector $a \in E$ and a unique vector $\gamma \in P$ such that $(a, c) \in \Gamma$ and $(Aa, \gamma) \in \Gamma'$.*

Proof. Using Lemma 1, we wish to solve the equations

$$Ma = c, \qquad Na = 0, \qquad \text{and} \qquad M^*\gamma + N^*\delta - Aa = 0.$$

We employ the Fredholm alternative theorem; the homogeneous adjoint system is

$$Mu_3 = 0, \qquad Nu_3 = 0, \qquad \text{and} \qquad M^*u_1 + N^*u_2 = Au_3.$$

Suppose u_1, u_2, and u_3 furnish a solution to the adjoint system; then $\langle Au_3, u_3 \rangle = 0$. A is HSD; therefore $Au_3 = 0$. Since Γ' is a confluence, we claim $u_1 = 0$, otherwise from condition (iv) of Lemma 1, we would have a vector $(0, u_1)$ in Γ'. To complete the proof, we note that $\gamma = 0$ whenever $c = 0$, and it follows that γ is uniquely determined by c. ■

Definition. Let Γ be a confluence and A an HSD matrix on E, then define $\Phi(A)$ by $\Phi(A)c = \gamma$, where c and γ are as in Theorem 2.

Theorem 2 shows that $\Phi(A)$ is a well-defined function on the set of HSD operators on E. Using the notation of Theorem 2, we see that $\langle \Phi(A)c, c \rangle = \langle \gamma, c \rangle$, but Γ is a confluence and hence $\langle \gamma, c \rangle = \langle Aa, a \rangle$. Since A is HSD and Γ is a confluence, we see that $\langle \Phi(A)c, c \rangle$ is real, positive, and 0 only when $\Phi(A)c$ is 0. Thus we have proved the following theorem.

Theorem 3. *$\Phi(A)$ is HSD if A is HSD.*

The concept of power is fundamental in electrical network theory. If an electrical network is represented by an HSD matrix A, then the power dissipated in the network is given by $\langle Aa, a \rangle$. Our next theorem relates the power of $\Phi(A)$ to the power of A.

Theorem 4. *Given a confluence Γ and the associated operator Φ, let c be an arbitrary vector, and let $\gamma = \Phi(A)c$; then for any $(a, c) \in \Gamma$, $\langle Aa, a \rangle \geqq \langle \gamma, c \rangle$, and moreover equality holds if and only if $(Aa, \gamma) \in \Gamma'$.*

Proof. From the definition of a confluence, equality will hold whenever $(Aa, \gamma) \in \Gamma'$.

In fact, by Theorem 2, there exists an a_0 such that $(a_0, c) \in \Gamma$ and $(Aa_0, \gamma) \in \Gamma'$. If $(a, c) \in \Gamma$, then since Γ is a vector space $(t, 0) \in \Gamma$ where $t = a - a_0$. Now we know that $(Aa_0, \gamma) \in \Gamma'$, so we have $\langle Aa_0, t \rangle = 0$. Using this fact we obtain the following: $\langle Aa, a\rangle = \langle Aa_0, a_0\rangle + \langle At, t\rangle = \langle \gamma, c\rangle + \langle At, t\rangle$. Therefore, we have the inequality. For the other half of the equality condition we see that if equality holds then $\langle At, t\rangle = 0$. This implies $At = 0$, or $Aa = Aa_0$, so that $(As, \gamma) = (Aa_0, \gamma) \in \Gamma'$. ■

Theorem 4 may be rewritten to obtain a minimization definition of Φ. Consider the following problem. Given c, define the quadratic form of $\Phi(A)$ at c by $\langle \Phi(A)c, c\rangle = \inf\{\langle Aa, a\rangle \mid (a, c) \in \Gamma\}$. Theorem 4 guarantees that the infimum is obtained.

Using the minimization expression, $\langle \Phi(A + B)c, c\rangle$ may be written as $\inf\{\langle (A + B)a, a\rangle \mid (a, c) \in \Gamma\}$. Separating the inner product $\langle (A + B)a, a\rangle$ and using the fact that the infimum of a sum is greater than the sum of the infimums, we have that

$$\langle \Phi(A + B)c, c\rangle \geqq \inf\{\langle Aa, a\rangle \mid (a, c) \in \Gamma\} + \inf\{\langle Ba, a\rangle \mid (a, c) \in \Gamma\}.$$

The right-hand side of this inequality is just $\Phi(A) + \Phi(B)$, and we have the following.

Theorem 5. $\Phi(A + B) \geqq \Phi(A) + \Phi(B)$.

We will now consider the relationship of Φ-operators which arise from different confluences. In order to do this, we must discuss Φ-operators applied to rank 1 matrices.

The reader should recall that if α is a vector, then $\alpha\alpha^*$ is a rank 1 HSD matrix. Moreover, any rank 1 HSD matrix A can be expressed uniquely (up to a scalar multiple) in the form $A = \alpha\alpha^*$; and the range of A is the subspace spanned by α. It is also relevant to observe that for vectors α and β $\langle \alpha, \beta\rangle = \beta^*\alpha$.

Lemma 3. *Let Γ be a confluence with dual confluence Γ', and let Φ be the operator induced by Γ. If $\alpha \in \operatorname{dom}(\Gamma')$, then $\Phi(\alpha\alpha^*) = \gamma\gamma^*$, for γ such that $(\alpha, \gamma) \in \Gamma'$, and if $\alpha \notin \operatorname{dom}(\Gamma')$, then $\Phi(\alpha\alpha^*) = 0$.*

Proof. Suppose that $(\alpha, \gamma) \in \Gamma'$; then using Theorem 4, for any $c \in P$,

$$\begin{aligned}
\langle \Phi(\alpha\alpha^*)c, c\rangle &= \inf\{\langle \alpha\alpha^*a, a\rangle \mid (a, c) \in \Gamma\} \\
&= \inf\{\langle \alpha^*a, \alpha^*a\rangle \mid (a, c) \in \Gamma\} \\
&= \inf\{|\langle a, \alpha\rangle|^2 \mid (a, c) \in \Gamma\} \\
&= |\langle c, \gamma\rangle|^2 \\
&= \langle \gamma\gamma^*c, c\rangle.
\end{aligned}$$

The penultimate equality follows from the definition of the dual confluence. Therefore we have that $\Phi(\alpha\alpha^*) = \gamma\gamma^*$. Next suppose that $\alpha \notin \operatorname{dom}(\Gamma')$. By Lemma 2 there is a vector $b \in \ker(\Gamma)$ such that $\langle b, \alpha\rangle = k \neq 0$. Given an arbitrary vector $c \in P$, choose a vector a such that $(a, c) \in \Gamma$. Then for all λ, since Γ is a subspace, the vector $(a + \lambda b, c) \in \Gamma$. However $\langle a + \lambda b, \alpha \rangle = \langle a, \alpha \rangle + \lambda k$, which can be made equal to 0 for an appropriate choice of λ. Thus the infimum is 0. ■

We have previously defined an order relation between matrices based on the quadratic form. In a similar manner, we may consider operations Φ_1 and Φ_2 derived from confluences Γ_1 and Γ_2, respectively. We define $\Phi_1 \leqq \Phi_2$ if $\Phi_1(A) \leqq \Phi_2(A)$ for all HSD matrices A. The following theorem gives necessary and sufficient conditions on the confluences Γ_1 and Γ_2 so that the corresponding Φ's satisfy $\Phi_1 \leqq \Phi_2$.

Theorem 6. *Let Φ_1 and Φ_2 be operations derived from the confluences Γ_1 and Γ_2, respectively. Then $\Phi_1 \leqq \Phi_2$ if and only if there is a constant k with $|k| \geqq 1$ such that for every vector $(a, c) \in \Gamma_2$, the vector $(a, kc) \in \Gamma_1$.*

Proof. Suppose that there is such a k. Let A be a HSD matrix; then given an arbitrary vector $c \in P$, there exists a vector a such that $(a, c) \in \Gamma_2$, and $(Aa, \gamma) \in \Gamma_2'$. By hypothesis, $(k^{-1}a, c) \in \Gamma_1$, so that by Theorem 4

$$\langle \Phi_1(A)c, c \rangle \leqq \langle Ak^{-1}a, k^{-1}a \rangle = |k|^{-2}\langle Aa, a \rangle = |k|^{-2}\langle \Phi_2(A)c, c \rangle.$$

Since $|k| \geqq 1$, it follows that $\Phi_1(A) \leqq \Phi_2(A)$ for all A.

Conversely, suppose that $\Phi_1(A) \leqq \Phi_2(A)$ for all HSD matrices A. We first show that if $(\alpha, \gamma_1) \in \Gamma_1'$ with $\gamma_1 \neq 0$, then $(\alpha, k\gamma_1) \in \Gamma_2'$. By Lemma 3 $\Phi_1(\alpha\alpha^*) = \gamma_1\gamma_1^*$. Moreover $\Phi_2(\alpha\alpha^*) \geqq \Phi_1(\alpha\alpha^*)$, and thus $\Phi_2(\alpha\alpha^*) \neq 0$. Using Lemma 3 again, $(\alpha, \gamma_2) \in \Gamma_2'$ for some γ_2, and $\Phi_2(\alpha\alpha^*) = \gamma_2\gamma_2^*$. Thus $\gamma_1\gamma_1^* \leqq \gamma_2\gamma_2^*$. Therefore $\gamma_2 = k\gamma_1$ where $|k| \geqq 1$ by Theorem 1 of [32].

Conceivably k depends on α. We now show that in fact k is independent of α. Let $(\alpha, \gamma) \in \Gamma_1'$ and $(\beta, \theta) \in \Gamma_1'$. We know that there exist k_1 and k_2 so that $(\alpha, k_1\gamma) \in \Gamma_2'$ and $(\beta, k_2\theta) \in \Gamma_2'$. We must consider two cases. First let γ and θ be independent. Since Γ_2' is a subspace, and since we know that $(\alpha + \beta, k_3(\gamma + \theta)) \in \Gamma_2'$ for some k_3, it follows that

$$(0, k_3(\theta + \gamma) - k_1\gamma - k_2\theta) \in \Gamma_2'.$$

But Γ_2' is a confluence, and we must have that $(k_3 - k_2)\theta + (k_3 - k_1)\gamma = 0$. Thus we must have that $k_1 = k_2 = k_3$ because θ and γ are linearly independent.

If θ and γ are linearly dependent and both nonzero, then so are $k_1\gamma$ and $k_2\theta$. Thus there exist λ_1 and λ_2, not both zero, such that $\lambda_1 k_1\gamma + \lambda_2 k_2\theta = 0$. Let $\sigma = \lambda_1\alpha + \lambda_2\beta$ then $\lambda_1(\alpha, k_1\gamma) + \lambda_2(\beta, k_2\theta) = (\sigma, 0)$, and this vector is in Γ_2'. Therefore by Lemma 3, $\Phi_2(\sigma\sigma^*) = 0$, and hence $\Phi_1(\sigma\sigma^*) = 0$. Again by Lemma 3, we have that since $(\sigma, \lambda_1\gamma + \lambda_2\theta) \in \Gamma_1'$. Then $\lambda_1\gamma + \lambda_2\theta = 0$. If $k_1 \neq k_2$ this contradicts the definition of λ_1 and λ_2.

We now know that if $(\alpha, \gamma) \in \Gamma_1'$, and $\gamma \neq 0$, then $(\alpha, k\gamma) \in \Gamma_2'$. We still need to show that if $(\alpha, 0) \in \Gamma_1'$, then $(\alpha, 0) \in \Gamma_2'$. To see this, consider some

$(\beta, \gamma) \in \Gamma'_1$, with $\gamma \neq 0$. Then $(\beta, k\gamma) \in \Gamma'_2$. Since Γ'_1 is a subspace, $(\alpha + \beta, \gamma) \in \Gamma'_1$, so that $(\alpha + \beta, k\gamma) \in \Gamma'_2$. Now Γ'_2 is a subspace, so we have that $(\alpha + \beta, k\gamma) - (\beta, k\gamma) = (\alpha, 0) \in \Gamma'_2$, as required.

We have proved the result for the dual confluences, the theorem however refers to the confluences themselves. To complete the proof, consider a vector $(a, c) \in \Gamma_2$. Then for any vector $(\alpha, \gamma) \in \Gamma'_1$, we have $(\alpha, k\gamma) \in \Gamma'_2$. By the definition of a confluence it follows that $\langle a, \alpha \rangle = \langle c, k\gamma \rangle = \langle k^*c, \gamma \rangle$. Therefore $(a, k^*c) \in \Gamma_1$. Since $|k^*| = |k| \geqq 1$, the theorem is proved. ∎

As a special case of Theorem 6, we have the following result. If for all HSD matrices A, $MAM^* \leqq NAN^*$, where M and N are the same size matrices, then $N = kM$ for some constant k with $|k| \geqq 1$.

3. SHORTED OPERATOR

One of the most interesting examples of a Φ-operator induced by a confluence is known as the shorted operator. Let S be a subspace of E and let P be isomorphic to S. For convenience, we will use the same symbols to represent vectors in S and P. We let the confluence Γ be the subspace consisting of all vectors of the form $(c + t, c)$ where t is in $S^\perp$. If A is an HSD matrix on E then we have that for any arbitrary c in P, $\langle \Phi(A)c, c \rangle = \inf\{\langle A(c + t), (c + t) \rangle | t \in S^\perp\}$. The operator $\Phi(A)$ on P can be identified with an operator on S; we call this operator $\Phi(A)$ also. The *shorted operator*, denoted $S(A)$, is defined to be the operator $\Phi(A) \oplus 0$ on E, where $\Phi(A)$ acts on S and 0 acts on S.

The shorted operator may also be characterized as the maximal element in an appropriate set of matrices; we show this in the next theorem.

Theorem 7. *Let $\Psi(A, S) = \{B | B$ is HSD, $B \leqq A$, and range of B is contained in $S\}$; then $S(A)$ is the supremum of $\Psi(A, S)$.*

Proof. First we show that $S(A)$ is in $\Psi(A, S)$. Clearly $S(A)$ is HSD, and the range of $S(A)$ is contained in S. Now given an a in E, let $a = s + t$ with s in S and t in $S^\perp$, then

$$\begin{aligned} \langle S(A)a, a \rangle &= \langle S(A)s, s \rangle = \inf\{\langle A(s + u), (s + u) \rangle | u \in S^\perp\} \\ &\leqq \langle A(s + t), (s + t) \rangle = \langle Aa, a \rangle; \end{aligned}$$

thus $S(A)$ is in $\Psi(A, S)$. Now consider an arbitrary matrix B in $\Psi(A, S)$. For any a in D, let $a = s + t$ as above, then

$$\langle S(A)a, a \rangle = \langle S(A)s, s \rangle = \langle A(s + u), (s + u) \rangle \qquad \text{for some} \quad u \text{ in } s^\perp.$$

Since B is in $\Psi(A, S)$,

$$\langle A(s + u), (s + u) \rangle \geqq \langle B(s + u), (s + u) \rangle = \langle Bs, s \rangle = \langle Ba, a \rangle.$$

Therefore $S(A) \geqq B$. ∎

The following theorem is a dual to Theorem 7; the proof is given in [12].

Theorem 8. *Let $\Upsilon(A, S) = \{QAQ^* \mid Q^2 = I$, and the range of Q is $S\}$; then $S(A)$ is the infimum of $\Upsilon(A, S)$.*

The Q's in Theorem 8 are projections onto S. One may show that for finite dimensional spaces, the infimum is obtained for the projection Q given by

$$Q = \begin{bmatrix} I & -A_{12}A_{22}^{-1} \\ 0 & 0 \end{bmatrix},$$

where A is partitioned,

$$A = \begin{bmatrix} A_{11} & A_{12} \\ A_{21} & A_{22} \end{bmatrix},$$

where $A_{11}: S \to S$, $A_{12}: S \to S$, etc.

Two explicit representations are available for the shorted operator, and these will be given without proof; the first is due to Krein [46], and the second is from [12].

Theorem 9. *Let T be the subspace of all vectors x such that $A^{1/2}x$ is in S, and let Q_T be the orthogonal projection onto T; then $S(A) = A^{1/2}Q_T A^{1/2}$.*

Theorem 10. *Let the matrix A be partitioned into blocks A_{11}, A_{12}, etc. as above; then $S(A) = A_{11} - A_{12}A_{22}^{-1}A_{21}$ on S and 0 on $S^{\perp}$.*

In each case, the explicit representation formula is established by showing that the expression for $S(A)$ furnishes a supremum for $\Psi(A, S)$. In contrast, we have established the existence of the supremum without any reference to an explicit formula.

Many other interesting properties of $S(A)$ are discussed in [4, 12]. In particular, we can prove directly from Theorem 7 that $S(A + B) \geqq S(A) + S(B)$ (this is a special case of Theorem 5) with equality if and only if the range of $A - S(A) + B - S(B)$ is contained in the orthogonal complement of S. This equality condition is important in network applications (see [27]). Another important property of the shorted operator is the following continuity result, which also follows from Theorem 7; a proof is given in [12].

Theorem 11. *If E_n is a sequence of HSD matrices monotonically decreasing to 0, then $S(A + E_n)$ monotonically decreases to $S(A)$.*

In addition to the shorted operator being continuous, it is differentiable when A_{22} is invertible.

Theorem 12. *If A_{22} is invertible, then $S(A)$ is differentiable at A.*

Proof. Let E be a suitably partitioned HSD matrix such that $A_{22} + E_{22}$ is invertible. We note that $A_{22} + E_{22}$ is invertible if E is sufficiently small. Letting Q be the projection from Theorem 8, it is a direct matrix calculation to show that

$$\begin{aligned} S(A + E) &= S(A) + QEQ^* \\ &\quad - (A_{12}A_{22}^{-1}E_{22} - E_{12})(A_{22} + E_{22})^{-1}(E_{22}A_{22}^{-1}A_{21} - E_{21}). \end{aligned}$$

The latter term is quadratic in E, so the desired linear approximation is given by QEQ^*. ■

As we have seen, the shorted operator is a Φ-operator. In fact, it seems, from the definition of the shorted operator, that it is a very simple Φ-operator. We now show that, in some sense, it is the most general Φ-operator. Our result is that every Φ-operator may be explicitly written in terms of the confluence basis matrices and the shorted operator.

Theorem 13. *Let Γ be a confluence with basis matrices K and L; then if we let $G^* = [K^*\ L^*]$, the induced Φ-operator may be written as follows $\Phi(A) = S(GAG^*)$.*

Proof. Let A be an HSD matrix and let c be in P, then from Theorem 2, there exists an a and γ such that $(a, c) \in \Gamma$ and $(Aa, \gamma) \in \Gamma'$. Using Lemma 1, part (iii), we have that $KAa = \gamma$ and $LAa = 0$, and using Lemma 1, part (iv), we know there exists a d so that $K^*c + L^*d = a$. Eliminating a yields

$$\begin{bmatrix} \gamma \\ 0 \end{bmatrix} = \begin{bmatrix} K \\ L \end{bmatrix} A [K^* \quad L^*] \begin{bmatrix} d \\ d \end{bmatrix}.$$

It is then immediate from Theorem 10 and the definition of G that $\gamma = S(GAG^*)c$. Since c is arbitrary, we are done. ■

We have now completed our analysis of Φ-operators. We note that we have not attempted to be all inclusive, but only present the fundamental properties of Φ-operators. We refer the reader to [14–17, 63] for further properties and applications of these operators.

In the next sections, we describe areas in which confluences, and therefore Φ-operators, arise naturally.

4. SCALAR NETWORKS

Our first example of a confluence arises from the study of electrical networks defined on graphs. Let G be a graph with distinguished edges—called ports, and other edges—called branches. We will assume that the set of ports contain neither a circuit nor a cocircuit of G. This assumption ensures that we have the proper current behavior in the network. Our network is also assumed to be resistive; by resistive we mean that the only elements present in the edges of G are positive resistors.

The space of all of the allowable current flows in G, governed only by Kirchhoff's law, forms a confluence Γ. The restriction on the circuits and cocircuits guarantees that the conditions for a confluence are satisfied. It is also easy to verify that the set of all allowable voltage drops in G, governed only by Kirchhoff's voltage law, is a confluence. Moreover it is the dual confluence Γ'. Our fundamental existence theorem, Theorem 2, yields a function Φ from the space of positive operators on the branches to the space of positive operators on the ports. Using the terminology of electrical network theory, the operator Φ maps the branch impedance matrix to the port impedance matrix.

As one application of our theory of Φ-operators, we see that we may use the explicit representation formula of Theorem 13 to compute the port impedance matrix of a network. For the case of a 1-port, this computation is similar to one given by Cederbaum in [27].

As another application of our theory, we describe a variational problem that may be solved using the derivative formula of Theorem 12. We are interested in the variation in the port impedance matrix with respect to a variation in branch impedance matrix. Bott and Duffin [23] consider this type of variational problem where they restrict the variation in the branch impedance matrix to be a variation in only one branch. We will restrict ourselves to this kind of variation, however, we consider the n-port case, whereas Bott and Duffin consider the 1-port case.

Assume that we have a diagonal HSD matrix A—this represents the resistances in the branches. We then know that $\Phi(A)$ is the port impedance matrix. Our question is if a_{ii} (the ith diagonal element of A) varies slightly, how does $\Phi(A)$ vary? Our answer to this question is contained in Theorem 14 below. We must first make a technical assumption. We need to ensure that the branch currents are uniquely determined by the port currents. Duffin defines such a concept for networks; he calls it positively connected [22, 37]. In our formulation, we assume that LAL^* is invertible. This condition forbids any internal zero resistance branch circuits. Since the columns of L^* are independent, if A is invertible, then our condition is satisfied.

Lemma 4. *Given a network and the associated confluence Γ, and given an HSD matrix A, if LAL^* is invertible, then*

(i) *for any port current $c \in P$, the associated branch current $a \in E$ such that $(a, c) \in \Gamma$ and $(As, \gamma) \in \Gamma'$ is given by $a = Rc$, with*

$$R = K^* - L^*(LAL^*)^{-1}(LAK^*);$$

(ii) *the Φ induced by Γ satisfies $\Phi(A + E) = \Phi(A) + R^*ER + o(\|E\|)$.*

Proof. For (i), as in the proof of Theorem 13, we have $KAa = \gamma$, $LA_a = 0$, and there is a d so that $K^*c + L^*d = a$. Since LAL^* is invertible, we may eliminate d and obtain $a = Rc$, with R given as above. For part (ii) we make use of Theorem 12; using the shorted operator definition, we may expand $\Phi(A + E)$ as follows

$$\Phi(A + E) = S\left(\begin{bmatrix} K \\ L \end{bmatrix} A[K^* \quad L^*] + \begin{bmatrix} K \\ L \end{bmatrix} E[K^* \quad L^*]\right)$$

$$= S\left(\begin{bmatrix} K \\ L \end{bmatrix} A[K^* \quad L^*]\right) + Q\begin{bmatrix} K \\ L \end{bmatrix} E[K^* \quad L^*]Q^* + o(\|E\|),$$

where Q is given by

$$Q = \begin{bmatrix} I & -KAL^*(LAL^*)^{-1} \\ 0 & 0 \end{bmatrix}.$$

The first term is $\Phi(A)$ and the second term may be simplified to obtain R^*ER. ■

The matrix R^* is termed the modified circuit matrix by Cederbaum [27]; Thulasiraman and Murti [62] have extensively studied this matrix. Returning to our variational problem, let E be a diagonal matrix with $E_{11} = \varepsilon$, and $E_{ii} = 0$ for $i \neq 1$. We will consider the variation in branch 1 for convenience.

Theorem 14. *If A is a diagonal matrix, and LAL^* is invertible, then if the vector a represents the branch current obtained from a port current of $c = (1, 0, \ldots, 0)$, we have $\Phi(A + E) = \Phi(A) + \varepsilon\langle a, a\rangle + o(|\varepsilon|)$.*

Proof. This is a restatement of part (ii) of Lemma 4 where R^*ER has been rewritten $R^*\varepsilon c^*cR$, which is εa^*a or $\varepsilon\langle a, a\rangle$. ■

Letting $\varepsilon \to 0$, we see that the expression $[\Phi(A + E) - \Phi(A)]/\varepsilon$ approaches $\langle a, a\rangle$, and this extends the results of Bott and Duffin to n-ports.

Our development of a confluence from a graphical network is related to the work of Seshu and Reed [60] and Bryant [25]. The book by Penfield *et al.* [56] also contains a similar type of construction in the case of a network with no ports.

Not all confluences arise from graphs, the confluence with basis elements $(1, 1, 1)$ and $(1, -1, 1)$ is an example. However every confluence may be thought of as arising from a matroid. Matroids were first described by Whitney [67], and further considered by Minty [49] and Tutte [65]. Matroids may be thought of as generalized graphs, and Bruno and Weinberg have shown that an electrical network may be defined on a matroid [24]. We have also considered matroid networks; in [16], we show that many of the classical concepts for graphical networks—flows, joint resistance, extremal length (see [1, 36])—extend to matroid confluences.

5. NETWORK INTERCONNECTIONS AND OPERATOR NETWORKS

The interconnection of operator networks also gives rise to confluences. The concept of an operator network may be approached from two different points of view. One may consider a graph with operators in the branches, or one may consider the interconnection of two or more n-port networks. These two problems are closely related and both may be analyzed using our confluence theory. We will focus on the interconnection of two n-port networks for the purpose of illustration.

An n-port network is a black box with $2n$ external terminals divided into pairs. Each pair is termed a port. We assume that only resistors are hidden inside the black boxes. We assume that current flows into a box from one terminal of each port and out of the box from the other terminal of the same port. We measure the voltage drop across each port. We also assume that an HSD matrix exists to represent the current–voltage relationship—this matrix is termed the impedance matrix for the n-port. For example, consider two n-port networks connected in series. The series connection is accomplished by connecting each port of network 1 in series with the corresponding port of network 2. The impedance matrix of the combined network will be the ordinary sum of the impedance matrices of the component networks.

In terms of confluences, we may write that the Φ induced by the series confluence applied to a block diagonal matrix with blocks A and B yields

$$\Phi\left(\begin{bmatrix} A & 0 \\ 0 & B \end{bmatrix}\right) = A + B.$$

As we vary the type of interconnections between the component networks, we produce other confluences. Each of these confluences induces a new Φ-operator applied to block diagonal matrices. The diagonal blocks correspond to the impedance matrices of the component networks.

Instead of using the Φ notation we may also write $A * B$ for the matrix of the combined network. With this notation, we may think of the interconnections of 2 n-port networks as inducing HSD matrix operations. In [7], we and Duffin consider the $*$ operation—we term it the router sum.

As we have just seen, one router operation is ordinary addition. We will now briefly discuss a few other interesting router operations. We refer the reader to [64] for a more detailed survey of the development of these operations.

If the two n-port networks are connected in parallel, then the associated confluence induces an operation termed parallel addition—this is denoted $A : B$. Erickson [43] first considered the parallel sum in the scalar case. The extension to the matrix case is done by Anderson and Duffin in [5]. If the component matrices are invertible then the parallel sum is given by the formula $A : B = (A^{-1} + B^{-1})^{-1}$. In the case where the inverses do not exist, an alternate expression is needed. The appropriate expression is $A : B = A(A + B)^{+}B$. We are using the symbol C^{+} to denote a generalized inverse of C; see [2] or [57]. However, since the range of B is contained in the range of $A + B$, for HSD A and B, the generalized inverse of $A + B$ only acts on its range, and therefore is a true inverse.

Another operation, hybrid addition, arises from the hybrid connection of n-ports. Hybrid addition is a combination of series and parallel additions. Duffin *et al.* [39] use the hybrid sum to obtain an elegant solution to the n-port synthesis problem. The mathematical theory of the hybrid router is presented in [42].

As our last special case of a router, we mention cascade addition, denoted $A \circ B$. The cascade sum arises from the cascade connection confluence. One of the more interesting questions concerning cascade addition is a fixed point problem—given an HSD A, find an HSD X such that $A \circ X = X$. We will discuss this further in a later section. Cascade addition is not commutative, whereas parallel, hybrid, and obviously series additions are commutative. The cascade sum is further described in [10].

We therefore see that the class of router operations is quite general and that it contains a number of interesting special cases. We will now summarize some of the basic properties of any router operation. Further details and proofs may be found in [7].

To facilitate our discussion of the router sum, we introduce the following change in notation. Instead of representing members of a confluence by (a, c), we will use the notation (a, b, c). We partition the branch portion E into two parts—these parts represent the ports of the two component networks. Our confluence Γ is a set of tripletons (a, b, c) which form a subspace, and such that for any c, there exists an a and b so (a, b, c) is in Γ, and if $(0, 0, c)$ is in Γ then $c = 0$. In this setting, Γ is a subspace of $E_1 \oplus E_2 \oplus P$, where $E_1 \oplus E_2$ corresponds to the E in our original confluence definition.

Given a confluence Γ and the associated router operation $*$, the following inequalities hold. Some of these results are a restatement of Theorems 3 and 4.

Theorem 15. *If A, B, C, and D are HSD matrices, then*

(i) $\langle A * Bc, c\rangle \leqq \langle Aa, a\rangle + \langle Bb, b\rangle$ *for all* $(a, b, c) \in \Gamma$. *Moreover, for every c there exists a and b so that equality holds.*

(ii) *Series-router inequality*:

$$(A + B) * (C + D) \geqq (A * C) + (B * D).$$

(iii) *Parallel-router inequality*:

$$(A : B) * (C : D) \leqq (A * C) : (B * D).$$

In Theorem 6 we characterized the relationship between Φ-operators arising from different confluences. A similar result for routers would be interesting. Unfortunately a complete characterization is not available. The following however does hold, we present it without proof.

Theorem 16. *Let Γ_1 and Γ_2 be confluences with associated router operations $*_1$ and $*_2$. If for constants k_1 and k_2 with $|k_i| \leqq 1$, (a, b, c) in Γ_2 implies that $(k_1 a, k_2 b, c)$ is in Γ_1, then $A *_1 B \leqq A *_2 B$ for all HSD A and B.*

We conjecture that the converse holds. However, we can only prove the result with the added assumption that for any c in P, the a and b such that (a, b, c) is in Γ_1 are unique. We note that the series confluence satisfies this condition. In the case of equality of the router sums, the appropriate version of this conjecture is proved by Bott and Duffin [23].

In our discussion of routers, we have tacitly assumed that ideal isolation transformers are present at all of the ports of our networks. The analysis of the interconnections of networks without the use of transformers is quite difficult. For example, the series connection is not even a commutative operation without transformers. The transformerless parallel connection is analyzed by Lempel and Cederbaum [47] and Murti and Thulasiraman [53]. Using confluences, in [13], we also consider this type of parallel connection. Our result may be summarized by saying that if the transformerless parallel connection impedance matrix exists, then the power in the transformerless impedance matrix is less than the power in the impedance matrix with transformers. Another way to state this fact is to say that since the transformers restrict the current flows, the power goes down without them.

6. INFINITE DIMENSIONAL AND FREQUENCY DEPENDENT CONFLUENCES

So far, we have only considered confluences arising from finite resistive networks; as we have seen, this corresponds to HSD impedance matrices. By dropping either of these restrictions on the network, different problems are encountered. If we allow our network to contain gyrators, inductors, or capacitors, we lose the HSD condition of our impedance matrix. On the other hand if we allow our underlying graph to be infinite, we no longer have matrices, but we must now consider operators on a Hilbert space.

With gyrators present in the network, the matrices are no longer symmetric, however the quadratic form is still positive, since gyrators are passive [61]. Anderson [3] and Lewis and Newman [48] considered matrices which correspond to resistive networks containing gyrators. Anderson characterized these matrices and he shows that many of the results of parallel addition extend to this case. Duffin and Morley further develop the network application of these matrices in [40].

If the network contains inductors or capacitors, then the impedance matrix will be positive real. A positive real matrix is a matrix function of a complex variable—termed the frequency. We refer the reader to [21] or [66] for a further discussion of positive real matrices. Since the confluence is determined by the topology of the network and not by the elements contained in the branches, the confluence theory still applies. In particular, there exists a Φ-operator, and in [14], we show that a Φ-operator applied to a positive real matrix yields a positive real matrix. However, we lose the natural partial ordering of the HSD case, and the inequality results do not apply.

The concept of a confluence may be extended to infinite dimensional cases in many different ways. For example, one may consider an infinite graph with resistors or positive operators in each of the branches. Flanders [45] and Zemanian [68–70] establish many classical results for infinite networks. Dolezal [30] expands this work to infinite networks which contain internal current or voltage sources. A generalization of an infinite network is a Hilbert port, which is an infinite network with a positive real impedance operator; Dolezal and Zemanian [29–31] have extensively studied Hilbert ports.

We now turn to a different method of extending the theory of confluences to Hilbert space. What is the router sum of two Hilbert space operators? In terms of networks, we are asking for the impedance operator of the interconnection of two infinite resistive networks. Fillmore and Williams [44] first answered this question for the case of the parallel router. In an effort to simplify the Fillmore and Williams theory, we showed that the shorted operator extends to Hilbert space [12]. (Krein [46] had previously considered the shorted operator in Hilbert space without referring to the network interpretation.) Since every router can be written using the shorted operator, the full theory of routers extends to Hilbert space.

Another example of an infinite network is provided by infinite ladders. An infinite ladder can be viewed as the infinite sequential connection of given networks. If the same network is connected with itself in a ladder, then a fixed point problem arises naturally. In [8] a fixed point problem for parallel addition was considered—find X such that $X = A : X + W$. The equation arose from the study of a linear time series, but the network interpretation is obvious. Another example of a ladder fixed point problem is provided by the infinite cascade connection of a given n-port network with itself. Let A denote an invertible HSD matrix, then we wish to solve the equation $A \circ X = X \circ A = X$. In [17], we have shown that there exists a unique solution to this fixed point problem. We also know that the fixed point X is a block diagonal matrix, however, we have been unable to determine an explicit form for the diagonal blocks—except in the simple case of $n = 2$.

7. GENERALIZATIONS AND RELATED WORK

In the previous sections, we have presented basically our development of an operator algebra derived from electrical network theory. In this section, we survey the results of others in this area, and discuss generalizations and related work.

Parallel addition was the first network induced operation defined. As we mentioned previously, Fillmore and Williams [44] first generalized the parallel sum to Hilbert space. Nishio and Ando [54] characterize parallel addition by showing that it is the only operation which satisfies a set of continuity, inequality, and scalar case conditions. Morley [52] expands the theory of the parallel sum.

Many of our results are based on the assumption that the operators are positive. If this assumption is relaxed, then other types of results are possible. Rao and Mitra [57] consider the parallel sum of arbitrary matrices. Their theory extends to hybrid addition (see [50]). Duffin and Morley [41] also consider network related operations on classes of operators other than the positive class. They have investigated a new partial ordering for operators and have obtained many interesting results. At present, there seems to be no physical interpretation of this ordering.

The question of the subtraction of routers was first considered for the parallel router and the hybrid router in [6]. Subsequently, Pekarey and Smul'tan [55] extended parallel subtraction to Hilbert space; they also rediscovered many results concerning parallel addition. Unitary routers will be defined below, but here we note that in [11], we and Reynolds considered the problem of unitary router subtraction.

In many ways the shorted operator is the most fundamental concept in our theory of confluences and Φ-operators (routers). It is not surprising that it has been the most extensively studied of all of the network operations. Another reason for the great interest in the shorted operator is that the shorted operator is a special case of the Schur complement. We refer to the paper by Cottle [28] for a thorough analysis of the theory and applications of the Schur complement. We also note that Carlson *et al.* [26] consider this operation. Ando [19] has added to the theory of the Schur complement by characterizing those matrices which have a generalized Schur complement.

As another illustration of the shorted operator, we refer to a paper by Rosenberg [59]. Rosenberg shows that given a subspace S, every invertible HSD matrix A may be written as $E + F$, where the range of E is contained in S, and the range of F is contained in the orthogonal complement of S. It is shown in [4] that in fact $E = S(A)$, and moreover the result is true for HSD operators. Ando has generalized this result; in [18], he considers more general splittings and terms them Lebesgue decompositions.

The first case of a nontrivial multiplicative inequality for a network operation is given by Ando for the shorted operator [54]. We present it next, with a simplified proof.

Theorem 17. *For an HSD A, $S(A^2) \leqq (S(A))^2$.*

Proof. Let A be an invertible HSD operator. Let $F = A - S(A)$ and let Q be the projection of Theorem 8. A simple computation yields that $QF = FQ^* = 0$. Since $A = F + S(A)$ we have that $A^2 = F^2 + FS(A) + S(A)F + (S(A))^2$, and therefore $QA^2Q^* = (S(A))^2$. By Theorem 8, we then have that $S(A^2) \leqq (S(A))^2$. The noninvertible case follows by taking limits and applying the continuity result of Theorem 11. ■

Recall that every router $*$ may be written as

$$A * B = S\left(G\begin{bmatrix} A & 0 \\ 0 & B \end{bmatrix}G^*\right).$$

In terms of the confluence basis, $G^* = [K^* \; L^*]$. We term a router $*$ a unitary router if G is a unitary matrix ($GG^* = I$); see [17]. An interesting problem is to characterize those confluences which generate a unitary router. A simple computation using Theorem 17 shows that $A^2 * B^2 \leqq (A * B)^2$ for any unitary router. In particular, it is easy to show that $2(A : B)$, twice the parallel sum, is a unitary router, and therefore we see that $A^2 : B^2 \leqq 2(A : B)^2$.

Ando has observed that $2(A : B)$ is the harmonic mean of A and B. This complements the fact that $(A + B)/2$ is the arithmetic mean. One obvious problem is to define the geometric mean of A and B. If A and B commute then $(AB)^{1/2}$ is the natural definition. Ando [20] defines the geometric mean $G(A, B)$, for HSD A and B, as the maximum HSD X so that $\left[\begin{smallmatrix} A & X \\ X & B \end{smallmatrix}\right]$ is HSD. He then shows that the geometric mean satisfies the appropriate inequality, $(A + B)/2 \geqq G(A, B) \geqq 2(A : B)$. Moreover, if A and B commute, then $G(A, B) = (AB)^{1/2}$. We do not know if $G(A, B)$ can be written as a router.

Throughout this paper, we have dealt with matrices and operators. It is therefore natural to consider the behavior of our operations with respect to the classical matrix and operator quantities—norm, trace, and determinant. In [5], the following inequalities for parallel addition are given:

$$\|A : B\| \leqq \|A\| : \|B\|;$$

$$\operatorname{tr}(A : B) \leqq \operatorname{tr}(A) : \operatorname{tr}(B);$$

$$\det(A : B) \leqq \det(A) : \det(B).$$

The right-hand sides of these inequalities are scalar parallel sums. To obtain these types of inequalities for hybrid or cascade addition, one must define generalized operations. We will illustrate using the norm. Partition A into blocks as before A_{11}, A_{12}, etc. Define $|||A|||$ as a 2 by 2 diagonal matrix with diagonal elements equal to $\|A_{11}\|$ and $\|A_{22}\|$. Then for hybrid addition, one has $|||A * B||| \leqq |||A||| * |||B|||$. The generalized norm was first defined in [63]. Duffin and Morley also use this quantity (see [40] or [51]).

As the last extension area, we will briefly discuss the application of this work to nonlinear networks. For an HSD operator A, the quadratic form is a real valued convex function, and this function represents the power dissipated. As we have seen, using the minimum formulation of Theorem 6, we may define a Φ-operator or a router in terms of the quadratic form. One may pose the same problem for arbitrary convex functions. In [9], we have considered this problem, and have shown that with the appropriate technical assumptions, the theory of routers extends to this case. Rockafellar [58] had previously defined the parallel and hybrid sums for convex functions. He used different terminology and did not refer to the network interpretation. Finally, we mention that, using a different approach, Duffin [33, 34] has also developed the theory of nonlinear networks, and in fact has defined a nonlinear shorted operator.

REFERENCES

1. L. V. ALFORS AND L. SARIO, "Riemann Surfaces," Princeton Univ. Press, Princeton, New Jersey, 1960, pp. 214–228.
2. A. ALBERT, "Regression and the Moore–Penrose Pseudoinverse," Academic Press, New York, 1972.
3. W. N. ANDERSON, JR., Series and Parallel Addition of Operators, Ph.D. Thesis, Carnegie-Mellon Univ., Pittsburgh, Pennsylvania, 1968.
4. W. N. ANDERSON, JR., Shorted operators, *SIAM J. Appl. Math.* **20** (1971), 576–594.
5. W. N. ANDERSON, JR. AND R. J. DUFFIN, Series and parallel addition of matrices, *J. Math. Anal. Appl.* **26** (1969), 576–594.
6. W. N. ANDERSON, Jr., R. J. DUFFIN, AND G. E. TRAPP, Parallel subtraction of matrices, *Proc. Nat. Acad. Sci. U.S.A.* **69** (1972), 2531–2532.
7. W. N. ANDERSON, JR., R. J. DUFFIN, AND G. E. TRAPP, Matrix operations induced by network connections, *SIAM J. Control Optim.* **13** (1975), 446–461.
8. W. N. ANDERSON, JR., G. D. KLEINDORFER, P. R. KLEINDORFER, AND M. B. WOODROOFE, Consistent estimates of the parameters of a linear system, *Ann. Math. Statist.* **40** (1969), 2064–2075.
9. W. N. ANDERSON, JR., T. D. MORLEY, AND G. E. TRAPP, Fenchal duality for nonlinear networks, *IEEE Trans. Circuits and Systems* **CAS-25** (1978), 762–765.
10. W. N. ANDERSON, Jr., D. F. REYNOLDS, AND G. E. TRAPP, Cascade addition of matrices, *Proc. West Virginia Acad. Sci.* **46** (1974), 185–192.
11. W. N. ANDERSON, JR., D. F. REYNOLDS, AND G. E. TRAPP, Subtraction of network induced matrix operations, *Proc. West Virginia Acad. Sci.* **47** (1975), 212–215.
12. W. N. ANDERSON, JR. AND G. E. TRAPP, Shorted operators II, *SIAM J. Appl. Math.* **28** (1975), 160–171.

13. W. N. ANDERSON, JR. AND G. E. TRAPP, Inequalities for the parallel connection of resistive n-port networks, *J. Franklin Inst.* **299** (1975), 305–313.
14. W. N. ANDERSON, JR. AND G. E. TRAPP, An algebra of operator networks, *Proc. Internat. Symp. Oper. Theory Networks Systems, Montreal*, pp. 5–15, Western Periodicals Co., North Hollywood, California, 1975.
15. W. N. ANDERSON, JR. AND G. E. TRAPP, A class of monotone operator functions related to electrical network theory, *Linear Algebra Appl.* **15** (1976), 53–67.
16. W. N. ANDERSON, JR. AND G. E. TRAPP, Algebraic properties of networks on matroids, *Proc. 20th Midwest Symp. Circuits Systems, Texas Tech Univ.*, Western Periodicals Co., North Hollywood, California, 1977. pp. 384–390.
17. W. N. ANDERSON, JR. AND G. E. TRAPP, Matrix operations induced by network connections II, *Bull. Calcutta Math. Soc.* (to appear).
18. T. ANDO, Lebesgue-type decomposition of positive operators, *Acta Sci. Math. (Szeged)* **38** (1976), 253–260.
19. T. ANDO, Commutativity formulas for generalized Schur completements, *Linear Algebra Appl.* (to appear).
20. T. ANDO, Topics on operator inequalities, Lecture Notes, Sapporo, Japan, 1978.
21. H. BART, M. A. KAASHOEK, AND D. C. LAY, Relative inverses of finite meromorphic operator functions, *Math. Ann.* **218** (1975), 199–210.
22. S. BHARGAVA AND R. J. DUFFIN, Network models for maximization of heat transfer under weight constraints, *J. Networks* **2** (1972), 285–299.
23. R. BOTT AND R. J. DUFFIN, On the algebra of networks, *Trans. Amer. Math. Soc.* **74** (1953), 99–109.
24. J. BRUNO AND L. WEINBERG, Generalized networks: Networks embedded on a matroid, Part I, II, *J. Networks* **6** (1976), 53–94, 231–272.
25. P. R. BRYANT, The algebra and topology of electrical networks, *Proc. Inst. Electr. Engrs., Part C* **108** (1961), 215–229.
26. D. CARLSON, E. HAYNSWORTH, AND T. MARKHAM, A generalization of the Schur complement by means of the Moore–Penrose pseudoinverse, *SIAM J. Appl. Math.* **26** (1974), 169–175.
27. I. CEDERBAUM, On equivalence of resistive n-port networks, *IEEE Trans. Circuit Theory* **CT-12** (1965), 338–344.
28. R. W. COTTLE, Manifestation of the Schur complement, *Linear Algebra Appl.* **8** (1974), 189–211.
29. V. DOLEZAL, Hilbert networks I, *SIAM J. Control Optim.* **12** (1974), 755–778.
30. V. DOLEZAL, Nonlinear networks with current sources and Tellegens theorem, *SIAM J. Math. Anal.* **8** (1977), 473–485.
31. V. DOLEZAL AND A. H. ZEMANIAN, Hilbert networks II—Some qualitative properties, *SIAM J. Control Optim.* **13** (1975), 153–161.
32. R. G. DOUGLAS, On majorization, factorization, and range inclusion of operators in Hilbert space, *Proc. Amer. Math. Soc.* **17** (1966), 413–416.
33. R. J. DUFFIN, Nonlinear networks I, *Bull. Amer. Math. Soc.* **52** (1946), 833–838.
34. R. J. DUFFIN, Nonlinear networks III, *Bull. Amer. Math. Soc.* **55** (1949), 119–129.
35. R. J. DUFFIN, An analysis of the Wang algebra of networks, *Trans. Amer. Math. Soc.* **93** (1959), 114–131.
36. R. J. DUFFIN, The extremal length of a network, *J. Math. Anal. Appl.* **5** (1962), 200–215.
37. R. J. DUFFIN, Optimum heat transfer and network programming, *J. Math. Mech.* **17** (1968), 759–768.
38. R. J. DUFFIN, Network models, *in* "Mathematical Aspects of Electrical Network Analysis," *Amer. Math. Soc.*, Providence, Rhode Island (1971), pp. 65–91.
39. R. J. DUFFIN, D. HAZONY, AND N. MORRISON, Network synthesis through hybrid matrices, *SIAM J. Appl. Math.* **14** (1966), 390–413.

40. R. J. DUFFIN AND T. D. MORLEY, Inequalities induced by network connections, Part II—Hybrid connections, *J. Math. Annal. Appl.* **67** (1979), 215–231.
41. R. J. DUFFIN AND T. D. MORLEY, Operator networks treated by Sylvester unisignants, preprint.
42. R. J. DUFFIN AND G. E. TRAPP, Hybrid addition of matrices—A network theory concept. *J. Appl. Anal.* **2** (1972), 241–254.
43. K. E. ERICKSON, A new operation for analyzing series—Parallel networks, *IEEE Trans. Circuit Theory* **CT-6** (1959), 124–126.
44. P. A. FILLMORE AND J. P. WILLIAMS, On operator ranges, *Adv. in Math.* **7** (1971), 254–281.
45. H. FLANDERS, Infinite networks I—Resistive networks, *IEEE Trans. Circuit Theory* **CT-18** (1971), 326–331.
46. M. G. KREIN, The theory of selfadjoint extensions of semibounded Hermitian operators and its applications I, *Mat. Sb.* **20** (62) (1947), 431–495. (In Russ.; Engl. summary.)
47. A. LEMPEL AND I. CEDERBAUM, Parallel interconnection of n-port Networks, *IEEE Trans Circuit Theory* **CT-14** (1967), 274–279.
48. T. LEWIS AND T. NEWMAN, Pseudoinverses of positive semidefinite matrices, *SIAM J. Appl. Math.* **16** (1968), 701–703.
49. G. MINTY, On the axiomatic foundations of the theories of directed linear graphs, electrical networks, and network programming, *J. Math. Mech.* **15** (1966), 485–520.
50. S. K. MITRA AND G. E. TRAPP, On hybrid addition of matrices, *Linear Algebra Appl.* **10** (1975), 19–35.
51. T. D. MORLEY, Operator functions induced by network connections, Ph.D. Thesis, Carnegie-Mellon Univ., Pittsburgh, Pennsylvania, 1976.
52. T. D. MORLEY, Parallel summation, Maxwell's principle and the infimum of projections, preprint.
53. V. C. K. MURTI AND K. THULASIRAMAN, Parallel connections of n-port networks, *Proc. IEEE*, **55** (1967), 1216–1217.
54. K. NISHIO AND T. ANDO, Characterizations of operations derived from network connections, *J. Math. Anal. Appl.* **53** (1976), 539–549.
55. E. L. PEKAREY AND JU. L. SMUL'TAN, Parallel addition and parallel subtraction of operators, *Izv. Akad. Nauk SSSR* **10** (1976), 351–370.
56. P. PENFIELD, JR., R. SPENCE, AND S. DUINKER, "Tellegen's Theorem and Electrical Networks," Research Monograph, No. 58, MIT Press, Cambridge, Massachusetts, 1970.
57. C. R. RAO AND S. K. MITRA, "Generalized Inverse of Matrices and Its Applications," Wiley, New York, 1971.
58. R. T. ROCKAFELLAR, "Convex Analysis," Princeton Univ. Press, Princeton, New Jersey, 1970.
59. M. ROSENBERG, Range decomposition and generalized inverse of non-negative Hermitian matrices, *SIAM Rev.* **11** (1969), 568–571.
60. S. SESHU AND M. B. REED, "Linear Graphs and Electrical Networks," Addison-Wesley, Reading, Massachusetts, 1961.
61. B. D. H. TELLEGEN, The gyrator, a new network element, *Philips Res. Rep.* **3** (1948), 81–101.
62. K. THULASIRAMAN AND V. C. K. MURTI, Modified cut-set matrix of an n-port network, *Proc IEE* **115** (1968), 1263–1268.
63. G. E. TRAPP, "Algebraic Operations Derived from Electrical Networks," Ph.D. Thesis, Carnegie-Mellon Univ., Pittsburgh, Pennsylvania, 1970.
64. G. E. TRAPP, Matrix algebra derived from network theory, *in* "Network and Signal Theory," Peter Peregrinus, London, 1973, pp. 89–94.
65. W. T. TUTTE, Lectures on matroids, *J. Res. Nat. Bur. Standards Sect. B* **69** (1965), 1–48.
66. L. WEINBERG AND P. SLEPIAN, Positive real matrices, *J. Math. Mech.* **9** (1960), 71–84.

67. H. Whitney, On the abstract properties of linear dependence, *Amer. J. Math.* **57** (1935), 509–533.
68. A. H. Zemanian, Infinite networks of positive operators, *Circuit Theory and Appl.* **2** (1974), 69–78.
69. A. H. Zemanian, The complete behavior of certain infinite networks under Kirchoff's node and loop laws, *SIAM J. Appl. Math.* **30** (1976), 278–295.
70. A. H. Zemanian, Infinite electrical networks, *Proc. IEEE* **64** (1976), 6–17.

W. N. Anderson, Jr.
DEPARTMENT OF MATHEMATICS
EAST TENNESSEE STATE UNIVERSITY
JOHNSON CITY, TENNESSEE

G. E. Trapp
DEPARTMENT OF COMPUTER SCIENCE
WEST VIRGINIA UNIVERSITY
MORGANTOWN, WEST VIRGINIA

Pulse Control in Dynamic Systems

DOV HAZONY

Of concern are problems in engineering and physics for which the principal signals are pulses. A synthesis procedure is provided for generating a piecewise constant pulse, with steps of arbitrary widths, evoking a time limited response in an *RLC* network. The number of steps is the degree of the network plus one. Many examples are shown.

1. INTRODUCTION

Of concern are pulse management and control in dynamic systems. Unless special conditions are met, it is very likely that the pulse response of a dynamic system will have tails or asymptotes that stretch out to infinity. Accordingly, an antenna may expand the transmitted signal and a communication channel would cause signals to overlap. A filter in the path of the signal may serve well to scrub off the noise, but a long tail is also added. The problem also exists in electromechanical systems, where a transducer keeps pulsating long after the original excitation. Similarly, a stepping motor continues to jitter for an extended time.

As the signal travels through the system, natural resonances are stimulated. This excitation continues until the resonances die down. A properly designed pulse, however, will deexcite the system as it exits. In this way the system would be excited only for the duration of the pulse.

Gerst and Diamond posed this problem in [1]. They also suggested several solutions. Philipp Dines and the author continued the work and showed that it is also possible to control some aspects of the output.

These works are reviewed and extended. It will be seen that a number of degrees of freedom may be added which need be no larger than the degree of the network plus one.

The main part of the proposal will be dedicated to the definition of the problem and to the development of the relevant tools. It will be shown that there is a wide variety of shapes that can be used to evoke a time limited response of any dynamic system. Moreover, bipolar pulses may also be used with very beneficial results. Finally, extensions of these ideas will be discussed as well as some important engineering applications.

ISBN 0-12-178150-X

2. THE FINITE LAPLACE TRANSFORM

Consider a dynamic system where the input is $e_i(t)$, the output is $e_0(t)$, and they have the Laplace transforms E_1 and E_2 governed by the relationship

$$E_2 = t_{21} E_1, \tag{1}$$

where $t_{21}(s)$ represents the transfer function of an initially relaxed dynamic system.

The problem at hand is to obtain an input e_i such that the output is confined to the interval 0, T. This defines a Laplace transform

$$E_2 = \int_0^T e_0(t) e^{-st}\, dt, \tag{2}$$

which is denoted as a *finite Laplace transform.*

The following are examples of finite Laplace transforms:

$$\frac{1 - \exp(-s)}{s}, \quad \frac{1 + \exp(-\pi s)}{s^2 + 1}, \quad \frac{1 - \exp[-(s+1)]}{s+1}. \tag{3}$$

Note that these functions have their poles canceled by zeros of the numerators. Hence they are entire functions. It has been proved by Gerst and Diamond [1] that:

Theorem 1. *If P_i and D are polynomials in s with the P_i's of lower degree than D, and if the a_i's are nonnegative real numbers, then*

$$G(s) = \frac{1}{D} \sum_{i=1}^{k} P_i \exp(-a_i s) \tag{4}$$

is a finite Laplace transform of length a ($=$ max a_i) if, and only if, it is entire.

In what follows we shall insist that both E_1 and E_2 shall satisfy the conditions of Theorem 1.

3. EXAMPLES

The main approach will be introduced by examples. Consider the one-pole low-pass filter shown in Fig. 1.

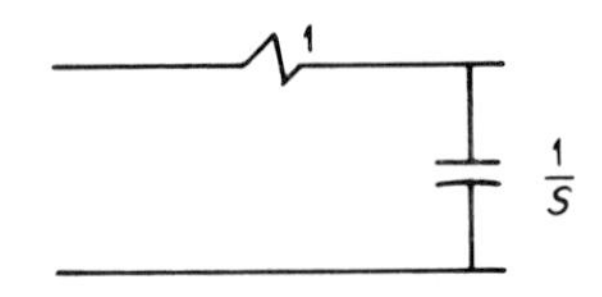

Fig. 1. $t_{21} = (E_2/E_1)|_{I_2=0} = 1(s+1)$.

The output will be a finite Laplace transform if it is of the form ($a > 0$)

$$E_2 = \frac{1 - \exp[-a(s + 1)]}{s + 1}.$$

Hence the input is $1 - \exp[-a(s + 1)]$, which is made up of two delta functions in the time domain. Following Gerst and Diamond the input is enhanced:

$$E_1 = \frac{[1 - \exp(-as)]\{1 - \exp[-a(s + 1)]\}}{s}. \tag{5}$$

This produces a stepwise constant input with output shown in Fig. 2.

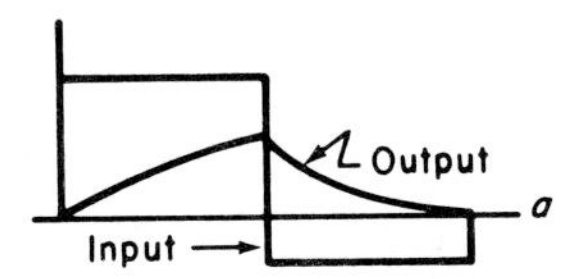

Fig. 2. A response to Eq. (5).

A stronger response is obtained when the input is bipolar. For example, when

$$E_1 = \left[1 - 2\left(\frac{1 + e^a}{2}\right)^{-s} + e^{-as}\right] \Big/ s \tag{6}$$

the response is shown in Fig. 3. The broken line represents Fig. 2. It is seen that the output is significantly higher for a bipolar input. Consider now the case of two complex poles:

$$t_{21} = \frac{2}{s^2 + 2s + 2} = \frac{2}{(s + i + j)(s + i + j)}.$$

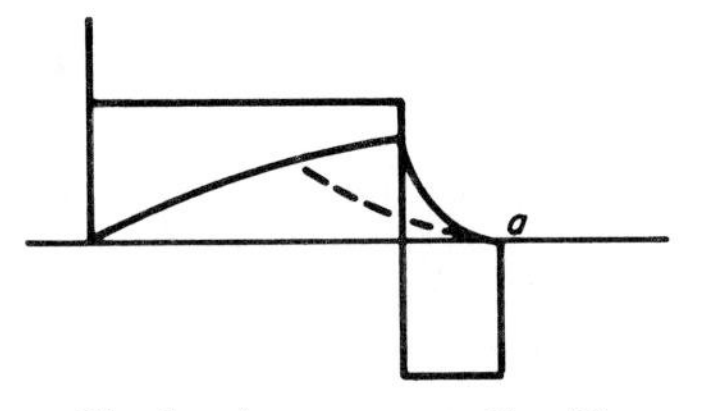

Fig. 3. A response to Eq. (6).

This is the voltage transfer function of the low-pass filter shown in Fig. 4. One possible input would have the Laplace transform

$$\frac{1 - \exp(-as - a)}{s + 1}$$

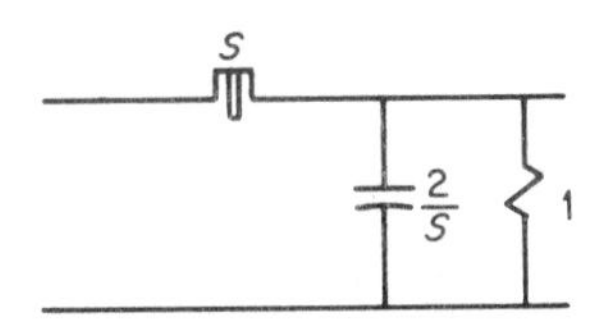

Fig. 4. Low-pass filter.

provided of course that the numerator is zero at $(s + 1 \pm j) = 0$, which are the zeros of the denominator. This restricts $a_{\min}$ to be 2π.

This example is suggested by Gerst and Diamond. To remove this last restriction on the interval "a" they suggested the following:

$$E_1 = \frac{\{1 - \exp[-(a/3)s]\}\{1 - \exp[-(a/3)(s + 1 + j)]\} \times \{1 - \exp[-(a/3)(s + 1 - j)]\}}{s}. \tag{7}$$

Again the first term, $\{1 - \exp[-(a/3)s]\}/s$, is inserted to make the input piecewise constant. The other terms are zero at $(s + 1 \pm j) = 0$. Gerst and Diamond call their method *zero insertion.* Note that a is arbitrary and can be very small. The fact that "a" is arbitrary permits optimization schemes. These can be carried out in respect to efficiency, shape, or output energy.

4. DEVELOPMENT FOR DEGREE TWO

Applying the above technique to the filter

$$t_{21} = (\alpha^2 + \beta^2)/[(s + \alpha)^2 + \beta^2], \tag{8}$$

Gerst and Diamond determined the following input–output pair where T is the pulse duration:

$$E_i(s) = \frac{(1 - e^{-Ts/3})(1 - e^{-T(s+\alpha-j\beta)/3})(1 - e^{-T(s+\alpha+j\beta)/3})}{s}.$$

$$E_o(s) = \frac{(\alpha^2 + \beta^2)(1 - e^{-Ts/3})(1 - e^{-T(s+\alpha-j\beta)/3})(1 - e^{-T(s+\alpha+j\beta)/3})}{s[(s + \alpha)^2 + \beta^2]}. \tag{9}$$

Let $u(t - t_0) = 1$ for $t \geqq t_0$ and zero elsewhere. The following is the time response of the above:

$$e_i(t) = u(t) - \left(1 + 2e^{-T\alpha/3} \cos \frac{T\beta}{3}\right) u\left(t - \frac{T}{3}\right) + \left(e^{-2T\alpha/3} + 2e^{-T\alpha/3} \cos \frac{T\beta}{3}\right)$$
$$\times u\left(t - \frac{2T}{3}\right) - e^{-2T\alpha/3} u(t - T),$$

$$e_o(t) = r(t) - \left(1 + 2e^{-T\alpha/3}\cos\frac{T\beta}{3}\right)r\left(t - \frac{T}{3}\right) + \left(e^{-2T\alpha/3} + 2e^{-T\alpha/3}\cos\frac{T\beta}{3}\right)$$
$$\times\, r\left(t - \frac{2T}{3}\right) - e^{-2T\alpha/3}r(t - T),$$

where

$$r(t) = \left[1 - \frac{(\alpha^2 + \beta^2)^{1/2}}{\beta}e^{-\alpha t}\cos\left(\beta t - \tan^{-1}\frac{\alpha}{\beta}\right)\right]u(t). \tag{10}$$

Thus the method has one arbitrary constant. It is piecewise constant with *three equally spaced steps*. The following input function was proposed in [2]:

$$E_1 = N_1/s,$$
$$N_1 = 1 - \frac{\sin b}{\sin(b - a)}\exp[-a(s + \alpha)/\beta] + \frac{\sin a}{\sin(b - a)}\exp[-b(s + \alpha)/\beta], \tag{11}$$

where a and b are roots of

$$f = \frac{\cos x - \exp(-x\alpha/\beta)}{\sin x} = \text{const.} \tag{12}$$

Setting $s = 0$, or $-\alpha \pm j\beta$, gives $N_1 = 0$. Typical input–output plots (for $\alpha = \beta = 1$) are shown in Fig. 5.

A study of f [Eq. (12)] shows that it is multivalued for $x > \pi$. Hence the resulting minimum pulse length is larger than π. The following function [N_1 in Eq. (13)] has no length limit. Moreover, it has *three degrees of freedom*. The function will be derived in Section 6.

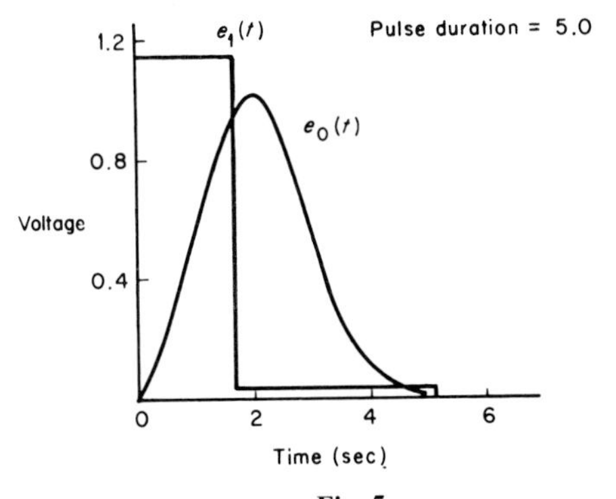

Fig. 5.

Thus

$$E_1 = N_1/s: \tag{13}$$

$$N_1 = 1 + A\exp[-a(s+\alpha)/\beta] + B\exp[-b(s+\alpha)/\beta] + C\exp[-c(s+\alpha)/\beta],$$

$$AD = \sin(b-c) + \sin c\exp(-b\alpha/\beta) - \sin b\exp(-c\alpha/\beta),$$

$$BD = \sin(c-a) + \sin a\exp(-ca/\beta) - \sin c\exp(-a\alpha/\beta),$$

$$CD = \sin(a-b) + \sin b\exp(-a\alpha/\beta) - \sin a\exp(-b\alpha/\beta),$$

$$D = \sin(a-c)\exp(-b\alpha/\beta) + \sin(c-b)\exp(-a\alpha/\beta) + \sin(b-a)\exp(-c\alpha/\beta).$$

This function (N_1) is zero at the origin and at $s = -\alpha \pm j\beta$. *The break points a, b, and c are perfectly arbitrary.*

5. BIPOLAR EXAMPLES

A bipolar pulse is relatively easy to produce either by switching or by hard amplification. An example for degree one was discussed in Section 3. In what follows it will be shown that it is possible to obtain a time limiting bipolar pulse for the degree 2 network. An example of a higher degree will be developed at the end of this section leading into a discussion on pulse compressors.

Without loss in generality let $\alpha = \beta = 1$ and $c = \pi$ in Eq. (13). This gives

$$N_1 = 1 - c_1\exp(-as) + c_2\exp(-bs) - c_3\exp(-\pi s),$$

$$C_4C_1 = \sin b\,[1 + \exp(-\pi)]\exp(-a),$$

$$C_4C_2 = \sin a\,[1 + \exp(-\pi)]\exp(-b), \tag{14a}$$

$$C_4C_3 = [\sin(a-b) + \sin b\exp(-a) - \sin a\exp(-b)]\exp(-\pi),$$

$$C_4 = \sin(b-a)\exp(-\pi) + \sin b\exp(-a) - \sin a\exp(-b).$$

This function simplifies considerably when

$$\sin a\exp a = \sin b\exp b = K.$$

Accordingly,

$$N_1 = 1 - C_1[\exp(-as) - \exp(-bs)] - \exp(-\pi s), \tag{14b}$$

$$C_1 = \frac{\sin b\,(1 + \exp\pi)\exp(-a)}{\sin(b-a)}.$$

A plot of $\sin x\exp x$ versus x is shown in Fig. 6.

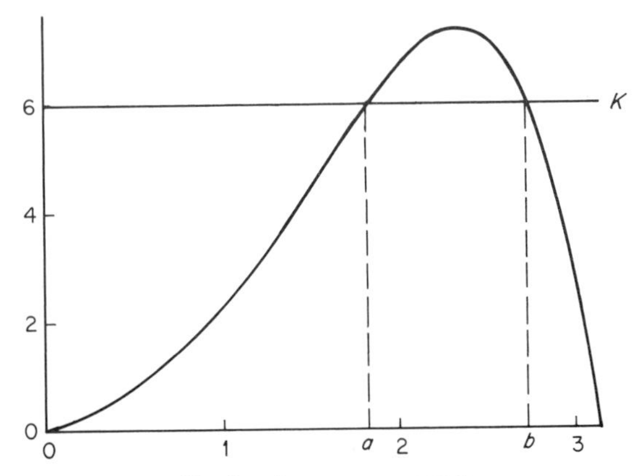

Fig. 6. sin x exp x versus x.

When $K \to 0$ the coefficients a and b approach 0 and π, respectively. Then C_1 [Eq. (14b)] approaches unity. On the other hand, the condition $b - a$ approaches zero (near $x = 3\pi/4$) makes C_1 approach infinity. It follows that C_1 may assume any value $1 < C_1 < \infty$. Making $C_1 = 2$ gives:

$$E_1 = N_1/s = [1 \exp(-as) + 2 \exp(-bs) - \exp(-\pi s)]/s \tag{15}$$

representing the bipolar pulse.

A computer plot of output signals corresponding to Eqs. (9) and (15) is shown in Fig. 7. It is seen that the response to the bipolar input is significantly higher. It follows from this development that any bilevel signal ($e_{\text{in}} = A$ or $-B$) can be designed in regards to break points to produce time limited responses.

The corresponding inputs are shown in 10 : 1 scale. The energy delivered to the resistor (Fig. 4) is about double for the bipolar input.

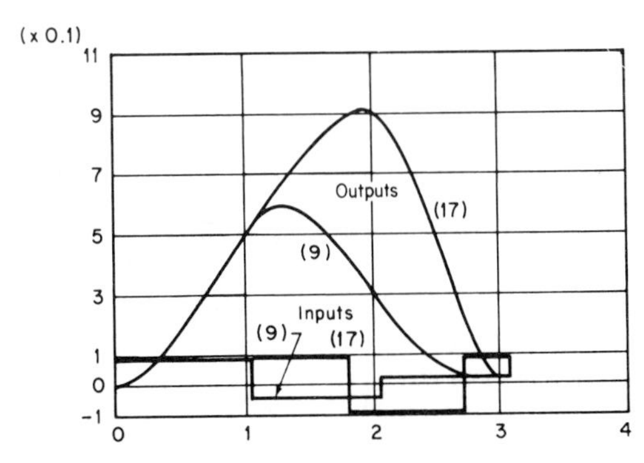

Fig. 7. Outputs corresponding to Eqs. (9) and (17).

The bipolar pulse break points can be readily obtained for some important problems. Take for example a pulse whose transform has multiple zeros at the origin:

$$\begin{aligned} N_1 &= (1 - 2e^{-s} + e^{-2s})/s, \\ N_2 &= (1 - e^{-2s})N_1, \\ &\vdots \\ N_n &= (1 - e^{-s}) \prod_{i=1}^{n} (1 - e^{-is})/s, \end{aligned} \tag{16}$$

Here n is the number of zeros at the origin of N_n/s. Clearly the number of break points of $L^{-1}(N/s)$ increases exponentially with n. To avoid this rapid increase in breakpoints a minimal solution will be derived for a quadruple zero at the origin. Let $f_4(t)$ denote a bipolar pulse whose transform $F_4(s)$ is given by

$$sF_4(s) = 1 - 2e^{-as} + 2e^{-bs} - 2e^{-cs} + 2e^{-ds} - e^{-s},$$

where

$$0 < a < b < c < d < 1.$$

This function has a high order zero of degree five iff

$$\begin{aligned} a - b + c - d &= -\tfrac{1}{2}, \\ a^2 - b^2 + c^2 - d^2 &= -\tfrac{1}{2}, \\ a^3 - b^3 + c^3 - d^3 &= -\tfrac{1}{2}, \\ a^4 - b^4 + c^4 - d^4 &= -\tfrac{1}{2}. \end{aligned} \tag{17}$$

Solving this set gives $a = \frac{1}{8}(3 - \sqrt{5})$, $b = \frac{1}{8}(5 - \sqrt{5})$, $c = \frac{1}{8}(3 + \sqrt{5})$, and $d = \frac{1}{8}(5 + \sqrt{5})$. Similarly we solve for the break points of three other bipolar signals. This is shown in Fig. 8.

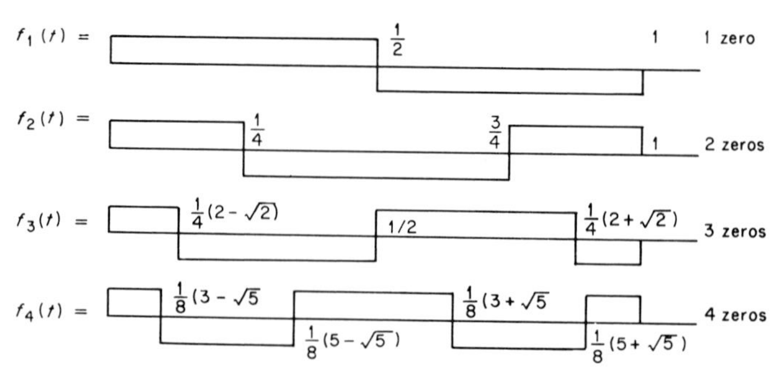

Fig. 8. Pulses whose transforms have multiple origin zeros. The break points are $a_i = \frac{1}{2}(1 - \cos i\pi/n + 1)(i = 0, 1, \ldots, n + 1)$.

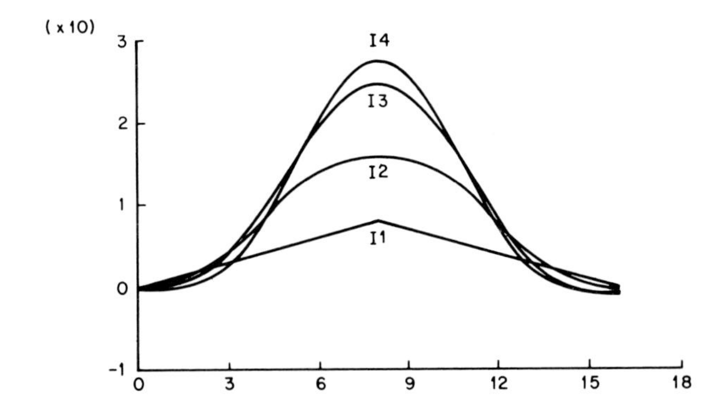

Fig. 9. Comparison between quadruple integration of $f_4(t)$ ($I4$), triple integration of $f_3(t)$ ($I3$), double integration of $f_2(t)$ ($I2$), and one integration of $f_1(t)$ ($I1$).

The fourth order zero at the origin of $F_4(s)$ makes $F_4(s)/s^4$ a finite Laplace transform. This makes $\iiiint f_4(t)\,dt$ time limited. Figure 9 shows a plot of this integral (I4) as well as plots of $\iiint f_3(t)\,dt$ (I3), $\iint f_2(t)\,dt$ (I2), and $\int f_1(t)\,dt$ (I1). *It is seen that these integrators serve as pulse compressors.* Moreover, since replacing s by s/k retains the finite Laplace transform property of $k^4F_4(s)/s^4$ it follows that *these compressors are effective irrespective of time scale change.*

The characteristics of the compression can be further enhanced by judicious addition of the output of a double integrator to that of the quadruple integrator. Both integrators are operating on $f_4(t)$. This is shown in Fig. 10, where IF represents four integrations less two. Both in Figs. 9 and 10 the time scale has been increased from 1 to 16 seconds.

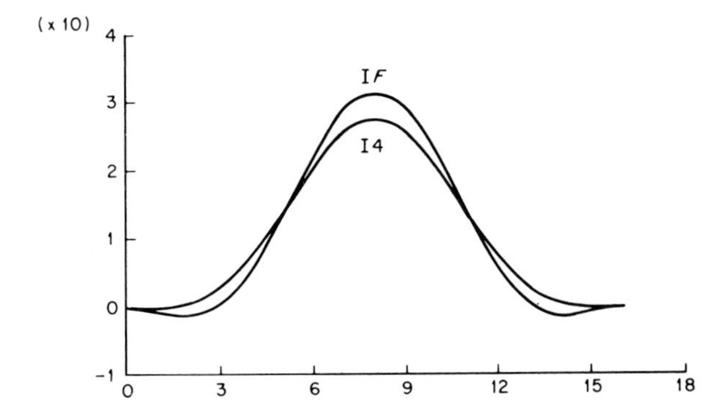

Fig. 10. Comparison between quadruple integration of $f_4(t)$ ($I4$) and $I4$ less double integration of $f_4(t)$.

6. THE MAIN THEOREM

Given an RLC network of degree n, it is possible to find a stepwise constant pulse, with $n + 1$ steps of arbitrary lengths, which will evoke a time limited network response.

The proof is by induction. It will be shown that new poles can be canceled by the introduction of new steps. Let N_1 in Eq. (11) be written as

$$F_{ab} = 1 - \frac{\sin b}{\sin(b-a)} \exp[-a(s+\alpha)/\beta] + \frac{\sin a}{\sin(b-a)} \exp[-b(s+\alpha)/\beta],$$

$$F_{bc} = 1 - \frac{\sin c}{\sin(c-b)} \exp[-b(s+\alpha)/\beta] + \frac{\sin b}{\sin(c-b)} \exp[-c(s+\alpha)/\beta]. \tag{18}$$

Clearly both F_{ab} and F_{bc} are zero at $(s + \alpha \pm j\beta) = 0$. Let $F(0)$ denote $F(s)|_{s=0}$. Then the following function, F_{abc},

$$F_{abc} = \frac{F_{ab}}{F_{ab}(0)} - \frac{F_{bc}}{F_{bc}(0)} \tag{19}$$

is zero at $s = 0$, and $-\alpha \pm j\beta$. In fact N_1 [Eq. (13)] is the same as $F_{abc}/F_{abc}(\infty)$. It follows that this process will generate a zero for any additional real pole of the network. To generate an additional complex pole, say at s_0, proceed in two steps:

Let $U = \text{Re } F(s_0)$, $V = \text{Im } F(s_0)$, and

$$G_{abc} = \frac{U_{bc}F_{ab} - U_{ab}F_{bc}}{U_{bc}V_{ab} - U_{ab}V_{bc}}, \qquad G_{bcd} = \frac{U_{cd}F_{bc} - U_{bc}F_{cd}}{U_{cd}V_{bc} - U_{bc}V_{cd}}. \tag{20}$$

Thus both G_{abc} and G_{bcd} are zero at $s = -\alpha \pm j\beta$ and are equal to j at $s = s_0$. Therefore

$$F_{abcd} = \frac{G_{abc} - G_{bcd}}{G_{abc}(\infty) - G_{bcd}(\infty)} \tag{21}$$

is zero at $s = -\alpha \pm j\beta$ as well as at s_0 and $\bar{s}_0$. Moreover, to produce a zero at the origin we add one more point:

$$F_{abcde} = \frac{F_{bcde}(0)F_{abcd} - F_{abcd}(0)F_{bcde}}{F_{bcde}(0)F_{abcd}(\infty) - F_{abcd}(0)F_{bcde}(\infty)}.$$

Similarly, any number of zeros would be generated. This completes the proof.

7. CONCLUSION

A synthesis procedure is developed for producing a piecewise constant pulse having an arbitrary number of zeros in the entire left plane. This pulse can be used to elicit a time limited response in any dynamic system. Examples are shown where a powerful response can be achieved by use of bipolar inputs.

REFERENCES

1. I. Gerst and J. Diamond, The elimination of intersymbol interference by input signal shaping, *Proc. IRE* (1961), 1195–1203.
2. P. Dines and D. Hazony, Optimization of time limited outputs in band limited channel, *Proc. 13th Allerton Conf.* (1975), 515–522.

This work was supported in part by U.S. Air Force Contract No. AFOSR76-2886B.

DEPARTMENT OF ELECTRICAL ENGINEERING AND APPLIED PHYSICS
CASE WESTERN RESERVE UNIVERSITY
CLEVELAND, OHIO

MOS Neuristor Lines†

R. W. NEWCOMB

An MOS–RC line is considered which has all the properties of neuristors, as defined by Crane. For this the nonlinear diffusion equations describing distributed structures are derived and discussed. Experimental evidence is presented as well as extensions.

Voy a nombrar las cosas, los sonoros
Altos que ven el festejar del viento [1].

1. INTRODUCTION

In his Ph.D. dissertation Crane [2, 3] introduced the concept of the neuristor, this being [2, p. iii] "defined as a device having the form of a one-dimensional channel along which signals may flow, the signals taking the form of propagating discharges having the following properties: 1. Threshold of stimulability, 2 uniform velocity of propagation, 3. attentuationless propagation, 4. refractory period following the passage of a discharge, after which the neuristor can support a discharge." At the time, although various possibilities were put forth, no physical realizations existed in electronic form, and, thus, the name "heuristor" was coined [4].

However, as the concept became known, a number of electronic realizations appeared [5–26] as well as characterizations of such and related electronic schemata [27–56]. And, since the neuristor, by its definition above, possesses many of the properties of nerve axons and cells, their various models [57–81] are neuristor-like devices. It should be emphasized, though, that the neuristor was conceived not as a model of the nerve axon but as a device for computer construction, for which it has been shown capable of carrying out all digital logic functions [2, 82–84]. Consequently, to be useful in the context of computer construction, neuristors compatible with large scale integrated circuit technology should be available; this is the major shortcoming associated with all of the devices and models referenced above.

† This English translation from the Spanish is printed with permission of Academia de Ciencias de Cuba from Lineas MOS de tipo neuristor, *Communicaciones*, to appear.

ISBN 0-12-178150-X

With this in mind, under the joint Polish–American "Active Microelectronic Systems" program, Dr. Wilamowski and his colleagues developed a bipolar transistor RC neuristor line [85, 86] fully compatible with standard integrated circuit technology. A slightly different circuit was developed using CMOS (complementary metal oxide silicon) technology on the American side of the program [87] where the possibility of frequency coding of repetitive neuristor pulses was observed [88, 89] and checked on the Polish circuit [90]. Therefore, these and similar circuits merit further investigation in which case we here discuss a closer variant of the Polish circuit using CMOS devices, giving primarily experimental evidence to support its operation.

Con la mirada inmóvil del verano
Mi cariño sabrá de las veredas [1].

2. CIRCUITS

In the following paragraph we introduce the basic neuristor circuit we wish to discuss here, giving physical reasoning on its operation and experimental verification of its important properties. From these some valuable extensions are put forth.

2.1. Basic Circuit

The basic circuit to be introduced here is shown in Fig. 1 which represents a cascade of identical sections beyond the source. Each section of Fig. 1 is to be considered as a passive $R_l - C_l$ two-port "line" section with an active, nonlinear, and (most importantly) dynamic one-port load across its second port. For comparison one section of the circuit of Wilamowski *et al.* [85] from which it stems is shown in Fig. 2; as can be seen the major difference is the substitution of the MOS transistors for the bipolar ones. In Fig. 1 the connection node numbers are those of the pins on the MC-14007CP, CMOS transistor, package used for all experimental verifications given in this paper. Figure 3 shows typical pulses obtained experimentally, as discussed below, the top trace being the periodic input voltage, with negative pulses, and the bottom trace being the line voltage response at pin 3 of the fifth section of an eight section line of the form of Fig. 1.

The principle of operation is quite similar to that for Fig. 2 and will now be reviewed. For this reference to Fig. 4, which shows the line input (at the top) and voltages v_1, v_2, v_3 at pins 1, 2, 3 of the fifth stage and at pin 3 of the fourth stage, will prove helpful. In the resting state, when $v_{in} = 0$, capacitor C_2 is discharged with zero voltage across it while capacitor C_l is fully charged to the bias voltage V_B; no current flows and, hence, the voltages v_1 and v_{10} of pins 1 and 10 sit at ground potential while pins 2, 3, and 12 sit at the bias

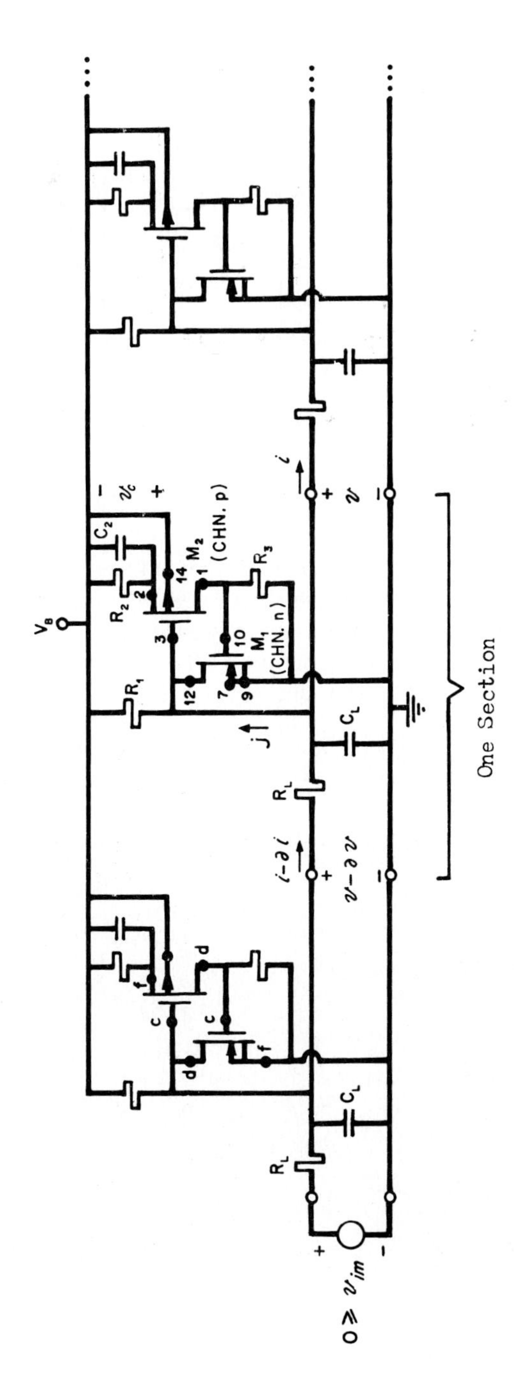

Fig. 1. Basic neuristor line circuit for negative pulse transmission.

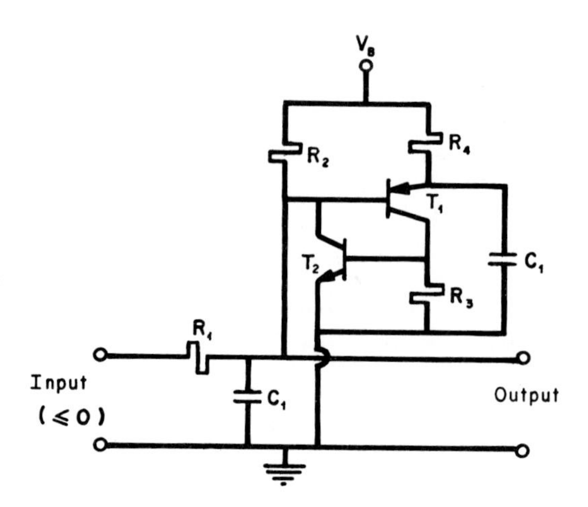

Fig. 2. One section of circuit of Wilamowski *et al.* [85].

potential V_B. From these observations we see that with zero input the gate-to-source voltage of the n-channel MOS transistor M_1 is given by $v_{gs_1} = v_{10} - v_9 = 0$ and similarly for the p-channel transistor M_2, $v_{gs_2} = v_3 - v_2 = 0$. In other words, under zero input conditions both M_1 and M_2 are turned off (i.e., zero drain current flows) since necessary requirements for their respective conduction are $v_{gs_1} > V_P$ and $v_{gs_2} < -V_P$, where $V_P > 0$ is the n-channel pinch-off voltage of the complementary MOS pair (this voltage being about 2 volts for our MC14007CP devices). With the application of a

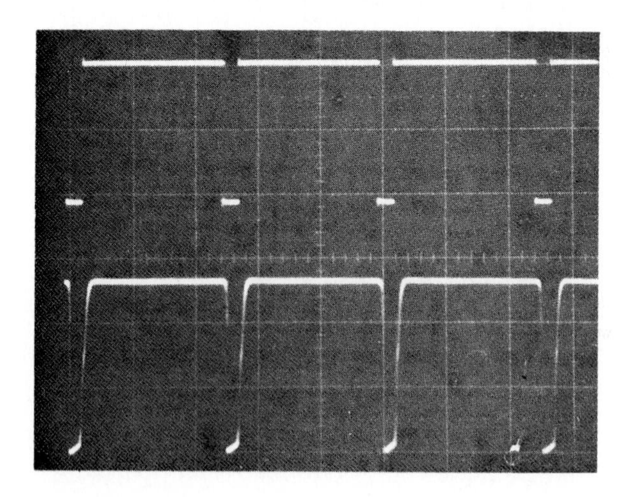

Fig. 3. Typical pulses of Fig. 1 (10 μsec/div). Top: Input voltage (2 V/div.; 0 = top graticule). Bottom: Voltage at pin 3 of fifth stage (5 V/div.; 0 = bottom graticule)

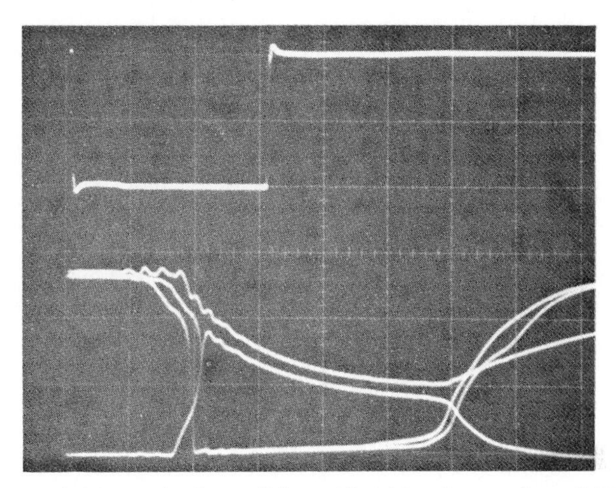

Fig. 4. Expanded internal voltages (0.5 μsec/div.). Top: Input voltage (2 V/div.). Bottom four traces—Internal voltages (5 V/div.): Top at left—v_2, fifth stage; Upper middle at left—v_3, fifth stage; Lower middle at left—v_3, fourth stage; Bottom at left—v_1, fifth stage.

sufficiently large negative input pulse, v_3 is lowered while v_2 can only slowly lower, being held by the capacitor C_2; with proper choice of parameters this turns on M_2. Sufficient current, as the activated source–drain current of M_2, flowing through R_3 raises the potential v_1 turning on M_1, a delay occurring due to the intrinsic delay in transferring the signal through M_2, and, more controllably, through the necessity of discharging C_l (through R_l and R_1 effectively in parallel) in order to lower v_3 to turn on M_2. The added current through R_1 due to the activation of M_1 further lowers v_3, this being the actual cause of a very sharp drop in the voltage v_3 [see Fig. 4 where this phenomena is seen at about 1 μsec in the fifth stage output (the input to this fifth section occurring as the fourth stage output at about 0.9 μsec)]. Consequent to this sharp drop in v_3, C_2 continues to charge up, this charging taking place following the MOS nonlinear (square-law) drain current characteristic until $v_3 - v_2$ approaches $-V_P$ (see Fig. 4 at about 2.9 μsec) during which time a quick change through the linear operation region of M_2 can occur as M_2 turns off. After a sufficient drop in v_1 a similar change in transistor M_1 takes place, it also turning off (see at about 3.4 μsec in the fifth stage output of Fig. 4). Here the state of refractoriness is fixed by the state of discharge of C_2, while the possibility for oscillations with a steady input exists if the situation is such that M_2 turns on again with sufficient discharge of C_2 (as occurs in Fig. 8 below).

Experiments were carried out to illustrate some of these points with oscilloscope traces shown in Figs. 3–9 (taken from a Tektronix 5403 with

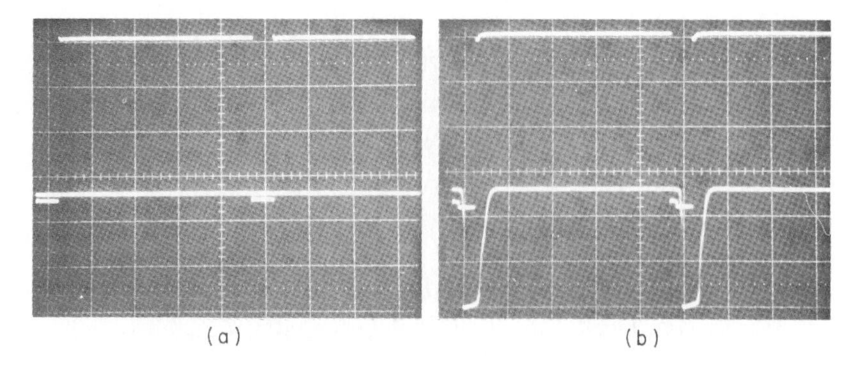

Fig. 5. Amplitude threshold (5 μsec/div.). (a) Input amplitude just below threshold. (b) Input amplitude just above threshold: Tops—Input (0.5 V/div.); Bottoms—Voltage at fifthstage (5 V/div.).

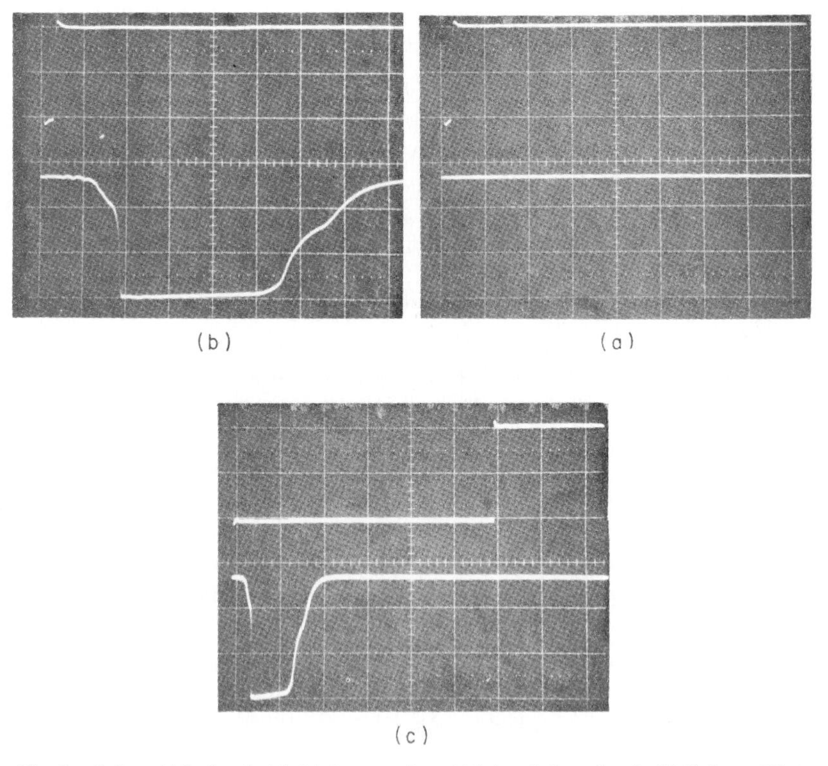

Fig. 6. Pulse width threshold. (a) Input pulse width just below threshold (0.5 μsec/div.). (b) Input pulse width just above threshold (0.5 μsec/div.). (c) Independence from input pulse width above threshold (2 μsec/div.): Tops—input (2 V/div.); Bottoms—voltage at fifth stage (5 V/div.).

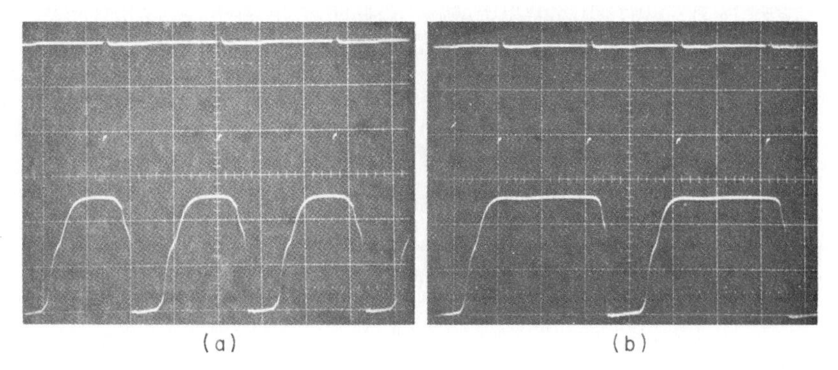

Fig. 7. Refractory period (2 μsec/div.). (a) Input repetition rate just prior to refractoriness. (b) Input repetition rate just after refractoriness. Tops—input (2 V/div.); Bottoms—voltage at fifth stage (5V/div.).

5A48 dual trace and 5B42 delaying time base plug-ins and a 013-0090-00 $\times$ 1 probe). In all of these the upper trace is the periodic negative going applied voltage pulse (from a HP 8002 pulse generator) while the lower trace is generally the voltage on the line (pin 3) at the fifth section of an eight section line (with a resistor R_l to ground at the right). The circuit element values were $R_l = R_1 = R_2 = 5.6$ kΩ, $R_3 = 2.7$ kΩ, $C_l = 0$ (except 50 pF and 100 pF in the last two parts of Fig. 9), $C_2 = 270$ pF, and $V_B = 13$ V (from a HP 721A power supply). In all of the traces, the very top and bottom grid lines show ground potential.

Commenting specifically on these traces, Fig. 5 shows that a pulse is excited when the input amplitude is above 1.85 V. Figure 6 shows that when the input pulse width is larger than 0.17 μsec (at 4 volts amplitude) an output is triggered; it also shows that this output is independent of the input pulse width when longer than its threshold width. Figure 7 illustrates that the refractory period is 5.5 μsec, since when input pulses occur more frequently some cannot trigger outputs. The interesting pulse repetition rate control by input amplitude is illustrated in Fig. 8 where the control is easy to be exercised over the range $-5.2\text{ V} > v_{\text{in}} > -7.3$ V. Finally, Fig. 9 shows the delay as it changes due to changes in C_l and R_l, the first two leftmost curves using only the intrinsic input capacitance of the transistor circuit (seemingly this intrinsic capacitance is about 10 pF being close to the parallel combination of the two gate-to-source capacitances of about 5 pF each). Experimentation also showed the delay per stage to be about identical for the middle five stages, of the eight stage line, being also as illustrated between the fourth and fifth stage in Fig. 4 (thus, about 0.1 μsec/stage).

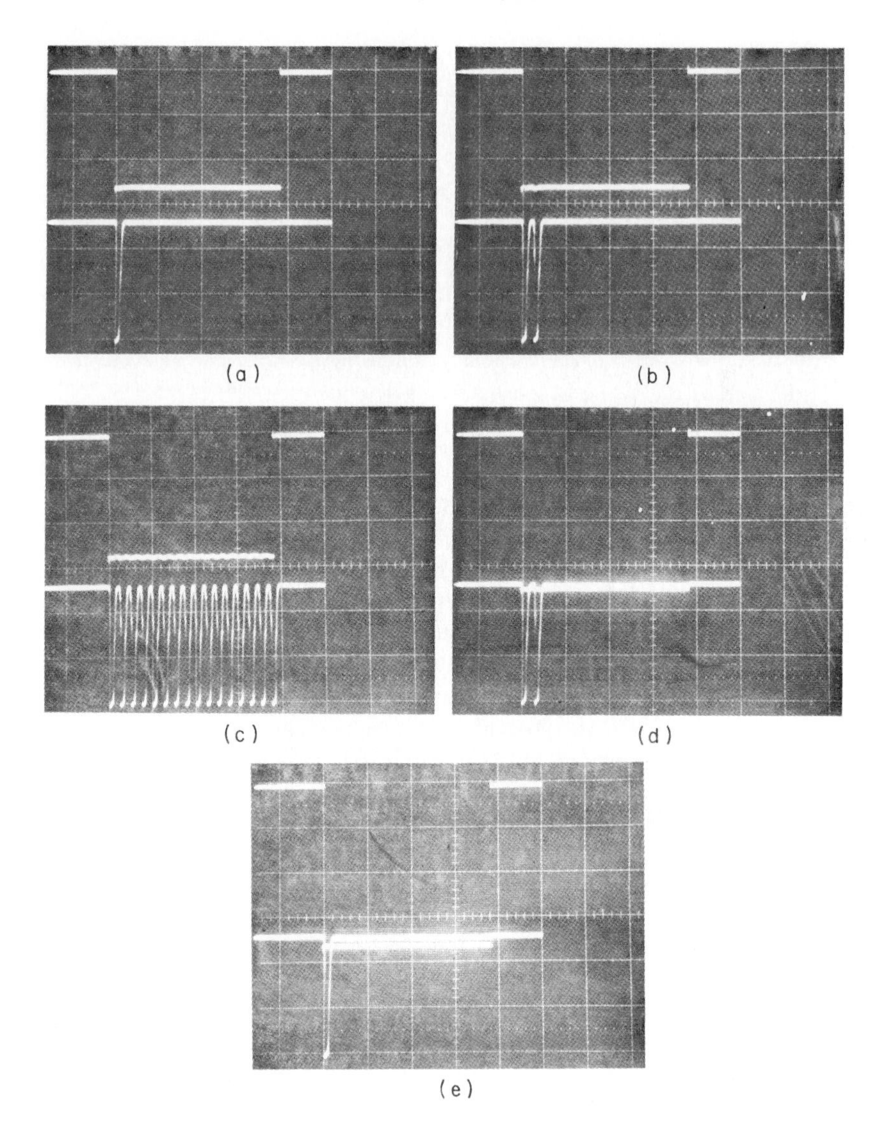

Fig. 8. Voltage controlled Pulse repetition rate (20 μsec/div.). (a) Input for single pulse output. (b) Input for double pulse output. (c) Input for continuous pulse generation. (d) Larger input for double pulse output. (e) Larger input for single pulse output: Tops—input (2 V/div.); Bottoms—voltage at fifth stage (5 V/div.).

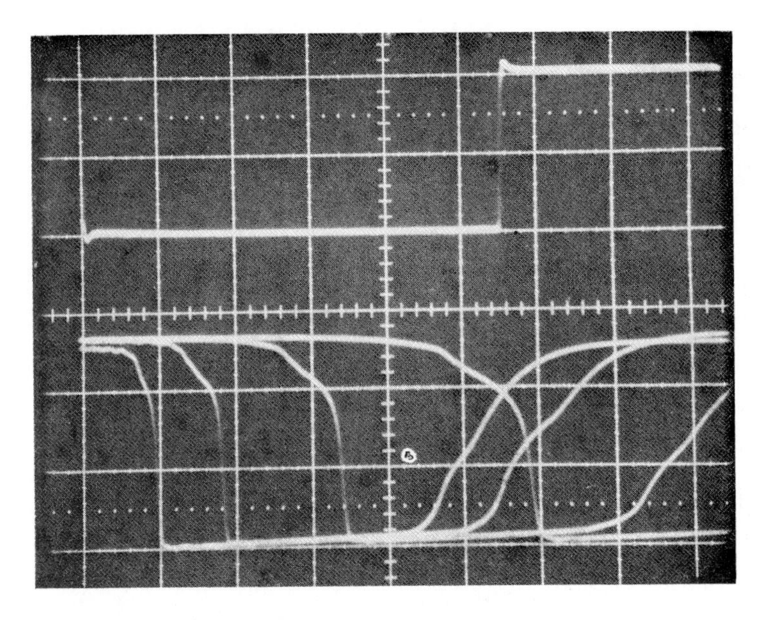

Fig. 9. Delay control by R_l and C_l (0.5 μsec/div.; V_B = 13 V). Top: input (2 V/div.). Bottom (5 V/div.): Leftmost—R_l = 2.7 kΩ, C_l = 0 pF; Second—R_l = 5.6 kΩ, C_l = 0 pF; Third—R_l = 5.6 kΩ, C_l = 50 pF; Rightmost—R_l = 5.6 kΩ, C_l = 100 pF.

2.2. Modifications and Other Circuits

By using the symmetries of CMOS devices, Fig. 1 is readily modified to transmit positive going pulses, as shown in Fig. 10. If one desires a line transmitting positive and negative pulses, one naturally thinks of realizing these two lines, of Figs. 1 and 10, as a simultaneous line, but the existence of two opposite resting levels on C_l, which one would like to be common, initially negates this attempt. However, interchanging ground and the bias points while reversing the sign of the bias voltages in these two types of "one-port loads" brings the resting level on C_l to zero. Doing this, while sharing R_1, gives a line, a section of which is shown in Fig. 11, which will transmit both positive and negative pulses. As a comment on its operation, we mention that it has proven rather touchy to operate, requiring rather close coordination between bias and input voltages, apparently due to substrate couplings. As a further comment we note that Fig. 11 is a nice configuration for complete integration since all capacitors have a common terminal at ground. In order to have identical positive and negative pulse characteristics the complementary transistors of the two parts (p-channel of upper circuit with n-channel of lower circuit, for example) most

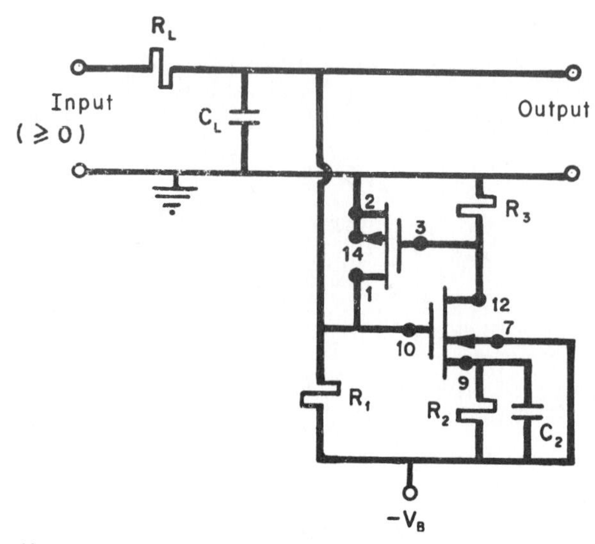

Fig. 10. One section of Fig. 1 modified for positive pulse transmission.

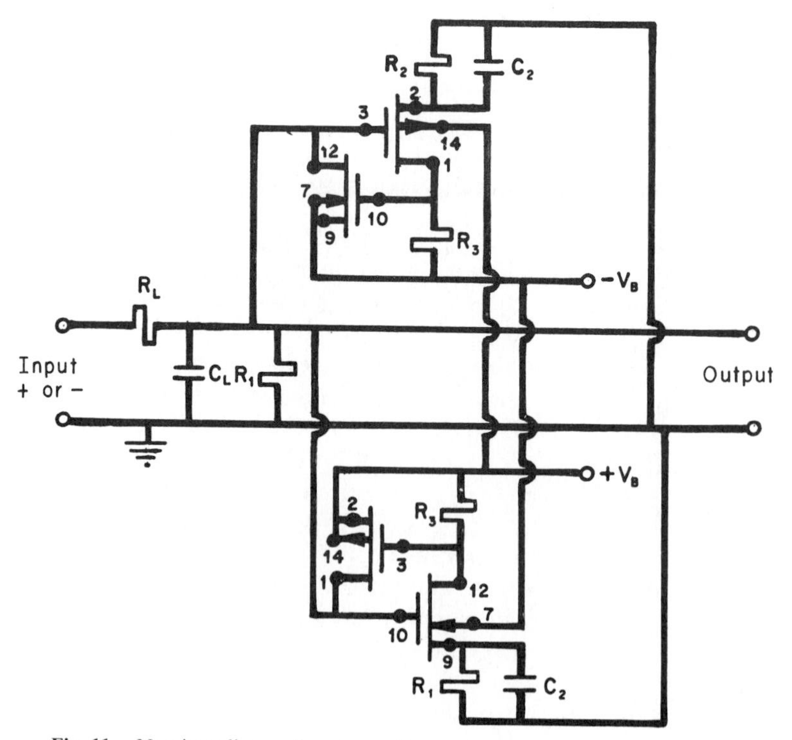

Fig. 11. Neuristor line section for positive and negative pulse transmission.

conveniently should be well matched, something still relatively hard to accomplish in integrated circuit technology though.

The circuits discussed to this point have all been of the nature of loaded lines, and, thus, most convenient for consideration of distributed realizations. A different configuration is shown in Fig. 12 of a previously developed but unpublished circuit which satisfies all of the neuristor definition conditions.

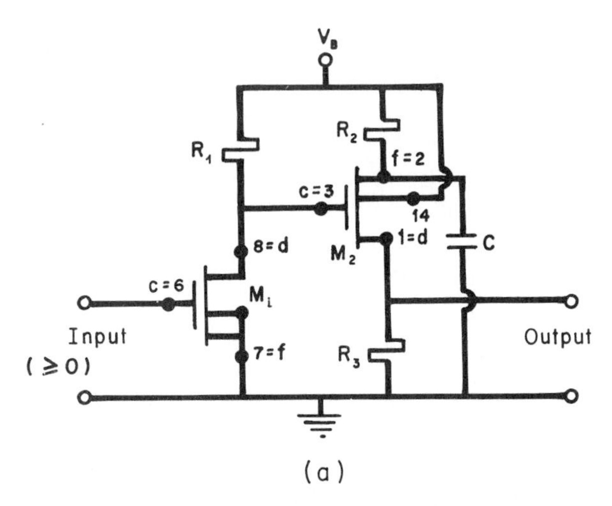

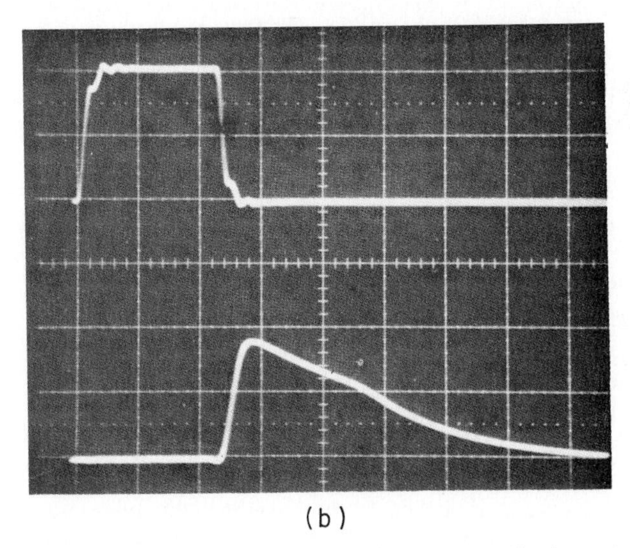

(b)

Fig. 12. Simple unilateral neuristor line section and response. (a) Circuit—typical values: $R_1 = R_2 = 20$ kΩ, $R_3 = 2.7$ kΩ, $V_B = 9$ V, $C = 100$ pF. (b) Response (2 V/div., 0.2 μsec/div.): Upper trace—input (zero at fifth trace); Lower: fifth stage output (zero at first trace).

Here the n-channel input transistor M_i merely acts as an inverter, which also gives unilateral transmission and isolation, while the p-channel M_2 gives the nonlinear and active structure with feedback necessary for pulse shaping. This section then transmits positive pulses and unilaterally from left to right. Probably the circuit of Fig. 12 is the simplest neuristor available, though the absence of real gain in the feedback path precludes it from having the voltage controlled repetition rate property, or so it seems. But this is remedied by inserting a feedback transistor, M_1 in the same relation to M_2 we had in Fig. 1, as per our previously published circuit [87], now studied extensively by Kulkarni-Kohli [88]. Further, the circuit can be made to transmit positive and negative pulses through a conjunction similar to that used to obtain Fig. 11, and probably more successfully due to the isolation of the stages (positive results are contained in [88, p. 30]).

Y la pobreza del lugar, y el polvo
Sitios de piedra decidida y limpia [1].

3. MATHEMATICAL ANALYSIS

In this section we show that by standard circuit analysis reasonable results can be obtained, as the delay per stage or the describing differential equations. However, we will see that the "state of the art" of modern mathematics is not to the point that analytic characterizations can be obtained from the nonlinear partial differential equations describing distributed MOS neuristors.

On all of the lumped circuits presented here it is a relatively straightforward matter to perform a computer aided analysis using standard models for the MOS transistors. This has been carried out [88] to good agreement on the derivative [87] of Fig. 12, from which a simple accurate model of the MOS transistor has been introduced by Kulkarni-Kohli [88, p. 36]; this will be used at Eqs. (2) and (4) below. However, due to the availability in the U.S. of CMOS packages at very low cost (about three transistor pairs for 0.30 U.S. dollars), it has been actually easier to carry out the experiments discussed earlier. Nevertheless, for design purposes it is important to have design equations for parameters of interest. Since for envisaged uses the delay per stage is probably the most important parameter to be controlled, we develop this before continuing to more abstract topics.

Applying circuit theory to any stage, but the first, of Fig. 1 we can roughly calculate the time delay through it as follows. The stage is activated by turning on M_2 which is accomplished by discharging C_l by an amount V_P. Roughly, this capacitor sees an effective resistance R_e of R_1 in parallel with R_l (since the previous state is assumed activated, which effectively grounds the coupling R_l and because the unactivated MOS transistors appear as open circuits).

The discharge law of the single time–constant circuit will then be $v_3(t) = V_B - V_B \exp(t/R_e C_l)$; at the turn on time t_1 of M_2 we have $v_3(t_1) \approx V_B - V_P$, or on solving

$$t_1 \approx R_e C_l \ln(V_B/V_P) \qquad \text{with} \quad R_e \approx (R_1 R_l)/(R_1 + R_l). \tag{1a}$$

Added to t_1 to obtain the actual delay t_d is the intrinsic delay t_i of each transistor (the time when drain current starts to flow after the gate–source voltage reaches its turn-on level). Thus

$$t_d = t_1 + 2t_i. \tag{1b}$$

Numerically we can check against Fig. 4 by using $t_i \approx 10$ nsec, $C_l \approx 10$ pF, $R_1 = R_l = 5.6\ \mathrm{k\Omega}$, $V_B = 13$ V, $V_P = 2v$, from which Eqs. (1) give $t_d \approx 0.07$ μsec while Fig. 4 shows about 0.1 μsec of delay per stage; Eq. (1a) is rough, but exhibits the important design parameters (see also Fig. 9). Other such formulas can be worked out for the remaining neuristor characteristics [85], though again these give more in the way of design insight than accurate determinations.

More challenging is the analysis of truly distributed circuits for which we will set up the pertinent partial differential equations for the MOS loaded lines; for these, as will become apparent, computer aided analysis is much more pertinent in the absence of analytic mathematical techniques by which to proceed.

First we consider characterization of the MOS devices, for which we introduce a function $F(\cdot, \cdot, \cdot)$ of three variables

$$F(x, y, z) = \beta(x - z)^2\{1 - \exp[-Ky/(x - z)]\}u(x - z), \tag{2}$$

where β and K are positive constants and $u(\cdot)$ is the unit step function

$$u(x) = \begin{cases} 1 & \text{if} \quad x > 0, \\ 0 & \text{if} \quad x \leq 0. \end{cases} \tag{3}$$

Then, as shown in a curve fitting manner by Kulkarni-Kohli [88, p. 36], matched n- and p-channel CMOS transistors are accurately described, respectively, by

$$i_{ds} = F(v_{gs}, v_{ds}, V_P), \qquad \text{n-channel}, \qquad V_P > 0, \tag{4a}$$

$$i_{ds} = -F(-v_{gs}, -v_{ds}, -V_{P_p}), \qquad \text{p-channel}, \qquad V_{P_p} = -V_P, \tag{4b}$$

where i_{ds} is the drain–source current, etc. (for the MC14007CP package, $K \approx 5$, $V_P \approx 2$ V, $\beta \approx 0.64 \times 10^{-3}$ A/V^2). Using these relationships we can turn to an analysis of a section of Fig. 1.

We have, on summing currents at node 12,

$$j = R_1^{-1}(v - V_B) + i_{ds_1}, \tag{5a}$$

and from Eq. (4a)

$$\begin{aligned} i_{ds_1} &= F(v_{gs_1}, v_{ds_1}, V_P) = F(v_1, v, V_P) \\ &= F(-R_3 i_{ds_2}, v, V_P). \end{aligned} \tag{5b}$$

From Eq. (4b), using the capacitor C_2 voltage v_c as defined in Fig. 1,

$$\begin{aligned} i_{ds_2} &= -F(-v_{gs_2}, -v_{ds_2}, V_P) \\ &= -F(-[v - (v_c + V_B)], [v_c + V_B + R_3 i_{ds_2}], V_P). \end{aligned} \tag{5c}$$

To proceed, Eq. (5c) needs to be solved for the i_{ds_2} which occurs on both of its sides and this solution substituted into Eq. (5b); such a solution always exists, as is seen by plotting the left side and the right side [using Eq. (2)] of Eq. (5c) and noting that for fixed values of v, v_c, R_3, V_B, V_P, K, and β there is one and only one intersection. Thus, we can functionally obtain

$$i_{ds_2} = G(v, v_c). \tag{6a}$$

Analytically it is difficult to give a more explicit formula for $G(\cdot, \cdot)$, but note that in our circuit the second argument, $y = [v_c + V_B + R_3 i_{ds_2}]$, in Eq. (5c) never goes negative in which case the unit step is $u(y) = 1$; if further $K = \infty$ then we obtain

$$i_{ds_2} = G(v, v_c) = -\beta(v_c + V_B - V_P - v)^2 u(v_c + V_B - V_P - v), \tag{6b}$$

which is a useful square-law approximation to the actual current.

Continuing, a straightforward substitution of Eq. (6a) into (5b) and then into (5a) yields

$$j = R_1^{-1}(v - V_B) + F(-R_3 G(v, v_c), v, V_P), \tag{7}$$

which gives the relationship of the one-port terminal current j in terms of the one-port terminal voltage and internal capacitor (state-variable) voltage v_c, where also we observe that in the resting state, $v = V_B$, $v_c = 0$, we have $j = 0$.

Next the dynamical equation for v_c is found by summing currents at node 2:

$$i_{ds_2} = R_2^{-1} v_c + C_2 \dot{v}_c \qquad (\text{where} \quad \dot{x} = dx/dt) \tag{8a}$$

or, from Eq. (6a)

$$C_2 \dot{v}_c = -R_2^{-1} v_c + G(v, v_c). \tag{8b}$$

Now consider Fig. 1 to be a distributed circuit with a section of length ∂x, in which case we can set up partial differential equations describing the line as follows. Across $R_l = r_l\, \partial x$ we have $(v - \partial v) - v = R_l(i - \partial i)$ and through $C_l = c_l\, \partial x$ we have $-\partial i = (c_l\, \partial x)(\partial v/\partial t) + j$ where from Eq. (7) we can express the current j as $j = f(v, v_c)\,\partial x$ with $f(V_B, 0) = 0$. Similarly from Eq. (8b) we can write, using $C_2 = c_2\,\partial x$, $(c_2\,\partial x)(\partial v_c/\partial t) = g(v, v_c)\,\partial x$ where $g(V_B, 0) = 0$. The equations of interest for a distributed realization will then be (on taking the limit as ∂x tends to 0)

$$(\partial^2 v/\partial x^2) = r_l c_l(\partial v/\partial t) + r_l f(v, v_c), \tag{9a}$$

$$(\partial v_c/\partial t) = c_2^{-1} g(v, v_c) \tag{9b}$$

with $v(x, t)$ having $v(0, t) = v_{in}(t)$ specified. Equations of this form are known as nonlinear diffusion equations in mathematics [91]. Stimulated by the problems in neural modeling, there has recently been considerable interest in such equations by mathematicians who have treated either special cases, often of very simple functions f and g, [92–112], or very general situations [113–121].

The main interests in Eqs. (9) have to do with traveling wave solutions, that is solutions which are only a function of $z = x - vt$ where v is the velocity of travel; we note that then

$$(\partial/\partial x) = -v^{-1}(\partial/\partial t) = (d/dz) = {}', \tag{10}$$

which leads to so called "dynamic steady state" [54, p. 175] state variable equations

$$v' = w, \tag{11a}$$

$$w' = -v^{-1} r_l c_l w + r_l f(v, v_c), \tag{11b}$$

$$v_c' = -c_2^{-1} v g(v, v_c). \tag{11c}$$

For these we know $v = V_P$, $v_c = w = 0$ define an equilibrium point, corresponding to the resting state of our line. Now it is known that if f and g were linear the (parabolic) set of partial differential equations (9) have no nonresting state traveling wave solution [110, p. 882], consequently the question of existence of solutions of Eqs. (11) is not easily dismissed. The real problem, which has as yet not been analytically solved in general [110, p. 886], is that of determining the velocity v of wave propagation, under appropriate boundary conditions, those treated so far(for very special f and g) representing either a single traveling pulse or a periodic train of traveling pulses [112–122]. However, using the time-delay found at Eq. (1) we can of course estimate the velocity v for our MOS circuits.

Although not much more can really be said analytically on the form of solutions one can try to linearize Eqs. (11). Linearization is most easily accomplished when in the active region as one need use only the square law behavior of the MOS transistors [as at Eq. (6b)]. Because of step function discontinuities, though, this is only tractable when linearization occurs about the excited state, of most practical interest for stability considerations of the transmitted pulse, but this excited state must first be analytically assumed.

Nevertheless, in the abstract mathematical sense probably most progress has been made in this direction [114–119].

> Las portales profundos, las mamparos
> Cerrados a la sombra y al silencio [1].

4. DISCUSSION

The MOS transistor being of simple construction and having nonlinear active voltage-controlled current source characteristics has seemed to be a fundamental building block for electronic circuits. Here we have illustrated its versatility in the realization of neuristor lines of various configurations. In fact the basic realization of either Fig. 1 or Fig. 2 has appeared most naturally to be implemented with MOS devices once its principle of operation, using currents controlled by voltages, is understood.

But what we have presented here also seems to be just a start in the area. To be sure, we are able to construct neuristor lines, as per Fig. 11, which transmit positive and negative pulses. Nevertheless, the practical lack of complete symmetry in constructing CMOS pairs makes the actual operation more difficult than for the single signed pulse lines. This lack of symmetry was ignored, for simplicity, in our setting up of the mathematical analysis of the basic line, since at Eq. (4) we assumed that the same parameters β, K, V_P described both the n-channel and the p-channel devices, which is not quite true. But correcting for this disymmetry on the basic line is straightforward. And, although we have seen that setting up describing equations, as Eqs. (9), is straightforward, it is not an easy matter to carry out an analysis of these equations, in which case most of the design to date has been carried out empirically either numerically on the computer or experimentally in the laboratory. In short we lack a complete mathematical theory of neuristor lines.

Besides just being a start in the mathematical sense, what we present is also a start in the physical sense in that physical constructs to this date are all by way of lumped sections while totally distributed integrated circuit neuristor lines, especially following Fig. 11, should be quite possible. Likewise since MOS transistors can readily realize resistors and capacitors, another promising direction of investigation, even in the lumped section case, is to obtain neuristor lines constructed solely from MOS devices. Other improvements, as electronically controlled delay, seem possible. Further, the material presented has covered only neuristor lines themselves. Although these are sufficient to realize all components for making computers, there is another class of circuits, called "neural junctions" [123–126], which can be used to combine neuristors in possibly new and interesting configurations. And, since neuristors are level detecting pulse generators, they can be considered for use wherever such are desired, as for example in delta modulators, which we presently have under investigation.

In the References we have listed a number of the more pertinent references, without thought to much selectivity. Consequently, for those wishing to get a firmer start in the field we recommend that, besides Crane's basic though somewhat general paper [3], the immediately following paper of the same issue by Nagumo *et al.* [9], still seems to us the specific paper of most interest. For those most interested in the mathematics, Hastings' tutorial paper [110] is quite understandable and covers the primary points of interest. As time goes on we believe the theory of "catastrophes" [127] will be seen to be more and more applicable. And for those wishing more, possibly relevant, works we add references [128–208], a few of which we have as yet not been able to locate. In the near future we do hope to make available an annotated collection of the references.

Finally, we believe that in the above we have shown again the possibility of making neuristors which can be constructed in integrated circuit form. Improvements are needed as mentioned but as illustrated by the various oscilloscope traces, the neuristor properties are obtained, and, thus, the nickname "heuristor" can now be taken as a play on the creator's shortened forename!

Y nombraré las cosas, tan despacio
Que cuando pierda el paraíso de mi calle
Pueda llamalas de pronto con el alba [1].

REFERENCES

1. E. Diego, Voy a nombrar las cosas, *in* "Nombrar las Cosas," Bolsilibros Unión, Havana, 1973, pp. 32–33.
2. H. D. Crane, Neuristor Studies Tech. Rep. No. 1506–2, Stanford Electron. Lab., 1960.
3. H. D. Crane, Neuristor—A novel device and system concept, *Proc. IRE* **50** (1962), 2048–2060.
4. H. D. Crane, The neuristor, *IRE Trans. Electron. Comput.* **EC-9** (1960), 370–371.
5. H. D. Crane and A. Rosengreen, Results from Experimental Relay Neuristor Lines, Interim Rep. No. 1, SRI Project 3286, 1961.
6. H. D. Crane, The neuristor, *Internat. Solid-State Circuits Confer., 1961.*
7. A. J. Cote, Jr., A neuristor prototype, *Proc. IRE* **49** (1961), 1430–1431.
8. A. J. Cote, Jr., Simulating nerve networks with four layer diodes, *Electronics* **34** (1951), 51–53. [Reprinted *in* "Tunnel-Diode and Semiconductor Circuits" (J. M. Carroll, ed.), McGraw-Hill, New York, 1963, pp. 294–296.]
9. J. Nagumo, S. Arimoto, and S. Yoshizawa, An active pulse transmission line simulating nerve axon, *Proc. IRE* **50** (1962), 2061–2070.
10. A. C. Scott, Neuristor propagation on a tunnel diode loaded transmission line, *Proc. IEEE* **51** (1963), 240.
11. A. Rosengreen, Experimental neuristor gives nerve-like propagation, *Electronics* **36** (1963), 25–27.

12. S. Noguchi, Y. Kumagai, and J. Oizumi, General considerations on the neuristor circuits, *Rep. Res. Inst. Electr. Comm., Tohoku Univ.* **14** 4 (1963), 155–184.
13. A. Ambroziak, Semidistributed neuristor line using unijunction transistors, *Solid-State Electron.* **7** (1964), 259–265.
14. S. Yoshizawa and J. Nagumo, A bistable distributed line, *Proc. IEEE* **52** (1964), 308.
15. R. H. Mattson, A neuristor realization, *Proc. IEEE* **52** (1964), 618–619.
16. A. J. Cote, Jr., Neuristor propagation in long-tunnel diodes, *Proc. IEEE* **53**, (1965), 164–165.
17. H. Kunov, Controllable piecewise linear lumped neuristor realization, *Electron. Lett.* **1** (1965), 134.
18. J. Nagumo, S. Yoshizawa, and S. Arimoto, Bistable transmission lines, *IEEE Trans. Circuit Theory* **CT-12** (1965), 400–412.
19. A. Hayasaka and J. Nishizawa, Pulse characteristics of the distributed Esaki diode, *Electron. Comm. Japan* **49** (1966), 123–133.
20. R. D. Parmentier, Recoverable neuristor propagation on superconductive tunnel junction strip lines, *Solid-State Electron.* **12** (1969), 287–297.
21. J.-I. Nishizawa and A. Hayasaka, Two-line neuristor with active element in series and in parallel, *Internat. J. Electron.* **26** (1969), 437–469.
22. B. M. Wilamowski, H. Yokogawa, and J.-I. Nishizawa, Neuristor propagation in low impedance line, *Internat. J. Electron.* **29**, (1970), 101–137.
23. M. B. Waldron, Syneuristor: A Device and Concept, Tech. Rep. No. 6560–22, Stanford Electron. Lab., 1971.
24. B. S. Borisov, F. F. Zolotarev, and B. B. Shamshev, An investigation of the excitation propagation velocity in a semidistributed p-n-p-n structure neuristor line, *Radio Engrg. Electron. Phys.* **16** (1971), 526–529.
25. G. F. Belova and Y. A. Parmenov, Investigation of the coupling between neuristor-line elements having p-n-p-n structures, *Radio Engrg. Electron. Phys.* **17** (1972), 1923–1926.
26. B. M. Wilamowski, Analysis of the neuristor line, *Bull. Acad. Polon. Sci.* **20** (1972), 38[127]–44[134].
27. M. C. Pease, Analytic Proof of Operability of a Second-Order Neuristor Line, Interim Rep. No. 3, Stanford Res. Inst. Project 3286, 1961.
28. A. C. Scott, Analysis of nonlinear distributed systems, *IRE Trans. Circuit Theory* **CT-9** (1962), 192–195.
29. J. Nishizawa, Studies on progressing-wave type of Esaki diode, *Electron. Sci.* **12** (1962), 4?
30. T. Janus, Moving source model of neuristor triggering, *Proc. IEEE* **51** (1963), 1049–1050.
31. T. M. Il'inova and R. V. Khokhlov, Wave processes in lines with nonlinear shunt resistance, *Radio Engrg. Electron. Phys.* **8** (1963), 1864–1972.
32. A. C. Scott, Distributed device applications of the superconducting tunnel junction, *Solid-State Electron.* **7** (1964), 137–146.
33. A. C. Scott, Steady propagation on nonlinear transmission lines, *IEEE Trans. Circuit Theory* **CT-11** (1964), 146–154.
34. Y. I. Vorontsov, Certain properties of delay lines containing tunnel diodes, *Radio Engrg. Electron. Phys.* **9** (1964), 478–483.
35. M. W. Green and H. D. Crane, Modes of Neuristor Propagation—A Study in Distributed Active Processes, Stanford Res. Inst., Final Rep., Contract Nonr-3212(00), 1964.
36. Y. I. Vorontsov, Velocity of propagation of stationary signals in lines having nonlinear resistance, *Radio Engrg. Electron. Phys.* **9** (1964), 1414–1416.
37. T. M. Il'inova, Interaction of waves in a distributed line with nonlinear parallel loss, *Radio Engrg. Electron Phys.* **9** (1964), 1728–1735.

38. Y. N. Vorontsov, On the interaction between oppositely traveling signals in lines containing a nonlinear resistance, *Radio Engrg. Electron Phys.* **9** (1964), 1812–1814.
39. T. Pavlidis, A new model for simple neural nets and its application in the design of a neural oscillator, *Bull. Math. Biophys.* **27** (1965), 215–229.
40. I. Richer, Pulse propagation along certain lumped nonlinear transmission lines, *Electron. Lett.* **1** (1965), 135–136.
41. S. Aono, Y. Kumagai, S. Noguchi, and J. Oizumi, Fundamental characteristics of the electronic neuristor line, *Rept. Res. Inst. Electr. Comm., Tohoku Univ.* (1965), 21–43?
42. V. F. Zolotarev and V. I. Stafeyev, Soviet Patent No. 258,374, cl. 21a′, 32/40, 1966.
43. L. Y. Wei, A new theory of nerve conduction, *IEEE Spectrum* **3** (1966), 123–127.
44. Y. I. Vorontsov and I. V. Polyskov, Investigation of undamped signals in lines with a nonlinear resistance, *Radio Engrg. Electron. Phys.* **11** (1966), 1449–1456.
45. L. A. Logunov, I. V. Polykov, and V. N. Serebryakov, Distributed tunnel diodes, *Radio Engrg. Electron Phys.* **12** (1967), 149–152.
46. H. Kunov, On recovery in a certain class of neuristors, *Proc. IEEE* **55** (1967), 428–429.
47. Y. I. Vorontsov, M. I. Kozhevnikova, and I. V. Polyakov, Wave processes in active RC-lines, *Radio Engrg. Electron. Phys.* **12** (1967), 644–648.
48. R. D. Parmentier, Stability analysis of neuristor waveforms, *Proc. IEEE* **55** (1967), 1498–1499.
49. R. J. Buratti and A. G. Lindgren, Neuristor waveforms and stability by the linear approximation, *Proc. IEEE* **56** (1968), 1392–1393.
50. R. D. Parmentier, Neuristor waveform analysis by Lyapunov's second method, *Proc. IEEE* **56** (1968), 1607–1608.
51. I. Richer, The switch-line: A simple lumped transmission line that can support unattenuated propagation, *IEEE Trans. Circuit Theory* **CT-13** (1968), 388–392.
52. A. G. Lindgren and R. J. Buratti, Stability of waveforms on active nonlinear transmission lines, *IEEE Trans. Circuit Theory* **CT-16** (1969), 274–279.
53. R. D. Parmentier, Neuristor analysis techniques for nonlinear distributed electronic systems, *Proc. IEEE* **58** (1970), 1829–1837.
54. A. C. Scott, "Active and Nonlinear Wave Propagation in Electronics," Wiley (Interscience), New York, 1970.
55. A. Weyns, Information Transmission in the Nervous System, Thesis Dept. Elektrotech., Katholieke Univ. Leuven, 1973. (In Flem.)
56. J. A. Koslosick, K. L. Landt, H. C. S. Hsuan, and K. E. Lonngren, Properties of solitary waves as observed on a nonlinear dispersive transmission line, *Proc. IEEE* **62**, (1974), 578–581.
57. R. S. Lillie, Transmission of activation in passive metals as a model of the protoplasmic on nerve type of transmission, *Science* **48** (1918), 51–60?
58. R. S. Lillie, The passive iron wire model of protoplasmic and nervous transmission and its physiological analogues, *Biol. Rev.* **11** (1936), 181–209.
59. A. M. Weinberg, Nerve conduction with distributed capacitance, *J. Appl. Phys.* **10** (1939), 128–134.
60. K. Yamagiwa, A model for the synapse, *Japan Med. J.* **2** (1949), 38–46?
61. A. L. Hodgkin and A. F. Huxley, A quantitative description of membrane current and its application to conduction and excitation in nerve, *J. Physiol.* (*London*) **117**, (1952), 500–544.
62. W. H. Freygang, Jr., Some functions of nerve cells in terms of an equivalent network, *Proc. IRE* **47** (1959), 1862–1869.
63. J. W. Moore, Electronic control of some bioelectric membranes, *Proc. IRE* **47**, (1959), 1869–1880.

64. W. R. Ashby, "Design for a Brain," 2nd Ed., Chapman & Hall, London, 1960.
65. K. S. Cole, The advance of electrical models for cells and axons, *Biophys. J.* 2 (1962), 101–119.
66. C. M. Wiley, Neural networks simulate body functions, *in* "Tunnel-Diode and Semiconductor Circuits" (J. M. Carroll, ed.), McGraw-Hill, New York, 1963, p. 270.
67. R. F. Reiss, ed., "Neural Theory and Modeling," Proceedings of the 1962 Ojai Symposium, Stanford Univ. Press, Stanford, California, 1964.
68. L. L. Anderson, A nerve model for experimental purposes, *Med. Electron. Biol. Engng.* **3** (1965), 315–316?
69. R. FitzHugh, A kinetic model of the conductance changes in nerve membrane, *J. Cell. Compar. Physiol.* **66** (1965), 111–117.
70. R. FitzHugh, An electronic model of the nerve membrane for demonstration purposes, *J. Appl. Physiol.* **21** (1966), 305–308.
71. L. D. Harmon and E. R. Lewis, Neural modeling, *Physiol. Rev.* **46** (1966), 513–591.
72. H. Kunov, Nonlinear Transmission Lines Simulating Nerve Axon, Thesis, Electron. Lab., Tech. Univ. of Denmark, Lyngby, 1966.
73. R. Suzuki, Mathematical analysis and application of iron-wire neuron model, *IEEE Trans. Bio-Med. Engrg.* **BME-14** (1967), 114–124.
74. R. FitzHugh, Motion picture of nerve impulse propagation using computer animation, *J. Appl. Physiol.* **25** (1968), 628–630.
75. E. W. Pottala, An Electronic Model Neuron with Multiple Input Capability, Ph.D. Thesis, Univ. of Maryland, College Park, 1970.
76. G. Roy, A simple electronic analog of the squid: The NEUROFET, *IEEE Trans. Biomed. Engrg.* **BME-19** (1972), 60–63.
77. K. N. Leibovic, "Nervous System Theory," Academic Press, New York, 1972.
78. J. S. Albus, Theoretical and Experimental Aspects of a Cerebellar Model, Ph.D. Thesis, Univ. of Maryland, College Park, 1972.
79. M. A. Parkhideh, Nonlinear Network Analysis for Neural Models, Ph.D. Thesis, Stanford Univ., Stanford, California, 1973.
80. M. A. B. Brazier, D. O. Walter, and D. Schneider, eds., Neural Modeling, Res. Rep. No. 1, Brain Inform. Serv., Univ. of California, Los Angeles, 1973.
81. R. M. Gulrajani, F. A. Roberge, and P. A. Mathieu, A field-effect transistor analog for the study of burst-generating neurons, *Proc. IEEE* **65** (1977), 807–809.
82. B. M. Wilamowski, A novel concept of neuristor logic, *Internat. J. Electron.* **33**, (1972), 659–663.
83. B. M. Wilamowski, O możliwósci budowy neurystorowych logicznych (Possibilities of constructing neuristor logic network), *Rozprawy Elektrotech.* **19** (1973), 273–280.
84. Z. Czarnul and B. Bialko, Selected neuristor logic circuits using single neuristor line sections (to appear).
85. B. M. Wilamowski, Z. Czarnul, and M. B. Bialko, Novel inductorless neuristor line, *Electron. Lett.* **11**, (1965), 355–356.
86. Z. Czarnul, M. Bialko, and R. W. Newcomb, A neuristor-line pulse train selector, *Electron. Lett.* **12** (1976), 205–206.
87. C. Kulkarni-Kohli and R. W. Newcomb, An integrable MOS neuristor line, *Proc. IEEE* **65** (1976), 1630–1632.
88. C. Kulkarni-Kohli, An Integrable MOS Neuristor Line: Design, Theory and Extensions, Ph.D. Thesis, Univ. of Maryland, College Park, 1977.
89. C. Kulkarni-Kohli and R. W. Newcomb, Voltage controlled oscillations in the MOS neural line, *Proc. IEEE Midwestern Symp. Circuits Systems.* pp. 134–138, 1977.
90. Z. Czarnul and M. Bialko, Utilization of a single inductorless neuristor line section as a voltage-to-frequency convertor, *Electron. Lett.* **13** (1977), 251–252.

91. C. Conley and J. Smoller, Remarks on traveling wave solutions of non-linear diffusion equations, *in* "Structural Stability, the Theory of Catastrophes and Applications in the Sciences" (P. Hilton, ed.), Lecture Notes in Mathematics, No. 525, Springer-Verlag, Berlin and New York, 1976, pp. 77–89.
92. A. Kolmogoroff, I. Petrovsky, N. Piscounoff, Etude de l'équation de la diffusion avec croissance de la quantité de matiere et son application a un probleme biologique, *Moscow Univ., Bull. Math. Méch., Ser. Internat. Sect. A.1* **6** (1937), 1–25.
93. F. Offner, A. Weinberg, and G. Young, Nerve conduction theory: Some mathematical consequences of Bernstein's model, *Bull. Math. Biophys.* **2** (1940), 89–103.
94. A. M. Weinberg, On the formal theory of Nerve conduction, *Bull. Math. Biophys* **2** (1940), 127–133.
95. R. FitzHugh, Mathematical models of threshold phenomena in the nerve membrane, *Bull. Math. Biophys.* **17** (1955), 257–278.
96. K. S. Cole, H. A. Antosiewicz, and P. Rabinowitz, Automatic computation of nerve excitation, *J. Soc. Indust. Appl. Math.* **3** (1955), 153–172.
97. K. S. Cole, H. A. Antosiewicz, and P. Rabinowitz, Automatic computation of nerve excitation, correction, *J. Soc. Indust. Appl. Math.* **6**, (1958), 196–197.
98. R. FitzHugh and H. A. Antosiewicz, Automatic computation of nerve excitation—Detailed corrections and additions, *J. Soc. Indust. App. Math.* **7** (1959), 447–458.
99. R. FitzHugh, Thresholds and plateaus in the Hodgkin–Huzley nerve equations, *J. Gen. Physiol.* **43** (1960), 867–896.
100. H. M. Lieberstein, On the Hodgkin–Huxley partial differential equation, *Math. Biosci.* (1967), 45–69.
101. H. M. Lieberstein, Numerical studies of the steady-state equations for a Hodgkin–Huxley model. *Math. Biosci.* **1** (1967), 181–211.
102. V. M. Eleonskii, Stability of simple stationary waves related to the nonlinear diffusion equation, *Soviet Phys.—JETP* **26** (53), (1968), 382(592)–384(597).
103. R. FitzHugh, Mathematical models of excitation and propagation in nerve, *in* "Biological Engineering" (H. P. Schwan, ed.), McGraw-Hill, New York, 1969, pp. 1–85.
104. H. P. McKean, Jr., Nagumo's equation, *Adv. in Math.* **4** (1970), 209–223.
105. N. H. Sabah and R. A. Spangler, Repetitive response of the Hodgkin–Huxley model for the squid giant axon, *J. Theoret. Biol.* **29** (1970), 155–171.
106. H. Cohen, Nonlinear diffusion problems, *in* "Studies in Applied Mathematics" (A. H. Taub, ed.), Prentice-Hall, Englewood Cliffs, New Jersey, 1971, pp. 27–64.
107. S. P. Hastings, On a third order differential equation from biology, *Quart. J. Math., Oxford Ser.* (2) **23** (1972), 435–448.
108. J. Rinzel and J. B. Keller, Traveling wave solutions of a nerve conduction equation, *Biophys. J.* **13** (1973), 1313–1337.
109. S. Hastings, The existence of periodic solutions to Nagumo's equation, *Quart. J. Math. Oxford Ser.* (2) **25** (1974), 369–378.
110. S. P. Hastings, Some mathematical problems from neurobiology, *Amer. Math. Monthly*, **82**, (1975), 881–895.
111. S. P. Hastings, On the existence of homoclinic and periodic orbits for the FitzHugh–Nagumo equations, *Quart. J. Math. Oxford Ser.* (2) **27** (1976), 123–134.
112. J. Rinzel, Nerve signaling and spatial stability of wave trains, *in* "Structural Stability, the Theory of Catastrophes and Applications in the Sciences" (P. Hilton, ed.), Lecture Notes in Mathematics, No. 525, Springer-Verlag, Berlin and New York, 1976, pp. 127–142.
113. S. D. Eidel'man "Parabolic Systems," North-Holland Publ., Amsterdam, 1969.
114. J. Evans and N. Shenk, Solutions to axon equations, *Biophys. J.* **10** (1970), 1090–1101.
115. J. W. Evans, Nerve axon equations: I, Linear approximations, *Indiana Univ. Math. J.* **21** (1972), 877–885.

116. J. W. Evans, Nerve axon equations: II, Stability at rest, *Indiana Univ. Math. J.* **22** (1972), 75–90.

117. J. W. Evans, Nerve axon equations: III, Stability of the nerve impulse, *Indiana Univ. Math. J.* **22**, (1972), 577–593.

118. J. W. Evans, Errata: Nerve axon equations: II & III, *Indiana Univ. Math. J.* **25** (1976), 301.

119. J. W. Evans, Nerve axon equations: IV, The stable and the unstable impulse, *Indiana Univ. Math. J.* **24** (1975), 1169–1190.

120. C. Conley, On traveling wave solutions of non-linear diffusion equations, *in* "Dynamical Systems, Theory and Applications" (J. Moser, ed.), Lecture Notes in Physics, Springer-Verlag, Berlin and New York, 1975, 498–510.

121. G. A. Carpenter, "Homoclinic, Heteroclinic, and Periodic Solutions of Autonomous Systems, with Applications to Nerve Impulse Equations," Report, Dept. Math., MIT, Cambridge, Massachusetts, 1975.

122. G. A. Carpenter, Nerve impulse equations, *in* "Structural Stability, the Theory of Catastrophes and Applications in the Sciences" (P. Hilton, ed.), Lecture Notes in Mathematics, No. 525, Springer-Verlag, Berlin and New York, 1976, pp. 58–76.

123. S. R. Gutman, Neuronal network discriminating local stimuli, *Biophysics* **19** (1974), 136–143.

124. M. V. L. Bennett, ed., "Synaptic Transmission and Neuronal Interaction," Raven, New York, 1974.

125. N. DeClaris, Neural type junctions as network elements, *Proc. IEEEE Midwestern Symp. Circuits Systems*, pp. 165–170, 1977.

126. N. DeClaris, Neural-type junctions: A new circuit concept, *Proc. IEEE Midwestern Symp. Circuits Systems*, pp. 268–271, 1976.

127. R. Thom, "Structural Stability and Morphogenesis, an Outline of a General Theory of Models," Benjamin, New York, 1975.

128. E. D. Adrian, "The Mechanism of Nervous Action, Electrical Studies of the Neurone," New Ed., Univ. of Pennsylvania Press, Philadelphia, 1959.

129. D. J. Aidley, "The Physiology of Excitable Cells," Cambridge Univ. Press, London and New York, 1971.

130. R. L. Beurle, Properties of a mass of cells capable of regenerating pulses, *Philos. Trans. Roy. Soc. London Ser. B*, **240** (1956), 55–95?

131. M. A. B. Brazier, "A History of the Electrical Activity of the Brain, The First Half-Century," Pitman, London, 1961.

132. H. A. Blair, On the intensity–time relations for stimulation by electric currents. *J. General Physiol.* **15**, (1932), 709–755.

133. M. A. B. Brazier, "The Electrical Activity of the Nervous System," 2nd Ed., Macmillan, New York, 1966.

134. D. R. Brillinger, Measuring the association of point processes: A case history, *Amer. Math. Monthly* **83** (1976), 16–22.

135. T. M. O. Burrows, K. A. Campbell, E. J. Howe, and J. Z. Young, Condition velocity and diameter of nerve fibres of cephalopods, *J. Physiol.* (*London*) **179** (1965), 39P–40P.

136. G. Carpenter, Travelling Wave Solutions of Nerve Impulse Equations, Ph.D. Thesis, Univ. of Wisconsin, Madison, 1974.

137. W. K. Chandler, R. FitzHugh, and K. S. Cole, Theoretical stability properties of a space-clamped axon, *Biophys. J.* **2** (1962), 105–127.

138. Y. A. Chizmadzhev, V. S. Markin, and A. L. Muler, Conformational model of excitable cell membranes—II, Basic equations, *Biophysics* **18** (1973), 70–76.

139. K. S. Cole, "Membranes, Ions and Impulses," Univ. of California Press, Berkeley, 1968.

140. C. C. Conley, On the existence of bounded progressive wave solutions of the Nagumo equation (in preparation).

141. H. D. Crane, Possibilities for signal processing in axon systems, *in* "Neural Theory and Modeling" (R. F. Reiss, ed.), Stanford Univ. Press, Stanford, California, 1964, pp. 138–153.
142. D. L. Dietmeyer, Bounds on the period of oscillatory activity in randomly interconnected networks of neuron-like elements, *IEEE Trans. Comput.* **C-17** (1968), 578–591.
143. J. C. Eccles, "The Physiology of Nerve Cells," Johns Hopkins Press, Baltimore, Maryland, 1957.
144. J. C. Eccles, "The Physiology of Synapse," Springer-Verlag, Berlin and New York, 1964.
145. G. P. Findlay, Studies of action potentials in the vacuole and cytoplasm of nitella, *Austral. J. Biol. Sci.* **12** (1959), 412–426.
146. U. F. Franck, Models for biological excitation processes, *Progr. Biophys.* **6** (1956), 171–206.
147. R. FitzHugh, Impulses and physiological states in theoretical models of nerve membrane, *Biophys. J.* **1** (1961), 445–466.
148. R. FitzHugh, Theoretical effect of temperature on threshold in the Hodgkin–Huxley nerve model, *J. General Physiol.* **49** (1966), 989–1005.
149. R. FitzHugh, Computation of impulse initiation and saltatory conduction in a myelinated nerve fiber, *Biophys. J.* **2** (1962), 11–21.
150. H. Fukutome, H. Tamura, and K. Sugata, An electric analogue of the neuron, *Kybernetik* **2** (1963), 28–32.
151. R. Granit, ed., "Muscular Afferents and Motor Control," Proceedings of the First Nobel Symposium, Almqvist & Wiksell, Stockholm, 1966.
152. L. Goldman and J. S. Albus, Computation of impulse conduction in myelinated fibers: Theoretical basis of velocity–diameter relation, *Biophys. J.* **8** (1968), 596–607.
153. J. S. Griffith, A Field theory of neural nets: I, Derivation of Field equations, *Bull. Math. Biophys.* **25** (1963), 111–120.
154. J. S. Griffith, A Field theory of neural nets: II, Properties of the Field equations, *The Bulletin of Mathematical Biophysics*, **27** (1965), 187–195.
155. R. Granit, "Mechanisms Regulating the Discharge of Motoneurons," Thomas, Springfield, Illinois, 1972.
156. L. D. Harmon, Studies with artificial neurons, I, Properties and functions of an artificial neuron, *Kybernetik* **1** (1961), 89–101.
157. L. D. Harmon, Neural analogs, *Proc. Spring Joint Comput. Conf., San Francisco* **21** (1962), 153–158.
158. J. Z. Hearon, Application of results from linear kinetics to the Hodgkin–Huxley equations, *Biophys. J.* **4** (1964), 69–75.
159. A. V. Hill, Excitation and accommodation in nerve, *Proc. Roy. Soc. London Ser. B*, **119** (1936), 305–355.
160. F. F. Hiltz, Artificial neuron, *Kybernetik* **1** (1963), 231–236.
161. A. L. Hodgkin, "The Conduction of the Nervous Impulse," Thomas, Springfield, Illinois, 1964.
162. A. S. Householder, A theory of steady-state activity in nerve-fiber networks: I, Definitions and preliminary lemmas, *Bull. Math. Biophys.* **3** (1941), 63–69.
163. A. S. Householder, A theory of steady-state activity in nerve-fiber networks: II, The simple circuit, *Bull. Math. Biophys.* **3** (1941), 105–112.
164. A. S. Householder, A neural mechanism for discrimination: II, Discrimination of weights, *Bull. Math. Biophys.* **2** (1940), 1–13.
165. T. M. Il'inova and R. V. Khokhlov, Wave processes in lines with nonlinear shunt resistance, *Radio Engrg. Electron. Phys.* **8** (1963), 1864–1972.
166. T. M. Il'inova, Interaction of waves in a distributed line with a nonlinear parallel loss, *Radio Engrg. Electron. Phys.* **9** (1964), 1728–1735.

167. M. H. JACOBS, "Diffusion Processes," Springer-Verlag, Berlin and New York, 1967.
168. F. JENIK AND H. HOEHNE, Uber die Impulsverarbeitung eines mathematischen Neuronmodelles, *Kybernetik* **3** (1966), 109–128.
169. D. JUNGE, "Nerve and Muscle Excitation," Sinauer Associates, Sunderland, Massachusetts, 1976.
170. B. KATZ, "Nerve, Muscle, and Synapse," McGraw-Hill, New York, 1966.
171. B. I. KHODOROV, "The Problem of Excitability. Electrical Excitability and Ionic Permeability of the Nerve Membrane," Plenum, New York, 1974.
172. B. N. Kholodenko, Networks of excitable elements—IV, Networks of elements without refractoriness in continuous time, *Biophysics* **19** (1974), 325–330.
173. H. D. LANDAHL, A note on mathematical models for the interaction of neural elements, *Bull. Math. Biophys.* **23** (1961), 91–97.
174. L. LAPICQUE, "L'Excitabilité en Fonction du Temps," Presses Univ. de France, Paris, 1926.
175. J. LEVINSON AND L. D. HARMON, Studies with artificial neurons, III: Mechanisms of flicker-fusion, *Kybernetik* **1** (1961), 107–117.
176. E. R. LEWIS, An electronic model of neuroelectric point processes, *Kybernetik* **5** (1968), 30–46.
177. R. J. MACGREGOR AND R. M. OLIVER, A model for repetitive firing in neurons, *Kybernetik* **16** (1974), 53–64.
178. A. M. MONNIER, "L'Excitation Electrique des Tissues," Hermann, Paris, 1934.
179. P. P. NELSON, Un modèle de neurone, *Bull. Math. Biophys.* **24** (1962), 159–181.
180. J. G. NICHOLLS AND D. VAN ESSEN, The nervous system of the leech, *Sci. Amer.* **230** (1974), 38–48.
181. F. OFFNER, Circuit theory of nervous conduction, *Amer. J. Physiol.* **126** (1939), P594.
182. V. F. PASTUSHENKO, V. S. MARKIN, AND Y. A. CHIZMADZHEV, Uniform modes of work of neurone networks—I, Steady Modes, *Biophysics* **19** (1974), 131–136.
183. V. F. PASTUSHENKO, V. S. MARKIN, AND Y. A. CHIZMADZHEV, Homogeneous modes of work of the neurone networks — II, Periodic solutions and stability of steady states, *Biophysics* **19** (1974), 319–324.
184. W. F. PICKARD, On the propagation of the nervous impulse down medullated and unmedullated fibers, *J. Theoret. Biol.* **11** (1966), 30–45.
185. G. I. POLIAKOV, "Neuron Structure of the Brain," Harvard Univ. Press, Cambridge, Massachusetts, 1972.
186. M. I. RABINOVICH, Self-oscillations in a transmission line with tunnel diodes, *Radio Engrg. Electron. Phys.* **11** (1966), 1271–1275.
187. N. RASHEVSKY, Note on the mathematical biophysics of temporal sequences of stimuli, *Bull. Math. Biophys.* **3** (1941), 89–92.
188. N. RASHEVSKY, "Mathematical Biophysics," *3rd Rev. Ed. in 2 Vols.*, Dover, New York, 1960.
189. J. RINZEL, Travelling Wave Solutions of a Nerve Conduction Equation, Ph.D. Thesis, Courant Inst., New York, 1973.
190. W. A. H. RUSHTON, Initiation of the propagated disturbance, *Proc. Roy. Soc. London Ser. B* **124** (1937), 210–243?
191. C. P. Sandbank, Integrated tunnel diode circuits, *in* "Solid Circuits and Microminiaturization (G. W. A. Dummer, ed.), Macmillan, New York, 1964, pp. 221–230.
192. A. C. Scott, Electrophysics of a nerve fiber, *Rev. Modern Phys.* **47** (1975), 487–533.
193. B. I. H. SCOTT, Electricity in plants, *Sci. Amer.* **211** (1962), 107–114.
194. D. R. SMITH AND C. H. DAVIDSON, Maintained activity in neural nets, *J. Assoc. Comput. Mach.* **9** (1962), 268–279.
195. C. J. SWIGERT, A mode control model of a neuron's axon and dendrites, *Kybernetik* **7**, (1970), 31–41.

196. I. Tasaki, Conduction of the nerve impulse, *in* "Handbook of Physiology: Neurophysiology" (J. Field, ed.), Vol. I, Amer. Physiol. Soc., Washington, D.C., 1959.
197. W. A. Van Bergeijk, Studies with artificial neurons, II: Analog of the external spiral innervation of the cochlea, *Kybernetik* **1** (1961), 102–107.
198. L. Y. Wei, Possible origin of action potential and birefringence change in nerve axon, *Bull. Math. Biophys.* **33** (1971), 521–537.
199. L. Y. Wei, Quantum theory of time-varying stimulation in nerve, *Bull. Math. Biol.* **35** (1973), 359–374.
200. L. Y. Wei, Dipole mechanisms of electrical, optical and thermal energy transductions in nerve membrane, *Ann. N.Y. Acad. Sci.* **227** (1974), 285–293.
201. L. Y. Wei, Dipole theory of heat production and absorption in nerve axon, *Biophys. J.* **12** (1972), 1159–1170.
202. A. M. Weinberg, The equivalence of the conduction theories of Rashevsky and Rushton, *Bull. Math. Biophys.* **2** (1940), 61–64.
203. J. W. Woodbury and W. E. Crill, On the problem of impulse conduction in the atrium, *Proc. Internat. Symp., Nervous Inhibitions*, Pergamon, New York, 1961.
204. N. L. Wulfsohn and A. Sances, Jr., eds., "The Nervous System and Electric Currents," Proceedings of the Annual Conference of the Neuro-Electric Society, Plenum, New York, Vol. 1, 1970; Vol. 2, 1971.
205. E. C. Zachmanoglou and D. W. Thoe, "Introduction to Partial Differential Equations with Applications," Williams & Wilkins, Baltimore, Maryland, 1976.
206. C. L. Chao, Wave propagation in a nonlinear transmission line and its application, *IEEE NEREM Rec.* **11** (1969), 164–165.
207. S. A. Reible and A. C. Scott, Pulse propagation on a superconductive neuristor, *J. Appl. Phys.* **46** (1975), 4935–4945.
208. V. A. Skorik, V. I. Stafeev, and G. I. Fursin, Some properties of n-p-i neuristor structures with a drift active coupling, *Soviet Phys.—Semiconduct.* **9**, (1975), 803–804.
209. A. Avila, Lluves tropicales, *in* "Los Trovaderos del Pueblo" (selected by S. Feijoo), Vol. 1, Univ. Central de las Villas, Santa Clara, Cuba, 1960, p. 316.

This research was supported in part by U.S. National Science Foundation (NSF) Grant Nos. ENG-03227 and GF-42178 and under the joint Polish–American "Active Microelectronic Systems" program. Any opinions, findings, conclusions, or recommendations are those of the author and do not necessarily reflect views of NSF.

Y después de fuerte aguaje
Viene la apacible calma;
En el desierto la palma
Doblando va su follaje [209].

At this point the author wishes to acknowledge the interest, assistance, and cooperation of his colleagues on the Polish–American program which has made this work possible as well as his Cuban colleagues, especially Dr. J. Altshuler without whom this would not have been undertaken and E. Diego whose outlook gave it spirit. For A. Lavendero and the interest and search assistance of A. Pfaffenberger, appreciation and acknowledgment.

ELECTRICAL ENGINEERING DEPARTMENT AND
THE APPLIED MATHEMATICS PROGRAM
UNIVERSITY OF MARYLAND
COLLEGE PARK, MARYLAND

Some Models in the Social and Behavioral Sciences Based on Proportioning Networks

A. H. ZEMANIAN

Four mathematical models relating to business investments, marketing systems in underdeveloped economies, collective farms, and interpersonal relationships in families are described. They are all based on the apparently new concept of a proportioning network.

1. INTRODUCTION

Proportioning networks comprise an apparently new kind of flow network. Until recently, they did not seem to have been discussed in the literature despite the fact that they arose in natural ways as mathematical models of several systems in the social and behavioral sciences. The objective of this paper is to describe four such models. Mathematical analyses of proportioning networks appear elsewhere [3–5].

2. INVESTMENTS AND RETURNS

The basic principle behind proportioning networks is common in the business world; namely, when businessmen invest in a venture, the profit ensuing is divided among them in proportion to their investments. This is figuratively illustrated by the star graph of Fig. 1. The end nodes represent the businessmen and have the indices $1, 2, \ldots, j, \ldots, m$ except for the index k, which is reserved for the central node. The central node represents the business venture, which for the sake of definiteness we assume to be a nineteenth-century whaling expedition. The quantity $S_j(t)$ assigned to the jth end node represents the dollar value of the resources the jth investor puts into the venture at the time t. The corresponding investment is represented by the flow $s_{jk}(t) = s_j(t)$ and may be money or the ship or the Captain's nautical abilities, all of which are assigned dollar values.

We let $t + 1$ represent the time just after the whaling expedition has been completed, when all costs and debts have been paid and the profit $s_j(t + 1)$ is to be distributed among the investors. That profit is in general a function

ISBN 0-12-178150-X

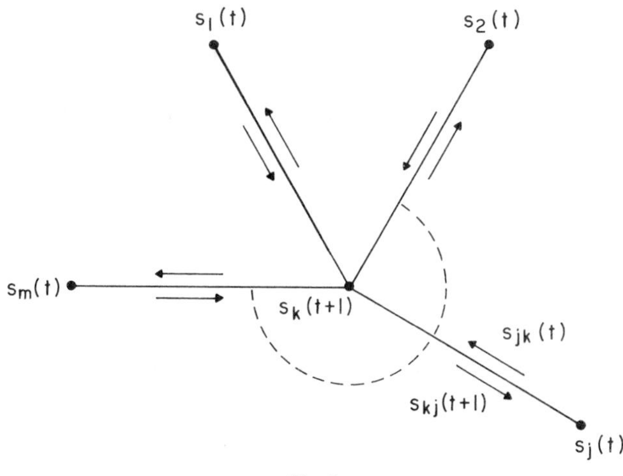

Fig. 1

of many variables, in particular, of the various flows coming into the central node at time t. By the customary proportionality rule, the jth investor receives the return

$$s_{kj}(t+1) = [s_{jk}(t) / \textstyle\sum_{i \in \mathrm{adj}(k)} s_{ik}(t)] s_k(t+1), \qquad t = 1, 3, 5, \ldots, \tag{2.1}$$

at time $t + 1$. Here, adj(k) is the "adjacency of k," that is, the index set of all nodes adjacent to the kth node. In this example and indeed throughout this paper all node values are restricted to positive values and all flows to non-negative values.

Let us assume that the investors come together once again to fund a new whaling expedition. To model this we let the jth investor have at time $t + 2$ a resource $s_j(t + 2)$, which in general is not equal to but may be a function of the return $s_{kj}(t + 1)$ or alternatively is an exogeneously given quantity. This leads to the investment $s_{jk}(t + 2) = s_j(t + 2)$. Upon repeating the above process many times, we obtain the two time series $\{s_{jk}(t)\}$ and $\{s_{kj}(t + 1)\}$, where $t = 1, 3, 5, \ldots$, for the two flows in the branch connecting nodes j and k, the second time series being a result of the rule (2.1). Figure 1 is an example of a proportioning network, albeit a very simple one.

3. A MARKETING NETWORK THAT OCCURS IN CERTAIN THIRD-WORLD COUNTRIES

We turn now to a situation that leads to proportioning networks with more complicated graphs than the star graph of the preceding example. This model was discussed in detail in [3], and so our present exposition will be brief.

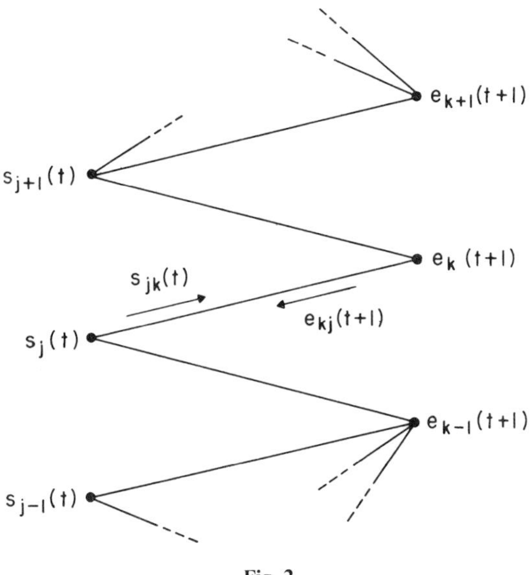

Fig. 2

In a number of African countries, such as Kenya, Nigeria, and Sierra Leone, farmers do not bring the staple food commodity directly to the urban centers but instead sell it in widely distributed and periodically open farmers' markets to traders; the traders are the ones who transport that commodity to the urban centers and resell it in wholesale markets for retail distribution [2]. This is schematically represented in Fig. 2. The nodes on the left of the bipartite graph represent the farmers' markets and are indexed by $\ldots, j-1, j, j+1, \ldots$. The nodes on the right represent the urban centers and are indexed by $\ldots, k-1, k, k+1, \ldots$. Also, $t = 1, 3, 5, \ldots$ denotes the days on which the farmers' markets are open. The node value $s_j(t)$ represents the amount of goods available in the jth farmers' market on market day t, and the node value $e_k(t+1)$ is the "intended order," that is, the total amount of goods the traders operating into the kth urban center will buy in the various farmers' markets on market day $t+2$ if the prices in the farmers' markets remain at their "normal" values. The latter for a particular farmers' market is the average price therein for the commodity under ordinary conditions.

Under a number of not too unreasonable assumptions on the behavior of the traders, it can be shown that the intended order $e_{kj}(t+1)$ of those traders operating between the jth farmers' market and the kth urban center is that proportion of $e_k(t+1)$ as was their supply $s_{jk}(t)$ of the commodity as a proportion of the total supply brought in on day t. That is,

$$e_{kj}(t+1) = \frac{s_{jk}(t)}{\sum_{i \in \mathrm{adj}(k)} s_{ik}(t)} e_k(t+1). \tag{3.1}$$

(In [3], we denoted $e_{kj}(t+1)$ by $O_{jk}(t+1)$.) Actually, $e_k(t+1)$ is a function of the demand behavior of the consumers in the kth urban center and the supply behavior of the traders delivering to that center, and therefore so too is $e_{kj}(t+1)$.

Similar assumptions on the behavior of the traders in the farmers' markets leads to the following proportioning rule for the flow $s_{jk}(t+2)$:

$$s_{jk}(t+2) = \frac{e_{kj}(t+1)}{\sum_{m \in \mathrm{adj}(j)} e_{mj}(t+1)} s_j(t+2). \tag{3.2}$$

The supply $s_j(t+2)$ of goods in the jth farmers' market may be exogenously given for each $t+2$ if the supply functions for the farmers are perfectly inelastic. Otherwise, $s_j(t+2)$ will depend on both the supply behavior of the farmers and the demand behavior of the traders in the jth farmers' market.

After an initial set of flows $s_{jk}(1)$ is assumed, the nonlinear recursion equations (3.1) and (3.2) determine a time series in each of the flows $e_{kj}(t+1)$ and $s_{jk}(t+2)$ for $t = 1, 2, 3, \ldots$. (In order to prevent any denominator from disappearing, we assume here that for each urban center at least one incoming flow is positive at $t = 1$.) We have once again a "proportioning network," so-called because of the proportioning rules (3.1) and (3.2).

The marketing systems in the aforementioned countries are more sluggish than the marketing systems of the developed countries because of the absence of market news [2]. This is reflected in the behavior of our model. A disturbance in one market, such as a sudden shortfall in the supply of goods in a particular farmers' market, is transmitted only by the trading activity and therefore diffuses in a step-by-step fashion throughout the marketing network. Thus, if the shortest path in Fig. 2 between two markets has n branches, then a disturbance in one of those markets at, say, time t cannot produce a perturbation in price in the other market until at least time $t+n$. Several other results concerning this model are also established in [3].

4. AN IDEALIZED COLLECTIVE FARM AND ITS ASSOCIATED PRIVATE PLOTS

Collective-farm workers in the Soviet Union have the privilege of operating private plots whose vegetable, dairy, and meat products they may keep for themselves either for consumption or for sale. Although these private plots comprise but a very small fraction of the total cultivated land of the collective farms, they are very productive and yield a substantial part of the income of collective-farm workers [1, pp. 278–280]. Thus, the private plots not only divert labor from the collective farms but also create some tension in that they are private operations attached to collectivized agriculture. A model

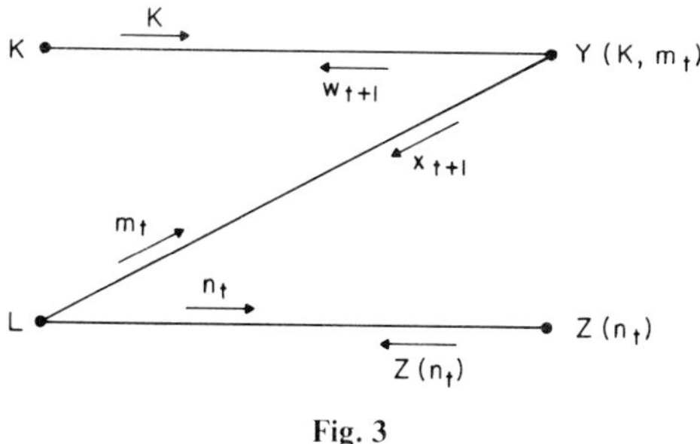

Fig. 3

of this dichotomy can be obtained by applying in a precise way some basic tenets of Marxian economics. It assumes that a collective farm divides its yield between the state and the collective-farm workers in a certain ideal way and leads to the proportioning network of Fig. 3 [6]. This then is an example of the use of proportioning networks within the framework of Marxian economics. Without making value judgments concerning the real-world ramifications of Marxism, we present here a summary of the key ideas in [6].

In the present example, $t = 1, 3, 5, \ldots$ is the index for the growing seasons. In order to obtain a tractable model, we aggregate all farm inputs per growing season for a given collective farm and its associated private plots into just two inputs, capital K and labor L. In accordance with Marx's labor theory of value, both K and L are measured in units of labor value per year and therefore can be added without mixing diverse units of measurement. K represents only that capital that is used up during a growing season, such as fertilizer, fuel, pesticides, and repair materials to maintain fixed assets.

During growing season t, the collective-farm workers allocate the labor m_t to the collective farm and the labor n_t to their private plots. Thus,

$$L = m_t + n_t. \tag{4.1}$$

Y is the production function for the collective farm and yields the output $Y(K, m_t)$ at time $t + 1$, which represents the end of the growing season. Z is the production function for the aggregate of all the private plots associated with the collective farm. Unlike the collective farm, the private plots are highly labor intensive and use up very little capital. (Many hoes and much muscle, but hardly any machinery.) We idealize this situation by taking Z to be a function of n_t alone. Since the collective-farm workers keep all of the yield from their private plots, $Z(n_t)$ is their return from those operations.

Next, we assume that there is no exploitation of the collective-farm workers by the state (i.e., by all the other workers in our idealized socialist state), or the converse. This dictates that the return w_{t+1} to the state and the return x_{t+1} to the collective-farm workers at the end of the growing season be given by the following proportioning rules.

$$w_{t+1} = [K/(K + m_t)]Y(K, m_t), \tag{4.2}$$

$$x_{t+1} = [m_t/(K + m_t)]Y(K, m_t). \tag{4.3}$$

To complete this proportioning-network model, we have to assign a proportioning behavior to the collective-farm workers as they allocate their labor between the collective farm and their private plots from year to year. Actually, it is equivalent to assume that the collective-farm workers as a group adjust their labor allocations to match the ratio of their returns from the two farm operations at the end of the preceding growing season. That is,

$$m_{t+2}/n_{t+2} = x_{t+1}/Z(n_t), \tag{4.4}$$

This is not unreasonable behavior, but there are virtually no data available to verify this or any other kind of behavior. Our model therefore indicates how the collective-farm and private-plots dichotomy would behave if the farm workers allocated their labor by the proportioning rule (4.4) and if Y and Z were known in our ideal Marxian economy. (In [6], we merely assumed that Y and Z had certain general shapes and that the collective-farm workers adjusted their labor allocations toward that operation yielding the better rate of return, without assuming how much the adjustment was. Nevertheless, a variety of conclusions could still be drawn from these less restrictive assumptions.)

Upon combining (4.4) with (4.1), we obtain

$$m_{t+2} = \frac{x_{t+1}}{x_{t+1} + Z(n_t)} L \tag{4.5}$$

and

$$n_{t+2} = \frac{Z(n_t)}{x_{t+1} + Z(n_t)} L, \tag{4.6}$$

which are the proportioning rules needed to complete our model.

5. ATTENTION-DEMANDING BEHAVIOR

The examples given so far all lead to bipartite proportioning networks. Nonbipartite proportioning networks also arise. One example is a system of spatially separated markets in a certain kind of barter economy [4, Sect. 2]. Another occurs with letter writing; when corresponding with friends, one tends to respond more often and with longer letters to those who send more and longer letters. We can idealize this situation by assuming that every

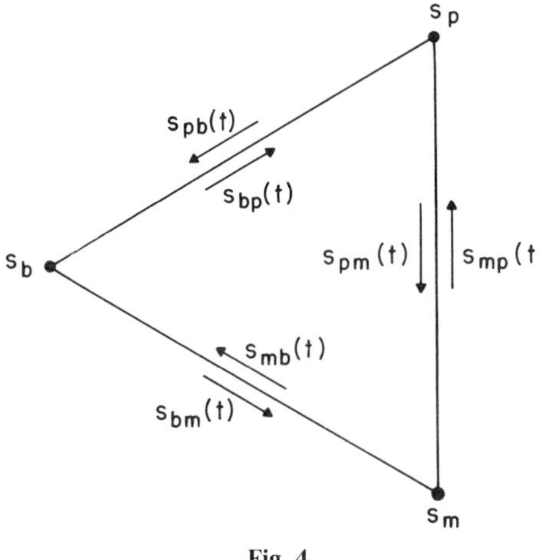

Fig. 4

person responds to his friends in proportion to the correspondence received from them, obtaining thereby a honbipartite proportioning network. A third example, one that we now describe, models attention-demanding behavior within a family in a rather unrealistic but suggestive—and perhaps entertaining—way. With this example we shall also indicate the various kinds of limiting behavior proportioning networks can have.

The simplest nonbipartite case occurs for a family of two parents and one child as indicated in Fig. 4. The subscript p stands for "Popa," m for "Moma," and *b* for "Beautiful Baby Boy" a.k.a. "The Brat." For simplicity's sake, we assume that day by day each person has a fixed amount of attention to be given to the family; these amounts are the node values s_p, s_m, and s_b. The amount of attention given by one individual to another on day *t*, where now $t = 1, 2, 3, \ldots$ is the flow from the node for the first individual to the node for the second; for instance, $s_{pm}(t)$ is the attention Popa gives to Moma during day *t*.

We idealize our model still further by assuming that only love and kindness are exchanged between individuals (no abuse or hateful behavior) and that the amount of such behavior from one individual toward another may vary from a negligible amount up to the total attention the first individual has available to give to the family. We assume in fact that attention-demanding and attention-providing behavior are essentially the same thing, that an appropriate unit can be devised for measuring the duration and intensity of such behavior, and that the numerical values thus assigned to the nodes (and flows) are all positive (respectively, nonnegative).

Our final assumption is the proportionality rule: If the attention Popa receives from Moma on day *t* is $x\%$ of the total attention he receives on that day from both Moma and Baby, then he responds by providing $x\%$ of $s_{\rm p}$ to Moma on day $t + 1$; similar rules are to hold between every ordered pair of individuals. This is not too unreasonable, for one does tend to respond to those that are paying attention to them and to ignore those that are ignoring them. Thus, we have

$$s_{jk}(t + 1) = \frac{s_{kj}(t)}{\sum_{i \in \mathrm{adj}(j)} s_{ij}(t)} s_k, \tag{5.1}$$

where $j, k = p, m, b$. We need merely require that for each j there exists at least one i for which $s_{ij}(1) \neq 0$ in order to obtain a time series for each flow where $t = 1, 2, 3, \ldots$.

We shall use this example to introduce numerical illustrations of the kinds of limiting behavior proportioning networks can have. Let us assume that Popa is away at work during much of the day, Baby is often out playing but is home more than Popa, and Moma—not being a liberated woman—is engrossed with kitchen and child. As a result, we might have $s_{\rm p} = 3$, $s_{\rm b} = 4$, and $s_{\rm m} = 5$. With these node values a stable situation is possible wherein everyone receives as much love and attention from each of the other members of the family as is given. This is indicated in Fig. 5 wherein every flow remains constant with respect to time [as can be seen by applying (5.1)], and the flows in opposite directions in any given branch are the same. We call this a balanced state [4] and view it as a happy family relationship.

Another kind of limiting state, which we call an oscillating state, is shown in Fig. 6. We view this as a situation wherein some tension exists within the family, for each individual keeps switching the allocation of attention between the other two family members. For instance, on day t Moma, who has been rather worn down by Baby's demands all day long, has a more than usual need for Popa's affection, but her advances are thwarted by Popa's unconcern toward her coupled with a jolly coddling of Baby. So, the next day

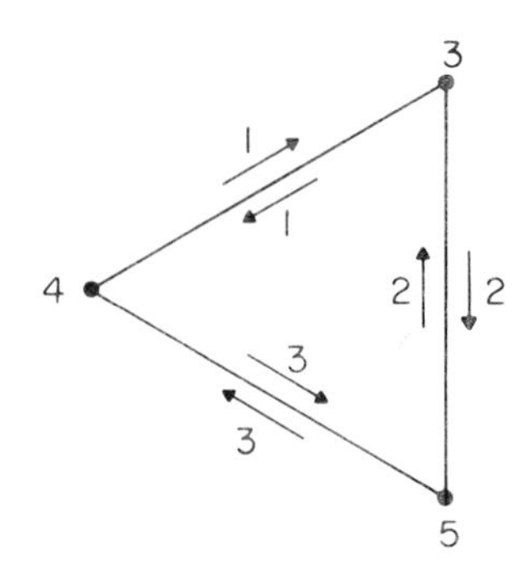

Fig. 5

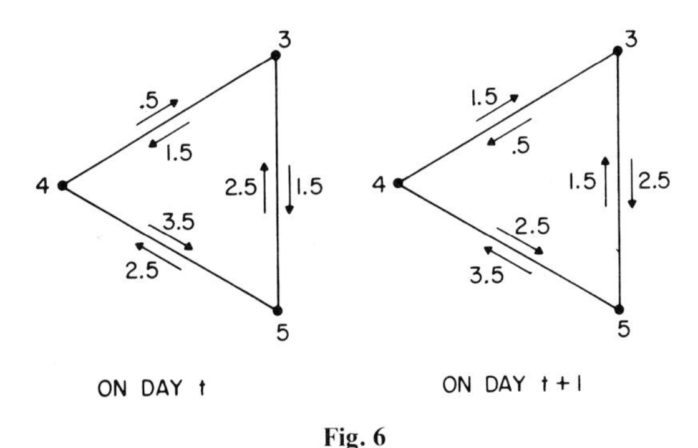

Fig. 6

Moma decides to give Popa a taste of his own medicine by switching most of her attention to Baby who was demanding it. Popa on the other hand tries to make amends by catering to Moma and thereby paying less attention to Baby. Baby however has started paying more attention to Popa who coddled him and less to Moma who ignored him. On the third day the situation reverts back to that of the first day for similar reasons. A continuation of this argument leads to a periodic variation of period 2, as can be seen by applying (5.1) to the numbers indicated in Fig. 6. This argument is certainly of questionable realism, but it makes a nice story and leads to some challenging mathematical problems [5].

Still another kind of unvarying state is the steady state, a generalization of the balanced state. The triangle network for our family has exactly three steady states that are not balanced states; they are shown in Figs. 7, 8, and 9. In every case the time series for each flow is constant with respect to time, but in at least one branch the flows in opposite directions are different. We might interpret Fig. 7 as the case where the relationship between Popa and Moma has degenerated to the point where they are no longer talking to each other. They are instead giving all their attention to Baby, who cannot adequately respond to either parent. In each of Figs. 8 and 9, a different pair of family members are no longer communicating.

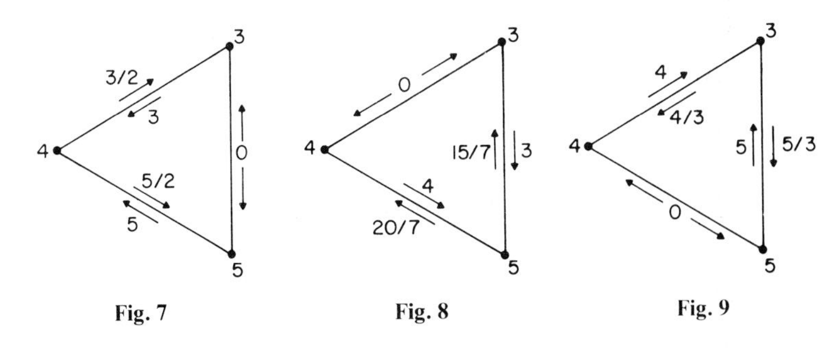

Fig. 7 **Fig. 8** **Fig. 9**

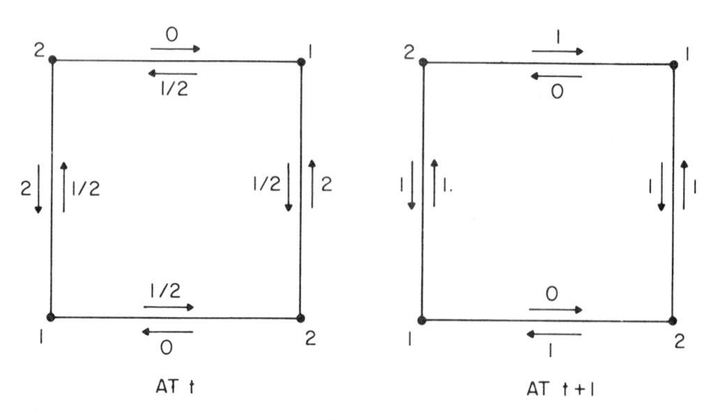

Fig. 10

Bipartite proportioning networks have yet another kind of limiting behavior that appears to be different from balanced, oscillating, and steady states. An example is indicated in Fig. 10. But, a closer inspection shows that it is a combination of two different steady states. More specifically, because of the bipartite structure, the flows on the inside of the square are independent of the flows on the outside of the square, and conversely. That is, we may alter the inner flows without altering the outer flows, and conversely. Let us replace the outer flows at time t by the outer flows at time $t + 1$ and replace the outer flows at $t + 1$ by the outer flows at t. This yields two different steady states. We conjecture that balanced, oscillating, and steady states, and the combination of steady states just illustrated exhaust all possible periodic responses for proportioning networks. To put it another way, we suspect that proportioning networks cannot have periodic responses other than those of period 1 (the constant responses: balanced and steady states) and of period 2 (oscillating states and the combined steady states illustrated by Fig. 10).

6. FINAL REMARKS

The three limiting cases just described are examined in [3–5]; they show that proportioning networks have a rich variety of limiting behaviors. On the other hand, not much information is presently available concerning the general dynamic behavior of proportioning networks; some results along these lines will appear in a future paper.

There are a number of ways the idea of a proportioning network can be extended. One way is to replace the arithmetic mean, which is the key concept behind proportioning networks, by other kinds of means such as the geometric or harmonic means. Another way is to make the flows at time t depend upon the flows at several prior instances of time by taking a weighted mean. Alternatively, one can replace discrete time by continuous time and use a memory function in a convolution-type integral to obtain the flows at time t through an integration of the flows over a prior interval of time. Furthermore, stochastic proportioning networks, wherein the node values and/or the initial flows are given by probability density functions, can also be considered.

REFERENCES

1. B. F. Johnson and P. Kilby, "Agricultural and Structural Transformation," Oxford Univ. Press, London and New York, 1975.
2. W. O. Jones, The structure of staple food marketing in Nigeria as revealed by price analysis, *Food Res. Inst. Stud. Agric. Econom., Trade, Dev.* **8** (1968), 95–123.
3. A. H. Zemanian, Proportioning networks: A model for a two-level marketing system in an underdeveloped economy, *SIAM J. Appl. Math.* **33** (1977), 619–639.
4. A. H. Zemanian, The balanced states of a proportioning network, *SIAM J. Appl. Math.* **34** (1978), 597–610.
5. A. H. Zemanian, The steady and oscillating states of a proportioning network, *SIAM J. Appl. Math.* **35** (1978), 496–507.
6. A. H. Zemanian, "An Idealized Collective Farm in a Marxian Economy, *Tech. Rep. No. 319, Coll. Engrg.*, State Univ. of New York at Stony Brook. 1978.

This work was supported by NSF Grant MCS7505268.

DEPARTMENT OF ELECTRICAL SCIENCES
STATE UNIVERSITY OF NEW YORK AT STONY BROOK
STONY BROOK, NEW YORK

Part III

MATHEMATICAL PROGRAMMING

A Helly-Type Theorem and Semiinfinite Programming

A. BEN-TAL

E. E. ROSINGER

A. BEN-ISRAEL

The equivalence between semiinfinite convex programming (in R^n) and certain finite subprograms (with at most n constraints) is established using a Helly-type theorem due to V. Klee.

1. INTRODUCTION

A *semiinfinite program* is a mathematical program of the form

$$\text{(P)} \qquad \begin{aligned} &\inf \quad f(\mathbf{x}) \\ &\text{s.t.} \quad g(\mathbf{x}, t) \leqq 0, \qquad t \in T, \end{aligned}$$

where $\mathbf{x} \in R^n$ and T is an infinite set, which henceforth is assumed compact.

Semiinfinite programs were studied by John [15], Charnes *et al.* [3–5], Duffin and Karlovitz [8], Krabs [18], Gehner [11, 12], and others.

Under suitable assumptions, we can associate with (P) a finite subprogram

$$(\mathrm{P} \cdot \tau) \qquad \begin{aligned} &\inf \quad f(\mathbf{x}) \\ &\text{s.t.} \quad g(\mathbf{x}, t) \leqq 0, \qquad t \in \tau, \end{aligned}$$

where τ is a finite subset of T, containing at most n elements.

If (P) is convex, this association becomes an equivalence in the sense that (P) and certain of its finite subprograms $(\mathrm{P} \cdot \tau)$ have the same optimal solutions. This equivalence is already evident in the classical characterizations of best Tchebychev approximations [6].

If (P) is a nonconvex program, the associated finite subprograms $(\mathrm{P} \cdot \tau)$ yield necessary conditions satisfied by the optimal solutions [15, Theorem 1].

The correspondence $(\mathrm{P}) \leftrightarrow (\mathrm{P} \cdot \tau)$ has been established, following John [15], by several authors, including Pshenichnyi [19, Chapter V, Corollary 2

ISBN 0-12-178150-X

to Theorem 5.1], Krabs [18, Part III, Section 3.4], Laurent [14, Part II, Section 2.7], and Gehner [11, 12].

In the above derivations, the association between (P) and $(\mathrm{P}\cdot\tau)$ is obtained indirectly from optimality considerations, by using Carathéodory's theorem [2, 7].

Thus, for example, Pshenichnyi [19] replaced (P) by

$$(\tilde{\mathrm{P}}) \qquad \begin{aligned} &\inf \quad f(\mathbf{x}) \\ &\text{s.t.} \quad g(\mathbf{x}) \leqq 0, \end{aligned}$$

where

$$g(\mathbf{x}) \triangleq \sup\{g(\mathbf{x}, t) : t \in T\}.$$

Since g is, in general, nondifferentiable [even if the $g(\cdot, t)$ are], the optimality condition for $(\tilde{\mathrm{P}})$ involves the subgradient set of g, given by Valadier [21] as

$$\begin{aligned} \partial g(\mathbf{x}) &= \operatorname{cl}\operatorname{conv} \bigcup \{\partial g(\mathbf{x}, t) : t \in T(\mathbf{x})\} \\ &= \operatorname{conv} \bigcup \{\nabla g(\mathbf{x}, t) : t \in T(\mathbf{x})\}, \end{aligned}$$

under suitable differentiability assumptions, where

$$T(\mathbf{x}) = \{t \in T : g(\mathbf{x}) = g(\mathbf{x}, t)\}.$$

Using the Carathéodory theorem, an optimal $\mathbf{x}^*$ can then be characterized (under the Kuhn–Tucker condition) by the existence of $n + 1$ points $t_i \in T(\mathbf{x}^*)$ and $n + 1$ scalars $\lambda_i \geqq 0$, $i = 1, \ldots, n + 1$, satisfying

$$\nabla f(\mathbf{x}^*) + \sum_{i=1}^{n+1} \lambda_i \nabla g(\mathbf{x}^*, t_i) = 0.$$

Under a suitable constraint qualification this is precisely the necessary and sufficient condition for $\mathbf{x}^*$ to be an optimal solution of $(\mathrm{P}.\tau)$, where $\tau = \{t_i : i = 1, \ldots, n + 1\}$.

The purpose of this paper is to develop the association between (P) and $(\mathrm{P}.\tau)$ directly, using a Helly-type theorem [7, 13], relating the intersection of an infinite family of convex sets to the intersections of its finite subfamilies of cardinality $\leqq n + 1$. This derivation is more elementary than the previous ones mentioned above, and it does not require differentiability. In our approach, optimality considerations enter after (P) has been reduced to a finite $(\mathrm{P}.\tau)$, permitting the use of finite optimality theory.

In Section 2 we adapt a Helly-type theorem of Klee [16, 17] for use in semiinfinite programming (Lemmas 2.2 and 2.3). Semiinfinite systems of strict convex inequalities are the subject of Corollaries 2.4 and 2.5.

In Section 3 we establish the equivalence between semiinfinite convex programs and certain finite subprograms (Theorem 3.1).

This Helly-type approach can also be used in other semiinfinite problems, e.g., semiinfinite linear inequalities [14], where Fenchel's Helly-type theorem ([10, Result 45, p. 101] or [20, Theorem 21.3]) should be applied, and Tchebychev approximations (e.g. [11, 12]). Linear equality constraints pose no additional difficulty in our Helly-type approach, since they just reduce the dimension n of the relevant space.

2. SOME CONSEQUENCES OF KLEE'S THEOREM

A family Γ of sets in R^n is called 0-*closed* if and only if every set in Γ is open and int $K \in \Gamma$ whenever K is the limit of a convergent sequence of sets of Γ [17].

Here by $\lim_{n\to\infty} K_n = K$ it is meant that

$$\bigcup_{1\leq m<\infty}\ \bigcap_{m\leq n<\infty} K_n = \bigcap_{1\leq m<\infty}\ \bigcup_{m\leq n<\infty} K_n = K.$$

The following Helly-type theorem for 0-closed families of convex sets will be used in the sequel.

Theorem 2.1. (Klee [17, (5.3)]). *Let Γ be a 0-closed family of convex sets. Then the intersection of all members of Γ is empty if and only if there are $n+1$ members of Γ whose intersection is empty.*

The following lemma permits taking Γ as a finite union of 0-closed families.

Lemma 2.2. *If Γ^1, Γ^2 are 0-closed families, so is their union.*

Proof. Let $\Gamma = \Gamma^1 \cup \Gamma^2$ and let $\{K_n\} \subset \Gamma$ be a convergent sequence with $K = \lim_n K_n$. Consider the subsequences $\{K_l^1\}$ and $\{K_m^2\}$ of $\{K_n\}$, consisting of its elements in Γ^1 and Γ^2, respectively. At most one of these subsequences is finite.

Assume that $\{K_l^1\}$ is infinite, then the inclusions

$$\limsup_l K_l^1 \subset \limsup_n K_n, \qquad \liminf_l K_l^1 \supset \liminf_n K_n$$

follow easily. But the sequence $\{K_n\}$ is converged to K; hence, the above inclusions will give

$$\limsup_l K_l^1 \subset K \subset \liminf_l K_l^1,$$

which by definition implies that $\lim K_l^1 = K$. Since Γ^1 is 0-closed, it follows that int $K \in \Gamma^1 \subset \Gamma$. ■

The following lemma establishes the 0-closedness of a certain family of open level sets.

Lemma 2.3. *Let T be a compact set, $D \subset R^n$ be convex with nonvoid interior, and let $g: D \times T \to R$ be a function such that:*

(A1) *For all $\mathbf{x} \in D$, $g(\mathbf{x}, t)$ is upper semicontinuous on T.*
(A2) *For all $t \in T$, $g(\mathbf{x}, t)$ is convex on D and has the level set*

$$K_t = \{\mathbf{x} \in D \,|\, g(\mathbf{x},\ t) < 0\}$$

nonvoid and open in D (e.g., for all $t \in T$, $g(\mathbf{x}, t)$ is upper semicontinuous on D).

Then the family of nonvoid open convex sets

$$\Gamma = \{K_t \,|\, t \in T\}$$

is 0-closed.

Proof. Let $\{K_{t_n}\}$ be any convergent sequence in Γ and let $\lim_n K_{t_n} = K$. We shall show that int $K \in \Gamma$, i.e., int $K = K_{t^*}$, for a certain $t^* \in T$. The compactness of T implies the existence of a subsequence $\{t_m\}$ in $\{t_n\}$ such that $\lim_m t_m = t^* \in T$. Then $\lim_m K_{t_m} = K$, due to the obvious inclusions $\limsup_m K_{t_m} \subset \limsup_n K_{t_n}$ and $\liminf_m K_{t_m} \supset \liminf_n K_{t_n}$. But

$$K_{t^*} \subset \text{int } K. \tag{1}$$

Indeed, assume $\mathbf{x} \in K_{t^*} \backslash K$. Then $g(\mathbf{x}, t^*) < 0$, while $\mathbf{x} \notin K_{t_l}$ for an infinite subsequence $\{t_l\}$ in $\{t_m\}$. Hence $g(\mathbf{x}, t_l) \geqq 0$, for every l. But $\lim_l t_l = t^*$; thus the condition (A1) will imply $g(\mathbf{x}, t^*) \geqq 0$, contradicting $g(\mathbf{x}, t^*) < 0$. Therefore $K_{t^*} \subset K$. The relation (1) will follow, since K_{t^*} is open, due to (A2). Now, we can prove the relation

$$\text{int } K = K_{t^*}. \tag{2}$$

Assume it is false. Then (1) implies

$$K_{t^*} \subset \text{int } K, \qquad K_{t^*} \neq \text{int } K,$$

which will imply the stronger relation

$$\text{int}(\text{int } K \backslash K_{t^*}) \neq \varnothing. \tag{3}$$

Indeed, take $\mathbf{z} \in \text{int } K \backslash K_{t^*}$. Since $\mathbf{z} \notin K_{t^*}$ and K_{t^*} is convex and open, there exists a hyperplane

$$H = \{\mathbf{x} \in R^n \,|\, h(\mathbf{x}) = a\}$$

such that $h(\mathbf{z}) = a$, while $h(\mathbf{x}) < a$ if $\mathbf{x} \in K_{t^*}$. Then int $K \cap \{\mathbf{x} \in R^n \,|\, h(x) > a\} \neq \varnothing$, since $\mathbf{z} \in \text{int } K$, which implies (3). Denote

$$H_t = \{\mathbf{x} \in D \,|\, g(\mathbf{x}, t) = 0\}.$$

We shall prove that

$$\text{int } H_t = \varnothing, \qquad \text{for all} \quad t \in T. \tag{4}$$

Indeed, assume $\mathbf{z} \in \operatorname{int} H_t$ for a certain $t \in T$ and take $\mathbf{w} \in K_t$. Define $\gamma: R^1 \to R^1$ by

$$\gamma(\lambda) = g(\lambda \mathbf{w} + (1 - \lambda)\mathbf{z}, t), \qquad \text{for} \quad \lambda \in R^1;$$

then $\gamma(1) = g(\mathbf{w}, t) < 0$, while $\gamma = 0$ in a neighborhood of $\lambda = 0$, contradicting (A2). Now, the relations (3) and (4) will result in

$$K \backslash (K_{t^*} \cup H_{t^*}) \neq \varnothing.$$

Taking $\mathbf{x} \in K \backslash (K_{t^*} \cup H_{t^*})$ it follows that $g(\mathbf{x}, t^*) > 0$. However, $\mathbf{x} \in K = \lim_m K_{t_m}$ implies $g(\mathbf{x}, t_m) < 0$ for each m except perhaps a finite number. Since $\lim_m t_m = t^*$, the above two inequalities contradict each other according to (A1). ■

By using Lemmas 2.2 and 2.3, Klee's theorem, Theorem 2.1, is applicable to semiinfinite systems of convex inequalities.

Corollary 2.4. *Let D be a subset on R^n with nonempty interior, and $\mathscr{I}$ a finite index set. For each $i \in \mathscr{I}$ let T^i be a compact set and let $g^i: R^n \times T^i \to R$ be a function satisfying* (A1) *and* (A2) *of Lemma* 2.3. *Then the system*

(S) $$g^i(\mathbf{x}, t) < 0, \qquad t \in T^i, \quad i \in \mathscr{I},$$

has no solution $\mathbf{x} \in D$ if and only if there is a finite subset

$$\tau \subset \bigcup_{i \in \mathscr{I}} T^i$$

containing at most $(n + 1)$ elements, such that the system

(S.τ) $$g^i(\mathbf{x}, t) < 0, \qquad t \in \tau \cap T^i, \quad i \in \mathscr{I}$$

has no solution in D.

Using a well-known theorem of alternatives for finite systems of convex inequalities [9], we get from Corollary 2.4 the following theorem of alternatives for infinite systems of convex inequalities.

Corollary 2.5. *The system* (S) *of Corollary* 2.4 *is inconsistent if and only if there is a finite subset*

$$\tau \subset \bigcup_{i \in \mathscr{I}} T^i$$

containing at most $(n + 1)$ elements, and a corresponding set of nonnegative "multipliers" $\{\lambda_t \geqq 0 : t \in \tau\}$ not all zero, such that

$$\sum_{i \in \mathscr{I}} \sum_{t \in \pi \cap T^i} \lambda_t g^i(\mathbf{x}, t) \geqq 0, \qquad \forall \mathbf{x} \in D.$$

A solvability theorem for semiinfinite systems of weak convex inequalities has been given by Bohnenblust *et al.* ([1, Lemma 1.5], see also [20, Theorem 21.3]).

3. SEMIINFINITE CONVEX PROGRAMS AND THEIR EQUIVALENT FINITE PROGRAMS

Consider the semiinfinite convex program

$$\text{(C)} \qquad \begin{aligned} &\inf \ f(\mathbf{x}) \\ &\text{s.t.} \ \ g^k(\mathbf{x}, t) \leqq 0, \quad t \in T^k, \quad k = 1, \ldots, m, \qquad \mathbf{x} \in D, \end{aligned}$$

where for $k = 1, \ldots, m$, T^k is a compact set, D is a convex subset of R^n with nonempty interior, and the following assumptions hold.

(A1) For each k, and for all $\mathbf{x} \in D$, $g^k(\mathbf{x}, \cdot)$ is an upper semicontinuous function on T^k.

(A2) For each k, and for all $t \in T^k$, $g^k(\cdot, t)$ is a lower semicontinuous convex function on D, and the set $\{\mathbf{x} : g^k(\mathbf{x}, t) < 0\}$ is open.

(A3) The objective function f is lower semicontinuous and convex on D.

(A4) (Slater Condition) The set

$$F^0 \triangleq \{\mathbf{x} \in D : g^k(\mathbf{x}, t) < 0, t \in T^k, k = 1, \ldots, m\},$$

is nonempty.

Theorem 3.1 *Let the convex program* (C) *satisfy* (A1)–(A4), *let* $\mathbf{x}$ *be a feasible solution of* (C) *and*

$$T^k(\mathbf{x}^*) \triangleq \{t \in T^k : g^k(\mathbf{x}, t) = 0\}, \qquad k = 1, \ldots, m,$$

$$\mathscr{P}^* \triangleq \{k : T^k(\mathbf{x}^*) \neq \varnothing\}.$$

Then $\mathbf{x}^*$ *is an optimal solution of* (C) *if and only if there is a finite set*

$$\tau^* \subset \bigcup_{k \in \mathscr{P}^*} T^k$$

containing at most n elements such that $\mathbf{x}^*$ *is an optimal solution of the finite convex problem*

$$(\text{C}.\tau^*) \qquad \begin{aligned} &\inf \ f(\mathbf{x}) \\ &\text{s.t.} \ \ g^k(\mathbf{x}, t) \leqq 0, \quad t \in \tau^* \cap T^k(x^*), \quad k \in \mathscr{P}^*, \qquad \mathbf{x} \in D. \end{aligned}$$

Proof. (i) We first show that $\mathbf{x}^*$ is optimal if and only if the system

$$\text{(B)} \qquad \begin{aligned} &f(\mathbf{x}) \ < f(\mathbf{x}^*) \\ &\text{s.t.} \ \ g^k(\mathbf{x}, t) < 0, \quad t \in T^k(\mathbf{x}^*), \quad k \in \mathscr{P}^*, \qquad \mathbf{x} \in D, \end{aligned}$$

has no solution. (See also [18, II.6.1].)

If. Suppose $\mathbf{x}^*$ is not optimal, i.e., there is a feasible $\bar{\mathbf{x}}$ with

$$f(\bar{\mathbf{x}}) < f(\mathbf{x}^*).$$

Let $\hat{\mathbf{x}} \in F^0$. Then the convex combination

$$\mathbf{x}(\lambda) = (1 - \lambda)\bar{\mathbf{x}} + \lambda\hat{\mathbf{x}}, \qquad 0 < \lambda < 1,$$

satisfies (B) for λ sufficiently small.

Only if. Suppose $\bar{\mathbf{x}}$ is a solution of (B). Then $\bar{\mathbf{x}}$ is a solution of

$$(\tilde{B}) \qquad \begin{aligned} & f(\mathbf{x}) < f(\mathbf{x}^*) \\ \text{s.t.} \quad & g^k(\mathbf{x}) < 0, \quad k \in \mathscr{P}^*, \qquad \mathbf{x} \in D, \end{aligned}$$

where

$$g^k(\mathbf{x}) \triangleq \sup\{g^k(\mathbf{x}, t) : t \in T^k\}.$$

Note that the supremum is attained (T^k is compact, and $g^k(\mathbf{x}, \cdot)$ is upper semicontinuous), so that

$$g^k(\mathbf{x}) = g^k(\mathbf{x}, t)$$

for some $t \in T^k$. Thus, in particular,

$$g^k(\mathbf{x}^*) < 0 \qquad \text{for} \quad k \notin \mathscr{P}^*.$$

Consider the convex combination

$$\mathbf{x}(\lambda) \triangleq (1 - \lambda)\mathbf{x}^* + \lambda\bar{\mathbf{x}}, \qquad 0 < \lambda < 1.$$

Then $\mathbf{x}(\lambda) \in D$, by the convexity of D. Further, by the convexity of g^k,

$$g^k(\mathbf{x}(\lambda)) \leqq (1 - \lambda)g^k(\mathbf{x}^*) + \lambda g^k(\bar{\mathbf{x}}) \begin{cases} < 0, & k \in \mathscr{P}^* \\ < 0, & k \notin \mathscr{P}^* \end{cases}$$

if λ is sufficiently small, and by the convexity of f,

$$f(\mathbf{x}(\lambda)) \leqq (1 - \lambda)f(\mathbf{x}^*) + \lambda f(\bar{\mathbf{x}}) < f(\mathbf{x}^*).$$

Thus, for λ sufficiently small, $\mathbf{x}(\lambda)$ is a feasible solution of (C), with

$$f(\mathbf{x}(\lambda)) < f(\mathbf{x}^*),$$

contradicting the optimality of x^*.

(ii) Denote $g^0(x, t) \triangleq f(x) - f(x^*)$, $T^0 \triangleq \{0\}$.

By a straightforward application of Corollary 2.4, we conclude that the system (B) is inconsistent if and only if there is a finite subset

$$\tau^* \subset \bigcup_{k \in \mathscr{P}^* \cup \{0\}} T^k$$

containing at most $n + 1$ elements such that the following finite system is inconsistent:

$$(\text{B}.\tau^*) \qquad g^k(\mathbf{x}, t) < 0, \qquad t \in \tau^* \cap T^k(\mathbf{x}^*), \quad k \in \mathscr{P}^* \cup \{0\}, \quad \mathbf{x} \in D.$$

We claim now that $0 \in \tau^*$. Indeed, if $0 \notin \tau^*$, then the nonexistence of a solution to the system

$$g^k(\mathbf{x}, t) < 0, \qquad t \in \tau^* \cap T^k(\mathbf{x}^*), \quad k \in \mathscr{P}^*,$$

will contradict (by the "if" part of Corollary 2.4) assumption (A4). We conclude that there is a finite subset $\tau^* \subset \bigcup_{k \in \mathscr{P}^*} T^k$ containing at most n elements such that the system

$$(\mathrm{B}.\tau^*) \qquad f(\mathbf{x}) < f(\mathbf{x}^*)$$
$$\text{s.t.} \quad g^k(\mathbf{x}, t) < 0, \quad t \in \tau^* \cap T^k(\mathbf{x}^*), \quad k \in \mathscr{P}^*, \qquad \mathbf{x} \in D,$$

has no solution.

(iii) Similarly to (i) we can now establish that the inconsistency of (B.τ^*) is equivalent to $\mathbf{x}^*$ being an optimal solution of ($P.\tau^*$). ■

REFERENCES

1. H. F. Bohnenblust, S. Karlin, and L. S. Shapley, Games with continuous, convex pay-off, "Contributions to the Theory of Games," Vol. 1, Princeton Univ. Press, Princeton, New Jersey, pp. 181–192.
2. C. Carathéodory, Uber den Variabilitalbereich der Koeffizienten von Potenzreihen, die gegebene Werte nicht annehem, *Math. Ann.* **64** (1907), 97–115.
3. A. Charnes, W. W. Cooper, and K. O. Kortanek, Duality in semi-infinite programs and some works of Haar and Caratheodory, *Management Sci.* **9** (1963), 209–228.
4. A. Charnes, W. W. Cooper, and K. D. Kortanek, On representations of semi-infinite programs which have no duality gaps, *Management Sci.* **12** (1965), 113–121.
5. A. Charnes, W. W. Cooper, and K. D. Kortanek, On the theory of semi-infinite programming and a generalization of the Kuhn–Tucker saddle point theorem for arbitrary convex functions, *Naval Res. Logist. Quart.* **16** (1969), 41–51.
6. E. W. Cheney, "Introduction to Approximation Theory," McGraw-Hill, New York, 1966.
7. L. Danzer, B. Grunbaum, and V. Klee, Helly's theorem and its relatives, *Convexity, Proc. Symp. Pure Math., Amer. Math. Soc.* **VII** (1963), 101–180.
8. R. J. Duffin and L. A. Karlovitz, An infinite linear program with a duality gap, *Management Sci.* **12** (1965), 122–134.
9. K. Fan, I. Glicksberg, and A. J. Hoffman, Systems of inequalities involving convex functions, *Proc. Amer. Math. Soc.* **8** (1957), 617–622.
10. W. Fenchel, Convex Cones, Sets and Functions, Lecture Notes, Princeton University, Princeton, New Jersey, 1951.
11. K. R. Gehner, Necessary and sufficient conditions for the Fritz John problem with linear equality constraints, *SIAM J. Control Optim.* **12** (1974), 140–149.
12. K. R. Gehner, Characterization theorems for constrained approximation problems via optimization theory, *J. Approx. Th.* **14** (1975), 51–76.
13. E. Helly, Uber Mengen konvexer Körper mit gemeinschaftlichen Punkten, *Jahresber Deutsch. Math.-Verein.* **32** (1923), 175–176.
14. R. G. Jeroslow and K. O. Kortanek, On semi-infinite systems of linear inequalities, *Israel J. Math.*
15. F. John, Extremum problems with inequalities as subsidiary conditions, "Studies and Essays, Courant Anniversary Volume," Wiley (Interscience), New York, 1948, pp. 187–204.

16. V. KLEE, The critical set of a convex body, *Amer. J. Math.* **75**, (1953), 178–188.
17. V. KLEE, Infinite-dimensional intersection theorems, *Convexity, Proc. Symp. Pure Math., Amer. Math. Soc.* **VII** (1963), 349–360.
18. W. KRABS, "Optimierung und Approximation," Teubner Studienbucher, Teubner, Stuttgart.
19. B. N. PSHENICHNYI, "Necessary Conditions for an Extremum," Dekker, New York, 1971.
20. R. T. ROCKAFELLAR, "Convex Analysis," Princeton Univ. Press, Princeton, New Jersey, 1970.
21. M. M. VALADIER, Sous-differentiels d'une borne superieure et d'une somme continue de fonctions convexes, *C. R. Acad. Sci., Ser. A* **268** (1969), 39–42.

Research partly supported by NSF Grant ENG 77-10126.

A. Ben-Tal and E. E. Rosinger
DEPARTMENT OF COMPUTER SCIENCE
TECHNION-ISRAEL INSTITUTE OF TECHNOLOGY
HAIFA, ISRAEL

A. Ben-Israel
DEPARTMENT OF MATHEMATICS
UNIVERSITY OF DELAWARE
NEWARK, DELAWARE 1971

Lagrange Dual Programs with Linear Constraints on the Multipliers

C. E. BLAIR

R. G. JEROSLOW

We consider Lagrange dual problems in which the vector of Lagrange multipliers, in addition to the usual nonnegativity conditions, satisfies certain linear or convex constraints. Of course, the maximum dual value, for such restricted classes of multipliers, need not equal the primal value, even if various "constraint qualifications" are appended.

We obtain expressions for this maximum "constrained dual" value, in terms of perturbation functions for the primal convex program, and value functions associated with the linear constraints. At least some information on the primal perturbation functions appears to be necessary in order to put upper or lower bounds on the "constrained dual" value. For the case that the constraints are, simply, that a positive weighted sum of the multipliers shall not exceed some given bound (plus, of course, the usual nonnegativities), only two values of an associated one-dimensional perturbation function is needed to obtain such upper and lower bounds. *Key words:* (1) Lagrangeans; (2) duality; (3) convexity.

Let S be a nonempty convex subset of a vector space and

$$f, g_1, \ldots, g_k : S \to R$$

be convex (not necessarily continuous) functions. An ordinary constrained convex optimization program (see, e.g. [7, Section 28]) is

$$\text{(P)} \qquad \begin{array}{ll} \text{minimize} & f(x) \\ \text{subject to} & g_i(x) \leqq 0 \quad \text{for} \quad 1 \leqq i \leqq k \quad \text{and} \quad x \in S. \end{array}$$

Throughout the paper, we assume that (P) is consistent.

For any $\lambda_1, \ldots, \lambda_k \geqq 0$ the Lagrange dual problem is

$$(\mathrm{D}_\lambda) \qquad \begin{array}{ll} \text{minimize} & f(x) + \sum_{i=1}^{k} \lambda_i g_i(x) \\ \text{subject to} & x \in S. \end{array}$$

ISBN 0-12-178150-X

The value $v(\mathrm{D}_\lambda)$ of (D_λ) clearly does not exceed the value $v(\mathrm{P})$ of (P) and it may be $-\infty$. One uses the unconstrained optimization problem (D_λ) to give information about the constrained one (P). If $v(\mathrm{D}) = \sup_{\lambda \geqq 0} v(\mathrm{D}_\lambda)$, then $v(\mathrm{D}) \leqq v(\mathrm{P})$.

In this paper, we study (D_λ) for those $\lambda \geqq 0$ satisfying certain conditions, for example (with $M \geqq 0$):

(a) $\sum_1^k \lambda_i \leqq M$; or
(b) $\lambda_i \leqq M$, $1 \leqq i \leqq k$; or, more generally,
(c) $A\lambda \leqq b$, where $b \in R^m$ and $A \in R^{m \times k}$.

In other words if T is a finite system of linear constraints on the λ_i, we are interested in $v(\mathrm{D}_T) = \sup\{v(\mathrm{D}_\lambda) \mid \lambda \text{ satisfies } T\}$. In an addendum, we show that certain of our results are valid for closed, convex sets as well as polyhedra.

We establish relations between $v(\mathrm{D}_T)$ and certain relaxations of (P). In the spirit of Duffin [4], Blair [1], Charnes *et al.* ([3], see also [2]). Duffin and Jeroslow [5], and Duffin and Karlovitz [6], we will use theorems about systems of linear inequalities systematically.

We begin with the situation in which $\{\lambda \mid A\lambda \leqq b,\ \lambda \geqq 0\}$ is bounded (i.e., a polytope). This case includes (1) and (2) above; and has simpler conclusions than the case of arbitrary A, which we treat later.

1. $\{\lambda \mid A\lambda \leqq b, \lambda \geqq 0\}$ IS BOUNDED

Lemma 1.† *Let $P = \{\lambda \mid A\lambda \leqq b,\ \lambda \geqq 0\}$ be nonempty and bounded (i.e., a polytope).*

There is no $\lambda \in P$ such that

$$f(x) + \sum_1^k \lambda_i g_i(x) \geqq L \tag{1}$$

for every $x \in S$, if and only if, there is a single point $\mathbf{x} \in S$ such that the finite system of linear inequalities in unknowns $\lambda_1, \ldots, \lambda_k$

$$(\mathrm{S}_\mathbf{x}) \qquad A\lambda \leqq b,$$

$$\sum_{i=1}^k g_i(\mathbf{x})\lambda_i \geqq L - f(\mathbf{x}),$$

$$\lambda \geqq 0$$

alone has no solution.

† Thomas L. Magnanti has pointed out to us, that this lemma can be derived from minimax theorems as, e.g., [7, Corollary 37.3.1], and he has provided an alternate proof of Theorem 1 by means of such minimax theorems.

Proof. The "if" part is immediate. We now prove the "only if" part.

Let $x \in S$ and denote, by $T_x \subset R^k$, the set of solutions to S_x. T_x is compact for every x, since it is closed by virtue of being a polyhedron, and bounded as it is a subset of P. Now $\bigcap_{x \in S} T_x = \varnothing$ if (1) fails. Therefore, there is a finite $F \subseteq S$ such that $\bigcap_{x \in F} T_x$ is empty, by standard compactness results. Then the system of linear inequalities

$$A\lambda \leqq b,$$

$$\sum_{i=1}^{k} g_i(x)\lambda_i \geqq L - f(x), \qquad \text{for all} \quad x \in F, \tag{2}$$

$$\lambda \geqq 0$$

has no solution.

By the Kuhn–Fourier theorem [8] there are nonnegative $w \in R^m$, $u_x \in R$ such that

$$w^t A - \sum_{x \in F} u_x(g_1(x), \ldots, g_k(x)) \geqq \mathbf{0} \quad \text{and} \quad w^t b + \sum_{x \in F} u_x(f(x) - L) < 0.$$

Since P is nonempty $u_x > 0$ for at least one $x \in F$. Let $N = \sum_{x \in F} u_x > 0$, and $\mathbf{x} = (1/N) \sum_{x \in F} u_x x$. Since $F \subseteq S$, clearly $\mathbf{x} \in S$. By convexity, $f(\mathbf{x}) \leqq (1/N) \sum_{x \in F} u_x f(x)$ and $g_i(\mathbf{x}) \leqq (1/N) \sum_{x \in F} u_x g_i(x)$. Therefore $w^t A - N[g_1(\mathbf{x}), \ldots, g_n(\mathbf{x})] \geqq \mathbf{0}$ and $w^t b + N[f(\mathbf{x}) - L)] < 0$, i.e., the system $(S_{\mathbf{x}})$ has no solutions. ■

Theorem 1. *Let P be as in the lemma. Let $V: R^k \to R$ be defined by*

$$V(u) = \max\{u\lambda \mid A\lambda \leqq b, \lambda \geqq 0\}. \tag{3}$$

Let $p: R^k \to R$ be defined by

$$p(u) = \inf\{f(x) \mid g_i(x) \leqq u_i \textit{ for } 1 \leqq i \leqq k \textit{ and } x \in S\}, \tag{4}$$

where we set $p(u) = +\infty$ if there is no $x \in S$ satisfying $g_i(x) \leqq u_i$ for $1 \leqq i \leqq k$, and $r + \infty = +\infty$ for all reals $r \in R$. Also, define

$$L = \inf_{u \in R^k} \{V(u) + p(u)\}. \tag{5}$$

If $L = -\infty$, then $v(\mathrm{D}_\lambda) = -\infty$ for all λ satisfying $\lambda \geqq 0$ and $A\lambda \leqq b$. In fact, if $L \in R \cup \{-\infty\}$, then $v(\mathrm{D}_\lambda) \leqq L$ for all λ with $\lambda \geqq 0$, $A\lambda \leqq b$.

Moreover, there exists $\boldsymbol{\lambda} \geqq 0$ with $A\boldsymbol{\lambda} \leqq b$ and $v(D_{\boldsymbol{\lambda}}) = L$.

Proof. First, suppose that $L \in R$.

Let $\lambda \in P$. We show that $v(\mathrm{D}_\lambda) \leqq L$. In fact, if $v(\mathrm{D}_\lambda) > L$, then for some $\varepsilon > 0$,

$$f(x) + \sum_{i=1}^{k} \lambda_i g_i(x) \geqq L + \varepsilon \qquad \text{for all} \quad x \in S. \tag{6}$$

Let $u \in R^k$ be arbitrary. If $p(u) \neq +\infty$, then $p(u) \in R$, since $p(u) = -\infty$ is ruled out by $L \in R$ [note that $V(u) \in R$ for all $u \in R^k$, by inspection of (3)]. For any $x \in X(u) = \{x \in S \mid g_i(x) \leqq u_i \text{ for } 1 \leqq i \leqq k\} \neq \varnothing$, (6) gives

$$f(x) + \sum_{i=1}^{k} \lambda_i u_i \geqq L + \varepsilon \tag{6$'$}$$

as all $\lambda_i \geqq 0$. From (6′), it follows at once, by taking the infimum over $x \in X(u)$, that

$$p(u) + \sum_{i=1}^{k} \lambda_i u_i \geqq L + \varepsilon. \tag{7}$$

Then invoking the definition (3) of $V(u)$, we have

$$p(u) + V(u) \geqq L + \varepsilon. \tag{8}$$

Also, if $p(u) = +\infty$, (8) is true. Therefore, (8) holds for all $u \in R^k$, contradicting the definition (5) of L.

Continuing with our assumption $L \in R$, note that, for any $x \in S$,

$$\begin{aligned} L - f(x) &\leqq L - p(g_1(x), \ldots, g_k(x)) \\ &\leqq V(g_1(x), \ldots, g_k(x)), \end{aligned} \tag{9}$$

using $u = (g_1(x), \ldots, g_k(x))$ in the definition (5) of L. From (9), (S_x) has a solution. By Lemma 1, a $\boldsymbol{\lambda}$, as described in Theorem 1, exists.

Finally, suppose that $L = -\infty$. Then we repeat the discussion, in Eqs. (6)–(8), with any finite value L' replacing $L + \varepsilon$ on the right-hand side in (6). Again, we obtain a contradiction. This proves that $v(\mathrm{D}_\lambda) = -\infty$ if $\lambda \geqq 0$ and $A\lambda \leqq b$. ■

The function p above is the usual perturbation function [7]. V is the criterion "value function" associated with a linear program, and as is well known, it can be written as the maximum of finitely many linear affine functions, each of which corresponds to an extreme point of the feasible region.

Theorem 1 tells us how good a value we can hope for in the program (D_λ) given that the λ must lie in P. Consider a family of polytopes $\mathrm{P}_\alpha = \{\lambda \mid A\lambda \leqq \alpha b,\ \lambda \geqq 0\}$ for $\alpha > 0$, $b \in R^m$, $b \geqq \mathbf{0}$. As α increases, P_α becomes larger. Given L we can ask what the smallest α is such that there exist $\lambda \in \mathrm{P}_\alpha$ with $v(\mathrm{D}_\lambda) \geqq L$. This is a "converse" of the question answered by Theorem 1.

Theorem 2. *Let* P_α *be polytopes as above. Define a* generalized perturbation function $G: R^m \to R$, *depending on* A, *by*

$$G(w) = \inf\{f(x) \mid \textit{subject to } x \in S \textit{ and } (g_1(x), \ldots, g_k(x)) \leqq w^{\mathrm{t}}A\}. \tag{10}$$

Define a modulus *for* $L \in R$, *by*

$$h_b(A, L) = \sup_{w \geqq 0} \frac{L - G(w)}{w^t b}, \tag{11}$$

where we conventionally assign $0/0 = 0$, $-(-\infty) = +\infty$, $L + \infty = +\infty$, $\alpha/0 = -\infty$ for $\alpha < 0$, $\alpha/0 = +\infty$ for $\alpha > 0$, $L - \infty = -\infty$, *and we set* $G(w) = +\infty$ *if there exists no* $x \in S$ *satisfying* $g_i(x) \leqq \sum_{j=1}^{M} w_j a_{ji}$ *for* $1 \leqq i \leqq k$.

Suppose that $\alpha \geqq 0$, $\alpha \in R$, $b \geqq 0$.

Then there exist $\lambda \in P_\alpha$ *such that* $v(D_\lambda) \geqq L$ *if and only if* $\alpha \geqq h_b(A, L)$. [*In particular, if* $h_b(A, L) = +\infty$, *then* L *is not obtainable for any* α*; and consequently also,* $v(D) \leqq L$ *if* $b > 0$.]

Proof. First we establish the "only if" part.

Suppose $f(x) + \sum \lambda_i g_i(x) \geqq L$ for all $x \in S$, i.e., $v(D_\lambda) \geqq L$, with $\lambda \in P_\alpha$. Let $w \geqq 0$. Choose $\varepsilon > 0$ and let x_0 be such that, assuming $G(w)$ finite, $(g_1(x_0), \ldots, g_n(x_0)) \leqq w^t A$ and $f(x_0) \leqq G(w) + \varepsilon$. Since $\lambda \in P_\alpha$, $L - G(w) - \varepsilon \leqq L - f(x_0) \leqq \sum \lambda_i g_i(x_0) \leqq w^t A\lambda \leqq \alpha(w^t b)$. Since ε was arbitrary, $L - G(w) \leqq \alpha(w^t b)$. The same result holds if $G(w) = +\infty$; and the same reasoning shows that $G(w) = -\infty$ cannot occur [as $v(D_\lambda) = -\infty$ for all $\lambda \geqq 0$].

If $w^t b \neq 0$, then $w^t b > 0$ (as $w \geqq 0$ and $b \geqq 0$), so $[L - G(w)]/(w^t b) \leqq \alpha$. If $w^t b = 0$, we have $L \leqq G(w)$, and so by our conventions $[L - G(w)]/(w^t b) \leqq \alpha$ as $\alpha \geqq 0$. Thus $h_b(A, L) \leqq \alpha$, as $w \geqq 0$ was arbitrary.

We now establish the "if" part. Let $\alpha \geqq h_b(A, L)$. We show the system (S_x), with b replaced with αb, has solutions for all $x \in S$ and apply Lemma 1.

Let $x \in S$. By the Kuhn–Fourier theorem, and the fact that P_α is nonempty, (S_x) has no solutions only if there is a $w \geqq 0$ such that $w^t A \geqq (g_1(x), \ldots, g_k(x))$ and $\alpha w^t b < L - f(x)$. Since $L - f(x) \leqq L - G(w)$, we obtain $\alpha w^t b < L - G(w)$. If $w^t b > 0$, we have $\alpha < (L - G(w))/w^t b$, which contradicts $\alpha \geqq h_b(A, L)$. If $w^t b = 0$, we have $G(w) < L$, hence $(L - G(w))/w^t b = +\infty$ by our conventions. Therefore, $h_b(A, L) = +\infty$, again contradicting $\alpha \geqq h_b(A, L)$. ■

For the case where some components of b are negative there is this result:

Theorem 3. *Let* $c \in R^m$, $c \geqq 0$, $P_\alpha^* = \{\lambda \mid \lambda \geqq 0,\ A\lambda \leqq b + \alpha c\}$, $\alpha \geqq 0$, $\alpha \in R$, *and suppose that* $P_\alpha \neq \varnothing$, *and* P_0 *is nonempty and bounded.*

Then there exist $\lambda \in P_\alpha^*$ *such that* $v(D_\lambda) \geqq L$ *iff*

$$\alpha \geqq \sup_{w \geqq 0} \frac{L - G(w) - wb}{wc}.$$

The proof is essentially the same as that for Theorem 2, which is a special case.

Theorem 2 for $\alpha = 1$ can be rephrased in terms of conjugate functions, if we restrict G to the domain $\{w \mid w \geqq 0\}$, and $P = \{\lambda \geqq 0 \mid A\lambda \leqq b\}$ is nonempty and bounded. Indeed, by Theorem 2, there exists $\lambda \in P$ with $v(\mathrm{D}_\lambda) \geqq L$ if and only if

$$1 \geqq [L - G(w)]/w^t b \qquad \text{for all} \quad w \geqq 0, \tag{12a}$$

or, equivalently,

$$w^t b \geqq L - G(w) \qquad \text{for all} \quad w \geqq 0, \quad \text{with} \quad w^t b > 0; \tag{12b}$$

and

$$L - G(w) \leqq 0 \qquad \text{for all} \quad w \geqq 0, \quad \text{with} \quad w^t b = 0. \tag{12c}$$

Now (12) is in turn equivalent to $w^t b \geqq L - G(w)$ for all $w \geqq 0$, i.e.,

$$G(w) - w^t(-b) \geqq L \qquad \text{for all} \quad w \geqq 0. \tag{13}$$

Finally, (13) is equivalent to $-G^c(-b) \geqq L$, where G^c is the conjugate function for G [7].

We now give some corollaries of the previous results, where the cases (a) and (b) referred to are those of the introduction.

Corollary 1. *There exist* $\lambda \geqq 0$ *such that* $v(\mathrm{D}_\lambda) \geqq L$ *and* (a) *is satisfied iff* $L \leqq \inf_{\varepsilon \geqq 0}\{M\varepsilon + h(\varepsilon)\}$, *where* $h(\varepsilon) = \inf\{f(x) \mid g_i(x) \leqq \varepsilon$ *for* $1 \leqq i \leqq k$ *and* $x \in S\}$.

Proof. We take A to be the $1 \times k$ matrix of ones and apply Theorem 1. For any $u \in R^k$ let $\alpha = \max\{0, \max\{u_i \mid 1 \leqq i \leqq k\}\}$. Then $V(u) = M\alpha$ and $p(u) \geqq p(\alpha, \ldots, \alpha)$ so $\inf_u\{V(u) + p(u)\} = \inf_{\varepsilon \geqq 0}\{M\varepsilon + h(\varepsilon)\}$. ■

We note that Corollary 1 also follows directly from Theorem 2, since, for a $1 \times k$ matrix of ones, $G = h$ where G is defined in (10) and h is defined in Corollary 1. Then the condition $w^t b \geqq L - G(w)$ for all $w \geqq 0$, of the discussion following Theorem 2, becomes $w_1 M \geqq L - h(w_1)$ for all $w_1 = \varepsilon \geqq 0$, and we obtain Corollary 1.

Corollary 2. *There exist* λ *such that* $v(\mathrm{D}_\lambda) \geqq L$ *and* (b) *is satisfied iff* $L \leqq \inf_{u \geqq 0}\{M(\sum u_i) + p(u)\}$.

Proof. Here A is an identity matrix and $V(u) = M \sum_{i=1}^k \max\{u_i, 0\}$. If $u_i' = \max\{u_i, 0\}$, $V(u') = V(u)$ and $p(u') \leqq p(u)$, so we need only look at those $u \geqq 0$ in taking $\inf_u\{V(u) + p(u)\}$. ■

Corollary 2 can also be obtained directly from Theorem 2; we omit the proof.

To give an application of these results in a computational setting, we take Theorem 2 for the case of one row ($m = 1$); in particular, we focus on the case (a) of the introduction for $M = 1$ (though the case of a constraint $\Sigma_1^k a_i \lambda_i \leqq 1$, for all $a_i > 0$, is similarly treated). Since a perturbation function

$h(\varepsilon)$ of Corollary 1 is involved in obtaining upper bounds on $v(D_\lambda)$, which already embodies a significant amount of information on the primal problem (P) [note that $h(0)$ alone is the value of (P)], our bounds will necessarily require some knowledge of the primal problem. It turns out, however, that to get bounds on $h_b(A, L)$ of Theorem 2 in this case, does not require a knowledge of all of h, but only of $h(\varepsilon_1)$ and $h(\varepsilon_2)$ for values $0 < \varepsilon_1 < \varepsilon_2$. The analysis follows in the next paragraphs; it is straightforward.

One trivially proves that G as defined in (10) is convex on its domain. By the paragraph following Corollary 1, so is h of Corollary 1. Now assume that $h(0)$ is defined [i.e., (P) has finite value $v(P)$] and $h(\varepsilon)$ is defined for at least one $\varepsilon > 0$. [From Theorem 2, if $h(\varepsilon) = -\infty$ for any $\varepsilon > 0$, then $v(D_\lambda) = -\infty$ for all $\lambda \geqq 0$.] By convexity, $h(\varepsilon)$ is then defined for all $\varepsilon \geqq 0$, and continuous for $\varepsilon > 0$; clearly, h is nonincreasing in $\varepsilon \geqq 0$. Therefore, the graph of h looks like the function drawn in Fig. 1.

In Fig. 1, we have drawn the case where $L\# = \sup\{h(\varepsilon) | \varepsilon > 0\} < h(0) = v(P)$, which can occur if h is discontinuous at its boundary point $\varepsilon = 0$ [of course, $L\# = h(0) = v(P)$ is forced if there is a Slater point for (P), for then $h(\varepsilon)$ is defined for some $\varepsilon < 0$ and 0 is an interior point of the domain of h, hence a point of continuity]. This need not be the case; if h is continuous at $\varepsilon = 0$, $L\# = h(0) = v(P)$. Also in Fig. 1, we have $h'(0) = -\infty$, where $h'(0)$ indicates the right directional derivative of h at 0, which exists as h is convex. This also need not be the case, for $h'(0) > -\infty$ is possible, and then [as

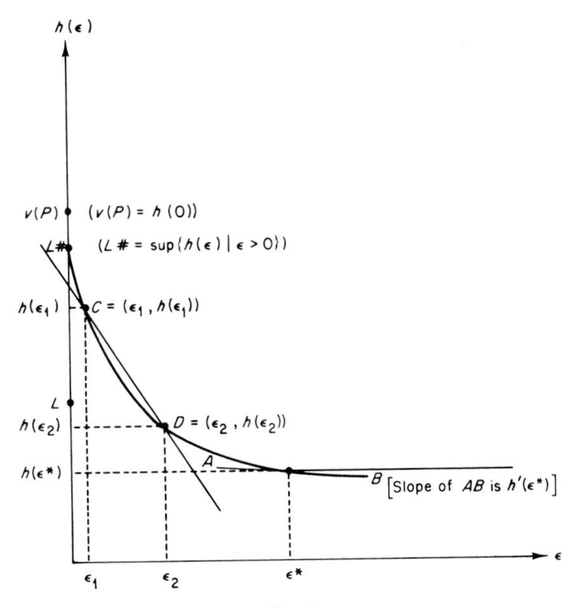

Fig. 1

$h'(0) \leqq 0$ by monotonicity] $h'(0)$ would be finite. Thus there are actually four possibilities at $\varepsilon = 0$ in drawing the graph of h [corresponding to $L\# < v(\mathrm{P})$ or $L\# = v(\mathrm{P})$, and $h'(0) = -\infty$ or $h'(0) > -\infty$], of which we have drawn only one. Furthermore, in our picture h has a horizontal asymptote, which simply means that the criterion function f of (P) is bounded below; but this need not be the case in general.

Note that P_α, as defined above Theorem 2, is always nonempty for all $\alpha \geqq 0$, for the constraint (a) with $M = 1$. By Theorem 2, there exists λ with $\lambda \geqq 0$ and $\sum_1^k \lambda_i \leqq \alpha$ if and only if $\alpha \geqq h_b(A, L)$. The supremum quotient on the right-hand side in (11) has an interesting interpretation; here $w = (w_1)$ and $b = (b_1) = (1)$.

For this quotient to be finite, one certainly requires $[L - G(0)]/(0 \cdot 1) = [L - h(0)]/0 < +\infty$, which, by the conventions in Theorem 2, necessitates $L \leqq h(0) = v(\mathrm{P})$. This fact is well known: the dual value $v(\mathrm{D})$ cannot exceed $v(\mathrm{P})$.

Moreover, in addition to $L \leqq v(\mathrm{P})$, for this quotient to be finite for $L = L\#$, requires that the right-sided directional derivative $g'(\varepsilon)$ of the function $g(\varepsilon)$ be finite at zero, where g is the continuous convex function defined by

$$g(\varepsilon) = \begin{cases} L\# & \text{if} \quad \varepsilon = 0 \\ h(\varepsilon) & \text{if} \quad \varepsilon > 0. \end{cases} \tag{14}$$

Indeed, the supremum quotient on the right-hand side in (11), for $w = (w_1) > 0$, is simply the negative of this right-sided directional derivative. Thus, when (as in Fig. 1) this directional derivative is $-\infty$, one has $v(\mathrm{D}_\lambda) < L\#$ for all $\lambda \geqq 0$ (as $\alpha = +\infty$ is forced, by Theorem 2); when this directional derivative is finite, there is $\lambda \geqq 0$ with $v(\mathrm{D}_\lambda) = L\#$ and in fact $\sum_1^k \lambda_i \leqq -g'(0) = h_1(A, L\#)$ (by Theorem 2). By the same reasoning, one cannot have $v(\mathrm{D}_\lambda) \geqq L$ for $\lambda \geqq 0$, if $L > L\#$, for then the right-hand side in (11) must be unbounded: this is a consequence of the well-known fact that $\lim\{p(\varepsilon_1, \ldots, \varepsilon_k) | \varepsilon_1, \ldots, \varepsilon_k \searrow 0^+\}$ is the value of the dual in (P).

Now the analysis of the previous paragraph extends to a point $\varepsilon^* > 0$, for this yields a convex program when translated to the origin [using convex constraints $g_i(x) - \varepsilon^* \leqq 0$, for $1 \leqq i \leqq k$]. Thus, the negative of the right-sided directional derivative of g at $\varepsilon = \varepsilon^*$ (which is that of h, by continuity at $\varepsilon = \varepsilon^*$) is a lower bound on $\sum_1^k \lambda_i$ for the translated program. By convexity of h, this directional derivative is not less than the directional derivative at $\varepsilon = 0$. Hence, a lower bound on $\sum_1^k \lambda_i$ for (P) is $-h'(\varepsilon^*)$ for any $\varepsilon^* > 0$. [Since $h'(\varepsilon)$ is defined also for $0 < \varepsilon < \varepsilon^*$, the same is true of the left-sided directional derivative, by convexity of h, and the left-sided derivative of course gives a better lower bound. In Fig. 1, we have drawn the case that h is differentiable at $\varepsilon = \varepsilon^*$.]

Now the directional derivative of a perturbation function can be hard to compute, but again the convexity of h simplifies the matter. In fact, if we know $h(\varepsilon_1)$ and $h(\varepsilon_2)$ for two values $0 < \varepsilon_1 < \varepsilon_2$, then a lower bound on $\sum_1^k \lambda_i$ is given by the negative of the slope of the secant line CD, to whit,

$$[h(\varepsilon_1) - h(\varepsilon_2)]/(\varepsilon_2 - \varepsilon_1). \tag{15}$$

Indeed, this secant line has slope not less than $h'(\varepsilon_1)$, and $-h'(\varepsilon_1)$ is a lower bound on $\sum_1^k \lambda_i$. Clearly, the lower bound given by (15) improves as ε_1 nears zero, and $\varepsilon_2 > \varepsilon_1$ is nearer to ε_1.

However, all the quantities (15) are lower bounds on $\sum_1^k \lambda_i$. If, in a computational setting, any of these bounds (15) is excessively large, one knows that the theoretically optimal dual value $L\# = v(0)$ cannot be computationally attained, and one will have to settle for a lower value $L < L\#$.

For $L < L\#$, the supremum quotient in Eq. (11) clearly gives the negative of the slope of a line through $(0, L)$ that is tangent to the graph of $h(\cdot)$, or, lacking such a point of tangency, asymptotic to it or parallel to an asymptote of it. Clearly, if the directional derivative at $\varepsilon = \varepsilon_2 > 0$ intersects the vertical axis in Fig. 1 at a point not below L, then ε_2 is at or to the left of this point of tangency, if it exists (if an asymptote exists, we view all $\varepsilon > 0$ as "to the left of" the "point of tangency"). Hence an *upper* bound on $\sum_1^k \lambda_i$ to attain $v(\mathrm{D}_\lambda) \geqq L$ is $-h'(\varepsilon_2)$, so another (generally weaker) upper bound is given by (15). Thus, if the value of (15) is computationally feasible, it yields a number L such that $v(\mathrm{D}_\lambda) \geqq L$ for some $\lambda \geqq 0$ with $\sum_1^k \lambda_i$ computationally feasible.

2. $\{\lambda \mid A\lambda \leqq b, \lambda \geqq 0\}$ MAY BE UNBOUNDED

One way of approaching the case treated in this section is by a simple reduction to the bounded case of the previous section. Again $P = \{\lambda \mid A\lambda \leqq b, \lambda \geqq 0\}$, $V(u)$ is as defined in Eq. (3), and $p(u)$ is defined as in Eq. (4). We shall also need the notations

$$P'_M = \left\{\lambda \mid A\lambda \leqq b, \lambda \geqq 0, \sum_1^k \lambda_i \leqq M\right\} \tag{16}$$

and

$$V'_M(u) = \max\{u\lambda \mid \lambda \in P'_M\}. \tag{17}$$

Here is a result one can obtain as an immediate application of Theorem 1.

Corollary 3. *Suppose that $P \neq \varnothing$, and set*

$$L^* = \lim_{M \to +\infty} \inf_{u \in R^k} \{V'_M(u) + p(u)\}. \tag{18}$$

If $L^* \in R \cup \{-\infty\}$, then $v(\mathrm{D}_\lambda) \leqq L^*$ for all λ with $\lambda \geqq 0$ and $A\lambda \leqq b$.

Moreover, if $L^* \in R$ and $L' < L^*$, there exists $\lambda \geqq 0$ with $A\lambda \leqq b$ and $v(\mathrm{D}_\lambda) \geqq L'$.

Finally, $L^* \in R \cup \{-\infty\}$.

Proof. Clearly, $V'_M(u)$ is monotone nondecreasing in M for any $u \in R^k$, hence the same is true of $\inf_{u \in R^k}\{V'_M(u) + p(u)\}$. Thus, the limit indicated in (18) exists.

If $L^* = -\infty$, then by monotonicity $\inf_{u \in R^k}\{V'_M(u) + p(u)\} = -\infty$ for all $M \geqq 0$. If $\lambda \geqq 0$ satisfies $A\lambda \leqq b$, then $\lambda \in P'_M$ for sufficiently large M, and Theorem 1 applies, showing $v(\mathrm{D}_\lambda) = -\infty$.

If $L^* \in R$, let $L\# > L^*$. By monotonicity, $\inf_{u \in R^k}\{V'_M(u) + p(u)\} < L\#$ for all M. If $\lambda \geqq 0$ satisfies $A\lambda \leqq b$, then $\lambda \in P'_M$ for sufficiently large M, and by Theorem 1 we have $v(P_\lambda) \leqq L\#$. Since $v(P_\lambda) \leqq L\#$ holds for all $L\# > L^*$, actually we have $v(\mathrm{P}_\lambda) \leqq L^*$ also.

Next, if $L' < L^*$, by monotonicity, for sufficiently large M,

$$\inf_{u \in R^k}\{V'_M(u) + p(u)\} > L'.$$

By Theorem 1, for M_0 sufficiently large, there is $\boldsymbol{\lambda} \in P'_{M_0} \subseteq P$ with $v(\mathrm{D}_\lambda) = \inf_{u \in R^k}\{V'_M(u) + p(u)\} > L'$.

Finally, by the last paragraph, if $L^* = +\infty$, then $v(\mathrm{D}_\lambda) > v(\mathrm{P})$ for some $\boldsymbol{\lambda} \in P$, since (P) is assumed consistent. However, as $v(\mathrm{D}) \leqq v(\mathrm{P})$, and $v(\mathrm{D}) = \sup\{v(\mathrm{D}_\lambda) | \lambda \geqq 0\}$, we have a contradiction. ■

The proof of Corollary 3 makes it evident, that there is $\boldsymbol{\lambda} \in P$ with $v(\mathrm{D}_\lambda) = L^*$ if and only if the limiting operation in (18) is vacuous, i.e., if and only if

$$L^* = \inf_{u \in R^k}\{V'_M(u) + p(u)\} \tag{19}$$

for all sufficiently large M.

It is clear that

$$V'_M(u) + p(u) \leqq V(u) + p(u) \tag{20}$$

for all $u \in R^k$, where one may interpret $\infty - \infty$ on the right-hand side of (20) to be $-\infty$. Taking the infimum over $u \in R^k$ and then applying the limiting operation to both sides of (20) yields

$$L^* \leqq \inf_{u \in R^k}\{V(u) + p(u)\}. \tag{21}$$

However, unlike the bounded case, strict inequality can occur in (21), i.e., one can have

$$\sup\{v(\mathrm{D}_\lambda) | \lambda \in P\} < \inf_{u \in R^k}\{V(u) + p(u)\} \tag{22}$$

as our next example shows.

Example. Let $P = \{\lambda \mid \lambda \geqq 0\}$. Then we have for $u \in R^k$, $u = (u_1, u_2)$,

$$V(u) = \begin{cases} 0 & \text{if } u_1 \leqq 0 \text{ and } u_2 \leqq 0 \\ +\infty & \text{if } u_1 > 0 \text{ or } u_2 > 0 \end{cases} \tag{23}$$

and

$$V'_M(u) = \begin{cases} 0 & \text{if } u_1 \leqq 0 \text{ and } u_2 \leqq 0 \\ M \max\{u_1, u_2\} & \text{if } u_1 > 0 \text{ or } u_2 > 0. \end{cases} \tag{24}$$

We calculate, using the fact that $p(u)$ is monotone nonincreasing in u,

$$\begin{aligned} \inf_{u \in R^k} \{V(u) + p(u)\} &= \inf\{p(u) \mid u_1 \leqq 0 \text{ and } u_2 \leqq 0\} \\ &= p(0) \end{aligned} \tag{25}$$

provided that $p(u)$ is never $-\infty$.

Now $p(0)$ is the value $v(\mathrm{P})$ of the primal problem (P), while $v(\mathrm{D}) = \sup\{v(\mathrm{D}_\lambda) \mid \lambda \geqq 0\}$ is the value of the dual. As is well known, one can have $v(\mathrm{D}) < v(\mathrm{P})$ even if $p(u)$ is never $-\infty$.

For example, in R^2 with $S = \{(w_1, w_2) \mid w_1 \geqq 0 \text{ and } w_2 \geqq 0\}$, $f(x) = f(w_1, w_2) = \max\{-1, -\sqrt{w_1 w_2}\}$ (recall that $\sqrt{w_1 w_2}$ is concave), $k = 1$, $g_1(x) = g_1(w_1, w_2) = w_1$, we have $v(\mathrm{P}) = p(0) = 0$. However, $p(u) = \inf\{\max\{-1, -\sqrt{w_1 w_2}\} \mid w_1 \leqq u\}$ is such that $p(u) = +\infty$ if $u < 0$, $p(u) = -1$ if $u > 0$ (since $-\sqrt{u w_2} \leqq -1$ for $w_2 > 0$ sufficiently large, whenever $u > 0$). Therefore $v(\mathrm{D}) = \lim\{p(u) \mid u \downarrow 0^+\} = -1 < v(\mathrm{P})$ [or one may demonstrate that $v(\mathrm{D}) = -1$ directly]. In this example, all functions are continuous on S.

Without introducing some form of "constraint qualification" which rules out examples of the type just given, there is another result, beyond that of Corollary 3, which involves a different analysis. First we need two preliminary lemmas.

Lemma 2. *Let $a^i \in R$, $b_i \in R$ for $i \in I$, $I \neq \varnothing$. Let Q be a polyhedron such that $Q \cap \{x \mid a^i x \geqq b_i, i \in I\} = \varnothing$ (I finite).*

Then there exist $c \in R^n$, $d \in R$ such that (i) $cx \geqq d$ for every $x \in Q$ and (ii) the system

$$\begin{aligned} a^i x &\geqq b_i, \qquad i \in I, \\ cx &\geqq d \end{aligned} \tag{26}$$

is inconsistent.

Proof. If $\{x \mid a^i x \geqq b_i, i \in I\}$ is empty we may take $c = 0$ and $d = 0$. Otherwise the separating hyperplane theorem trivially yields c, d such that (i) and (ii) hold, by strictly separating Q from $\{x \mid a^i x \geqq b_i, i \in I\}$. ■

Lemma 3. *Suppose that $P \neq \varnothing$.*

There is no $\lambda \in P$ with $v(\mathrm{D}_\lambda) \geqq L$ if and only if for every M, there is a $u \in R^k$ with $V(u) < +\infty$ and an $\mathbf{x} \in S$ such that the finite system of inequalities

$$\begin{aligned} u\lambda &\leqq V(u), \\ \sum_{i=1}^{k} g_i(\mathbf{x})\lambda_i &\geqq L - f(\mathbf{x}), \\ \sum_{i=1}^{k} \lambda_i &\leqq M, \\ \lambda &\geqq 0 \end{aligned} \tag{$27_{\mathbf{x},M,u}$}$$

is inconsistent.

Proof. The existence of $\lambda \in P$ such that $v(\mathrm{D}_\lambda) \geqq L$ is precisely equivalent to the consistency of this semiinfinite system for large M:

$$\begin{gathered} A\lambda \leqq b, \qquad \sum_{i=1}^{k} \lambda_i \leqq M, \\ \sum_{i=1}^{k} g_i(x)\lambda_i \geqq L - f(x), \qquad x \in S, \\ \lambda \geqq 0. \end{gathered} \tag{28}$$

We may apply compactness and Lemma 2 with $Q = P = \{\lambda \mid A\lambda \leqq b\}$ to obtain c, d such that for some finite set $F \subseteq S$,

$$\begin{gathered} c\lambda \geqq d, \qquad \sum_{i=1}^{k} \lambda_i \leqq M, \\ \sum_{i=1}^{k} g_i(x)\lambda_i \geqq L - f(x), \qquad x \in F, \\ \lambda \geqq 0 \end{gathered} \tag{28$'$}$$

is inconsistent. If we let $u = -c$ and replace $c\lambda \geqq d$ by $u\lambda \leqq V(u)$ (28$'$) remains inconsistent because $-d \geqq V(-c)$. Indeed, since $Q = P \neq \varnothing$ and $\lambda \in P$ implies $c\lambda \geqq d$, i.e., $(-c)\lambda \leqq -d$, we have $V(-c) \leqq -d$ at once.

Thus the following system is inconsistent:

$$\begin{aligned} u\lambda &\leqq V(u), \\ \sum_{i=1}^{k} g_i(x)\lambda_i &\geqq L - f(x), \qquad x \in F, \\ \sum_{i=1}^{k} \lambda_i &\leqq M, \\ \lambda &\geqq 0. \end{aligned} \tag{28$''$}$$

Since the polyhedron $Q_M = \{\lambda \geqq 0 | \sum_{i=1}^k \lambda_i \leqq M,\ u\lambda \leqq V(u)\}$ is either empty or nonempty and bounded, by Lemma 1 (with P replaced by Q_M) there is an $\mathbf{x}$ such that $(27_{\mathbf{x},M,u})$ is inconsistent. ■

As preliminary remarks to Theorem 4 just below, we note that always $V(u) > -\infty$ since $P \neq \varnothing$. Also, $V(u)$ is convex, as is well known [for example, note that $V(u) = \min\{\theta b | \theta A \geqq u,\ \theta \geqq 0\}$ if $V(u) < +\infty$, from duality; and a minimizing linear program is convex in its right-hand side u]. Clearly, $V(\alpha u) = \alpha V(u)$ for all $\alpha \geqq 0$, when $V(u) < +\infty$.

Lemma 3 can be rephrased, in terms of the perturbation function p, as follows.

Theorem 4. *Suppose $P \neq \varnothing$. Define*

$$h(\beta) = \inf_{\substack{u \in R^k \\ V(u) < +\infty}} \{V(u) + p(u + \beta e)\}, \tag{29}$$

where $e = (1, \ldots, 1)^t$ is a vector of ones. Let $J = \lim_{\beta \to 0+} h(\beta)$. Then

If $L < J$, there is $\lambda \in P$ with $v(\mathrm{D}_\lambda) \geqq L$. (30a)

If $L > J$, then $v(\mathrm{D}_\lambda) < L$ for every $\lambda \in P$. (30b)

There is $\lambda \in P$ with $v(\mathrm{D}_\lambda) = J$ if and only if $\lim_{\beta \to 0+} [h(\beta) - J]/\beta > -\infty$. (30c)

Proof. First we prove (30a) and the "if" part of (30c).

The function h is clearly nonincreasing. Also, h is convex. In fact, let $\beta_1, \beta_2 \geqq 0$ and $0 \leqq \lambda \leqq 1$ be given, and let $p_1 > h(\beta_1)$ and $p_2 > h(\beta_2)$. Then there exists u^1 with $V(u^1) < +\infty$, and u^2 with $V(u^2) < +\infty$, *such that*

$$p_1 > V(u^1) + p(u^1 + \beta_1 e),$$
$$p_2 > V(u^2) + p(u^2 + \beta_2 e).$$

Using the convexity of V and p, we obtain

$$\begin{aligned} \lambda p_1 + (1 - \lambda)p_2 &\geqq V(\lambda u^1 + (1 - \lambda)u^2) + p((\lambda u^1 + (1 - \lambda)u^2) \\ &\quad + (\lambda\beta_1 + (1 - \lambda)\beta_2)e) \\ &\geqq h(\lambda\beta_1 + (1 - \lambda)\beta_2). \end{aligned}$$

Since $p_1 > h(\beta_1)$ and $p_2 > h(\beta_2)$ are arbitrary, the last result above gives $\lambda h(\beta_1) + (1 - \lambda)h(\beta_2) \geqq h(\lambda\beta_1 + (1 - \lambda)\beta_2)$, and our proof of the convexity of h is complete.

In either of the two cases we are considering we can choose M_0 sufficiently large so that (i) $h(\beta) \geqq L - M_0\beta$ for all $\beta \geqq 0$ [with $L = J$ for 30c)] and (ii) there is a $\lambda \in P$ with $\sum \lambda_i \leqq M_0$.

We show that for any $x \in S$ and any u for which $V(u) < +\infty$, the system $(27_{\mathbf{x},M_0,u})$ is consistent by using the Kuhn–Fourier theorem [8].

If $(27_{\mathbf{x}, M_0, u})$ were inconsistent there would be multipliers α, δ, $\beta \geqq 0$ such that

$$\alpha u - \delta(g_1(\mathbf{x}), \ldots, g_k(\mathbf{x})) + \beta e \leqq 0. \tag{31a}$$

$$\alpha V(u) + \delta(f(\mathbf{x}) - L) + \beta M_0 < 0. \tag{31b}$$

If $\delta = 0$, (31a) and (31b) would imply that $P \cap \{\lambda | \sum \lambda_i \leqq M_0\} = \varnothing$. This would contradict property (ii) of M_0. So we may assume $\delta > 0$. If we replace α by α/δ and β by β/δ we may assume $\delta = 1$. By (31a) and the monotonicity of p, $p(\alpha u + \beta e) \leqq p(g_1(\mathbf{x}), \ldots, g_k(\mathbf{x})) \leqq f(\mathbf{x})$. But

$$\begin{aligned} 0 &> \alpha V(u) + (f(\mathbf{x}) - L) + \beta M_0 \text{ [by (31b)]} \\ &\geqq V(\alpha u) + p(\alpha u + \beta e) - L + \beta M_0 \\ &\geqq h(\beta) - L + \beta M_0 \geqq 0 \end{aligned}$$

[by property (i) of M_0]. This contradiction $0 > 0$ establishes that there are no $\alpha, \beta, \delta \geqq 0$ satisfying (31a) and (31b) so $(27_{\mathbf{x}, M, u})$ has a solution.

By Lemma 3, there is a $\lambda \in P$ with $v(\mathrm{D}_\lambda) \geqq L$. This completes the proof of (30a) and the "if" part of (30c).

Now we prove (30b) and the "only if" part of (30c). Let $\lambda \in P, \sum \lambda_i = M$. We wish to show $v(\mathrm{D}_\lambda) < L$ [where $L = J$ for (30c)].

In either of the two cases now treated, for any M there exists $\varepsilon > 0$ so that $h(\varepsilon) + \varepsilon M < L$. Choose u so that $V(u) + p(u + \varepsilon e) + \varepsilon M < L$. By definition of p, there is an $\mathbf{x} \in S$ such that $V(u) + f(x) + \varepsilon M < L$ and $(g_1(\mathbf{x}), \ldots, g_k(\mathbf{x})) \leqq u + \varepsilon e$. Since $L > V(u) + f(\mathbf{x}) + \varepsilon M \geqq u\lambda + f(\mathbf{x}) + \varepsilon(\sum \lambda_i) = f(\mathbf{x}) + (u + \varepsilon e)\lambda \geqq f(\mathbf{x}) + \sum \lambda_i g_i(\mathbf{x})$, we have $v(\mathrm{D}_\lambda) < L$. ■

Our next and final characterization of $\sup\{v(\mathrm{D}_\lambda) | \lambda \in P\}$, for the case of $P \neq \varnothing$ and P possibly unbounded, is a direct application of Theorem 3. First we define the altered perturbation function, depending on A, by

$$\begin{aligned} H(w, w_{m+1}) = \inf\{f(x) | &\text{subject to } x \in S \\ &\text{and } (g_1(x), \ldots, g_k(x)) \leqq w_t A + w_{m+1} e\} \end{aligned} \tag{32}$$

where $(w, w_{m+1}) \in R^{m+1}$, $w \in R^m$, and $e = (1, \ldots, 1)^t$ is again the vector of ones [compare (32) with (10)]. Now (32) is easily seen to be the previous perturbation function of type (10) for the constraint system

$$\begin{bmatrix} A\lambda \\ \sum_1^k \lambda_i \end{bmatrix} \leqq \begin{bmatrix} b \\ \gamma \end{bmatrix} + \alpha \begin{bmatrix} 0 \\ 1 \end{bmatrix} \tag{33}$$

in which a row $\sum_1^k \lambda_i \leqq 1$ has been added to the constraints of P to obtain a bounded polyhedron, and $\gamma \in R$.

Theorem 5. *Suppose that $P \neq \varnothing$ and $b \geqq 0$.*
There exists $\lambda \in P$ such that $v(D_\lambda) \geqq L$ if and only if the supremum

$$\sigma = \sup_{w, w_{n+1} \geqq 0} [L - H(w, w_{n+1}) - wb]/w_{n+1} \tag{34}$$

is finite. When this supremum is finite, in fact there is $\lambda \in P$ with $v(D_\lambda) \geqq L$ and $\sum_1^k \lambda_i = \sigma$; and there is no $\lambda \in P$ with $v(D_\lambda) \geqq L$ and $\sum_1^k \lambda_i < \sigma$.

Proof. We apply Theorem 3 with $\left[\begin{smallmatrix} b \\ \gamma \end{smallmatrix}\right]$ replacing b, and with $\left[\begin{smallmatrix} 0 \\ 1 \end{smallmatrix}\right]$ replacing c, where $\gamma = \min\{\sum_1^k \lambda_i \mid \lambda \in P\}$. ■

It is worth noting that, in Theorems 4 and 5 [the latter with H as defined in (34)], the vector e can be replaced by any vector with all components strictly positive. In fact, the same proofs are valid with this change.

3. ADDENDUM

We wish to show here that Theorems 1 and 4 hold for arbitrary closed convex sets as well as for polyhedra.

Theorem 4′. *Let K be a nonempty closed convex set contained in the non-negative orthant. If we define $V(u) = \max\{u\lambda \mid \lambda \in K\}$ and p, h as in Theorem 4, then the conclusions (30a)–(30c) hold with P replaced by K.*

The proof of Theorem 4 and Lemmas 3 can be repeated with P replaced by K throughout.

Theorem 1′. *Assume K is a nonempty, compact convex set. Then the conclusions of Theorem 1 hold with P replaced by K and V defined as in Theorem 4′.*

Proof. (Outline) It is possible to prove this result using the same methods as in Theorem 1.

For variety, we show how Theorem 4′ can be used to obtain the main result, namely the existence of $\lambda \in K$ with

$$v(D_\lambda) = \inf_{u \in R^K}\{V(u) + P(u)\} \equiv Q.$$

Let J be defined as in Theorem 4′. If K is bounded $V(u) < +\infty$ for all $u \in R^K$. If $\beta > 0$, $p(u + \beta e) \leqq p(u)$ so $Q \geqq J$. However $V(u) + p(u + \beta e)$ is close to $V(u + \beta e) + p(u + \beta e)$ when β is small so $J \geqq Q$, hence $J = Q$. Now conclusion (30a) of Theorem 4′ implies the existence of $\lambda^{(n)} \in K$ with $v(D_{\lambda^{(n)}}) \geqq Q - (1/n)$ for $n = 1, 2, \ldots$. Since K is compact, there is a limit point $\boldsymbol{\lambda} \in K$. Since $v(D_\lambda)$ is a closed, concave function of λ, we must have $v(D_\lambda) \geqq Q$. But (30b) says $v(D_\lambda) \leqq Q$, hence $v(D_\lambda) = Q$. ■

REFERENCES

1. C. E. Blair, Convex optimization and Lagrange multipliers, *Math. Programming* (to appear).
2. A. Charnes, W. W. Cooper, and K. O. Kortanek, On representations of semi-infinite programs which have no duality gaps, *Management Sci.* **12** (1965), 113–121.
3. A. Charnes, W. W. Cooper, and K. O. Kortanek, Duality, Haar programs, and finite sequence spaces, *Proc. Nat. Acad. Sci. U.S.A.* **68** (1962), 605–608.
4. R. J. Duffin, Convex analysis treated by linear programming, *Math. Programming* **4** (1973), 125–143.
5. R. J. Duffin and R. G. Jeroslow, Affine minorants and Lagrangean functions (in preparation).
6. R. J. Duffin and L. A. Karlovitz, An infinite linear program with a duality gap, *Management Sci.* **12** (1965), 122–134.
7. R. T. Rockafellar, "Convex Analysis," Princeton Univ. Press, Princeton, New Jersey, 1970.
8. J. Stoer and C. Witzgall, "Convexity and Optimization in Finite Dimensions I," Springer-Verlag, Berlin and New York, 1970.

This report was prepared as part of the activities of the Management Sciences Research Group, Carnegie-Mellon University, under Grant No. GP 37510 X1 of the National Science Foundation and contract N00014-75-0621 NRO47-048 with the U.S. Office of Naval Research.

C. E. Blair
DEPARTMENT OF BUSINESS ADMINISTRATION
UNIVERSITY OF ILLINOIS
URBANA, ILLINOIS

R. G. Jeroslow
DEPARTMENT OF MATHEMATICS
CARNEGIE-MELLON UNIVERSITY
PITTSBURGH, PENNSYLVANIA

and

COLLEGE OF INDUSTRIAL MANAGEMENT
GEORGIA INSTITUTE OF TECHNOLOGY
ATLANTA, GEORGIA

A View of Complementary Pivot Theory (Or Solving Equations with Homotopies)

B. C. EAVES

1. INTRODUCTION

Our purpose here is to give a brief, valid, and painless view of the equation solving computational method variously known as complementary pivot theory and/or fixed point methods. Our view begins in Section 2 with a bit of history and some of the successes of the method. In Section 3 the unique convergence proof of the method is elucidated with a riddle on ghosts. In Section 4 the general approach for solving equations with complementary pivot theory is encapsulated in the "homotopy principle." In Section 5 a simple example is used to illustrate both the convergence proof and the "homotopy principle." Rudiments of the general theory are stated and the "main theorem" is exhibited in an example in Section 6. Two representative complementary pivot algorithms are presented vis-a-vis the "homotopy principle and "main theorem" in Section 7. Finally, in Section 9 the principal difficulty of the method is discussed and some of the studies for dealing with this difficulty are mentioned.

2. A BIT OF HISTORY

If a point in time can be specified as the beginning of complementary pivot theory it is with the paper by Lemke and Howson [20]. In this paper a startling convergence proof was given for a finite algorithm for computing a Nash equilibrium of a bimatrix game. To understand their contribution, it was previously known that a Nash equilibrium existed via the Brouwer fixed point theorem and that exhaustive search offered a finite procedure for computing such an equilibrium. Furthermore, there is no theoretical proof that the Lemke–Howson algorithm has advantages over exhaustive search,

ISBN 0-12-178150-X

and, in fact, one can construct examples where the Lemke–Howson algorithm is no better than exhaustive search. However, the point is, as a practical matter, the Lemke–Howson algorithm versus exhaustive search enables one to solve bimatrix games with characteristic size of, say, one thousand versus thirty. The situation is analogous to that of Dantzig's simplex method and linear programs.

In the paper [19] Lemke specified what is now known as "Lemke's algorithm" and, thereby, showed that the convergence proof could be used for a much broader class of problems including quadratic programs.

The next steps came from Scarf in [26–28]. Using the convergence proof of Lemke and Howson he proved for the first time that a balanced game has a nonempty core and he described algorithms for computing a Brouwer fixed point and an equilibrium point for the general competitive equilibrium model. Once again, it has not been proved that the algorithm improved upon exhaustive search, but as a practical matter, problems could be solved that could not be solved before (note, for example, that Brouwer's theorem can be proved by repeated applications of Sperner's lemma and that only finitely many simplexes need be examined at each iteration in order to find a complete simplex).

There are now at least 200 papers in complementary pivot theory, and many very exciting developments have occurred. The convergence proof of Lemke and Howson is now understood to be intimately related to homotopy theory, a matter which is the crux of this paper. Many classical results have been given new "complementary pivot" proofs; to mention a few: Freidenfelds [9] and a connected set theorem of Browder, Kuhn [18] and the fundamental theorem of algebra, Garcia [11] and the last theorem of Poincaré, and Meyerson and Wright [23] and the Borsak–Ulman theorem.

A great deal of effort and ingenuity has been expended in making the complementary pivot algorithms more efficient; however, we shall not discuss these matters until the last section wherein we will describe the principal weakness of the complementary pivot algorithms.

Complementary pivot theory has been used to solve a number of specific problems; for instance, the following papers are concerned with complementary pivot theory applied to the solution of differential equations: Allgower and Jeppson [1, 2], Wilmuth [34], Cottle [4], Netravali and Saigal [24], and Kaneko [14]. Katzenelson's [15] algorithm, and subsequent developments thereof, for electrical network problems also fits comfortably into the framework of complementary pivot theory as discussed in this paper.

This paper is based upon Eaves and Scarf [8] and Eaves [6]. The reader might also want to consult Hirsch [12], Scarf [29], and Todd [31].

3. CONVERGENCE PROOF

The following riddle and its solution illustrate the extraordinary convergence proof of Lemka and Howson.

Ghost Riddle. I eased through the front door of the allegedly haunted house. Just as a ghost appeared, the front door slammed shut behind me. He spoke— "You are now locked inside our house, but it is your fate that except for this room which has one open door, every other room with a ghost has two open doors." I thought, "Is there a room without a ghost?"

Figure 1 spills the beans.

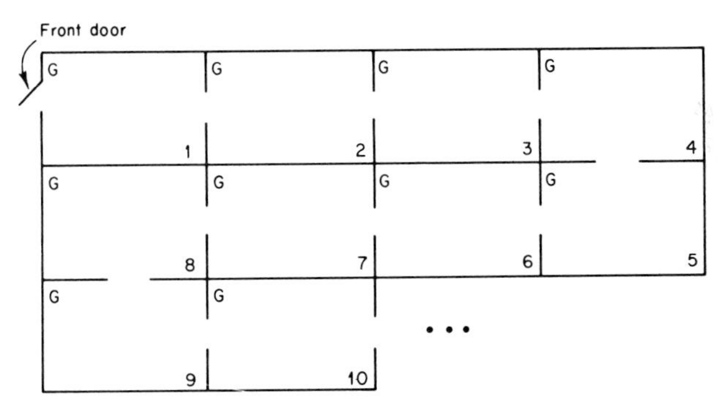

Fig. 1

Assume we are standing in room 1 with the front door closed. According to the riddle there is exactly one open door, so let us pass through the door into the adjoining room which we now call room 2. In room 2, assuming the presence of a ghost, there are two open doors available to us, one of which we just entered; so let us exit the other and enter room 3. We continue in this fashion to rooms 4, 5, etc. The essential property of this process is that no room is entered (i.e., numbered) more than once, which is to say, there is no cycling. A proof of this fact is available by assuming the contrary and examining the first room entered twice. Consequently, if there are only m rooms in the house, then the process must stop with m steps or less, and there is a room without a ghost. On the other hand, if the house is sufficiently haunted so as to have infinitely many rooms (George Dantzig supports this possibility) then either the process stops with a solution, that is, a room without a ghost, or it proceeds forever always entering new rooms.

In this isolated form the convergence proof appears uselessly simple. When we apply the convergence principle, the rooms will become pieces of linearity of some function and circumstances will not be quite so transparent.

4. HOMOTOPY PRINCIPLE

Now consider the continuous function $f: R^n \to R^n$ on n-dimensional Euclidean space and the system of equations $f(x) = y$. As a general procedure for solving such a system we offer the following.

Homotopy Principle. *To solve a system of equations, the system is first deformed to one which is trivial and has a unique solution. Beginning with the solution to the trivial problem a route of solutions is followed as the system is deformed, perhaps with retrogressions, back to the given system.*

Let us be more specific. First we introduce a family of problems

$$F(x, \theta) = y, \qquad 0 \leqq \theta \leqq 1,$$

where F is continuous in (x, θ), $F(\cdot, 0) = f$, and $F(x, 1) = y$ is a trivially solved system with a unique solution. We think of θ as deforming the given system $f(x) = y$ to the system $F(x, 1) = y$ with a unique solution, say x_1.

To obtain a solution to the given system $f(x) = y$ we follow the solution

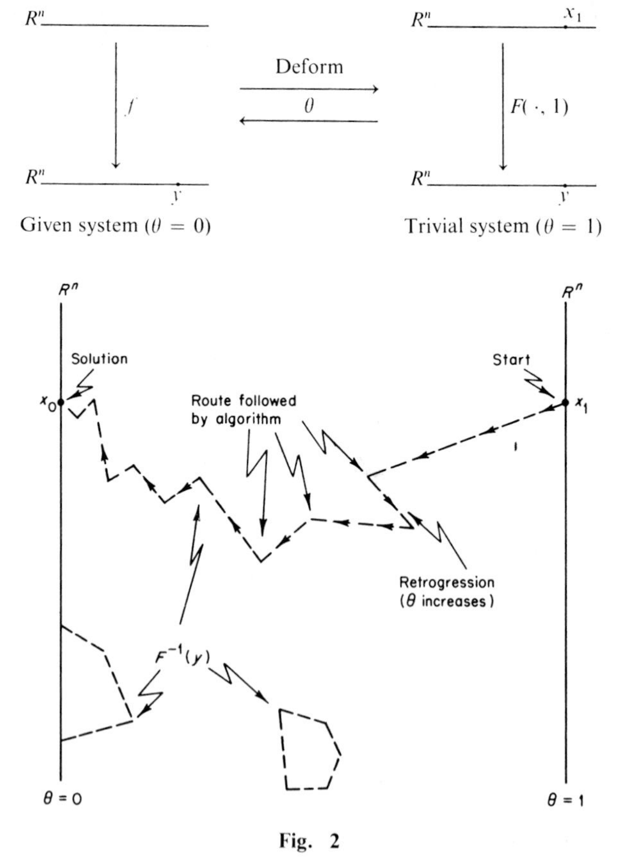

Fig. 2

of $F^{-1}(y)$ beginning with $(x_1, 1)$. Except for degenerate (rare) cases the component of $F^{-1}(y)$ that meets $(x_1, 1)$ is a route, that is, a path.

Assuming F is piecewise linear on $R^n \times (0, 1]$ Fig. 2 illustrates the situation quite well.

The algorithm begins with the point $(x_1, 1)$ and follows the route of $F^{-1}(y)$. Under various conditions it can be shown that the route eventually leads to $R^n \times 0$ and thus yields a solution of the given problem.

This principle is illustrated in Section 5 and given theoretical credence in Section 6.

5. AN EXAMPLE

In this section we exhibit the complementary pivot convergence proof and the "homotopy principle" by solving a system of piecewise linear equations. This particular system of equations was chosen for its pedagogical value; later we examine merely continuous functions.

Let S be an n-simplex in R^n with extreme points $s_0, s_1, \ldots, s_n$ (Fig. 3).

Let $\mathscr{S}$ be a collection of smaller n-simplexes which subdivides S (Fig. 4). By a vertex of $\mathscr{S}$ we mean a vertex of any element of $\mathscr{S}$.

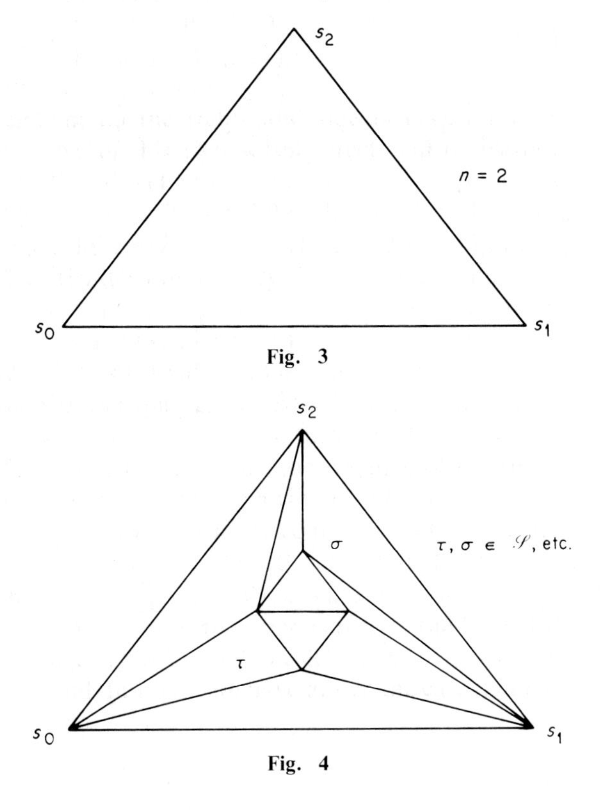

Fig. 3

Fig. 4

Now let $f: S \to S$ be a continuous function on the simplex with the following three properties.

(a) On each element of $\mathscr{S}$, f is linear, that is, affine.

(b) On the boundary of S, f is the identity, that is, $f(x) = x$ for x in ∂S.

(c) Vertices of the subdivision are carried by f into extreme points of S, that is, $f(\text{vertices}) \subset \{s_0, \ldots, s_n\}$.

Let y be some interior point of S and we consider the system of equations $f(x) = y$. Toward solving this system label the vertices of $\mathscr{S}$ according to the extreme point to which it is mapped; that is, define the labeling function l on the vertices by $l(v) = i$ if $f(v) = s_i$.

Assume that we thus obtain the diagram of Fig. 5. We call a simplex σ of $\mathscr{S}$ completely labeled only if all labels $0, 1, \ldots, n$ are present on its vertices. In Fig. 5 there is exactly one such, namely the upper right one.

Let us observe that solving $f(x) = y$ is equivalent, modulo solving a system of linear equations, to finding a completely labeled simplex. If a simplex τ of $\mathscr{S}$ has only labels $\{0, \ldots, n) \sim \{i\}$ then in view of the linearity of f on τ, f would map the entirety of τ into the face of S spanned by $\{s_0, \ldots, s_n\} \sim \{s_i\}$. Consequently, no point of τ would hit the interior point y. On the other hand, if a simplex τ of $\mathscr{S}$ is completely labeled then τ is mapped onto S, and here some point of τ hits y. So for the moment we focus on the task of finding a completely labeled simplex.

To execute the task of finding a completely labeled simplex we shall employ the "ghost riddle" as follows to obtain "Cohen's algorithm" (see [3]). We regard as a room a simplex of $\mathscr{S}$ and as an open door a face of a simplex of $\mathscr{S}$ with labels 0 and 1 [if $n > 2$ a face of a simplex of $\mathscr{S}$ with $(0, \ldots, n-1)$]. By passing through open doors the path followed is indicated in Fig. 6 and terminates with a completely labeled simplex.

Prima facia this procedure works so smoothly it seems coincidental. But suppose that the n-simplex σ of $\mathscr{S}$ has vertices $(t_0, \ldots, t_{n-1})$ with labels

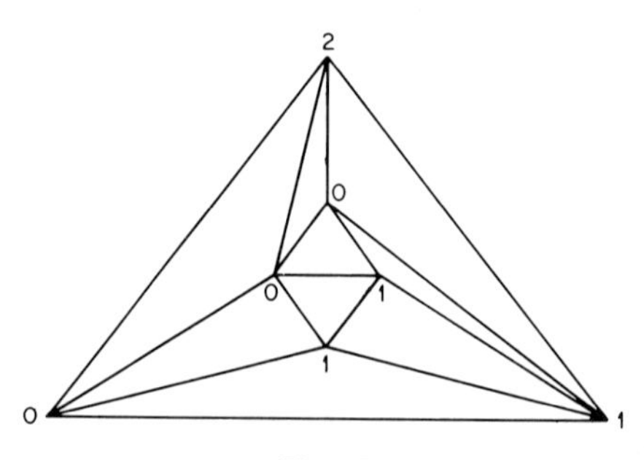

Fig. 5

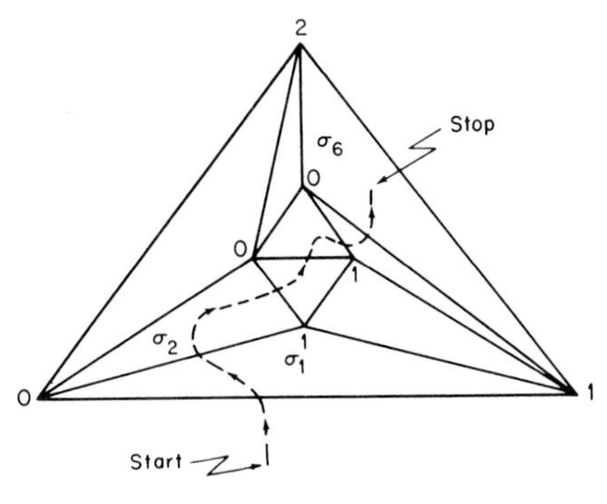

Fig. 6

$(0, \ldots, n-1)$. If the remaining vertex t_n has label n, then one has a completely labeled simplex. But, if the remaining labels is i for some $0 \leqq i \leqq n-1$, then there is exactly one other face of σ, namely, that spanned by $\{t_0, \ldots, t_n\} \sim \{t_i\}$, that has labels $\{0, \ldots, n-1\}$. Thus, if a room has at least one open door, then either it is the target (i.e., a completely labeled simplex) or it has exactly two open doors. Since only one door passes from outside the simplex S to the inside and since there are only finitely many simplexes, the procedure must terminate with a completely labeled simplex.

Our aim in the above exercise was principally to exhibit the convergence proof in action. Next we exhibit use of the "homotopy principle" and show that it yields the algorithm just given, "Cohen's algorithm," for solving $f(x) = y$.

Before applying the "homotopy principle" to solve $f(x) = y$ let us recall an elementary fact from linear algebra. Let $L: R^{n+1} \to R^n$ be a linear map of rank n from $(n+1)$-space to n-space. If y is any point in the range R^n, then $L^{-1}(y)$ is a line. Taking matters one step further, let σ be an $(n+1)$-cell, that is, a closed polyhedral convex set of dimension $n+1$. Let $L: \sigma \to R^n$ be a linear map where $L(\sigma)$ has dimension n. Then for most values of y in R^n either $L^{-1}(y)$ is empty or is a chord of σ whose endpoints lie interior to n-faces of σ (Fig. 7). There are a few values of y where $L^{-1}(y)$ meets an $(n-1)$-face of σ; these y's we call degenerate (or critical) but for convenience

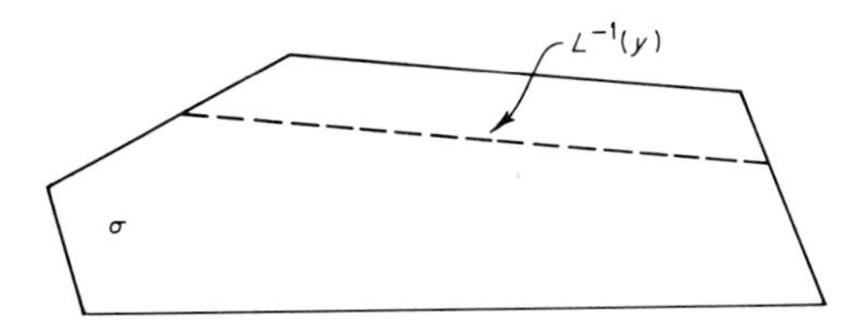

Fig. 7

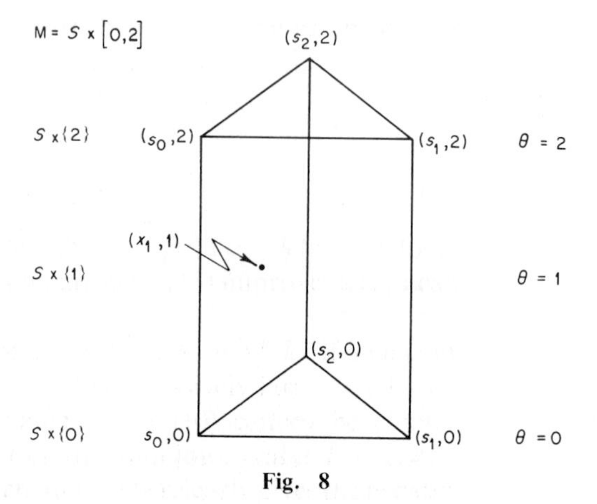

Fig. 8

we shall always assume that our y is regular, that is, not degenerate. There are measures for dealing with degenerate y's but the treatment for them will not be discussed in this paper (see Eaves [6]).

Given the system $f(x) = y$ we introduce a family of problems $F(x, 0) = y$. Let x_1 be any point in the interior of the face of S spanned by s_0 and s_1 (for $n > 2$ by $s_0, s_1, \ldots, s_{n-1}$) and define the homotopy, that is, function F by

$$F(x, \theta) = f(x) + (y - x_1)\theta$$

for $0 \leqq \theta \leqq 2$. So F carries points of the cylinder $M = S \times [0, 2]$ into R^n (Fig. 8).

We can subdivide the cylinder $M = S \times [0, 2]$ by letting $\mathscr{M}$ be the collection of cells of form $\sigma \times [0, 2]$ where σ is a cell of $\mathscr{S}$. Now observe that F is piecewise linear with respect to $\mathscr{M}$; that is, F is affine on each cell of $\mathscr{M}$.

At $\theta = 0$ the system $F(x, \theta) = y$ is the system $f(x) = y$. For $\theta > 0$ the system of particular interest is

$$(*) \qquad \begin{aligned} F(x, \theta) &= y, \\ (x, \theta) &\in \partial M, \\ \theta &> 0. \end{aligned}$$

The second condition requires that (x, θ) is in the boundary of M, that is, in the top, bottom, or some side of M. We argue that the system $(*)$ has exactly one solution, namely, $(x_1, 1)$.

If $\theta > 1$ then $y - \theta(y - x_1)$ is not in S and clearly $f(x) = y - \theta(y - x_1)$ can have no solution. If $0 < \theta < 1$ then $y - \theta(y - x_1)$ is interior to S and $f(x) = y - \theta(y - x_1)$ with $x \in \partial S$ can have no solution, since f is the identity on the boundary. If $\theta = 1$ our system becomes $f(x) = x_1$ with $x \in \partial M$,

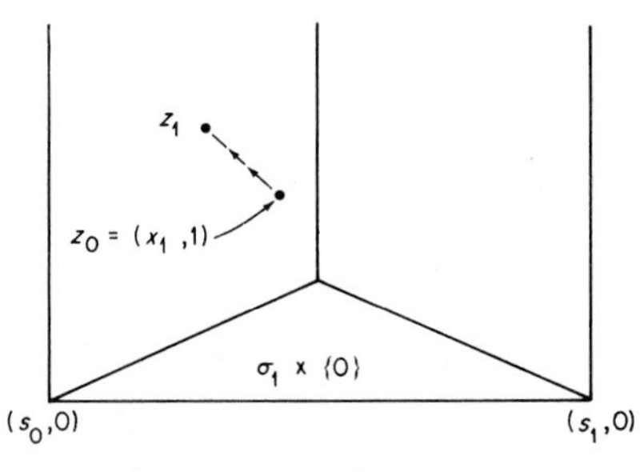

Fig. 9

and again since f is the identity on the boundary, clearly $x = x_1$ is the only solution.

So there is one solution $(x_1, 1)$ away from the bottom of M and we seek a solution on the bottom of M; this is precisely our desired situation. Assume that y is a regular value, that is to say, let us assume that $F^{-1}(y)$ does not meet any $(n - 1)$-faces of elements of $\mathcal{M}$. Given $(x_1, 1)$ and the cell $\sigma_1 \times [0, 2]$ of $\mathcal{M}$ which contains it we have a linear map from the $(n + 1)$-cell $\sigma_1 \times [0, 2]$ to R^n, so we apply our result from linear algebra. $F^{-1}(y) \cap (\sigma_1 \times [0, 2])$ is a chord of $\sigma_1 \times [0, 2]$.

Let $z_0 = (x_1, 1)$ be one end of the chord and z_1 the other (Fig. 9); calculating z_1 is just a matter of solving a linear system of equations. Next we go to the $(n + 1)$-cell $\sigma_2 \times [0, 2]$ of $\mathcal{M}$ that contains z_1 but not z_0 and repeat the procedure to get a chord $F^{-1}(y) \cap (\sigma_2 \times [0, 2])$ of $\sigma_2 \times [0, 2]$, etc. (Fig. 10). In this manner we continue to follow the route of $F^{-1}(y)$ beginning with $(x_1, 1)$. Observe that this route can have no forks. Eventually the route yields a point z_k in $S \times \{0\}$ and the system $f(x) = y$ is solved (Fig. 11).

What is the relation of the route of $F^{-1}(y)$ beginning at $(x_1, 1)$ and "Cohen's algorithm" applied to the problem? Well, they are in essence identical once the smoke has cleared. If the route of $F^{-1}(y)$ beginning with $(x_1, 1)$ is projected down to the base $S \times \{0\}$ of M we see that it passes through the same sequence $\sigma_1, \sigma_2, \ldots$ of rooms as "Cohen's algorithm"! In

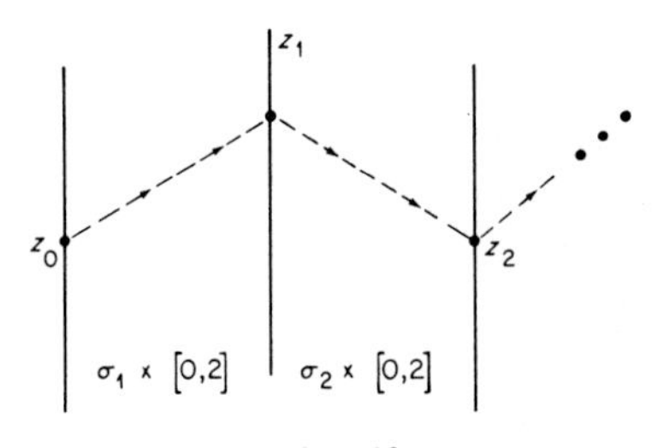

Fig. 10

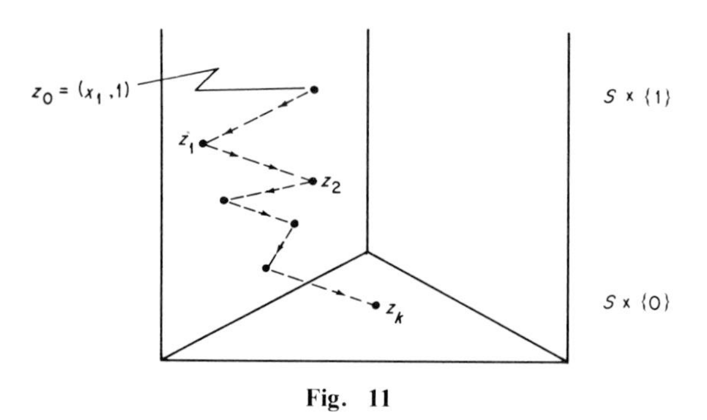

Fig. 11

this sense we regard "Cohen's algorithm" for $f(x) = y$ as that yielded by the "homotopy principle."

6. GENERAL THEORY

The general theory is to be sketched here. As the example of the previous section contains most of the ideas involved in the general theory, the conceptual step to the material presented here is small.

Cells are our building blocks. We define an m-cell to be a closed polyhedral convex set of dimension m.

Let $\mathscr{M}$ be a collection of m-cells and let M be the union of these cells. $(M, \mathscr{M})$ is defined to be a subdivided m-manifold if the following three conditions hold.

(a) Given any two cells of $\mathscr{M}$ either they do not meet or they meet in a common face.

(b) Any $(m - 1)$-face of a cell of $\mathscr{M}$ lies in at most two cells of $\mathscr{M}$.

(c) Given any point of M there is a neighborhood which meets only finitely many cells of $\mathscr{M}$.

Condition (a) prohibits construction such as

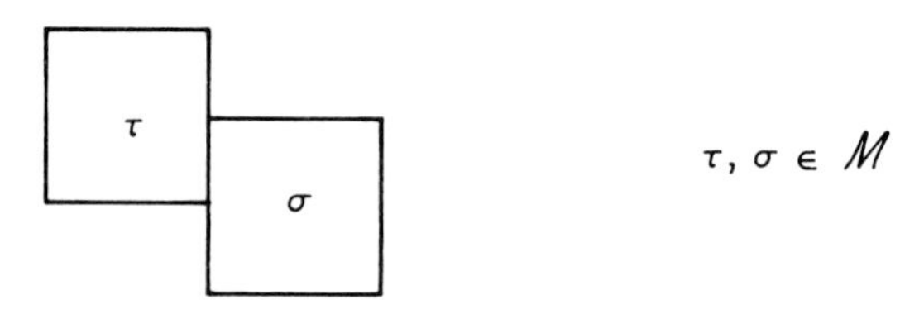

and requires constructions such as

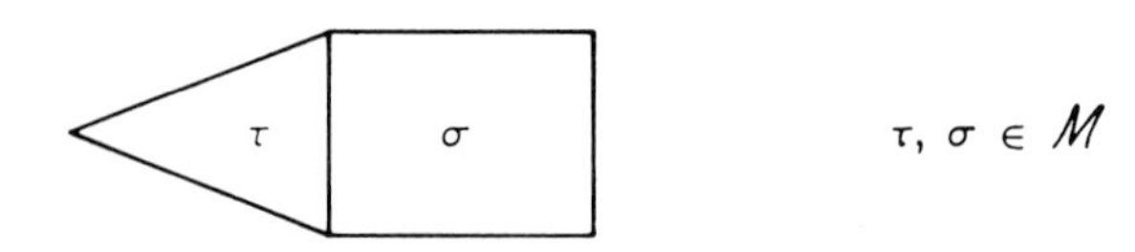

or

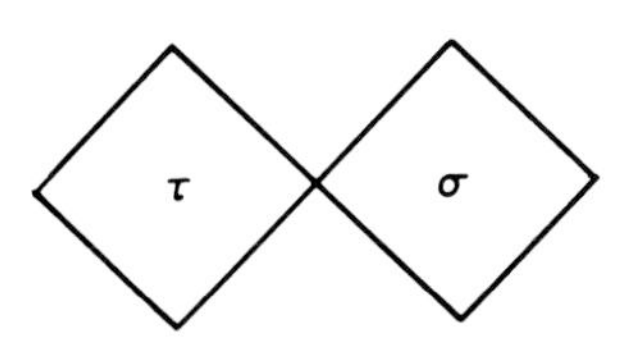

Condition (b) prohibits construction such as

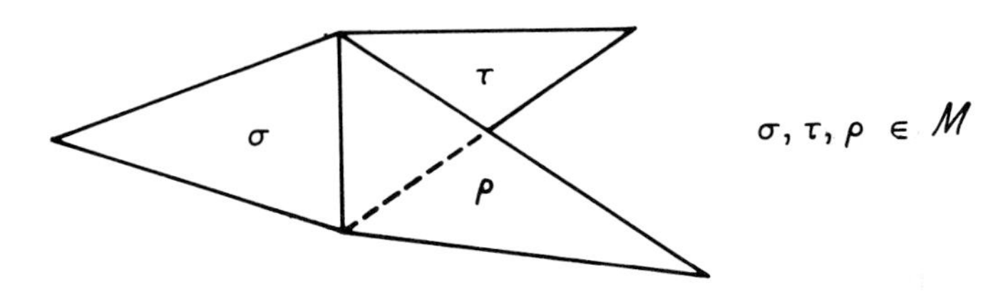

As an example of a 1-manifold we have

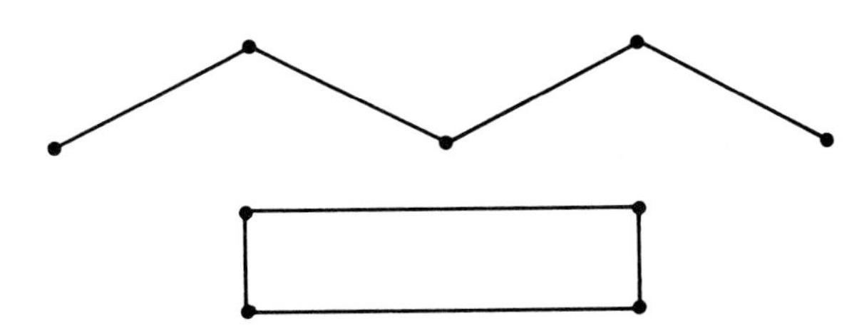

where $\mathscr{M}$ contains eight 1-cells. A 1-manifold is a disjoint collection of routes and loops; the previous example contains one route and one loop. Note that a route or loop of a 1-manifold contains no forks; a proof of this point requires essentially the "ghost argument" of Section 3. As examples of a subdivided 2-manifold we have $(S, \mathscr{S})$ of the previous section and of the surface of a cube where $\mathscr{M}$ is the set containing the top, bottom, and sides. We call M an m-manifold if for some $\mathscr{M}$, $(M, \mathscr{M})$ is a subdivided m-manifold.

By the boundary ∂M of a subdivided m-manifold we mean the union of all $(m - 1)$-faces of cells of $\mathscr{M}$ that lie in exactly one m-cell. Thus, for example, the boundary of the surface of a cube is empty. For further examples consider Fig. 12.

Let N be a 1-manifold and $(M, \mathscr{M})$ a subdivided m-manifold, where N is contained in M. We say that N is neat in $(M, \mathscr{M})$ if the following three conditions hold.

(a) N is closed in M.
(b) The boundary of N lies in the boundary of M.
(c) The collection of nonempty sets of form $N \cap \sigma$ with σ in $\mathscr{M}$ forms a subdivision of N.

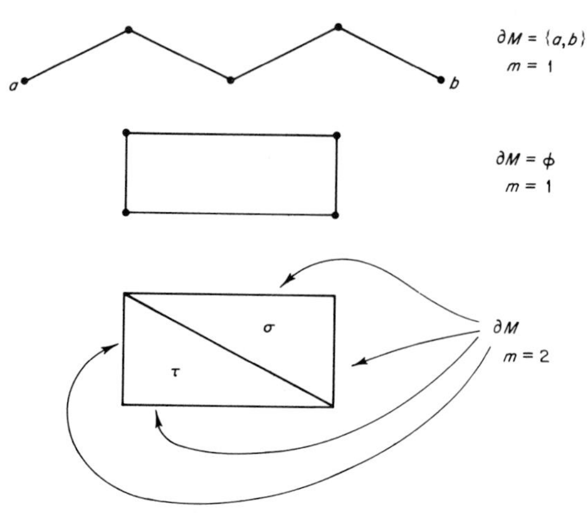

Fig. 12

In Fig. 13 we show a 1-manifold (which is a route) N neat in a subdivided 2-manifold M.

Now we can state the main theorem for regular values y.

Main Theorem. (*Regular Values*). *Let* $(M, \mathscr{M})$ *be an* $(n+1)$*-manifold and* $F: M \to R^n$ *be* $\mathscr{M}$ *piecewise linear. If* y *in* $F(M)$ *is regular, then* $F^{-1}(y)$ *is a* 1*-manifold neat in* $(M, \mathscr{M})$.

Recall that y is a regular value if $F^{-1}(y)$ does not meet any $(n-1)$-faces of cells of $(M, \mathscr{M})$. Note that for y regular, a loop of $F^{-1}(y)$ cannot meet the boundary of M.

For purposes of illustration of the theorem, consider the subdivided 2-manifold $(M, \mathscr{M})$ in R^2 shown in Fig. 14. Define $F = F_1F_2F_3F_4$ where the F_i's are defined in Fig. 15 together with the requirement that the F_i's are piecewise linear with respect to the indicated subdivisions. Let v represent any vertex. F_1 collapses the region A, F_2 flips C over to B, F_3 collapses the

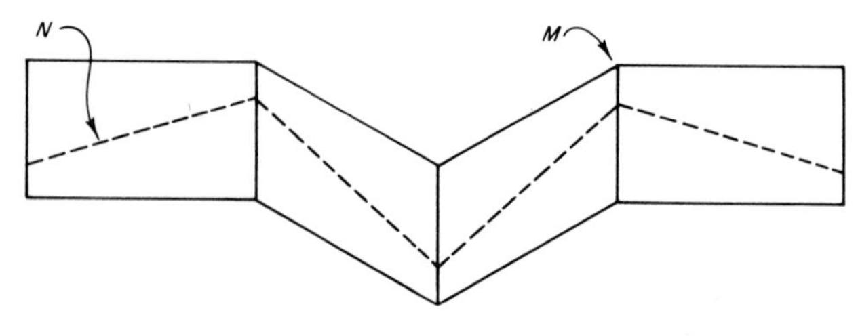

Fig. 13

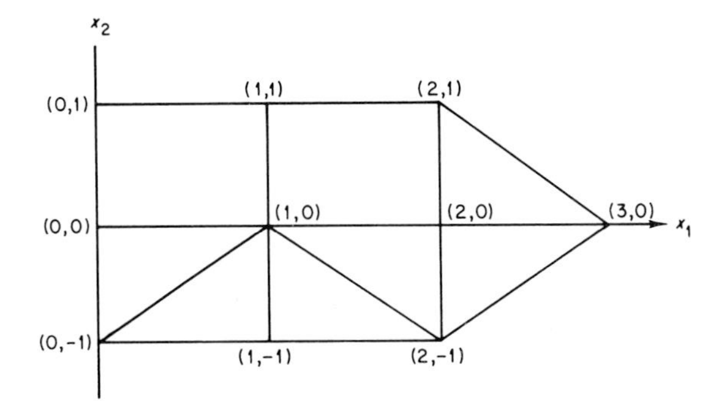

x_2
(0,1)
(1,1)
(2,1)
(0,0)
(1,0)
(2,0)
(3,0)
x_1
(0,-1)
(1,-1)
(2,-1)

Fig. 14

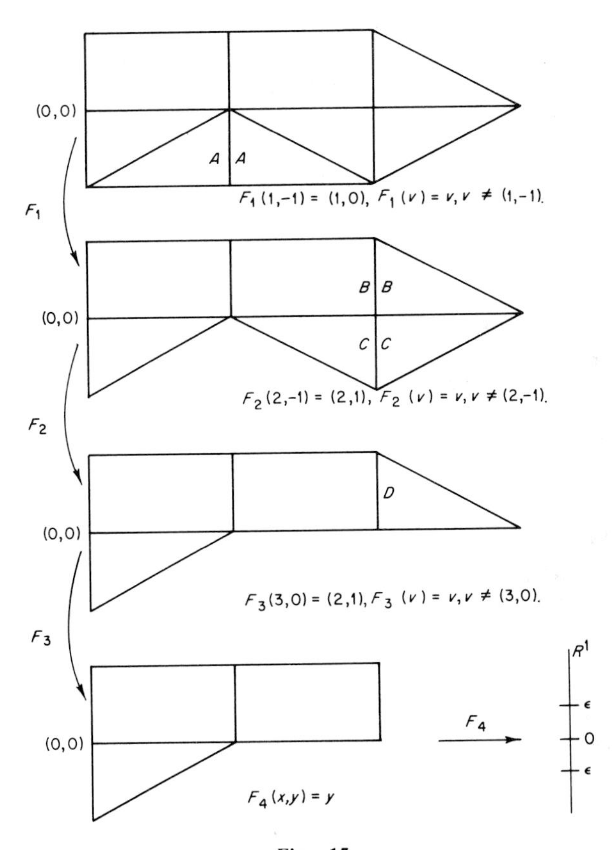

(0,0)
A
A
$F_1(1,-1) = (1,0),\ F_1(v) = v, v \neq (1,-1).$
F_1
(0,0)
B
B
C
C
$F_2(2,-1) = (2,1),\ F_2(v) = v, v \neq (2,-1).$
F_2
(0,0)
D
$F_3(3,0) = (2,1), F_3(v) = v, v \neq (3,0).$
F_3
R^1
ϵ
F_4
0
ϵ
(0,0)
$F_4(x,y) = y$

Fig. 15

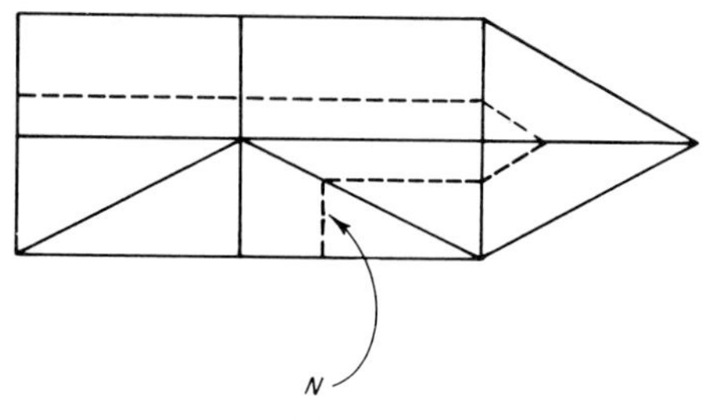

Fig. 16

region D, and F_4 projects to the vertical axis. So F_i^{-1} of any set can be discerned by inspection.

By calculating $F_4^{-1}(\varepsilon)$, $F_3^{-1}[F_4^{-1}(\varepsilon)]$, $F_2^{-1}[F_3^{-1}(F_4^{-1}(\varepsilon))]$, and finally $F^{-1}(\varepsilon) = F_1^{-1}[F_2^{-1}(F_3^{-1}(F_2^{-1}(\varepsilon)))]$ we can see that $F^{-1}(\varepsilon)$ is the 1-manifold (that is, route) N which is neat in $(M, \mathscr{M})$ (Fig. 16). $F^{-1}(y)$ will be a neat 1-manifold in $(M, \mathscr{M})$ for all y in $(-1, 0) \cup (0, 1)$. The values -1, 0, and 1 are degenerate and the reader might want to investigate F^{-1} of one or all of them, especially $F^{-1}(0)$.

The "main theorem" justifies the notion of following a route in $F^{-1}(y)$ in the "homotopy principle." That is, assuming y is a regular value, then in Section 5 we were following a 1-manifold (route) neat in $(M, \mathscr{M})$ to solve $f(x) = y$.

7. LEMKE'S ALGORITHM

The linear complementarity problem, which was first stated by Cottle, is:

Given an $n \times n$ matrix A and n-vector q find a z and w such that

$$Az - Iw = q,$$

$$z \geqq 0, \qquad w \geqq 0, \qquad z \cdot w = 0.$$

To solve such problems Lemke introduces an n-vector $d > 0$ and considers the augmented system

$$Az - Iw + d\theta = q,$$

$$z \geqq 0, \qquad w \geqq 0, \qquad \theta \geqq 0, \qquad z \cdot w = 0.$$

Lemke's algorithm proceeds by generating a path of solutions to the augmented system. To explain his algorithm from the perspective of the general theory first define $f_i : R^1 \to R^n$ for $i = 1, 2, \ldots, n$ by

$$f_i(x_i) = \begin{cases} A_i x_i & \text{if } x_i \geqq 0 \\ I_i x_i & \text{if } x_i \leqq 0, \end{cases}$$

where A_i and I_i are the ith columns of A and I, the identity, respectively. Define $f: R^n \to R^n$ by

$$f(x) = \sum_1^n f_i(x_i),$$

where

$$x = (x_1, \ldots, x_n).$$

The complementary problem is equivalent to $f(x) = q$ and the augmented system is equivalent to $f(x) + \theta d = q$ with $\theta \geqq 0$. So, define $F(x, \theta) = f(x) + \theta d$ as the homotopy; note that F is piecewise linear with respect to the orthants of $R^n \times R^1_+$. To solve $f(x) = q$, Lemke's algorithm follows the path of $F^{-1}(q)$ beginning with θ large and $x \leqq 0$. Observe that $F^{-1}(q)$, if q is regular, is a 1-manifold neat in $R^n \times R^1_+$ which is subdivided by its orthants. To show that the route of the algorithm yields a solution to $f(x) = q$ requires conditions on the matrix A, an issue that will not be treated here.

8. EAVES' ALGORITHM

Let $g: S \to \mathring{S}$ be a continuous function from a simplex S in R^n to its interior $\mathring{S}$. Let us compute a fixed point $x = g(x)$ of g, or equivalently, a zero of $f(x) \triangleq g(x) - x$. The first step is to subdivide the cylinder $M = S \times (0, 1]$ with $\mathscr{M}$ as indicated in Fig. 17. $S \times \{1\}$ should not be subdivided and the size of the simplexes of the subdivision should tend to zero uniformly as the simplexes near $S \times \{0\}$. We pause to note here that in a computer the subdivision exists only in the sense that there is a formula that can be used to generate portions of the subdivision as needed.

Let $F(x, \theta) = f(x)$ for all vertices (x, θ) of the subdivision. Define $F: M \to R^n$ by extending F to all of M in such a way that F is affine on the cells of $\mathscr{M}$; this extension is unique. Once again, in a computer F is only generated as needed. We now have the property that $F(\cdot, 1)$ is linear and

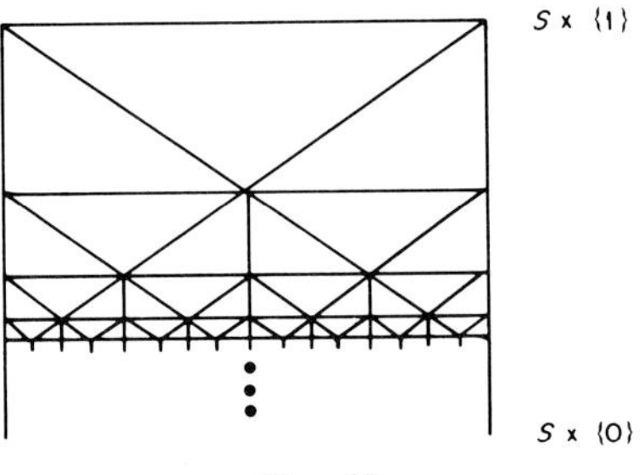

Fig. 17

$F(\cdot, t)$ tends to f as t tends to 0. $F(x, 1) = 0$ is a linear system and is easily shown to have a unique solution $(x_1, 1)$.

Beginning with the point $(x_1, 1)$ the route of $F^{-1}(0)$ is followed; $F^{-1}(0)$ is a neat 1-manifold in $(M, \mathscr{M})$ if 0 is regular. The x-component of the route tends (as the subdivision gets finer and finer) to a solution (or solutions) of $f(x) = 0$, that is, a fixed point (or fixed points) of g.

9. INTERNAL DEVELOPMENTS

The principal weakness of complementary pivot theory is simply that there are too many cells to traverse along the route of $F^{-1}(y)$. Many studies have improved the situation; let us mention those that seem to be the most important. The "restart" method of Merrill [22] permits, in effect, many cells to be skipped. In the presence of differentiability Saigal [25] has shown that Merrill's method can be used to obtain quadratic convergence; in addition, it becomes clear from his analysis that one should formulate the system of equations to be solved so that they are as smooth as possible; the routes to be followed are then less inclined to turn radically. In the absence of special structure Todd has shown that the simplexes (and cells) should be as round as possible, that is, not long slivers.

Kojima [17], and recently Todd [31], have shown how to use special structure of the function as linearity and separability to drastically reduce the number of cells. Van der Laan and Talman [33] have revived an idea of Shapley [30] which, in effect, enables one to move through some fraction of the cells more quickly; for lack of a better term this technique is often referred to a "variable dimension method." Garcia and Gould [10] have proposed an idea with similar affect. The works of Kellogg *et al.* [16] and Hirsch and Smale [13] avoid cells altogether by using differential homotopies (see Li [21]), but, then, following a differential path is more difficult than following a piecewise linear route; the trade-off is not yet understood.

REFERENCES

1. E. Allgower and M. Jeppson, The approximation of solution of nonlinear elliptic boundary value problems having several solutions, *in* "Numerische, Insbesondere Approximationstheoretische Behandlung von Funktionslgleichungen" (R. Ansorge and W. Törnig, eds.), Lecture Notes in Mathematics, Springer-Verlag, Berlin and New York, (to appear).
2. E. Allgower and M. Jeppson, Numerical solution of nonlinear boundary value problems with several solutions (to appear).
3. D. I. A. Cohen, On the Sperner lemma, *J. Combin. Theory* **2** (1967), 585–587.
4. R. W. Cottle, Complementarity and Variational Problems, TR SOL 74–6, Dept. Oper. Res., Stanford Univ., Stanford, California, 1974.
5. B. C. Eaves, Homotopies for computation of fixed points, *Math. Programming* **3** (1972), 1–22.

6. B. C. EAVES, A short course in solving equations with PL homotopies, *SIAM–AMS Proc.* **9** (1976), 73–143.
7. B. C. EAVES AND R. SAIGAL, Homotopies for computation of fixed points, *Math. Programming* **3** (1972), 225–237.
8. B. C. EAVES AND H. SCARF, The solution of systems of piecewise linear equations, *Math. Oper. Res.* **1** (1976), 1–27.
9. J. FREIDENFELDS, A set of intersection theorems and applications, *Math. Programming* **7** (1974), 199–211.
10. C. B. GARCIA AND F. J. GOULD, An Improved scalar Generated Homotopy Path for Solving $f(x) = 0$, Rep. No. 7633, Center Math. Stud. Business Econom., Univ. of Chicago, Chicago, Illinois, 1976.
11. C. B. GARCIA, A fixed point theorem including the last theorem of Poincaré, *Math. Programming* **8** (1975), 227–239.
12. M. W. HIRSCH, A proof of the nonretractability of a cell onto its boundary, *Proc. Amer. Math. Soc.* **14** (1963), 364–365.
13. M. W. HIRSCH AND S. SMALE, On Algorithms for Solving $f(x) = 0$, Dept. Math., Univ. of California, Berkeley.
14. I. KANEKO, A Mathematical Programming Method for the Inelastic Analysis of Reinforced Concrete Frames, TR 76–2, Dept. Indust. Engr., Univ. of Wisconsin, Madison, 1976.
15. J. KATZENELSON, An algorithm for solving nonlinear resistor networks, *Bell System Tech. J.* **44** (1965), 1605–1620.
16. R. B. KELLOGG, T.-Y. LI, AND J. YORKE, A Constructive Proof of the Brouwer Fixed Point Theorem and Computational Results, Univ. of Maryland and Univ. of Utah, 1975 (unpublished).
17. M. KOJIMA, On the Homotopic Approach to Systems of Equations with Separable Mappings, B-26, Dept. Inform. Sci., Tokyo Inst. of Tech., Tokyo, 1975.
18. H. W. KUHN, A new proof of the fundamental theorem of algebra, *Math. Programming*, **1** (1974), 148–158.
19. C. E. LEMKE, Bimatrix equilibrium points and mathematical programming, *Management Sci.* **11** (1965), 681–689.
20. C. E. LEMKE AND J. T. HOWSON, JR., Equilibrium points of bimatrix games, *SIAM J. Appl. Math.* **12** (1964), 413–423.
21. T.-Y. LI, Path Following Approaches for Solving Nonlinear Equations: Homotopy, Continuous Newton and Projection, Dept. Math., Michigan State Univ., East Lansing.
22. O. H. MERRILL, Applications and Extensions of an Algorithm that Computes Fixed Points of Certain Upper Semi-Continuous Point to Set Mappings, Ph.D. Thesis. Indust. Engr., Univ. of Michigan, Ann Arbor, 1972.
23. M. D. MEYERSON AND O. H. WRIGHT, A New and Constructive Proof of the Borsak–Ulam Theorem, *Proc. Amer. Math. Soc.* (to appear).
24. A. N. NETRAVALI AND R. SAIGAL, Optimum quantizer design using a fixed-point algorithm, *Bell System Tech. J.* **55**, (1976), 1423–1435.
25. R. SAIGAL, On the Convergence Rate of Algorithms for Solving Equations that are Based on Complementarity Pivoting, Bell Telephone Lab., Holmdel, New Jersey.
26. H. SCARF, The approximation of fixed points of a continuous mapping, *SIAM J. Appl. Math.* **15** (1967).
27. H. SCARF, The core of an N person game, *Econometrica* **35** (1967), 50–69.
28. H. SCARF, On the computation of equilibrium prices, *in* "Ten Economic Studies in the Tradition of Irving Fisher," Wiley, New York, 1967.
29. H. SCARF AND T. HANSEN, "Computation of Economic Equilibria," Yale Univ. Press, New Haven, Connecticut, 1973.
30. L. S. SHAPLEY, On balanced games without side payments, *in* "Mathematical Programming" (T. C. Hu and S. M. Robinson, eds.), Academic Press, New York, 1973, pp. 261–290.

31. M. J. TODD, "The Computation of Fixed Points and Applications," Springer-Verlag, Berlin and New York, 1976.
32. M. J. TODD, Exploiting Structure in Fixed Point Computation, Discussion Paper, Math. Res. Center, Univ. of Wisconsin, Madison.
33. G. VAN DER LAAN AND A. J. J. TALMAN, A New Algorithm for Computing Fixed Points, Free Univ., Amsterdam, 1978.
34. R. J. WILMUTH, The Computations of Fixed Points, Ph.D. Thesis, Dept. Oper. Res., Stanford Univ., Stanford, California, 1973.

DEPARTMENT OF OPERATIONS RESEARCH
STANFORD UNIVERSITY
STANFORD, CALIFORNIA

Selected Applications of Semiinfinite Programming

P. R. GRIBIK

Semiinfinite programming is one of the many subjects that have been advanced by Professor Duffin's research. In this paper, a few applications of semiinfinite programming are examined. The problem of meeting air quality standards throughout a region at least cost can be formulated as a semiinfinite program. The economic implications of such a model are discussed. The uses of semiinfinite programming in approximation is illustrated by an example problem. The geometric programming problem can be formulated as a semiinfinite program. This formulation has uses in the classification of geometric programs and in the development of solution techniques. Finally, the use of semiinfinite programming in developing solution techniques for optimal experimental design problems is examined.

1. INTRODUCTION

Semiinfinite programming has various uses in modeling and numerical analysis. Some of these uses will be illustrated by example problems. The first problem we will examine is an air pollution control model, and certain of its economic implications will be considered. Semiinfinite programming also has uses in approximation. These are exemplified by a problem arising from the design of a digital filter. Finally, we will show how it can be used to develop and improve solution techniques for geometric programming and regression experimental design. Before proceeding with these applications, we will define the classes of semiinfinite programs we will consider, and review some theory.

The most general problem we will consider is the *convex semiinfinite program.*

Program I

Find $V_{\mathrm{I}} = \inf G(y)$
for all $y \in R^n$ *subject to*

$$\sum_{r=1}^{n} y_r u_r(t) \geqq u_{n+1}(t) \qquad \text{for} \quad t \in S.$$

ISBN 0-12-178150-X

We will only consider Program I under the following assumptions:

(i) $G(\cdot)$ is a convex function on R^n that is continuously differentiable on its domain;

(ii) S is a compact subset of R^m;

(iii) $u_r(\cdot) \in C(S)$, the continuous functions on S, for $r = 1, 2, \ldots, n + 1$.

If $G(\cdot)$ is linear, we have an important special case, namely, the linear semiinfinite programming problem.

Program D

Find $V_D = \inf \sum_{r=1}^n y_r \mu_r$
for $y \in R^n$ *subject to*

$$\sum_{r=1}^{n} y_r u_r(t) \geqq u_{n+1}(t) \qquad \text{for} \quad t \in S.$$

We can write a dual program for Program D.

Program P

Find $V_P = \sup \sum_{t \in S} \lambda(t) u_{n+1}(t)$
subject to

$$\sum_{t \in S} \lambda(t) u_r(t) = \mu_r \qquad \text{for} \quad r = 1, 2, \ldots, n,$$

$$\lambda(t) \geqq 0 \qquad \text{for} \quad t \in S, \text{ and}$$

$$\lambda(\cdot) = 0 \qquad \text{for all but a finite number of } t \text{ in } S.$$

We will assume that (ii) and (iii) hold for Program D. If Krein's condition is also satisfied, i.e., there exist $c_1, \ldots, c_n$ which satisfy

$$\sum_{r=1}^{n} c_r u_r(t) > u_{n+1}(t) \qquad \text{for all} \quad t \in S,$$

the following properties can be shown ([17] or [21]).

Theorem 1. *If Program* D *satisfies* (ii) *and* (iii) *and Krein's condition, the following assertions hold*:

(1) $V_D = -\infty$ *if and only if Program* P *is inconsistent.*

(2) *If Program* P *is consistent, then* $V_P = V_D$.

(3) *If Program* P *is consistent, then its value is assumed for a* $\lambda(\cdot)$ *such that the set*

$$\Omega(\lambda) \equiv \{x \in S \mid \lambda(x) \neq 0\}$$

contains at most n *points.*

(4) *Let* y^* *be an optimal solution for Program* D *and let* $\lambda^*(\cdot)$ *be an optimal solution for Program* P. *Then the following* complementary slackness *conditions hold*:

$$\sum_{r=1}^{n} y_r^* u_r(t) = u_{n+1}(t) \qquad \text{for} \quad t \in \Omega(\lambda^*).$$

The references contain works dealing with the theory of semiinfinite programming [5, 17, 21, 24] and also with solution techniques [8, 14, 18, 20]. We will now proceed with illustrations of the uses of semiinfinite programming.

2. LEAST-COST STRATEGIES FOR AIR POLLUTION ABATEMENT

The first example we will consider is one in which the infinite system of linear inequalities arises naturally. Suppose that there are n sources of a chemically inert pollutant in a region, and that we wish to control the emissions so that the annual mean ground level pollution concentration at each point in the region satisfies some standard. Furthermore, suppose that we wish to find a control policy whose total cost is minimal. This problem has been studied by Gorr *et al.* [13] and Gustafson and Kortanek [19].

Using diffusion modeling and a source inventory, we can compute the following functions:

$u_r(x)$ = annual mean pollutant concentration at point x due to source r before control,
$u_0(x)$ = annual mean pollutant concentration at point x due to all sources not under the regulator's control.

We will also need the following definitions and data:

S = control region,
e_r = fraction by which source r is to reduce its output of pollutant (control variable),
$G_r(e_r)$ = cost of reducing by e_r at source r,
$\Psi(x)$ = maximum annual mean concentration of pollutant to be permitted at point x,
E_r = maximum amount we will require source r to reduce, $E_r \in (0, 1)$.

Since we assumed that the pollutant is chemically inert, we can use superposition to find the pollutant concentration at each point in x *after* reduction by $e_1, \ldots, e_n$, namely,

$$\sum_{r=1}^{n} (1 - e_r) u_r(x) + u_0(x).$$

If we define $\varphi(x) \equiv \sum_{r=0}^{n} u_r(x) - \Psi(x)$, the excess pollution at x before reduction, a least cost strategy would be a solution to the following program.

Program I*

Find $V_{I^*} = \inf \sum_{r=1}^{n} G_r(e_r)$
subject to

$$\sum_{r=1}^{n} e_r u_r(x) \geqq \varphi(x) \qquad \textit{for all} \quad x \textit{ in } S,$$

$$0 \leqq e_r \leqq E_r \qquad \textit{for} \quad r = 1, 2, \ldots, n.$$

In the following, we will assume that Program I* satisfies the following assumptions.

(i) $G_r(\cdot)$ is a continuously differentiable closed convex function for $r = 1, \ldots, n$.
(ii) $u_r(\cdot) \in C(S)$ for $r = 1, \ldots, n$, and $\varphi(\cdot) \in C(S)$.
(iii) S is compact.
(iv) There exist $\hat{e}_1, \ldots, \hat{e}_n$ such that

$$\sum_{r=1}^{n} \hat{e}_r u_r(x) > \varphi(x) \qquad \text{for all} \quad x \text{ in } S,$$

$$0 < \hat{e}_r < E_r \qquad \text{for} \quad r = 1, 2, \ldots, n.$$

Under these assumptions, Program I* is a special case of Program I. The properties of this program have been examined in the references given at the beginning of this section. We will be concerned with the information that can be obtained from a dual to Program I*. The case in which $G_r(\cdot)$ is linear for all r was studied by Samuelsson in [25].

Under assumptions (i)–(iv) of this section, Program I* will have optimal solutions; let $e_1^*, \ldots, e_n^*$ be one such solution. Also, one can show that the following program is dual to Program I*.

Program II

Find

$$V_{II} = \sup \sum_{x \in S} \lambda(x)\varphi(x) - \sum_{r=1}^{n} \lambda_r E_r + \left[\sum_{r=1}^{n} G_r(e_r^*) - \sum_{r=1}^{n} e_r^* \frac{dG_r(e_r^*)}{de_r}\right]$$

subject to

$$\sum_{x \in S} \lambda(x)u_r(x) - \lambda_r \leqq \frac{dG_r(e_r^*)}{de_r} \qquad \textit{for} \quad r = 1, \ldots, n,$$

$$\lambda(x) \geqq 0 \qquad \textit{for} \quad x \in S,$$

$$\lambda(x) = 0 \qquad \textit{for all but a finite number of } x \textit{ in } S,$$

$$\lambda_r \geqq 0 \qquad \textit{for} \quad r = 1, \ldots, n.$$

One can show that $V_{I^*} = V_{II}$ and that there is a $[\lambda^*(\cdot), \lambda_1^*, \ldots, \lambda_n^*]$ which is optimal for Program II. Furthermore, the following complementary slackness conditions hold:

$$\sum_{r=1}^{n} e_r^* u_r(x) = \varphi(x) \qquad \textit{for} \quad x \in \Omega(\lambda^*),$$

$$e_r^* = E_r \qquad \textit{if} \quad \lambda_r^* \neq 0,$$

$$\sum_{x \in S} \lambda^*(x)u_r(x) - \lambda_r^* = \frac{dG_r(e_r^*)}{de_r} \qquad \textit{if} \quad e_r^* \neq 0.$$

We will now proceed to develop an economic interpretation of the optimal solution of Program II. To this end, we will need the following perturbation of Program I*. Let $\Delta(\cdot) \in C(S)$ and $\delta \in R^n$.

Program I* (Δ; δ)

Find $V(\Delta; \delta) = \min \sum_{r=1}^{n} G_r(e_r)$
subject to

$$\sum_{r=1}^{n} e_r u_r(x) \geqq \varphi(x) + \Delta(x) \qquad \textit{for} \quad x \in S,$$

$$0 \leqq e_r \leqq E_r - \delta_r \qquad \textit{for} \quad r = 1, \ldots, n.$$

Obviously, Program I*(0; 0) is just Program I*.
The following lemma is easily proved.

Lemma 1. *Under assumptions* (i)–(iv) *of this section,* $V(\cdot\,;\cdot)$ *is a convex function on* $C(S) \times R^n$. *Also, there exists an* $\varepsilon > 0$ *such that if* $\max_{x \in S} |\Delta(x)| \leqq \varepsilon$ *and* $|\delta_r| \leqq \varepsilon$ *for* $r = 1, \ldots, n$ *then* $-\infty < V(\Delta; \delta) < \infty$.

Theorem 2. *An optimal solution of Program* II *is a subgradient of* $V(\Delta; \delta)$ *at* $\Delta = 0$, $\delta = 0$.

Proof. Let $[\lambda^*(\cdot), \lambda_1^*, \ldots, \lambda_n^*]$ be optimal for Program II, and let $e_1^*, \ldots, e_n^*$ be optimal for Program I*. For $\Delta \in C(S)$ and $\delta \in R^n$, define the linear program:

Program A(Δ; δ)

Find

$$VA(\Delta;\delta) = \min \sum_{r=1}^{n} G_r(e_r^*) + \sum_{r=1}^{n} y_r \frac{dG_r(e_r^*)}{de_r}$$

for $y \in R^n$ *subject to*

$$\sum_{r=1}^{n} y_r u_r(x) \geqq \Delta(x) \qquad \textit{for} \quad x \in \Omega(\lambda^*),$$

$$-y_r \geqq \delta_r \qquad \textit{if} \quad e_r^* = E_r,$$

$$y_r \geqq 0 \qquad \textit{if} \quad e_r^* = 0.$$

Its dual is

Program B(Δ; δ)

Find

$$VB(\Delta;\delta) = \max \sum_{r=1}^{n} G_r(e_r^*) + \sum_{\Omega(\lambda^*)} \gamma(x)\,\Delta(x) + \sum_{\{r \mid e_r^* = E_r\}} \gamma_r \delta_r$$

subject to

$$\sum_{\Omega(\lambda^*)} \gamma(x) u_r(x) = \frac{dG_r(e_r^*)}{de_r} \qquad \text{if} \quad 0 < e_r^* < E_r,$$

$$\sum_{\Omega(\lambda^*)} \gamma(x) u_r(x) - \gamma_r = \frac{dG_r(e_r^*)}{de_r} \qquad \text{if} \quad e_r^* = E_r,$$

$$\sum_{\Omega(\lambda^*)} \gamma(x) u_r(x) \leqq \frac{dG_r(e_r^*)}{de_r} \qquad \text{if} \quad e_r^* = 0.$$

Defining the value of an inconsistent program as $+\infty$, using the convexity of $G_r(\cdot)$ and the complementary slackness conditions, we have

$$V(\Delta;\delta) \geqq VA(\Delta;\delta).$$

Using linear programming duality

$$VA(\Delta;\delta) = VB(\Delta;\delta).$$

Using the complementary slackness conditions again, we have

$$VB(\Delta;\delta) \geqq \sum_{r=1}^{n} G_r(e_r^*) + \sum_{x \in S} \lambda^*(x)\Delta(x) + \sum_{r=1}^{n} \lambda_r^* \delta_r.$$

Using $V(0;0) = V_{I*}$ and since Δ was arbitrary in $C(S)$ and δ was arbitrary in R^n,

$$V(\Delta;\delta) \geqq V(0;0) + \sum_{x \in S} \lambda^*(x)\Delta(x) + \sum_{r=1}^{n} \lambda_r^* \delta_r$$

for all $\Delta \in C(S)$, $\delta \in R^n$. ■

Because of Theorem 2, we can make the following economic interpretation of an optimal solution of Program II, especially if $V(\Delta;\delta)$ has only one subgradient at $\Delta = 0$, $\delta = 0$. $\lambda^*(x)$ can be viewed as the marginal value of a unit reduction in pollution at point x given that we have implemented the reductions $e_1^*, \ldots, e_n^*$. Thus the dual program can be used to help make explicit the implicit values that were used to determine the standard $\Psi(\cdot)$. This could be of help in determining realistic standards that balance the value of clean air against the economic impact on an area of requiring its industry to reduce pollution. The λ_r^*'s can be viewed as the marginal disutility of not permitting source r to reduce more than E_r. Thus if $\lambda_r^* > 0$, we are saying that considerations other than the simple economic criterion we are using were used to set a limit on the reduction by source r.

3. APPROXIMATION

Semiinfinite programming techniques can be useful in solving many L_1 and L_∞ approximation problems. As an example, we will consider the problem of designing a two-dimensional digital filter with 0 phase whose frequency response is a good approximation to

$$F(\mu, \nu) = \begin{cases} 1 & \text{for} \quad \mu^2 + \nu^2 \leqq 1, \\ 0 & \text{for} \quad 1 < \mu^2 + \nu^2 \leqq 4, \\ \text{any value} & \text{for} \quad 4 < \mu^2 + \nu^2. \end{cases}$$

For the digital filter, we will consider one whose impulse response is given by

$$\begin{array}{lll} h(n,m) & \text{for} & -N \leqq n \leqq N, \quad -N \leqq m \leqq N \ (N \text{ fixed}), \\ & & m \text{ and } n \text{ integers}; \\ 0 & & \text{for all other } m \text{ and } n. \end{array}$$

To get 0 phase we will require that $h(n, m) = h(-n, -m)$. For this impulse response, the frequency response is

$$H(\mu, \nu) = h(0,0) + 2\sum_{n=1}^{N} h(n,0)\cos(n\mu)$$
$$+ 2\sum_{m=1}^{N} \sum_{n=-N}^{N} h(n,m)\cos(n\mu + m\nu).$$

Suppose that we wish to choose $h(\cdot, \cdot)$ so that

$$\max_{0 \leq \mu^2 + \nu^2 \leq 4} |F(\mu, \nu) - H(\mu, \nu)|$$

is minimized. Since $H(\cdot, \cdot)$ is continuous while $F(\cdot, \cdot)$ is not, we should not solve this problem but rather one where we permit a transition region at the discontinuity of $F(\cdot, \cdot)$.

If we choose $\delta > 0$ and set

$$\{(\mu, \nu) | 1 < \mu^2 + \nu^2 < 1 + \delta\}$$

as the transition region, the approximation problem can be formulated as the following problem:

minimize γ
subject to

$$|H(\mu, \nu) - 1| \leqq \gamma \quad \textit{for} \quad (\mu, \nu) \in \{(\mu, \nu) | \mu^2 + \nu^2 \leqq 1\},$$

$$|H(\mu, \nu) - 0| \leqq \gamma \quad \textit{for} \quad (\mu, \nu) \in \{(\mu, \nu) | 1 + \delta \leqq \mu^2 + \nu^2 \leqq 4\}.$$

In order to prevent $H(\cdot, \cdot)$ from taking too large or too small a value in the transition region, we will add the following constraints to the above program.

$$H(\mu, \nu) - 1 \leqq \gamma \quad \textit{for} \quad (\mu, \nu) \in \{(\mu, \nu) | 1 < \mu^2 + \nu^2 < 1 + \delta\},$$

$$-H(\mu, \nu) \leqq \gamma \quad \textit{for} \quad (\mu, \nu) \in \{(\mu, \nu) | 1 < \mu^2 + \nu^2 < 1 + \delta\}.$$

With these additional constraints, the program can be written as follows:

minimize γ
subject to

$$H(\mu, \nu) + \gamma \geqq 1 \quad \textit{for} \quad (\mu, \nu) \in \{(\mu, \nu) | \mu^2 + \nu^2 \leqq 1\},$$

$$-H(\mu, \nu) + \gamma \geqq -1 \quad \textit{for} \quad (\mu, \nu) \in \{(\mu, \nu) | \mu^2 + \nu^2 \leqq 1 + \delta\},$$

$$H(\mu, \nu) + \gamma \geqq 0 \quad \textit{for} \quad (\mu, \nu) \in \{(\mu, \nu) | 1 \leqq \mu^2 + \nu^2 \leqq 4\},$$

$$-H(\mu, \nu) + \gamma \geqq 0 \quad \textit{for} \quad (\mu, \nu) \in \{(\mu, \nu) | 1 + \delta \leqq \mu^2 + \nu^2 \leqq 4\}.$$

Since $H(\cdot, \cdot)$ is linear in the variables, namely the $h(n, m)$, this program is a linear semiinfinite program.

Another interesting use of semiinfinite programming in approximation and control is given by Glashoff and Gustafson in [10]. In this paper, they consider the problem of heating the ends of a rod so that the temperature in the rod at time T is close in the L_∞ norm to some desired temperature function.

4. GEOMETRIC PROGRAMMING

Semiinfinite programming has also been of use in developing improved cutting plane algorithms for prototype geometric programming problems. These problems can always be written in the following form.

Program G

Find $V_G = \inf t_1$
for $t \in R^m$ *subject to*

$$g_k(t) \leqq 1 \quad \text{for} \quad k = 1, \ldots, p,$$
$$t_j > 0 \quad \text{for} \quad j = 1, \ldots, m,$$

where

$$g_k(t) = \sum_{i=m_k}^{n_k} c_i t_1^{a_{i1}} t_2^{a_{i2}} \cdots t_m^{a_{im}}$$

with a_{ij} *a real constant for* $i = 1, \ldots, n_p$, $j = 1, \ldots, m$, *and* c_i *a positive constant for* $i = 1, \ldots, n_p$ *and* $m_1 = 1, m_2 = n_1 + 1, \ldots, m_p = n_{p-1} + 1$. *To simplify notation, define*

$$A = \begin{bmatrix} 1 & 0 & \cdots & 0 \\ a_{11} & a_{12} & \cdots & a_{1m} \\ \vdots & \vdots & & \vdots \\ a_{n_p 1} & a_{n_p 2} & \cdots & a_{n_p m} \end{bmatrix},$$

$$J[k] = \{m_k, m_k + 1, \ldots, n_k\},$$

and

$$\ln c = (0, \ln c_1, \ldots, \ln c_{n_p})^T.$$

In the following, we will assume that Program G is canonical and that A is of rank m.

Using a logarithmic transformation of variables, $\ln t_j = z_j$, Program G can be transformed into a convex program.

Program C*

Find $V_{C*} = \inf x_0$
for $z \in R^m$, $x_0 \in R$, $x \in R^n$; *subject to*

$$-Az + (x_0, x^T)^T = \ln c,$$

$$\sum_{i \in J[k]} e^{x_i} \leqq 1 \quad \text{for} \quad k = 1, \ldots, p.$$

Since A is of rank m,

$$-Az + (x_0, x^T)^T = \ln c$$

can be written in an equivalent form,

$$Iz + Hx = g \quad \text{with} \quad I \in R^{m\times m},\ H \in R^{m\times n_p},\ g \in R^m,$$
$$x_0 - d^T x = h \quad \text{with} \quad d \in R^{n_p},\ h \in R,$$
$$Dx = f \quad \text{with} \quad D \in R^{(n_p - m)\times n_p},\ f \in R^{n_p - m}.$$

Thus Program C* can be written in the following equivalent form.

Program C

Find $V_C = \inf d^T x$
for $x \in R^{n_p}$ *subject to*

$$\sum_{i\in J[k]} e^{x_i} \leqq 1 \qquad \text{for} \quad k = 1, \ldots, p,$$
$$Dx = f,$$
$$x_i \leqq 0 \qquad \text{for} \quad i = 1, \ldots, n_p.$$

In [4], Charnes *et al.* develop the following linear semiinfinite program that is equivalent to Program C.

Program S

Find $\inf d^T x$
for $x \in R^{n_p}$ *subject to*

$$\sum_{i\in J[k]} x_i e^{\beta_i} \leqq \sum_{i\in J[k]} \beta_i e^{\beta_i} \qquad \text{for all } \beta_{m_k}, \ldots, \beta_{n_k} \text{ which satisfy}$$
$$\sum_{i\in J[k]} \beta_i = 1 \quad \text{for} \quad k = 1, \ldots, p;$$
$$Dx = f;$$
$$x_i \leqq 0 \qquad \text{for} \quad i = 1, \ldots, n_p.$$

In [11], Gochet and Smeers contrasted the use of Kelley's cutting plane algorithm [22] on Program C with the use of the alternating algorithm of Gustafson and Kortanek [20] on Program S. Let $\bar{x} \in R^{n_p}$ satisfy

$$D\bar{x} = f,$$
$$\bar{x}_i \leqq 0, \qquad i = 1, \ldots, n_p.$$

If $\bar{x}$ is infeasible for Program C, and therefore Program S, and if k^* is the most violated constraint in Program C, Kelley's algorithm will generate the cut

$$-\sum_{i\in J[k^*]} x_i e^{x_i} \geqq -\sum_{i\in J[k^*]} \bar{x}_i e^{\bar{x}_i} + \sum_{i\in J[k^*]} e^{\bar{x}_i} - 1.$$

Gochet and Smeers show that the alternating algorithm will generate the cut

$$-\sum_{i\in J[k^*]} x_i e^{\bar{x}_i} \geqq -\sum_{i\in J[k^*]} \bar{x}_i e^{\bar{x}_i} + \left(\sum_{i\in J[k^*]} e^{\bar{x}_i}\right)\left(\ln \sum_{i\in J[k^*]} e^{\bar{x}_i}\right).$$

These cuts are parallel; but since

$$u \ln u > u - 1 \qquad \text{for} \quad u > 1,$$

the latter cut is deeper. (Also see [12].)

In [16], we show that similar results can be obtained if the central cutting plane algorithm for convex programs developed by Elzinga and Moore [7] is applied to Program C while the algorithm in [14] is applied to Program S. While these theoretical results do not mean that the algorithms based on the semiinfinite program are better, computational experience with the 16 test problems of Beck and Ecker [3] does indicate that they are better.

In the next section, we will discuss a problem for which the standard cutting plane algorithms may fail.

5. OPTIMAL EXPERIMENTAL DESIGN

The experimental design problem which we shall consider begins with the following model. Let X be a given compact set in some vector space, and let $f(\cdot) = (f_1(\cdot), \ldots, f_n(\cdot))^T$ be a continuous mapping of X into R^n. We consider random variables of the form

$$g(x) = \sum_{r=1}^{n} c_r f_r(x) + \Psi(x),$$

where $\Psi(x)$ is a random variable with $E\{\Psi(x)\} = 0$ and $E\{\Psi(x)^2\} = 1$ and where $c = (c_1, \ldots, c_n)^T$ is a vector of unknown constants. The vector c is to be estimated by making stochastically independent measurements of $g(\cdot)$ at various points in X. The problem is to allocate one's measurement resources so that the estimate of c is good.

Let the measurements be made at the points $x_1, \ldots, x_m$ in X with n_i independent measurements made at point x_i. Denote the result of the jth measurement at point x_i by g_{ij}. Set

$$N = \sum_{i=1}^{m} n_i,$$

$$M = \sum_{i=1}^{m} \frac{n_i}{N} f(x_i) f(x_i)^T,$$

$$b = \sum_{i=1}^{m} n_i f(x_i) \bar{g}_i,$$

$$\bar{g}_i = \frac{1}{n_i}\left(\sum_{j=1}^{n_i} g_{ij}\right).$$

If M is nonsingular, the best linear unbiased estimator of c is $\hat{c} = (1/N)M^{-1}b$ and the covariance matrix of $\hat{c}$ is $(1/N)M^{-1}$.

If we have N measurements to allocate to points in X, we may wish to choose the allocation so that $(1/N)M^{-1}$ is small in terms of the positive semidefinite partial order or in terms of some criterion function. Following Kiefer, Wolfowitz, and others (see [23]), programs of the following type are used to aid in this allocation.

Program P

Find $V_{\mathrm{P}} = \inf \Phi(M)$
subject to

$$M \in \Omega \equiv \left\{ M(\xi) \in R^{n \times n} \,\middle|\, M(\xi) = \int_X f(x) f(x)^T \, \xi(dx) \text{ with } \xi(\cdot) \text{ a probability measure on } X \right\}.$$

We consider Program P under the following additional assumptions:

(i) $\Phi(\cdot)$ is continuously differentiable on a nonempty open set $Q \subseteq R^{n \times n}$. [Define $\nabla\Phi(\cdot)$: $R^{n \times n} \to R^{n \times n}$ by $(\nabla\Phi(\cdot))_{ij} = \partial\Phi(\cdot)/\partial m_{ij}$.]
(ii) $Q \supset \{M \in \Omega | \Phi(M) < \infty\} \equiv \Omega_\infty$.
(iii) $\Omega - \Omega_\infty$ is a subset of the relative boundary of Ω.
(iv) $\Phi(\cdot)$ is a closed proper convex function on Ω.

Two important criterion functions satisfy these assumptions, namely, $\log \det(M^{-1})$ (D-optimality) and, with C a positive definite matrix, $\mathrm{tr}[CM^{-1}]$ (L-optimality).

One can easily show that Ω is a compact convex set. Thus Program P is a convex program and one would expect standard solution techniques to work. However, there are difficulties. Suppose one attempted to apply a feasible direction method such as the quasi-Newton or conjugate gradient method. In these methods, if one wants to achieve the theoretical convergence rate, an exact line search is usually required. Thus, if one is at a point $\overline{M} \in \Omega_\infty$, a direction of descent is chosen, say $\hat{M}$, and a new point in Ω_∞ is found by solving

$$\begin{array}{ll} \textit{minimize} & \Phi(\overline{M} + \alpha\hat{M}) \\ \textit{subject to} & \overline{M} + \alpha\hat{M} \in \Omega. \end{array}$$

Since evaluating $\Phi(M)$ for most $\Phi(\cdot)$ of interest usually requires finding M^{-1}, such a line search can be very difficult.

Fedorov [9], Atwood [1, 2], Tsay [26], and Wynn [27, 28] have avoided this difficulty by using a variant of the Frank–Wolfe algorithm. The simplest algorithm of this type is the following:

Step 0. Find an $M_0 \in \Omega_\infty$. Set $k = 1$.

Step 1. Find $x_k \in X$ which solves

$$\min_{x \in X} \operatorname{tr}[\nabla\Phi(M_{k-1}) f(x) f(x)^{\mathrm{T}}].$$

Step 2. If $\operatorname{tr}[\nabla\Phi(M_{k-1})[f(x_k)f(x_k)^{\mathrm{T}} - M_{k-1}]] \geqq 0$, stop; M_{k-1} is optimal. Otherwise find α_k which solves

$$\min_{0 \leqq \alpha \leqq 1} \Phi((1-\alpha)M_{k-1} + \alpha f(x_k)f(x_k)^{\mathrm{T}}).$$

Set $M_k = (1 - \alpha_k)M_{k-1} + \alpha_k f(x_k)f(x_k)^{\mathrm{T}}$.
Set $k = k + 1$ and go to Step 1. ■

Since $f(x)f(x)^{\mathrm{T}}$ is of rank one, the inverse of $((1-\alpha)M_{k-1} + \alpha f(x_k)f(x_k)^{\mathrm{T}})$ can be found from M_{k-1}^{-1} by using the rank one updating formulas. Thus the line search required in Step 2 can be relatively simple.

While this algorithm is easy to implement, it does have some drawbacks. For example, zigzagging is very likely to occur. Also, while $\lim_{k\to\infty} \Phi(M_k) = V_{\mathrm{P}}$, the convergence rate approaches zero as the optimum is approached. Atwood has made improvements to this algorithm which speed convergence. Edahl [6] has shown that two rank-one updates can be performed simultaneously, and this appears to help eliminate zigzagging. However, their convergence proofs still rely upon the algorithm as stated so one cannot say that the convergence rate behavior will be better.

Because of these difficulties, cutting plane solution techniques might be attractive. Unfortunately, the standard cutting plane techniques do not apply since $\Phi(\cdot)$ and $\nabla\Phi(\cdot)$ are not bounded on Ω for criteria of interest. However, it is possible to combine the central cutting plane algorithm of Elzinga and Moore with a semiinfinite programming variant [14] to obtain a workable algorithm.

One can characterize the set of optimal solutions for Program P as the set of solutions of an infinite system of linear inequalities (see [15]).

Theorem 3. *M^* in Ω is optimal for Program* P *if and only if*

$$\operatorname{tr}[\nabla\Phi(M)\, M^*] \leqq \operatorname{tr}[\nabla\Phi(M)\, M] \qquad \textit{for all} \quad M \in \Omega_\infty.$$

Choose r_1 and r_2 that satisfy $V_{\mathrm{P}} < r_1 < r_2 < \infty$, and define

$$\Omega_{(r_1 r_2)} \equiv \{M \in \Omega \mid r_1 \leqq \Phi(M) \leqq r_2\}.$$

Define a new program.

Program P*

minimize $\Phi(M)$
for $M \in \Omega$ *subject to*

$$\operatorname{tr}[\nabla\Phi(\overline{M})\, M] \leqq \operatorname{tr}[\nabla\Phi(\overline{M})\, \overline{M}] \qquad \textit{for} \quad \overline{M} \in \Omega_{(r_1, r_2)}.$$

Program P* is a convex semiinfinite program and because of Theorem 3, M^* is optimal for Program P if and only if it is optimal for Program P*. If at any stage k, a matrix M_k is generated for which $\Phi(M_k) > r_2$, we will generate a cut from one of the linear inequalities.

Step 0. Find a nonoptimal $\hat{M}_0 \in \Omega_\infty$. Set $r_1 = \Phi(\hat{M}_0)$.
Choose b, r_2 such that $b < V_{\mathrm{P}} < r_1 < r_2 < \infty$.
Choose $\beta \in (0, 1)$. Let SD_0 be the program

maximize σ
subject to $z + \sigma \leqq r_1, \qquad z \geqq b, \qquad M \in \Omega.$ ■

Set $k = 1$.

Step 1. Let (M_k, z_k, σ_k) be optimal for SD_{k-1}. If $\sigma_k = 0$, stop; $\hat{M}_{k-1}$ is optimal.

Step 2. Use deletion rules, if desired. Call the resulting program SD_{k-1}.

Step 3. (i) If $\Phi(M_k) > r_2$, find an $\overline{M}_k \in \Omega_{(r_1 r_2)}$ such that

$$\operatorname{tr}[\nabla\Phi(\overline{M}_k)\, M_k] \geqq \operatorname{tr}[\nabla\Phi(\overline{M}_k)\, \overline{M}_k].$$

Add the constraint

$$\operatorname{tr}[\nabla\Phi(\overline{M}_k)\, M] + \sum_{i=j}^{n} \sum_{j=1}^{n} [\nabla\Phi(\overline{M}_k))_{ij}^2]^{1/2}\sigma \leqq \operatorname{tr}[\nabla\Phi(\overline{M}_k)\, \overline{M}_k]$$

to Program SD_{k-1}. Set $\hat{M}_k = \hat{M}_{k-1}$.
(ii) If $\Phi(M_k) < r_2$ and $\Phi(M_k) - z_k > 0$, add the constraint

$$\Phi(M_k) + \operatorname{tr}[\nabla\Phi(M_k)\,(M - M_k)] - z + \left[1 + \sum_{i=1}^{n} \sum_{j=1}^{n} (\nabla\Phi(M_k))_{ij}^2\right]^{1/2} \sigma \leqq 0$$

to Program SD_{k-1}. Set $\hat{M}_k = \hat{M}_{k-1}$.
(iii) If $\Phi(M_k) - z_k \leqq 0$, add the constraint

$$z + \sigma \leqq \Phi(M_k)$$

to Program SD_{k-1}. Set $\hat{M}_k = M_k$.

In any case, call the resulting Program SD_k. Set $k = k + 1$ and go to Step 1.

Deletion Rule 1. *Delete any constraint generated by Step* 3(iii) *if* $\Phi(M_k) \leqq z_k$.

Deletion Rule 2. *Delete any constraint generated by Step* 3(i) *or* (ii) *if the constaint was*

(a) *generated at Iteration* $j, j < k$;
(b) $\sigma_k < \beta\sigma_j$;
(c) *the constraint was not tight at* (M_k, z_k, σ_k).

One can prove the following theorem.

Theorem 4. *If the algorithm does not terminate, the limit points of the sequence* $\{\hat{M}_k\}_{k=0}^{\infty}$ *are optimal for Program* P. *Furthermore, between distinct elements of the sequence, there is a linear rate of convergence in objective function value.*

All the cuts to be added in Step 3(ii) and (iii) are easy to compute. The only fairly difficult one is in Step 3(i). However, since $\Phi(\cdot)$ is convex, a suitable $\overline{M}_k$ can be found by a crude bisecting line search on

$$\{(1-\alpha)\hat{M}_0 + \alpha M_k | 0 \leqq \alpha \leqq 1\}.$$

If $\hat{M}_0$ has previously been used to generate a cut, the following procedure may be used.

Step I. Set $A = \hat{M}_0, B = M_k$.

Step II. Let $C = \frac{1}{2}A + \frac{1}{2}B$.
If $r_1 \leqq \Phi(C) \leqq r_2$, set $\overline{M}_k = C$; stop.
If $\Phi(C) < r_1$, set $A = C$.
If $\Phi(C) > r_2$, set $B = C$.
Repeat Step II.

Also, as we proceed, enough constraints should be added after a certain point so that Step 3(i) will no longer be used. Computational experience with this algorithm and certain variants will be given in future papers.

REFERENCES

1. C. L. Atwood, Sequences converging to D-optimal designs of experiments, *Ann. Math. Statist.* **1** (1973), 342–352.
2. C. L. Atwood, Convergent design sequences, for sufficiently regular optimality criteria, *Ann. Statist.* **4** (1976), 1124–1138.
3. P. A. Beck and J. G. Ecker, Some Computational Experience with a Modified Convex Simplex Algorithm for Geometric Programs, Rep. ADTC–12–20, Rensselaer Polytech. Inst., Troy, New York, 1972.
4. A. Charnes, W. W. Cooper, and K. O. Kortanek, Semi-infinite programming, differentiability and geometric programming, Part I: With examples and applications in economics and management science, *J. Math. Sci. India, R. S. Varma Memorial Vol.* **6** (1971), 19–40.

5. R. J. DUFFIN, Infinite programs, *Ann. of Math. Stud.*No. 38 (1956), 157–176.
6. R. EDAHL, A Modified Steepest Descent Algorithm for Continuous D-Optimal Designs, Tech. Rep., Sch. Urban Public Affairs, Carnegie–Mellon Univ., Pittsburgh, Pennsylvania, 1978.
7. J. ELZINGA AND T. G. MOORE, A central cutting plane algorithm for the convex programming problem, *Math. Programming* **8** (1975), 134–145.
8. K. FAHLANDER, Computer Programs for Semi-Infinite Optimization, TRITA-NA-7312, Swed. Inst. Appl. Math., Stockholm, 1973.
9. V. V. FEDOROV, "Theory of Optimal Experiments," Academic Press, New York, 1972.
10. K. GLASHOFF AND S.-Å. GUSTAFSON, Numerical treatment of a parabolic boundary-value control problem, *J. Optim. Theory Appl.* **19** (1976), 645–663.
11. W. GOCHET AND Y. SMEERS, On the Use of Linear Programs to Solve Prototype Geometric Programs, CORE Discussion Paper No. 7229, 1972.
12. W. GOCHET, K. O. KORTANEK, AND Y. SMEERS, Using Semi-Infinite Programming in Geometric Programming, CORE Discussion Paper No. 7304, 1973.
13. W. L. GORR, S.-Å. GUSTAFSON, AND K. O. KORTANEK, Optimal control strategies for air quality standards and regulatory policy, *Environ. Plann.* **4** (1972), 183–192.
14. P. R. GRIBIK, A central cutting plane algorithm for semi-infinite programming problems, *in* "Semi-Infinite Programming" (R. Hettich, ed.), Springer–Verlag, Berlin, 1979, pp. 66–82.
15. P. R. GRIBIK AND K. O. KORTANEK, Equivalence theorems and cutting plane algorithms for a class of experimental design problems, *SIAM J. Appl. Math.* **32** (1977), 232–259.
16. P. R. GRIBIK AND D. N. LEE, A comparison of two central cutting plane algorithms for prototype geometric programming problems, *Methods Oper. Res.* (to appear).
17. S.-Å. GUSTAFSON, Nonlinear systems in semi-infinite programming, *in* "Solutions of Nonlinear Equations," (G. B. Byrnes and C. A. Hall, eds.), Academic Press, New York, 1973.
18. S.-Å. GUSTAFSON, On computational applications of the theory of the moment problems, *Rocky Mt. J. Math.* **4** (1974), 227–240.
19. S.-Å. GUSTAFSON AND K. O. KORTANEK, Mathematical models for air pollution and control: Numerical determination of optimizing abatement policies, *in* "Models for Environmental Pollution Control" (R. A. Deininger, ed.), Ann Arbor Sci. Publ., Ann Arbor, Michigan, 1973, pp. 251–265.
20. S.-Å. GUSTAFSON AND K. O. KORTANEK, Numerical solution of a class of semi-infinite programming problems, *NRLQ* **20** (1973), 477–504.
21. S.-Å. GUSTAFSON, K. O. KORTANEK AND H. M. SAMUELSSON, On Dual Programs and Finite-Dimensional Moment Cones, Series in Numerical Optimization and Pollution Abatement, Rep. No. 8, Sch. Urban Public Affairs, Carnegie–Mellon Univ., Pittsburgh, Pennsylvania, 1973.
22. J. KELLEY, The cutting plane method for solving convex programs, *J. SIAM* **8** (1960), 703–712.
23. J. KIEFER, General equivalence theory for optimum designs (approximate theory), *Ann. Statist.* **2** (1974), 849–879.
24. K. O. KORTANEK, Constructing a perfect duality in infinite programming, *Appl. Math. Optim.* **3** (1977), 357–372.
25. H. M. SAMUELSSON, A Note on the Duality Interpretation of an Air Pollution Abatement Model, Series in Numerical Optimization and Pollution Abatement, Rep. No. 7, Sch. Urban Public Affairs, Carnegie–Mellon Univ., Pittsburgh, Pennsylvania, 1972.
26. J.-Y. TSAY, The Iterative Methods for Calculating Optimal Experimental Designs, Ph.D. Thesis, Purdue Univ., Lafayette, Indiana, 1974.
27. H. P. WYNN, The sequential generation of D-optimal designs, *Ann. Math. Statist.* **41** (1970), 1655–1664.

28. H. P. WYNN, Results in the theory and construction of D-optimal experimental designs (with discussion), *J. Roy. Statist. Soc. Ser. B* **34** (1972), 133–147.

COMPUTER SYSTEM AND SERVICES DEPARTMENT
PACIFIC GAS AND ELECTRIC COMPANY
SAN FRANCISCO, CALIFORNIA

Inequalities and Approximation

WILLIAM W. HAGER

In this survey, we discuss a class of inequalities related to the following topics:
(1) Error estimates for variational inequality and optimal control approximation
(2) Stability for mathematical programs
(3) Solution regularity for optimal control problems
The material discussed in Sections 4–8 will be developed more fully in forthcoming papers [6–8].

1. ERROR ESTIMATES FOR QUADRATIC COST

Consider the problem

$$\text{minimize} \quad \{J(v) = a(v, v) + l(v) : v \in K\}, \tag{1.1}$$

where K is a convex subset of the Banach space $\mathcal{U}$, $l(\cdot)$ is a bounded linear functional on $\mathcal{U}$, and $a(\cdot, \cdot)$ is a symmetric, bounded bilinear form on $\mathcal{U}$. Suppose that there exists a solution $u \in K$ to (1.1), and let $K^h \subset \mathcal{U}$ be an approximation to K. No assumptions are made regarding K^h; in particular, it need not be convex. If $u^h \in K^h$ solves the problem

$$\text{minimize} \quad \{J(v) : v \in K^h\}, \tag{1.2}$$

we shall estimate the error $u - u^h$ in terms of energy $a(u - u^h, u - u^h)$.

Since K is convex and $J(\cdot)$ is differentiable, we have the standard variational inequality [12].

$$DJ[u](v - u) \geqq 0 \qquad \text{for all} \quad v \in K, \tag{1.3}$$

where

$$DJ[u](v) = 2a(u, v) + l(v). \tag{1.4}$$

Expanding $J(\cdot)$ about u gives us

$$J(u^h) = J(u) + DJ[u](u^h - u) + a(u^h - u, u^h - u). \tag{1.5}$$

Moreover, (1.3) implies that

$$\begin{aligned} DJ[u](u^h - u) &= DJ[u](u^h - v) + DJ[u](v - u) \\ &\geqq DJ[u](u^h - v) \end{aligned} \tag{1.6}$$

ISBN 0-12-178150-X

for all $v \in K$. On the other hand, since u^h minimizes $J(\cdot)$ over K^h, we have

$$J(u^h) \leqq J(v^h) = J(u) + DJ[u](v^h - u) + a(v^h - u, v^h - u) \quad (1.7)$$

for all $v^h \in K^h$. Finally combining (1.5)–(1.7), we get

$$a(u^h - u, u^h - u) \leqq DJ[u](v - u^h) + DJ[u](v^h - u) + a(v^h - u, v^h - u) \quad (1.8)$$

for all $v \in K$ and $v^h \in K^h$.

Special cases of (1.8) are the following:

(i) $K^h \subset K$. Choosing $v = u^h$ in (1.8) gives

$$a(u^h - u, u^h - u) \leqq DJ[u](v^h - u) + a(v^h - u, v^h - u) \quad (1.9)$$

for all $v^h \in K^h$.

(ii) $K = \mathscr{U}$. Hence (1.3) implies that $DJ[u](v) = 0$ for all $v \in K$ and (1.8) yields

$$a(u^h - u, u^h - u) \leqq a(v^h - u, v^h - u) \quad (1.10)$$

for all $v^h \in K^h$.

A classical application of (1.8) is the obstacle problem [3, 4] where we have

$$\begin{aligned} \mathscr{U} &= H_0^1(\Omega), \\ K &= \{v \in \mathscr{U} : v \geqq \psi \text{ on } \Omega\}, \\ J(v) &= \int_\Omega [|\nabla v|^2 - 2fv]. \end{aligned} \quad (1.11)$$

Here $\Omega \subset R^2$ is a bounded open set, $f \in L^2(\Omega)$, $H^m(\Omega)$ is the standard Sobolev space consisting of functions whose derivatives through order m are square integrable on Ω, $H_0^1(\Omega) \subset H^1(\Omega)$ is the subspace consisting of functions vanishing on $\partial\Omega$, and $\psi \in H^2(\Omega)$ is the given obstacle. If $\partial\Omega$ is sufficiently regular, there exists a solution $u \in H^2(\Omega)$ for problem (1.1). To simplify the exposition we assume that Ω is a polygon although this restriction is easily removed [3].

Let $S^h \subset H_0^1(\Omega)$ denote a piecewise linear subspace that satisfies the standard interpolation bound

$$\|g - g^{\mathrm{I}}\|_{H^k} \leqq ch^{2-k} \quad (1.12)$$

for all $g \in H^2(\Omega)$ and $k = 0, 1$, where h denotes the diameter of the biggest triangle in the triangulation of Ω and c denotes a generic constant that is independent of h. Finally, we define the set

$$K^h = \{v^h \in S^h : v^h \geqq \psi^{\mathrm{I}} \quad \text{on} \quad \Omega\}. \quad (1.13)$$

Integrating by parts $DJ[u](v)$ given by (1.4) gives us

$$DJ[u](v) = \langle w, v\rangle, \qquad w = -2(\Delta u + f), \tag{1.14}$$

where $\langle\cdot,\cdot\rangle$ denotes the $L^2(\Omega)$ inner product. Furthermore, using the variational inequality, it can be shown that

$$w \geqq 0, \qquad w(u - \psi) = 0 \qquad \text{almost everywhere on } \Omega. \tag{1.15}$$

We now substitute $v = u$ and $v^h = u^{\mathrm{I}}$ into (1.8); applying (1.14) leads to

$$\begin{aligned} &DJ[u](u - u^h) + DJ[u](u^{\mathrm{I}} - u) \\ &= \langle w, u^{\mathrm{I}} - u^h\rangle \\ &= \underbrace{\langle w, \psi^{\mathrm{I}} - u^h\rangle}_{\leqq 0} + \underbrace{\langle w, \psi - \psi^{\mathrm{I}}\rangle}_{= O(h^2)} + \underbrace{\langle w, u - \psi\rangle}_{= 0} + \underbrace{\langle w, u^{\mathrm{I}} - u\rangle}_{= O(h^2)} \\ &\leqq ch^2 \end{aligned} \tag{1.16}$$

since $w \geqq 0$ by (1.15), $\psi^{\mathrm{I}} - u^h \leqq 0$ by (1.13), $\|\psi - \psi^{\mathrm{I}}\|_{H^0} = O(h^2) = \|u - u^{\mathrm{I}}\|_{H^0}$ by (1.12), and $\langle w, u - \psi\rangle = 0$ by (1.15). Similarly, we have

$$a(u - u^{\mathrm{I}}, u - u^{\mathrm{I}}) \leqq \|u - u^{\mathrm{I}}\|_{H^1}^2 = O(h^2). \tag{1.17}$$

Combining (1.8), (1.16), and (1.17), we obtain the estimate

$$a(u - u^h, u - u^h) \leqq O(h^2). \tag{1.18}$$

Using quadratic elements and sharper regularity results established by Brézis [1, 2], it can be shown that

$$\|u - u^h\|_{H^1} = O(h^{1 \cdot 5 - \varepsilon}) \qquad \text{for any} \quad \varepsilon > 0 \quad [3].$$

2. ERROR ESTIMATES FOR DIFFERENTIABLE COST

Now let us consider the equation: Find $u \in H_0^1(\Omega)$ such that

$$u''(x) = e^{u(x)} \qquad \text{for all} \quad x \in \Omega \tag{2.1}$$

where $\Omega = (0, 1)$ and let $u \in H^2(\Omega)$ denote the solution. Defining the functional

$$J(v) = \int_\Omega [(v')^2 + 2e^v], \tag{2.2}$$

(2.1) is equivalent to the variational problem

$$\text{minimize } \{J(v) : v \in K\} \tag{2.3}$$

where $K \equiv H_0^1(\Omega)$. Letting $S^h \subset H_0^1(\Omega)$ denote the space of continuous, piecewise linear polynomials with h = maximum grid interval, we select $K^h = S^h$ and consider the approximation (1.2).

The estimate (1.8) no longer applies due to the e^v term included in $J(v)$. To generalize our earlier results, suppose that $J : \mathscr{U} \to R$ is differentiable; hence, (1.3) holds [12]. Moreover, suppose that there exists $\alpha > 0$ such that

$$J(v) - J(w) - DJ[w](v - w) \geqq \alpha\|v - w\|^2 \tag{2.4}$$

for all $v, w \in \mathscr{U}$ where $\|\cdot\|$ denotes the norm on $\mathscr{U}$. Consequently, (1.5) can be replaced by

$$J(u^h) \geqq J(u) + DJ[u](u^h - u) + \alpha\|u - u^h\|^2. \tag{2.5}$$

In addition, define the parameter

$$c(v, w) = \frac{J(v) - J(w) - DJ[w](v - w)}{\|v - w\|^2} \tag{2.6}$$

for all $v \neq w$. Hence (1.7) can be replaced by

$$J(u^h) \leqq J(u) + DJ[u](v^h - u) + c(v^h, u)\|v^h - u\|^2 \tag{2.7}$$

for all $v^h \in K^h$. Combining (2.5), (1.6), and (2.7), we get

$$\alpha\|u^h - u\|^2 \leqq DJ[u](v - u^h) + DJ[u](v^h - u) + c(v^h, u)\|v^h - u\|^2 \tag{2.8}$$

for all $v \in K$ and $v^h \in K^h$.

Now let us apply (2.8) to our particular equation (2.1). Observe that

$$\begin{aligned} DJ[w](v) &= 2\int_\Omega [w'v' + e^w v] \\ &= 2\langle -w'' + e^w, v\rangle \end{aligned} \tag{2.9}$$

and

$$J(v) - J(w) - DJ[w](v - w) = \langle (v - w)', (v - w)'\rangle + \langle e^\gamma(v - w), v - w\rangle \tag{2.10}$$

by Taylor's theorem where $\gamma(x)$ lies between $v(x)$ and $w(x)$. Hence we have

$$(2.10)\begin{cases} \geqq \langle (v - w)', (v - w)'\rangle, \\ \leqq \langle (v - w)', (v - w)'\rangle[\exp\{\|v'\|_{L^2} + \|w'\|_{L^2}\} + 1], \end{cases} \tag{2.11}$$

since

$$\begin{aligned} \|v\|_{L^\infty} &\leqq \|v'\|_{L^2} && \text{for all} \quad v \in H_0^1(\Omega), \\ \gamma(x) &\leqq \|v\|_{L^\infty} + \|w\|_{L^\infty} && \text{for all} \quad x \in \Omega. \end{aligned} \tag{2.12}$$

Finally (2.1) implies that $DJ[u](v) = 0$ for all $v \in H_0^1(\Omega)$ and (2.8)–(2.11) yield for $v = u^h$ and $v^h = u^{\mathrm{I}}$

$$\langle (u - u^h)', (u - u^h)'\rangle \leqq ch^2. \tag{2.13}$$

3. ERROR ESTIMATES FOR NONDIFFERENTIABLE COST

Consider the Bingham fluid problem that is given by (1.1) with the choices

$$\mathscr{U} = H_0^1(\Omega) = K,$$
$$J(v) = \int_\Omega [|\nabla v|^2 + |\nabla v| - 2fv] \tag{3.1}$$

with $f \in L^2(\Omega)$. Letting $S^h \subset \mathscr{U}$ denote the piecewise linear subspace of Section 2, we again take $K^h = S^h$ and study the approximation (1.2).

Observe that (2.8) cannot be utilized since $J(\cdot)$ is nondifferentiable. To generalize (1.8) or (2.8), suppose that

$$J(v) = J_{\mathrm{n}}(v) + J_{\mathrm{d}}(v), \tag{3.2}$$

where $J_{\mathrm{n}}(\cdot)$ is convex but possibly nondifferentiable, and

$$J_{\mathrm{d}}(v) = a(v, v) + l(v). \tag{3.3}$$

If $u \in K$ solves (1.1), then the following variational inequality holds [12]:

$$DJ_{\mathrm{d}}[u](v - u) + J_{\mathrm{n}}(v) \geqq J_{\mathrm{n}}(u) \tag{3.4}$$

for all $v \in K$ where

$$DJ_{\mathrm{d}}[u](v) = 2a(u, v) + l(v). \tag{3.5}$$

Hence we have

$$J(u^h) = J_{\mathrm{n}}(u^h) + J_{\mathrm{d}}(u) + DJ_{\mathrm{d}}[u](u^h - u) + a(u^h - u, u^h - u). \tag{3.6}$$

Moreover, by (3.4), we find that

$$DJ_{\mathrm{d}}[u](u^h - u) \geqq DJ_{\mathrm{d}}[u](u^h - v) + J_{\mathrm{n}}(u) - J_{\mathrm{n}}(v) \tag{3.7}$$

for all $v \in K$. On the other hand, we observe that

$$J(u^h) \leqq J(v^h) = J_{\mathrm{n}}(v^h) + J_{\mathrm{d}}(u) + DJ_{\mathrm{d}}[u](v^h - u) + a(v^h - u, v^h - u) \tag{3.8}$$

for all $v^h \in K^h$. Finally the combination (3.6)–(3.8) yields

$$\begin{aligned} a(u^h - u, u^h - u) \leqq\ & J_{\mathrm{n}}(v^h) - J_{\mathrm{n}}(u) + J_{\mathrm{n}}(v) - J_{\mathrm{n}}(u^h) \\ & + DJ_{\mathrm{d}}[u](v^h - u) + DJ_{\mathrm{d}}[u](v - u^h) \\ & + a(v^h - u, v^h - u) \end{aligned} \tag{3.9}$$

for all $v \in K$ and $v^h \in K^h$.

Applying (3.9) to the Bingham fluid problem using $v = u^h$ and $v^h = u^{\mathrm{I}}$, we obtain

$$\begin{aligned} \|\nabla(u^h - u)\|_{H^0}^2 & \leqq \langle |(u^{\mathrm{I}} - u)|, 1\rangle - 2\langle \Delta u + f, u^{\mathrm{I}} - u\rangle + \|\nabla(u^{\mathrm{I}} - u)\|_{H^0}^2 \\ & \leqq ch. \end{aligned} \tag{3.10}$$

With a more careful analysis, one can establish the estimate

$$\|\nabla(u^h - u)\|_{H^0}^2 \leqq ch^{2-\varepsilon} \tag{3.11}$$

for any $\varepsilon > 0$. See Glowinski [5] for the details of (3.11).

4. PERTURBATIONS IN THE COST

In the previous sections, we studied the effect of replacing the constraint set K in (1.1) by an approximation K^h. Now let us consider the case where the constraint set is fixed, but the cost functional is permitted to depend on a parameter.

For example, consider the quadratic programs

$$\text{minimize} \quad \{u^{\mathrm{T}}R_j u + 2r_j^{\mathrm{T}}u : u \in K\} \tag{4.1}$$

for $j = 1, 2$ where $K \subset R^n$ is convex. Suppose that there exist solutions (u_1, u_2) to (4.1) associated with $j = 1, 2$, respectively. Hence the following variational inequality holds:

$$(R_j u_j + r_j)^{\mathrm{T}}(v - u_j) \geqq 0 \qquad \text{for all} \quad v \in K \tag{4.2}$$

and $j = 1, 2$.

Choosing $(j = 1, v = u_2)$ and $(j = 2, v = u_1)$ and adding the resulting relations yields:

$$(u_2 - u_1)^{\mathrm{T}}R_2(u_2 - u_1) \leqq (r_1 - r_2)^{\mathrm{T}}(u_2 - u_1) + u_1^{\mathrm{T}}(R_1 - R_2)(u_2 - u_1). \tag{4.3}$$

If the smallest eigenvalue, α, of R_2 is positive, (4.3) implies that

$$\alpha|u_2 - u_1| \leqq |r_1 - r_2| + |u_1||R_1 - R_2|, \tag{4.4}$$

where $|\cdot|$ denotes the Euclidean norm. If K is compact or R_1 is positive definite, then an a priori bound can be given for $|u_1|$; hence (4.4) implies that the solution to (4.1) depends Lipschitz continuously on the perturbation.

We introduce the following notation: If $g : R^{m_1} \times R^{m_2} \times \cdots \times R^{m_l} \to R$, we let $\nabla_j g$ denote the gradient of $g(y_1, \ldots, y_l)$ with respect to y_j where $y_k \in R^{m_k}$ for $k = 1, \ldots, l$.

The following generalization of (4.4) is easily established [7]:

Theorem 4.1. *Let $K \subset R^n$ be nonempty, closed, and convex, $E \subset R^m$, $f: R^n \times R^m \to R$ be differentiable in its first n arguments on $K \times R^m$, and assume that there exists $\alpha > 0$ such that*

$$f(y, \xi) \geqq f(x, \xi) + \nabla_1 f(x, \xi)(y - x) + \alpha|x - y|^2 \tag{4.5}$$

for all $x, y \in K$ and $\xi \in E$. Then for all $\xi \in E$, there exists a unique $x(\xi) \in K$ satisfying

$$f(x(\xi), \xi) = \min\{f(y, \xi) : y \in K\}; \tag{4.6}$$

and given $\bar{x} \in K$, we have

$$|x(\xi) - \bar{x}| \leqq |\nabla_1 f(\bar{x}, \xi)|/\alpha. \tag{4.7}$$

Also, if $\nabla_1 f(z, \cdot)$ is continuous for all $z \in K$, then $z(\cdot)$ is continuous on E, and if, moreover, $\nabla_2 \nabla_1 f(\cdot, \cdot)$ is continuous, then

$$|x(\xi_1) - x(\xi_2)| \leqq \frac{|\xi_1 - \xi_2|}{2\alpha} \max_{0 \leq s \leq 1} |\nabla_2 \nabla_1 f(x(\xi_2), \xi_1 + s(\xi_2 - \xi_1))|. \tag{4.8}$$

5. DUAL APPROXIMATIONS

One case where the perturbation parameter appears linearly in the cost of a program arises in the study of dual methods for mathematical programs. Consider the program

$$\text{minimize} \quad \{f(z) : z \in R^n, g(z) \leqq 0\} \tag{5.1}$$

where $g : R^n \to R^m$ and $f : R^n \to R$. The Lagrange dual function is given by

$$\mathscr{L}(\eta) = \inf\{f(z) + \eta^{\mathrm{T}} g(z) : z \in R^n\} \tag{5.2}$$

and the associated dual problem becomes

$$\sup\{\mathscr{L}(\eta) : \eta \in R^m, \eta \geqq 0\}. \tag{5.3}$$

Under suitable assumptions, there exist solutions η^* to (5.3) and z^* to (5.1). Moreover, z^* achieves the minimum in (5.2) for $\eta = \eta^*$.

Now define the set

$$K = \{\eta \in R^m : \eta \geqq 0\}, \tag{5.4}$$

and let K^h be an approximation to K. The following approximation to (5.3) is considered:

$$\sup\{\mathscr{L}(\eta) : \eta \in K^h\}. \tag{5.5}$$

If η^h solves (5.5) and z^h achieves the minimum in (5.2) for $\eta = \eta^h$, let us estimate $z^* - z^h$.

Referring to our development in Section 1–3, we see the need for an inequality of the form (2.4) and an estimate of the parameter c in (2.6). To attack these estimates, define the Lagrangian

$$\mathscr{L}(z, \eta) = f(z) + \eta^{\mathrm{T}} g(z) \tag{5.6}$$

and assume the following:

(i) $f, g \in C^2$;
(ii) there exists $\alpha > 0$ such that

$$\nabla_1^2 \mathscr{L}(z, \eta) > \alpha I \qquad \text{for all} \quad z \in R^n \quad \text{and} \quad \eta \in E.$$

where $E \subset R^m$ and ∇_1^2 denotes the Hessian of $\mathscr{L}(z, \eta)$ with respect to z.

Letting $z(\eta)$ denote the minimizing value of z in (5.2), we know that

$$\nabla_1 \mathscr{L}(z(\eta), \eta) = 0 \tag{5.7}$$

since the minimization is unconstrained. Moreover, by Theorem 4.1, $z(\cdot)$ is differentiable almost everywhere on E. Hence the chain rule and (5.7) give us

$$D\mathscr{L}[\mu](\eta) = \eta^{\mathrm{T}} g(z(\mu)) \tag{5.8}$$

for almost every $\mu \in E$. Motivated by relation (5.8) and our earlier development in Section 1–3, we study the quantity

$$\mathscr{L}(\eta_1) - \mathscr{L}(\eta_2) - g(z(\eta_2))^{\mathrm{T}}(\eta_1 - \eta_2). \tag{5.9}$$

The following result can be established [7]:

$$(5.9) \begin{cases} \leqq -(\alpha/2)|z(\eta_1) - z(\eta_2)|^2, \\ \geqq -|\nabla g(z(\eta_2))^{\mathrm{T}}(\eta_1 - \eta_2)|^2/\alpha. \end{cases} \tag{5.10}$$

Therefore, proceeding as in Section 2, we get the estimate:

$$\begin{aligned}(\alpha^2/2)|z(\eta^h) - z(\eta^*)|^2 \leqq \alpha g(z(\eta^*))^{\mathrm{T}}(\eta^h - \mu + \eta^* - \mu^h) \\ + |\nabla g(z(\eta^*))^{\mathrm{T}}|^2 |\mu^h - \eta^*|^2\end{aligned} \tag{5.11}$$

for all $\mu^h \in K^h$ and $\mu \in K$.

6. RITZ–TREFFTZ FOR OPTIMAL CONTROL

As an application of the results in Section 5, consider the following control problem:

$$\text{minimize} \quad \left\{ C(x, u) = \int_0^1 f(x(t), u(t), t)\, dt \right\}$$

subject to

$$\begin{aligned} &\dot{x}(t) = Ax(t) + Bu(t) \qquad \text{for almost every} \quad t \in [0, 1], \\ &x(0) = x_0, \\ &G(u(t), t) \leqq 0 \qquad \text{for all} \quad t \in [0, 1], \\ &x \in \mathscr{A}(R^n), \qquad u \in L^\infty(R^m), \end{aligned} \tag{6.1}$$

where x: $[0, 1] \to R^n$, u: $[0, 1] \to R^m$, and $\mathscr{A}$ denotes absolutely continuous functions. The Lagrange dual functional associated with (6.1) is given by

$$\begin{aligned}\mathscr{L}(p, \lambda) = \inf\{&C(x, u) + \langle p, \dot{x} - Ax - Bu\rangle \\ &+ \langle G(u), \lambda\rangle : x \in \mathscr{A}(R^n), u \in L^\infty(R^m), x(0) = x_0\}, \end{aligned}\tag{6.2}$$

and the dual problem becomes [9]

$$\begin{gathered}\sup\{\mathscr{L}(p, \lambda) : (p, \lambda) \in K\}, \\ K \equiv \{(p, \lambda) : p \in BV, p(1) = 0, \lambda \in L^1, \lambda \geqq 0\},\end{gathered}\tag{6.3}$$

where BV denotes the space of functions with bounded variation. Under suitable assumptions [7, 9], there exist solutions (x^*, u^*) to (6.1) and (p^*, λ^*) to (6.3). Moreover,

$$G(u^*(t), t)^{\mathrm{T}} \lambda^*(t) = 0 \qquad \text{for almost every} \quad t \in [0, 1] \tag{6.4}$$

and (x^*, u^*) achieve the minimum in (6.2) for $(p, \lambda) = (p^*, \lambda^*)$.

Now let $K^h \subset K$ denote the subset consisting of continuous, piecewise linear functions p such that $p(1) = 0$ and piecewise constant functions λ such that $\lambda \geqq 0$. As usual, the superscript h denotes the maximum grid interval. Consider the following approximation to (6.3):

$$\sup\{\mathscr{L}(p, \lambda) : (p, \lambda) \in K^h\}. \tag{6.5}$$

Suppose that (6.5) has the solution $(p^h, \lambda^h) \in K^h$ and that (x^h, u^h) achieve the minimum in (6.2) for $(p, \lambda) = (p^h, \lambda^h)$. Let us estimate the errors $(x^h - x^*)$ and $(u^h - u^*)$.

Assume that $f, G \in C^2$, the components of $G(\cdot, t)$ are convex for all $t \in [0, 1]$, and there exists $\alpha > 0$ such that

$$\nabla_1^2 f(z, t) > \alpha I \tag{6.6}$$

for all $z \in R^{n+m}$ and $t \in [0, 1]$. Integrating by parts in (6.2), it can be shown that [7]

$$\mathscr{L}(p, \lambda) = -p(0)^{\mathrm{T}} x_0 + \int_0^1 F(t)\, dt, \tag{6.7}$$

where

$$\begin{aligned}F(t) = \inf\{&f(x, u, t) - \dot{p}(t)^{\mathrm{T}} x \\ &-p(t)^{\mathrm{T}}(Ax + Bu) + G(u, t)^{\mathrm{T}} \lambda(t) : x \in R^n, u \in R^m\}.\end{aligned}\tag{6.8}$$

That is, $\mathscr{L}(p, \lambda)$ can be expressed as the integral of a pointwise minimum.

Moreover, with the identifications

$$\eta = \begin{bmatrix} \dot{p}(t) \\ p(t) \\ \lambda(t) \end{bmatrix}, \qquad z = \begin{bmatrix} x \\ u \end{bmatrix},$$

and

$$g(z) = \begin{bmatrix} -x \\ -Ax - Bu \\ G(u, t) \end{bmatrix}, \tag{6.9}$$

we can apply the results of Section 5 for each time t and integrate over $t \in [0, 1]$ to get

$$\begin{aligned}(\alpha/2)\{\|x^h - x\|_{L^2}^2 + \|u^h - u^*\|_{L^2}^2\} \\ \leqq \langle G(u^*), \lambda^* - \mu^h\rangle + C\{\|\dot{p}^* - \dot{q}^h\|_{L^2}^2 + \|p^* - q^h\|_{L^2}^2 + \|\lambda^* - \mu^h\|_{L^2}^2\}\end{aligned} \tag{6.10}$$

for all $(q^h, \mu^h) \in K^h$ satisfying $q^h(0) = p^*(0)$ where C depends on A, B, and $\nabla_1 G(u^*(\cdot), \cdot)$.

Now suppose that $(\dot{p}^*, \lambda^*)$ are Lipschitz continuous [6], select $q^h = p^{\mathrm{I}}$, and let $\mu^h = \lambda^{\mathrm{I}} =$ the piecewise constant function agreeing with the minimum value of λ^* on each grid interval. Since $G(u^*(t), t) \leqq 0 \leqq \lambda^*(t)$ for all $t \in [0, 1]$ and (6.4) holds, we conclude that

$$\langle G(u^*), \lambda^* - \lambda^{\mathrm{I}}\rangle = 0 \tag{6.11}$$

while the remaining terms on the right side of (6.10) are $O(h^2)$. To summarize,

$$\|x^h - x^*\|_{L^2}, \|u^h - u^*\|_{L^2} = O(h). \tag{6.12}$$

(For additional results in this area, see [7 and 11].

7. SEMIDUAL METHODS IN OPTIMAL CONTROL

In the previous section, we introduced a dual multiplier for both the differential equation and the control constraint. Now let us consider a semidual approach where a dual multiplier is only used for the differential equation. In particular, let us consider the problem

$$\text{minimize} \quad \left\{ C(x, u) = \frac{1}{2} \int_0^1 [x(t)^{\mathrm{T}} Q x(t) + u(t)^{\mathrm{T}} R u(t)] \, dt \right\}$$

subject to

$$\begin{aligned} \dot{x}(t) &= Ax(t) + Bu(t) \qquad \text{for almost every} \quad t \in [0, 1], \\ x(0) &= x_0, \\ G(u(t), t) &\leqq 0 \qquad \text{for all} \quad t \in [0, 1], \\ x \in \mathscr{A}(R^n), &\qquad u \in L^\infty(R^m). \end{aligned} \tag{7.1}$$

Define the set

$$U = \{u \in L^\infty(R^m) : G(u(t), t) \leqq 0 \text{ for all } t \in [0, 1]\},$$

and the semidual functional

$$\mathscr{L}(p) = \inf\{C(x, u) + \langle p, \dot{x} - Ax - Bu \rangle : x \in \mathscr{A}(R^n), x(0) = x_0, u \in U\}. \tag{7.2}$$

The semidual problem becomes

$$\sup\{\mathscr{L}(p) : p \in K\} \qquad \text{where} \quad K \equiv \{p \in BV : p(1) = 0\}. \tag{7.3}$$

Suppose that Q and R are positive definite; hence, it follows from [7, 9] that there exist solutions (x^*, u^*) to (7.1) and p^* to (7.3). Moreover, (x^*, u^*) achieve the minimum in (7.2) for $p = p^*$ and the following adjoint equation and minimum principle hold:

$$\dot{p}^*(t) = -A^{\mathrm{T}} p^*(t) + Q x^*(t) \qquad \text{for almost every} \quad t \in [0, 1], \tag{7.4}$$

$$M(u^*(t), p^*(t)) = \min\{M(u, p^*(t)) : u \in R^m, G(u, t) \leqq 0\} \quad \text{for almost every} \quad t \in [0, 1], \tag{7.5}$$

where

$$M(u, p(t)) \equiv \tfrac{1}{2} u^{\mathrm{T}} R u - p(t) B u.$$

Let $K^h \subset K$ be the space of continuous, piecewise linear polynomials, and consider the following approximation to (7.3):

$$\sup\{\mathscr{L}(p) : p \in K^h\}. \tag{7.6}$$

Suppose that p^h solves (7.6) and that $(x, u) = (x^h, u^h)$ achieves the minimum in (7.2) for $p = p^h$. We study the errors $(x^* - x^h)$ and $(u^* - u^h)$.

Integrating by parts in (7.2), it can be shown that

$$\mathscr{L}(p) = \mathscr{L}_1(p) + \mathscr{L}_2(p), \tag{7.7}$$

where

$$\mathscr{L}_1(p) = -\frac{1}{2}\int_0^1 (\dot{p}(t) + A^{\mathrm{T}}p(t))^{\mathrm{T}}Q(\dot{p}(t) + A^{\mathrm{T}}p(t))\,dt, \tag{7.8}$$

$$\mathscr{L}_2(p) = -p(0)^{\mathrm{T}}x_0 + \int_0^1 \gamma(p(t))\,dt, \tag{7.9}$$

$$\gamma(p(t)) \equiv \inf\{M(u, p(t)) : u \in R^m,\ G(u, t) \leqq 0\}. \tag{7.10}$$

Furthermore, it can be shown that (x^h, u^h) satisfy relations (7.4)–(7.5) with superscript $*$ replaced by h.

Since $\mathscr{L}_1(p)$ is a quadratic functional and $\mathscr{L}_2(p)$ is possibly nondifferentiable, we can apply the bound (3.9) choosing $v = p^h$ and $v^h = p^{\mathrm{I}}$:

$$\begin{aligned} -\mathscr{L}_1(p^h - p^*) \leqq\ & -\mathscr{L}_1(p^{\mathrm{I}} - p^*) \\ & + \langle \dot{p}^* + A^{\mathrm{T}}p^*, Q[\dot{p}^{\mathrm{I}} - \dot{p}^* + A^{\mathrm{T}}(p^{\mathrm{I}} - p^*)]\rangle \\ & + \mathscr{L}_2(p^*) - \mathscr{L}_2(p^{\mathrm{I}}). \end{aligned} \tag{7.11}$$

If there exists a continuous control $\bar{u}$ such that $G(\bar{u}(t), t) \leqq 0$ for all $t \in [0, 1]$, it follows from (7.4)–(7.5) that $\ddot{p}^* \in L^\infty$. Integrating by parts the $\dot{p}^{\mathrm{I}} - \dot{p}^*$ term appearing in (7.11) and applying the interpolation estimate $\|p^* + p^{\mathrm{I}}\|_{L^2} = O(h^2)$, we get:

$$\begin{aligned} -\mathscr{L}_1(p^h - p^*) &\leqq ch^2 + \mathscr{L}_2(p^*) - \mathscr{L}_2(p^{\mathrm{I}}) \\ &\leqq ch^2 + \int_0^1 (p^{\mathrm{I}}(t) - p^*(t))^{\mathrm{T}}Bu^{\mathrm{I}}(t)\,dt, \end{aligned} \tag{7.12}$$

where $u = u^{\mathrm{I}}(t)$ achieves the minimum in (7.10) for $p(t) = p^{\mathrm{I}}(t)$. Applying (4.7), $\|u^{\mathrm{I}}\|_{L^\infty}$ is bounded uniformly in h, and (7.12) gives us:

$$-\mathscr{L}_1(p^h - p^*) \leqq ch^2. \tag{7.13}$$

Since $p^h(1) = p^*(1) = 0$, we show in [10] that there exists a constant $\beta > 0$ such that

$$\beta\|p^h - p^*\|_{H^1}^2 \leqq -\mathscr{L}_1(p^h - p^*). \tag{7.14}$$

Therefore, we obtain

$$\|p^h - p^*\|_{H^1} \leqq ch. \tag{7.15}$$

Since (p^h, x^h) satisfy (7.4) with superscript $*$ replaced by h, (7.14) also implies that

$$\|x^* - x^h\|_{L^2} \leqq ch. \tag{7.16}$$

Finally we combine the inequality (7.14), relation (7.5) with superscript $*$ replaced by h, and the quadratic program stability bound (4.4) to get:

$$\|u^* - u^h\|_{L^2} \leqq ch. \tag{7.17}$$

8. LIPSCHITZ CONTINUITY FOR CONSTRAINED PROCESSES

In the previous sections, we studied the effect on a program of replacing a constraint set K by an approximation K^h, and the dependence of the solution to a program on a parameter appearing in the cost. Now we study stability (or more specifically Lipschitz continuity) in the abstract setting of a "constrained process."

A constrained process can be described as follows: Let $\mathscr{S}$ be a Banach space, $\mathscr{D}$ be a convex subset of a Banach space, $z: \mathscr{D} \to \mathscr{S}$, and $c: \mathscr{D} \to$ (power set of $\{1, \ldots, n\}$). Two examples of constrained processes are the following:

(1) A control problem such as (6.1) with optimal solution (x^*, u^*); we choose $\mathscr{D} = [0, 1]$, the time interval, $z(d) = (x^*(d), u^*(d))$, and $c(d) =$ indices of the binding constraints associated with $u^*(d)$.

(2) A mathematical program such as

$$\text{minimize} \quad \{f(x, \xi): x \in R^n, g(x, \xi) \leqq 0\}, \tag{8.1}$$

where $\xi \in R^m$ is a given parameter. Suppose that there exists a solution $x(\xi)$ to (8.1) for $\xi \in \mathscr{D} \subset R^m$, a convex subset. We then choose $z(d) = x(d)$ and $c(d) =$ indices of the binding constraints associated with $x(d)$.

Our goal is to estimate the Lipschitz constant for $z(\cdot)$. First consider the program (8.1), and let $\# c(d)$ denote the number of elements in the set $c(d)$. Recall that the Kuhn–Tucker conditions give us a system of $n + \# c(\xi)$ equations in the same number of unknowns: $x(\xi)$ and the dual multipliers associated with binding constraints. If $c(\xi_1) = c(\xi_2)$, then $|x(\xi_1) - x(\xi_2)|$ can often be estimated in terms of $|\xi_1 - \xi_2|$ using the implicit function theorem. Similar results apply to the control problem but with the Kuhn–Tucker conditions replaced by the Pontryagin minimum principle and the adjoint equation.

On the other hand, suppose that $c(\xi_1) \neq c(\xi_2)$. A key result on this subject is given in [6]; namely, a Lipschitz constant that is valid for compatible parameters (where the binding constraints agree) is also valid for noncompatible parameters. To be more precise, assume that $z(\cdot)$ is continuous, and $c(\cdot)$ has the following property: If $\{d_k\} \subset \mathscr{D}$, $d_k \to d \in \mathscr{D}$ as $k \to \infty$, and $I \subset c(d_k)$ for all k, then $I \subset c(d)$.

Given $d, e \in \mathscr{D}$, we define the segment

$$[d, e] = \{(1 - \lambda)d + \lambda e : 0 \leqq \lambda \leqq 1\}$$

and we say that (d, e) are *compatible* if $c(d) = c(e)$ and $c(\delta) \subset c(d)$ for all $\delta \in [d, e]$.

Theorem 8.1. *If γ satisfies*

$$\|z(d) - z(e)\|_{\mathscr{S}} \leqq \gamma \|d - e\|_{\mathscr{D}} \tag{8.2}$$

for all compatible data $(d, e) \in \mathscr{D} \times \mathscr{D}$, *then* γ *satisfies* (8.2) *for all data* $(d, e) \in \mathscr{D} \times \mathscr{D}$.

The application of Theorem 8.1 to derive Lipschitz continuity results for both mathematical programs and optimal control problems is given in [6].

REFERENCES

1. H. Brézis, Nouveaux théorèmes de régularité pour les problèmes unilatéraux, *Recontre Physiciens Théoriciens Mathématiciens, Strasbourg* **12** (1971).
2. H. Brézis, Seuil de régularité pour certain problèmes unilateraux, *C. R. Acad. Sci., Ser. A* **273** (1971), 35–37.
3. F. Brezzi, W. W. Hager, and P. A. Raviart, Error estimates for the finite element solution of variational inequalities, *Numer. Math.* **28** (1977), 431–443.
4. R. S. Falk, Error estimates for the approximation of a class of variational inequalities, *Math. Comp.* **28** (1974), 308–312.
5. R. Glowinski, Sur l'approximation d'une inéquation variationnelle elliptique, *RAIRO Anal. Numer.* **10**, No. 12 (1976), 13–30.
6. W. W. Hager, Lipschitz continuity for constrained processes, *SIAM J. Control Optim.* **17** (1979), 321–338.
7. W. W. Hager, Convex control and dual approximations, *Control Cybernet.* **8** (1979), 5–22.
8. W. W. Hager and G. Ianculescu, Semidual approximations in optimal control (to appear).
9. W. W. Hager and S. K. Mitter, Lagrange duality theory for convex control problems, *SIAM J. Control Optim.* **14** (1976), 843–856.
10. W. W. Hager, Rates of Convergence for Discrete Approximations to Problems in Control Theory, Ph.D. Thesis, Mass. Inst. of Tech., Cambridge, Massachusetts, 1974.
11. W. W. Hager, The Ritz–Trefftz method for state and control constrained optimal control problems, *SIAM J. Numer. Anal.* **12** (1975), 854–867.
12. J. L. Lions, "Optimal Control of Systems Governed by Partial Differential Equations" (transl. by S. K. Mitter), Springer-Verlag, Berlin and New York, 1971.

This research was supported in part by Office of Naval Research Grant N00014-76-C-0369, and the Naval Surface Weapons Center, Silver Springs, Maryland.

DEPARTMENT OF MATHEMATICS
CARNEGIE–MELLON UNIVERSITY
PITTSBURGH, PENNSYLVANIA

A Brief Survey of Complementarity Theory

C. E. LEMKE

A descriptive summary is presented of some of the developments over the past 15 years in the area of mathematical programming which has been variously called complementary pivoting, simplicial approximation methods, and fixed-point methods. Developments in both linear complementarity and the more general area are discussed, including the more recent "constructive" approaches based on a system of ordinary differential equations.

This contribution is a descriptive summary of some of the developments over the past 15 years in an area of mathematical programming which has often been described by the term complementary pivoting. However, other descriptives have also been used such as simplicial approximation methods, fixed-point methods, and more recently continuation methods or path-following methods. Primarily, however, the focus has been on *constructive* approaches to some mathematical problems. Here the word "constructive" is used to draw attention to the "constructive-proof" nature of the methods developed. A name is needed to focus on that at the same time includes some quite recent additions of "path-following" methods which I would like to include in the discussion. I shall therefore refer to the area under discussion as constructive approximation methods (CAM), meant only for this summary, or, broadly, to the developments as the "CAM-activity."

There are three sections: (1) preliminary considerations, (2) the linear complementarity problem, and (3) the more general CAM-activity.

Some material has been relegated to appendices.

The overall intent of this summary is to call attention to the total effort of the "CAM-activity," and hopefully to hint at its part in the mainstream of mathematics both past and present. But this is done from the author's personal perspective, hence with some bias, and, again, hopefully without too much factual distortion or omission.

1. PRELIMINARY CONSIDERATIONS

Roughly, the general models which have received attention during the CAM-activity are *fixed-point problems*, *complementarity problems*, (finding) solutions to *systems of equations*, and some types of *variational inequality*

Reproduced with permission of the publishers from "Variational Inequalities and Complementarity Problems (Theory and Applications)" (R. W. Cottla, F. G. Giannessi, and J. L. Lions, eds.). John Wiley and Sons Limited, Chichester, U.K., 1979.
ISBN 0-12-178150-X

problems. Within the scope of the CAM-activity, by now each of these has its own areas of theory and application. In this summary however, to provide a focus, *systems of equations* are stressed as a general model. Not too much will be lost by this stress since there are certain relevant equivalences between the four noted, and also since, in consideration of developments to the present, this seems to be the trend.

Further, two other terms tend to describe the CAM-activity, namely, *piecewise-linear systems* and the *homotopy approach*. To a large degree, we will be concerned with finding zeros to a (square) system of equations using the homotopy approach, and often when the functions defining the system are piecewise linear.

The scope of the CAM-activity over the past 15 years is evidenced primarily by the more than 200 publications, representing work of perhaps 100 persons concerned either with theory, models, applications, associated numerical analyses, generalizations or consolidations, or with computational aspects, inclusive of computer programs.

Only a few of these publications will be referenced here, and then generally only when needed. However, I shall draw heavily on the bibliography compiled by Eaves which accompanies his paper appearing in [B], and when only names are noted here, work referred to will most likely be referenced there. I shall also call attention to the three recent volumes [A, B, and C]. I shall make much reference to [C]. In particular, for history and a view of the future, the excellent "Introduction" therein, written by H. E. Scarf, is strongly recommended.

By now we more or less agree on what constitutes nonlinear programming (NLP). I will say that it consists of linear programming (LP), and the rest of it. In retrospect over the past 25 or so years there seems to be emerging the parallel that the linear complementarity problem (LCP) is to CAM as LP is to NLP. Much of the activity we are addressing does concern LCP, namely, the problem:

Given the matrix M $(n \times n)$, and column q $(n \times 1)$: Find x such that

$$y := q + Mx; \qquad x \geqq 0, y \geqq 0, \quad \text{and} \quad y^{\mathrm{T}}x = 0. \tag{1}$$

(The superscript T denotes matric transposition. The definitional equality ":=" means that the symbol on the left is being defined, "=:" that on the right.)

1.1. Some Equivalences

As the general model of focus, consider the following:

Given the column $F := [F_1, F_2, \ldots, F_n]$, where $F_i\colon D \to R$ and $D \subset R^n$:

$$\text{Find} \quad x \in D \qquad \text{such that} \quad F(x) = 0. \tag{2}$$

(Square brackets are used to denote a column.)

Generally F is at least continuous. The CAM-activity has been concerned with the finite-dimensional cases, but there have been exceptions in the following forms, and one can expect some effort on infinite-dimensional cases in the next few years.

The simple fact that $x \in D$ is a zero of F iff it is a fixed point of $G: G(x) := F(x) + x$ gives an equivalence of (2) and the fixed-point problem:

$$\text{(FP)} \qquad \text{Find} \quad x \in D \qquad \text{such that} \quad G(x) = x \tag{3}$$

(usually, $G: D \to D$). This equivalence suffices for our purposes since, in a given case, stated assumptions on D and F make the differences essentially notational.

The two others which have accounted for much of the CAM-activity are as follows:

The *complementarity problem*: Given $f: R^n \to R^n$:

$$\text{(CP)} \quad \text{Find} \quad x \geqq 0 \qquad \text{such that} \quad y := f(x) \geqq 0 \quad \text{and} \quad x^{\mathrm{T}} y = 0 \tag{4}$$

(equivalently: $x_i y_i = 0, i = 1, 2, \ldots, n$).

The *variational inequality problem*: Given $g: R^n \to R^n$ and $D \subset R^n$, find a "stationary point" of the pair (g, D); that is:

$$\text{(VIP)} \quad \text{Find} \quad x \in D, \qquad \text{such that for all} \quad y \in D: (y - x)^{\mathrm{T}} g(x) \geqq 0 \tag{5}$$

[or, x satisfies: $x^{\mathrm{T}} g(x) = \inf_D y^{\mathrm{T}} g(x)$].

Of course, CP is LCP when f is affine: $f(x) = q + Mx$. Some relations and equivalences comprise Appendix 1.

The most well-known and probably most applicable example to date of CP relates to the differentiable case of NLP:

$$\min_D h_0(x), \qquad \text{where} \quad D: = \{x : h(x) \leqq 0, x \geqq 0\}, \tag{6}$$

where h_0 and $h = [h_1, h_2, \ldots, h_m]$ are given and $D \subset R^n$.

Generalizing the classical method of Lagrange, the Karush–Kuhn–Tucker (KKT) conditions (which assume the form of CP) furnish necessary conditions for an x to yield a solution to (6) (more generally, a local minimum). Briefly described, writing $z := [x, u]$ (a column), where the u_i are the "Lagrange multipliers," if one forms the "Lagrangian"

$$L(z) := h_0(x) + u^{\mathrm{T}} h(x) \tag{7}$$

a necessary condition that x yield a solution to (6) is that (with a sufficient "constraint qualification" on D) for some $u \geqq 0$

$$D_x L(z) \geqq 0; \qquad -D_u L(z) \geqq 0 \quad \text{and} \quad x^{\mathrm{T}} D_x L(z) = 0 = u^{\mathrm{T}} D_u L(z), \tag{8}$$

which has the form (4) with $f(z) := [D_x L(z), -D_u L(z)]$.

2. THE LINEAR COMPLEMENTARITY PROBLEM

2.1. A Classical Example

Building on the works of Cottle, Dantzig, Kuhn, Tucker, and so many others, LCP (1) is more closely observable as a generalization of linear and quadratic programming than CAM is of, say, NLP, mainly because of the new thrusts of simplicial approximation methods.

As a classical example, the KKT conditions for the quadratic programming problem

$$\text{(QP)} \qquad \min(c^{\mathrm{T}}x + x^{\mathrm{T}}Qx/2), \qquad \text{where} \quad y := b + Ax, \quad u, x \geqq 0, \tag{9}$$

where $Q = Q^{\mathrm{T}}$, may be expressed in the form

$$\begin{bmatrix} y \\ v \end{bmatrix} := \begin{bmatrix} b \\ c \end{bmatrix} + \begin{bmatrix} 0 & A \\ -A^{\mathrm{T}} & Q \end{bmatrix}\begin{bmatrix} u \\ x \end{bmatrix}; \quad \begin{bmatrix} u \\ x \end{bmatrix}, \begin{bmatrix} y \\ v \end{bmatrix} \geqq 0 \quad \text{and} \quad \begin{bmatrix} u \\ x \end{bmatrix}^{\mathrm{T}} \begin{bmatrix} y \\ v \end{bmatrix} = 0. \tag{10}$$

QP (9) includes LP, namely the case $Q = 0$, in which case (10) gives the KKT conditions for LP. In particular, for the *convex* QP (which includes LP), namely, when Q is positive semidefinite (psd), any solution to (10) yields a solution to QP (as well as to its dual problem).

In this sense, LCP is a generalization of LP, which we repeat here for reference:

$$\text{(LP)} \qquad \min c^{\mathrm{T}}x, \qquad \text{where} \quad y := b + Ax \quad \text{and} \quad y, x \geqq 0. \tag{11}$$

2.2. Models and Applications of LCP

As of the present writing, LCP has found application in the following areas among others:

(1) LP and QP
(2) Bimatrix and polymatrix games
(3) Mechanics, plasticity, etc.,
(4) Economic equilibrium
(5) Approximate solutions to differential equations
(6) Classification of square matrices
(7) Solutions to systems $F(x) = 0$ (n equations in n unknowns)

We reference the names of some of the more prominent workers in these areas, roughly in chronological order.

(1) In a real sense, all of the proposed methods of QP, inclusive of LP, may be visualized in the context of the KKT conditions (10). The recent survey of QP by Cottle [L.6] is recommended. Emphasizing the roots of

LCP, many of the following are concerned exclusively with QP: Dantzig, Wolfe, Beale, Zoutendijk, van de Panne, Winston, Graves, Keller, Cottle, Cottle and Dantzig, and Sacher.

(2) Howson–Lemke, Howson, Eaves, and Shapley.

(3) In this area, see the papers by Cottle [L.1] and Kaneko [L.4].

(4) Dantzig and Manne, Hansen, T., Evers, J. J. M., and Engles, C., Eaves. It may be expected that additional such models of economic equilibrium will be developed.

(5) In particular, again see [L.1]. In many instances of this area of application, the LCP is based upon QP.

(6) A substantial proportion of the CAM-activity has consisted of contributions in this area. The development to date is reported in the forthcoming [L.2], which unifies the contributions. Some of the main contributors are Gale, Samelson–Thrall–Wesler, Cottle, Dantzig, Murty, Lemke, Eaves, Pang, Mangasarian, Chandraskekaran, Karamardian, Garcia, Saigal, Mohan, Perriera, and Evers. Of special note are the works on principal pivoting of A. W. Tucker and the comprehensive papers of Fiedler and Ptak.

(7) One of the most recommended papers dealing with developments over the past 15 years is the paper by Eaves in [B], which develops a theory for solving general PL systems $F(x) = 0$ and includes the methods of simplicial approximation. The paper builds upon a previous paper by Eaves and Scarf and is a forerunner of a book by Eaves and Saigal.

2.3 Complementary Pivoting for LCP

It is well known that the LP model (11) has many applications. Historically, the growth of the LP area, and indeed of the NLP area, was in large part due to the advent of the simplex method (SM) of G. B. Dantzig as, on the one hand, an efficient way of solving (or "processing") LP problems (for any data c, b, A), and, on the other hand, as a tool for "constructively" proving theorems in LP and related areas.

The SM may be described as generating a *sequence of pivots* on a linear system, namely, $y = b + Ax$, whose termination pivot gives the resolution of LP. To a large extent this is also the case for LCP. We develop this thought as an integral part of this summary.

In turn these sequences of pivots, viewed geometrically, generate a piecewise-linear path (PL path), comprised of a finite number of line segments.

This generation of a PL path is also characteristic of the simplicial approximation methods (SAM) of the CAM-activity, which we consider in Section 3, and on such paths each segment may be considered as generated by a "pivot."

A relevant summary of the process of pivoting is developed in Appendix 2.

In LP essentially all methods in use utilize pivoting (SM or a variant), as distinct from *iterative* (i.e., basically nonterminating) methods. To date this also holds for LCP.

In 1965 the author proposed a pivot scheme, which we shall call the complementary pivot scheme (CPS), essentially as a method for solving QP via the LCP (10). CPS is similar to the "principal pivot" methods (PPM) of Cottle and Dantzig, which had been proposed prior to CPS for the matrix M psd [such as the coefficient matrix in (10) when Q is psd], or when M is, more generally, a P-matrix (that is, M has positive principal minors).

In turn, the CPS was modeled on the Lemke–Howson pivot scheme published in 1964 which furnished the first "constructive" proof (by pivoting) of the existence of a Nash equilibrium point for a bimatrix game. Later Eaves pointed out that CPS itself could process the game case also.

In LP one seeks to minimize a functional [$c^{\mathrm{T}}x$ in (11)], and SM (like most methods of NLP) is a method of "successive improvement" in $c^{\mathrm{T}}x$, whereas the novelty of CP is that it does not share this property. LCP is not per se an "optimization" problem. Rather (but roughly) one deals with "equilibrium"-type settings (such as the KKT conditions, or the game setting generally).

To describe CPS briefly (while alluding to the more general "homotopy approach"), prior to pivoting one has selected an "artificial" column $d \geqq 0$ for which for some $z_0 > 0$ one has $\bar{q} := q + z_0 d > 0$. In the generation of the PL path, by pivoting, using the "augmented" linear system

$$w = (q + z_0 d) + Mz, \tag{12}$$

the "parameter" z_0 is systematically varied. Along the path the conditions (a) $\bar{q} \geqq 0$, and (b) $z^{\mathrm{T}}w = 0$ are maintained. A solution to LCP therefore is obtained for a point with $z_0 = 0$.

We note here, to expand upon later, that also in the methods of simplicial approximation used in generating solutions to "homotopy systems": $H(x_0, x) = 0$ in the cases when H is PL, one generates PL paths of zeros of H (on which the parameter x_0 varies) quite like that which we have just discussed (then the linear system is "huge," and a column is generated only when used). In general, the PL H is a (linear) approximation to the underlying H whose zeros are sought. This brief allusion to these methods accounts also for the use of the term "complementary pivoting." Shortly, we shall exhibit CPS itself as generating a zero of a system by this homotopy approach.

2.4. Classes of Matrices, LCP, and CPS

Whereas SM "processes" LP for any data, CPS is more "matrix dependent." For given M and q, CPS also terminates in a finite number of pivots and either in a solution to LCP (with $z_0 = 0$) or not (with $z_0 \neq 0$).

This fact has given rise to much effort directed toward identifying *classes* of matrices M about which, on the one hand, one can make statements regarding existence and/or uniqueness of solutions, and on the other hand, classes which can be "processed" by CPS: here meaning an M such that one can say that, for a given q, CPS either terminates in a solution to the LCP, or else determines that LCP is *infeasible* (meaning that already $w = q + Mz$, $w, z \geqq 0$ has no solution). Two associated classes have been defined:

$$Q := \{M : \text{for all } q, \text{ LCP has a solution}\}$$

and (13)

$$Q_0 := \{M : \text{for each } q \text{ such that LCP is feasible there exists a solution}\}.$$

Over the years, quite substantial subclasses of Q and Q_0 have been identified. Worthwhile *characterizations* of Q and Q_0 themselves have yet to be found. Such characterizations would definitely be of great interest.

Many classes of square matrices have been identified and studied in this regard, with results of interest in other areas also. A summary of the efforts involved is given in Appendix 3 for the interested reader. This is discussed further in the forthcoming [L.2].

2.5. Existence and Uniqueness Results in LCP

Many results regarding existence and uniqueness in LCP have accumulated since the identification of the LCP. In the large these are summarized in the Q_0 and Q columns in Table I of class inclusions in Appendix 3, some of which we isolate here for emphasis.

To begin, note the "total inclusion" of the strong class (D) of positive-definite matrices (pd) M. In this sense all of the classes studied (in the upper portion of Table I) involve generalizations of properties of (D) and (D_0).

In particular, (P) (P-matrices) have the distinction that: M is in (P) iff *for all* q, LCP has a *unique* solution. Much effort has been expended in finding alternate characterizations of this exceptional class.

Note that the *strong* classes (E), (C), (P), (D) [all subclasses of (E)] are in Q; i.e., have solutions *for all* q.

Next note the role of the class (S), and feasibility: if $M \in (S)$, then for all q, LCP is *feasible*, and conversely. In particular, if $M \in Q_0$, then $M \in Q$ iff also $M \in (S)$ [that is, $Q = Q_0 \cap (S)$].

(D_0) shows up most prevalently in the applications (such as in QP), mainly because $x^{\mathrm{T}}Mx$ is *convex* iff $M \in (D_0)$. $(D_0) \subset Q_0$ and in LCP has the distinction that if $M \in (D_0)$ then, for all q such that $w = q + Mz$ is nondegenerate, LCP has 0 or 1 solutions.

The class (Z) has a large role in LCP. $(Z) \subset Q_0$, and hence, for example, $(K) := (Z) \cap (S) \subset Q$; but in fact $(K) \subset (P)$. (K) is the well-known class of "Minkowski" or "Leontief" matrices.

Finally, with regard to the CPS, as we have noted, the question as to which matrices M can be "processed" by CPS initially motivated much of the study concerning classes. If we for the moment identify as (N) the class of matrices M such that for any given q, CPS (applied with any permissible $d \geqq 0$), if it does not terminate in a solution to LCP, will ascertain that LCP is infeasible for that q. Then, in particular, $(N) \subset Q_0$, and it has been shown that $(N) \neq Q_0$.

In particular, with reference to Table I, as shown by Eaves, the large class $E_0^+ \subset (N)$. This result has been extended somewhat by Garcia. Also, for example, $(Z) \subset (N)$, as shown by Saigal.

Since (N) has not been completely identified, it may be expected that there will be additional efforts in this direction.

2.6. CPS as a Simple Example of the Homotopy Approach

A simple, but potentially useful, formulation of LCP is given by Eaves [G.3], which we illustrate with an example:

$$\text{Write} \quad w = q + Mz \quad \text{as} \quad [M \quad I]\begin{bmatrix} z \\ -w \end{bmatrix} + q = 0.$$

Let O_r, for $r = 1, 2, \ldots, 2^n$ denote the orthants of R^n in some order. Here let A_i denote the ith column of matrix A. For each r, let B^r denote the $n \times n$ matrix such that, for $x \in \operatorname{int} O_r$,

$$(B^r)_i = \begin{cases} M_i & \text{iff} \quad x_i > 0 \\ I_i & \text{iff} \quad x_i < 0. \end{cases} \tag{14}$$

Define $F: R^n \to R^n$ by

$$F(x) := B^r x + q \qquad \text{iff} \quad x \in 0_r. \tag{15}$$

Then x satisfies $F(x) = 0$ iff x solves LCP.

As a simple example of the homotopy approach, Eaves gives a picture of LCP as follows: If $d \geqq 0$, and for some $x_0 > 0 : d + x_0 q > 0$, consider $H: [0, \infty] \times R^n \to R^n$ given by

$$H(x_0, x) := F(x) + x_0 d. \tag{16}$$

If the linear system $w = (q + x_0 d) + Mz$ is nondegenerate (see Appendix 2), the PL path of zeros of H (which contains a ray in the nonpositive orthant of R^n) corresponds exactly with the path followed by CPS (and this path may be visualized as a non-self-intersecting path in R^n, parametrized by x_0). This is illustrated in Fig. 1 for $q := [{}^{-6}_{-4}]$, $d := [{}^{1}_{1}]$, and the P-matrix (see

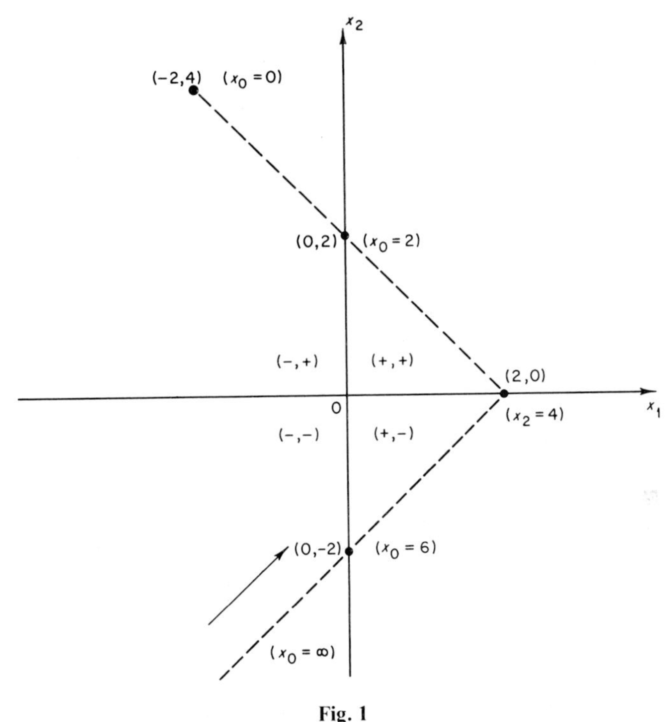

Fig. 1

Appendix 3) $M := \begin{bmatrix} 1 & 2 \\ 0 & 1 \end{bmatrix}$. In fact, this example is actually the case $n = 2$ of the example used by Murty to illustrate the "computational complexity" of CPS, which is not polynomial—the CPS takes $2^n - 1$ pivots; that is, the PL path pierces the interiors of all orthants in R^n.

2.7. Some Relations between LCP and LP

We note some results which tend to clarify distinctions between LP and the more general LCP.

First, since the KKT conditions for the LP problem (11) form an LCP problem with matrix $M \in (D_0)$, CPS can process the LP problem. When it does (and $d := e$, a column of 1's) it is precisely the "self-primal-dual method" discussed by Dantzig in his text "Linear Programming and Extensions" which uses either simplex-method or dual-simplex-method iterations depending upon the state. Some empirical results (due to Ravindran and others) seem to indicate that this technique applied to LP might be as efficient for general problems as the SM.

Secondly, there are large classes of matrices M (as yet incompletely identified) first and extensively studied by Mangasarian, then studied and augmented by Cottle and Pang, with the property that given M there is a readily available column p such that the solution to the LP

$$\min p^{\mathrm{T}}x, \qquad \text{where} \quad q + Mx \geqq 0, \quad x \geqq 0 \tag{17}$$

is also a solution to LCP with the same M and q.

Such classes are rather intimately related to the class (Z) which serves to yield the following relationship (Mohan) between CPS and SM; for CPS, again take $d := e$. Consider the SM applied to the LP, where $M \in (Z)$,

$$\min z_0, \qquad \text{where} \quad z_0 e + q + Mz \geqq 0, \quad z \geqq 0. \tag{18}$$

Then SM and CPS go through exactly the same sequence of pivots in the processing.

Finally, of a different nature is the following recent result (an example of which is contained in the model of Dantzig and Manne in [L.3], who were the first to apply it): If M is *copositive*, and the LP

$$\min q^{\mathrm{T}}z, \qquad \text{where} \quad q + Mz \geqq 0 \quad \text{and} \quad z \geqq 0, \tag{19}$$

has a *finite minimum*, then LCP defined by q and M has a solution which can be found by CPS.

3. CAM DEVELOPMENTS

Even from the vantage point of the present, there is no doubt that the timely publication by H. E. Scarf in 1967 of his constructive proof of the Brouwer fixed-point theorem provided the most impetus for the overall developments in CAM. It is also quite possible that Scarf is one of the few who could have anticipated the subsequent effects of his work. I again refer to his version of developments since 1967 in the "Introduction" article in [C].

In his proof of the Sperner lemma for the simplex, Scarf developed the notion of "primitive set," together with a pivotal algorithm for proceeding to a next primitive set. A sequence of primitive sets was thereby generated which terminated in a "completely-labeled" primitive set, which gave the desired approximation to a fixed point of the given function in the simplex. This initial use of the "primitive set" notion utilized the bimatrix-game setting, and the Howson–Lemke algorithm.

Following this, the process was "refined," by Scarf and Hansen and by Kuhn, toward the process of generating a sequence of "almost-completely labeled" members (simplices) of a standard simplicial subdivision of the simplex. This, in turn, has given rise to the many applications since of the constructive simplicial approximation (or "fixed-point") methods. Many of the early developments are described in Scarf's monograph "On the Computation of Economic Equilibria," 1973, written in collaboration with T. Hansen. Although almost all of these subsequent developments do not use per se Scarf's initial "primitive set" concept, that concept is in some respects more general and will quite probably form the setting for some future developments.

3.1. Simplicial Approximation Methods (SAM)

Uses of simplicial approximation subsequent to Scarf's initial contribution, peripheral developments [such as studies of appropriate simplicial subdivisions (SS)], and developments in related areas (such as the development of CP), have been many and varied. We shall shortly identify these in some detail.

Basic to all of these uses is the following:

One is given a SS, call it K^n, in R^n. Without loss of generality K^n is a subdivision of R^n itself, whose elements are simplices, each being the convex hull of some $n + 1$ *vertices* of K^n. Let K^0 denote the set of all vertices V of K^n. Each member of K^n has $n + 1$ *facets*, each facet being the convex hull of n of the $n + 1$ vertices defining the simplex. K^{n-1} is the collection of all facets of K^n. Each member of K^{n-1} is a facet of exactly two members of K^n. Two members of K^n are *adjacent* iff they share (intersect in) a common facet. Thus, each member of K^n has $n + 1$ *neighbors* (adjacent simplexes).

In each application a *sequence* of members of K^n is identified, such that adjacent members of the sequence are adjacent simplices, with no simplex appearing twice. Thus, adjacent members of the sequence differ in but one vertex. If S^{k-1}, S^k, S^{k+1} denote three consecutive members of the sequence, Fig. 2 illustrates the implicit "pivot" (or drop–add) process. In this sense, each pivot involves one "new" vertex, so that the sequence may be considered as a sequence of vertices.

Given a SS K^n, let $L\colon K^0 \to R^n$ be a function defined on K^0, with values in R^n. There is then a uniquely defined PL (continuous) function $L\colon R^n \to R^n$ that agrees with L on K^0 and is affine on each simplex of K^n.

In each of the applications of SAM, one has such a function to generate the sequence of simplices. (L is the "labeling" function.) Such an L generally has the property that it identifies a subset of K^n of "permissible" simplices; each permissible simplex has at most two neighbors which are permissible. (In Fig. 2 S^k has two permissible neighbors; namely, S^{k-1} and S^{k+1}. S^{k+1} might have only S^k as a permissible neighbor.) Given K^n and an L, a sequence can be started at some permissible S^0 which has *one* permissible simplex. It is then characteristic in all of the SAM applications that a *unique* sequence

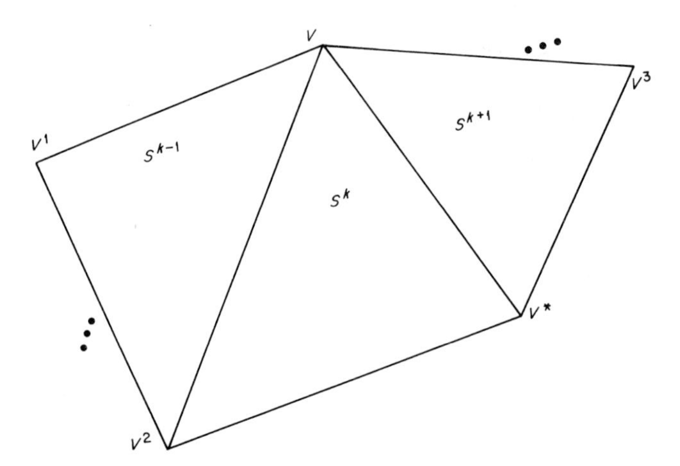

Fig. 2. Drop V^1—add V^*; drop V^2—add V^3.

is generated (which may be infinite) terminating if and only if a permissible simplex with one permissible neighbor is encountered. This is the feature inherent in the Howson–Lemke paper, subsequently utilized by Scarf.

From the computational point of view, one must be able to identify and conveniently store a current member of the sequence, such as S^k. Then, for the kth iteration, with V^* the vertex added on the previous iteration, one computes $L(V^*)$; subsequently determines the "pivot pair" (V^2, V^3); and then "updates" the stored S^k to S^{k+1}.

We discuss briefly each of the following main topics of development in the SAM area:

(1) Refined approximations and the homotopy approach.
(2) Algorithms.
(3) Simplicial subdivisions (triangulations).
(4) Theoretical results and uses.

At the present writing there are many indications that the general approach of SAM will find other, perhaps novel, applications, building upon the accumulated efforts to date. As we have noted, one may view most of the uses in terms of finding zeros of a system of equations but only for focus. To do proper justice, in a survey, to the scope of the CAM development in toto would require a much more comprehensive effort.

(1) Scarf's initial computation resulted in an approximation to a fixed point which, in particular, depended upon the fineness (mesh size) of the SS. In order to obtain an improved approximation, one would have to refine

the SS (smaller simplices), and *restart* the process. For some time people were concerned with this "restart problem." It is safe to say that this problem has been resolved, and not only with respect to approximating fixed points over the simplex. There were two notable results in this direction, each of which in turn has had value well beyond just resolving the "restart problem." The two techniques involved have been called the "sandwich method" and the "homotopy approach." Merrill is credited with initiating the sandwich method, and Eaves with initiating the homotopy approach. We shall develop the latter in some detail later. Roughly, in the sandwich method, using an SS one obtains an approximate solution to the given problem. Using a finer SS (generally a refinement of the previous), a next "iteration," starting from the previous approximation, leads to an improved approximation. Eaves initial paper utilizing the homotopy approach and concerned with fixed points over the simplex was, on the other hand, a method which "continually" improved the approximation in the sense of "moving through" a preassigned sequence of SSs, with "adjacent" SSs forming the boundary of an SS of one higher dimension. It was the first example, in SAM, of the (deliberate) generation of an infinite sequence of simplices which, in the limit, gave a fixed point. A sense in which each of these approaches is a special case of the other is discussed by Eaves in [G.3]. Subsequently others have utilized these ideas or provided variations. In this regard see again the "Introduction" [G.5] of Scarf. In particular, Eaves and Saigal developed a first extension of the Eaves result from the simplex to all of R^n. Other contributors include Gould, Fisher and Tolle, Kuhn, Lüthi, Wilmuth, and MacKinnon.

(2) Besides the more general algorithms included in (1) above, there have been algorithms (or methods) developed for specific problems, notably NLP, via the KKT conditions (several of quite recent vintage), or more generally the CP. But of course, in a broader sense, and in view of the fact that we are concerned with "constructive" methods, the designation "algorithm" may extend from a constructive proof for a class of theorems, to a detailed, computer-based method, incorporating a specific type of SS, designed for a specific problem (such as finding zeros of a polynomial), and possibly combining, in hybrid fashion, SAM computations with Newton-like steps. Such variations and more have been described. We note some of the contributors.

In NLP and CP, of particular note are Gould, Fisher, and Tolle, Kojima Karamardian, Scarf and Hansen, Mangasarian, Merrill, Freidenfelds, Lüthi, P. Reiser, Habetler and Price, and Garcia, and Saigal.

Finding fixed points and solving systems of equations, of particular note are Scarf and Hansen, Garcia, Garcia and Zangwill (general polynomial systems), Kuhn, and Kojima (roots of a polynomial).

(3) There are various reasons for studying different classes of SSs. Whereas in principle for most "theoretical" applications almost any SS may be used, in an actual application of say a SAM method (for example, for a computer program) there are practical considerations which require appropriate SS's (for example, relating to "storing" a current simplex, or to rates of convergence questions). Actually, it appears that there are very few kinds of SS's that have practical value. In any case, it also appears that more work needs to be done in this area. Work in this area has, in particular, been contributed by Scarf and Hansen, Kuhn, Eaves, Saigal, Todd, and Merrill. In particular, the survey by Saigal [G.8] is recommended.

(4) Perhaps one of the most impressive contributions of the SAM approach, again starting with Scarf's initial results, concerns the "constructive proof" aspect. By this means many of the classical results, e.g., dealing with fixed points or more generally with systems of equations, have been proven more simply, often generalized. Contributors in this regard include Scarf, Eaves, Garcia, Karamardian, Wilson, Moré, Shapley, Garcia–Gould, Kojima, Kuhn, Kojima–Meggido, Garcia–Lemke–Luthi, Charnes–Garcia–Lemke, and Garcia–Zangwill.

3.2. Solutions to Systems of Equations

We have noted that most of the CAM development may be viewed in terms of finding solutions to a system $F(x) = 0$ of n equations in n unknowns in a given region D in R^n. We have also noted that most of the recent SAM effort in this regard stresses the "homotopy approach." It appears, moreover, that in this setting the overlap between the CAM accomplishments and the more classical areas of algebraic or differential topology will be more readily identified. In particular, there is much indication of this in the very recent "path following" approaches we have mentioned and which we will make more note of shortly. In any case, we now identify by an example the homotopy approach which is contained in the very recent paper by Garcia and Zangwill [G.4].

If F is affine: $F(x) = b + Ax$, there is a unique solution iff the (constant) derivative $F'(x) = A$ is nonsingular. In particular, then, the zero is *isolated*. More generally, if, in a region D, at every zero of F, F' exists, is continuous, and is nonsingular, then, by the implicit-function theorem, the zeros are isolated. Then, for example, if D is compact it contains a finite number of zeros.

To locate zeros one sets up a homotopy: $H: R^{n+1} \to R^n$

$$H(x_0, x), \qquad \text{where} \quad H(0, x) =: F_0(x) \quad \text{and} \quad H(1, x) = F(x). \tag{20}$$

F_0 is some appropriately selected START function, with known zeros in D.

x_0 is the associated "parameter." Two simple forms, which to date are the two most widely used, are (16) and

$$H(x_0, x) := (1 - x_0)F_0(x) + x_0 F(x), \tag{21}$$

both linear in x_0. In almost all of the actual uses F_0 is affine with a unique zero.

Let us write $D^* := [0, 1] \times D$. Briefly, in D^*, with various choices or assumptions, one visualizes the set of zeros of H as consisting of a finite number of nonintersecting curves, all of whose endpoints are in bdry D^*. Thus, for $x_0 = 0$ they are the zeros of F_0, and for $x_0 = 1$ they are the (desired) zeros of F in D.

Generally, the SAM method is to (constructively) approximate one or more of these curves. For this one uses a SS of D^* and the PL approximation H^*, which agrees with H on the vertex set of the SS and is affine on each simplex of the SS. In particular, $F^*(x) := H^*(1, x)$ is the PL approximation to F in the SS of R^n induced by the SS of D^*. With the proper care, the zeros of H^* form PL paths—line segments in each simplex of the SS—whose endpoints with $x_0 = 1$ approximate zeros of F in D.

In a more recent series of developments, one more directly "follows a path" of zeros of H by some more classical (in any case different) approximating scheme. Such a scheme embodies the "constructive" element of such an approach.

Returning to D^* and the zeros of H, as in (21), we consider a development which is essentially in [G.4].

Suppose that D is compact, so that D^* is also, and that the derivative H' of H has rank n and is continuous at each point $x^* := (x_0, x)$ in D^*.

Visualize x^* on a path of zeros of H as a continuously differentiable function of a real parameter θ, so that $H(x^*(\theta)) = 0$ identically in θ. By the chain rule:

$$H'(x^*)x^{*\prime} = 0 \qquad [\text{where} \quad x^{*\prime} := x^{*\prime}(\theta) := dx^*(\theta)/d\theta]. \tag{22}$$

In the rk n $n \times n + 1$ matrix H', let H'_{-i} denote H' with its ith column deleted. The assumptions permit the assertion that the following system of $n + 1$ ordinary differential equations:

$$x_i^{*\prime} = (-1)^{i+1} \det H'_{-i}, \qquad i = 0, 1, \ldots, n, \tag{23}$$

satisfies (22), and hence its solution through some given zero of H is a "solution path" in the set where $H = 0$. It is then seen rather simply that each zero of H in D^* lies on a path of solutions to (23), and that therefore the set of zeros of H consists of a finite number of nonintersecting curves with continuous tangents $x^{*\prime}$, all of whose endpoints are in bdry D. This is a theorem in [G.4], and is illustrated in Fig. 3.

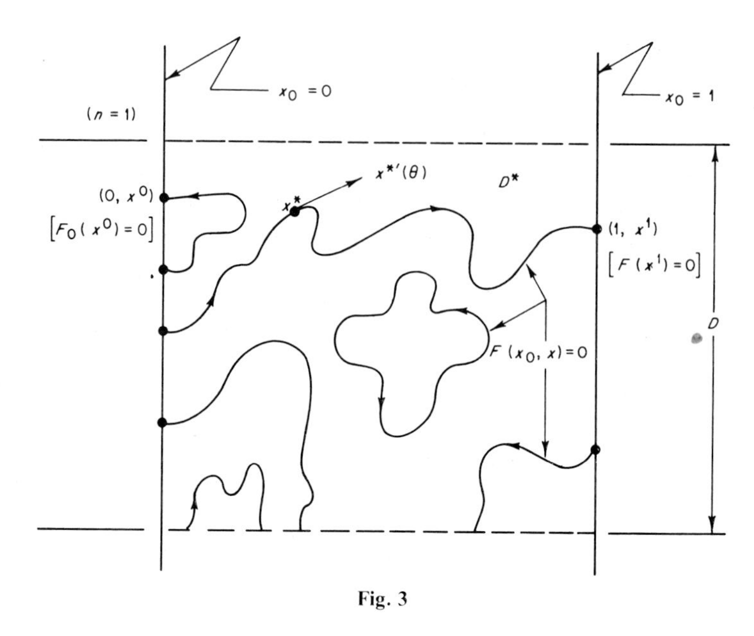

Fig. 3

To indicate how (23) comes about, for a fixed x^* write $A := H'(x^*)$. Since rk $A = n$, one has $Ac = 0$ for a column c such that $c^T c = 1$, and c is unique except for sign. Hence, $y = c$ is the unique solution to

$$\begin{bmatrix} c^T \\ A \end{bmatrix} y = \begin{bmatrix} 1 \\ 0 \end{bmatrix}. \tag{24}$$

Hence (from the adjoint, or using Cramer's rule)

$$c_i = k(-1)^{i+1} \det H'_{-i} \qquad \left(\text{where } 1/k = \det \begin{bmatrix} c^T \\ A \end{bmatrix} = \pm[\det AA^T]^{1/2} \right). \tag{25}$$

In particular, any continuous multiple of $c = c(x^*)$, such as defines the differential system in (23), gives a continuous measure of the tangent to a path through a given zero of H on which $H = 0$.

One objective of Garcia and Zangwill was to present a simplified development of many of the results of CAM *and* of more classical developments related to solving systems. In particular, they adopt the point of view that the situation illustrated in Fig. 3 is a "typical" one. That is, in particular, even though the above assumption of continuous differentiability is rather strong, they point out, using Sard's theorem and the Weierstrass approximation theorem, that their assumption holds "with probability one." Again, in particular, an important element in the SAM development has been that in many instances assuming merely that F is continuous suffices to obtain results (for example, the Brouwer fixed-point theorem), and without reference to differentiability. It is nonetheless interesting and of value to retain the picture in Fig. 3.

We next note the "pivotal" aspect of a typical SAM application. Visualize the cylindrical region D^* as above, and a SS of D^* (more generally, and without loss of generality, of $[0, 1] \times R^n$). The homotopy H is now PL with respect to H (hence, determined by the vertices of the SS), and one is concerned with the zeros of H in D^*. With some mild assumptions of nondegeneracy and "regularity," the picture is quite like that in Fig. 3. Again, with the zeros of $F_0(x) := H(0, x)$ known, it is first assumed that all zeros in D of $F_0(x)$ and of $F(x) := H(1, x)$ occur only in relative interiors of simplices in the $x_0 = 0, 1$ portions of bdry D^*. (Then it follows that H, in each such simplex containing a zero, has a nonsingular derivative there, and hence that the zeros of F_0 and F in D are finite in number.) Then, from each such zero, a PL path of zeros of H may be initiated. We illustrate a local piece of such a PL path in Fig. 4.

Corresponding to the simplex S^k is the abbreviated linear system

$$\begin{bmatrix} 0 \\ 1 \end{bmatrix} = y_0 \begin{bmatrix} H(V^0) \\ 1 \end{bmatrix} + y_1 \begin{bmatrix} H(V^1) \\ 1 \end{bmatrix} + y_2 \begin{bmatrix} H(V^2) \\ 1 \end{bmatrix}; \qquad \begin{bmatrix} y_0 \\ y_1 \\ y_2 \end{bmatrix} \geqq 0. \quad (26)$$

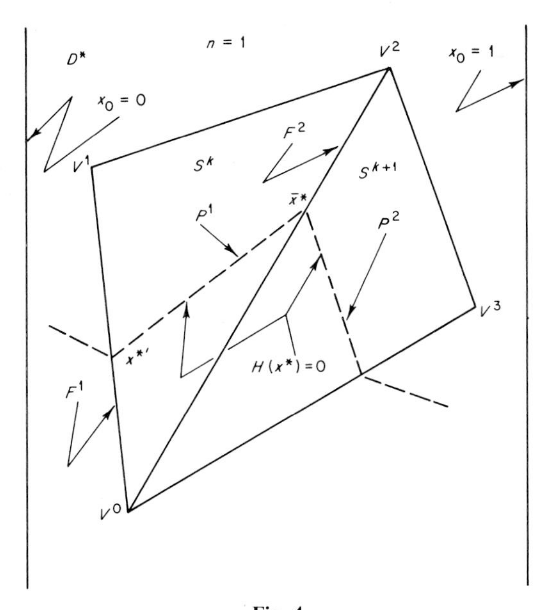

Fig. 4

[Since $n = 1$, the values $H(V)$ are scalars.] The segment P^1 is the set of all (feasible) solutions to (26). We are supposing that in the facet F^1 of $S^k x^{*\prime} = \bar{y}_0 V^0 + \bar{y}_1 V^1$ is the only zero of H, and that $x^{*\prime}$ is in the relative interior of F^1 (so that $\bar{y}_0, \bar{y}_1 > 0$). Thus, $[y_0, y_1, y_2] = [\bar{y}_0, \bar{y}_1, 0]$ is the solution to (26) with $y_2 = 0$. Introducing y_2 as basic (increasing y_2 from 0), we find that y_1 becomes 0 at $\bar{x}^*$. The matrix

$$\begin{bmatrix} H(V^0) & H(V^1) \\ 1 & 1 \end{bmatrix}$$

is nonsingular, and *by the pivot* process, the matrix

$$\begin{bmatrix} H(V^1) & H(V^2) \\ 1 & 1 \end{bmatrix}$$

is likewise nonsingular. We make the (nondegeneracy) assumption that the linear system (26) is nondegenerate. Thus, at $y_1 = 0$, $[y_0 \; y_1 \; y_2] = [y_0', 0, y_2']$, where $y_0', y_2' > 0$, so that $\bar{x}^* := y_0' V^0 + y_2' V^2$ is in the relative interior of F^2, the facet of S^k opposite V^1.

Now the local knowledge of the SS yields (uniquely) the vertex V^3 (assumed to be in D^*), and geometrically we have the "drop–add" operation: drop V^1–add V^3 leading from S^k to S^{k+1}. Then the computed column $[H(V^3) \; 1]$ is added to (26) (as "next pivot column"), and the situation is exactly as with S^k. Note that, therefore, with the assumption that as one proceeds each segment of zeros of H generated has its endpoints in relative interiors of facets of the SS, so that a unique path of zeros is generated, the $n \times n + 1$ matrix H' has rk n in each simplex used, so that all of the zeros in the sequence of simplices used comprise the path generated. In particular, therefore, since with the initial assumption, such a path may be initiated at any zero of F_0 or of F in D, there is a finite number of nonintersecting PL paths in D^* which contain the zeros of F_0 and F.

3.3. Continuation Methods

Of course, the problem of finding zeros of systems of equations has concerned us for many, many years, as has the existence problem. It is safe to say that the SAM approach is yet another tool, and a valuable one, which adds much to the area. In particular, the SAM approach, being *constructive* has attracted much attention. We again note that these methods derived from Scarf's constructive proof of the Brouwer theorem.

By "continuation" methods, one means methods which trace a path of zeros of a homotopy $H(x_0, x)$, based on a function $F(x)$, whose zeros are sought. These have also been called "Davidenko" methods (see, e.g. [G.2]), since they trace back at least as far as to a paper by Davidenko "On the Approximate Solution of a System of Nonlinear Equations" published in 1953.

In the past three years a series of papers [G1, G.2, G.6] indicate a development of "constructive" methods which appears on the one hand to have a computational potential paralleling that of the SAM approach, and on the other hand to indicate a larger area of fruitful (future) study which embraces both approaches. However, the extent of such a unification and the form it will take is a matter for the next few years to decide. It will no doubt have its bases in combinatorial and differential topology.

The series of papers in question was initiated by Kellogg *et al.* [G.6], which also concerned the Brouwer theorem with impressive computational results. We shall terminate this summary with a brief discussion of the approach, as it relates to our previous development.

Let $D \subset R^n$ be a bounded, open, convex set (such as the interior of a simplex), so that $\bar{D}$, the closure of D, is compact. Let $G: \bar{D} \to D$ be continuous. Brouwer's theorem states the existence of a fixed point; for some x^* in D, $G(x^*) = x^*$. Let C denote the set of fixed points of G in D.

There is the "classical" function $H:(\bar{D} - C) \to (\text{bdry } D)$ defined as follows: For $x \in \bar{D} - C$, $H(x)$ is the point where the ray from $G(x)$ through x meets bdry D. This is illustrated in Fig. 5 (D is the simplex).

Note that for $y \in \text{bdry } D$, $H(y) = y$.

Thus, for a function $\theta: (\bar{D} - C) \to R$

$$\begin{aligned} H(x) &= x + \theta(x)[G(x) - x] \\ &= [1 - \theta(x)]x + \theta(x)G(x), \qquad \text{where} \quad \theta(x) < 0. \end{aligned} \tag{27}$$

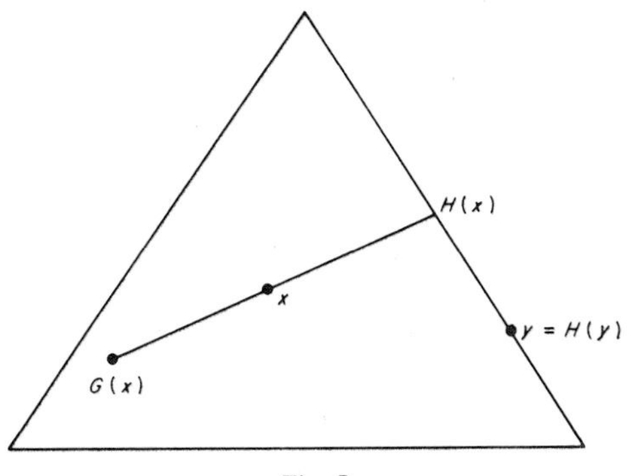

Fig. 5

The θ, hence H, is continuous. A classical *existence* proof is based on the fact that there is no continuous function $h: \bar{D} \to \text{bdry } D$ such that $H(y) = y$ when $y \in \text{bdry } D$—which would contradict the assumption that $C = \varphi$. In the following way (see [G.6]) this proof was rendered "constructive":

In 1959 Pontryagin [G.7] studied the homotopy properties of maps $\varphi: \bar{B}^n \to \text{bdry } B^n$ from a ball into its boundary. In particular, he used Sard's theorem (see below) and (quoting from [G. 6]) its corollary fact that for almost all $y \in \text{bdry } B^n$ the inverse image $\varphi^{-1}(y)$ is a finite union of non-intersecting closed curves (such as we see in Fig. 3).

Sard's theorem appears to be the source of the "a.e." (almost everywhere), or "probability one" statements currently being used. We give a (quite simplified) version which is the one people have used (e.g., Garcia–Zangwill [G.4]):

Theorem (Sard). *Let $D \subset R^{n+1}$ be open and bounded. Let $f: \bar{D} \to R^n$ have a continuous derivative f'. Then, if f has all second order partials, the set of all $y \in R^n$ such that for some point $x \in f^{-1}(y)$, $\text{rk } f'(x) < n$, has measure* 0.

If $\text{rk } f'(x) = n$ for all $x \in f^{-1}(y)$, y is a *regular* value of f (which, in particular, is true when $f^{-1}(y) = \varphi$). Thus, regularity holds a.e.

In 1963 Hirsch published a paper [G.9] wherein, in a very elegant way, he proved the fact noted after (27) and thereby the Brouwer theorem. With the assumption that no fixed points exist, he used the "standard retraction" (27) and, employing SS's and Sard's theorem, he investigated $H^{-1}(y)$, for a regular value y, and, exhausting the possibilities, obtained a contradiction.

The above ideas were used in [G.6]. From a regular point $y^0 \in \text{bdry } D$, that path of $H^{-1}(y^0)$ which contains y^0 is considered. Since H maps from R^n to an $(n-1)$-dimensional space, with the assumption that G has a continuous second derivative, by Sard's theorem the regularity of y^0 ensures that on that path $\text{rk } H'(x) = n - 1$, so that the path is a non-self-intersecting curve with a continuous tangent. If z is on the path and $z \in \text{bdry } D$, then $z = H(z) = H(y^0) = y^0$, but, with the rank condition, a piece of the path in some neighborhood of y^0 is an arc. Hence, the path cannot lead back to bdry D, and it is argued that the path must contain a point of C, as desired.

From the computational point of view, the authors set up a system of differential equations—*exactly* (23), but with $i = 0$ deleted, and, again, the "constructive" aspect is embodied in the numerical "path following" to a point of C, starting at y^0.

It is of much interest (see Scarf's relevant remarks in [G.5]) to note how the authors relate their approach to the "continuous" Newton method [compare the derivation of (23)]:

From (27), writing $F(x) := G(x) - x$,

$$H'(x) = H' = (I + \theta F') + F\theta'^{T} = \theta(F' + \theta^{-1}I) + F\theta'^{T}. \tag{28}$$

Since rk $H' = n - 1$, a $v \neq 0$, unique within a multiple, satisfies $H'v = 0$. Now, away from y^0, $H(x) - x$ is bounded away from 0 and equals θF. Hence, as F approaches 0 along the path, θ approaches $-\infty$. Assuming $(F' + \theta^{-1}I)$ is nonsingular [which is true if F' has no eigenvalues in $(1, \infty)$], with proper choice of the multiple, we may take the tangent $x' = v$ as

$$x' = -(F' + \theta^{-1}I)^{-1}F. \tag{29}$$

This is visualize as the basis for an approximation scheme. Near a fixed point (a zero of F) θ^{-1} is small, so that the calculations are as those applied to

$$x' = -F'^{-1}F, \tag{30}$$

the Newton equations. (For example, with "step size 1," the kth Newton iteration would be

$$x^{k+1} = x^k - [F'(x^k)]^{-1}F(x^k).) \tag{31}$$

As with the SAM development, this "path-following" technique has been extended to the applications to other, more general, problems, and apparently with some comparable numerical successes. We have tried to indicate not only the differences, but the similarities of these two general "constructive" approaches, and to suggest that there will be new developments in the near future embracing these two. Again, in this regard, the remarks of Scarf [G.5], especially as they regard the conjunction with the Newton-like approaches, are to be noted.

APPENDIX 1: SOME RELATIONS AND EQUIVALENCES

(i) In (5), in the case $D = R^n_+ := \{x : x \geqq 0\}$, it is readily seen (Karamardian) that (4) and (5) are equivalent. We shortly note a reverse result.

(ii) From the standpoint of analytic generalization, (5) seems to be most generalizable (see, e.g., Cottle [L.1]); namely, when V_1 and V_2 are vector spaces over R, $g: V_1 \to V_2$, and a bilinear form $B(V_1, V_2)$ replaces the scalar product $x^{T}y$.

(iii) CP has many equivalents of the form (2). Given f in (4), a simplest, but one often used to apply theoretical results, is obtained when an F, as in (2), is defined by

$$F_i(x) := \min(x_i, F_i(x)). \tag{A.1}$$

Then, for $D = R^n$, $F(x) = 0$ iff x is a solution to CP.

A more general transformation, displayed and used by Mangasarian (see his paper in [B]) is the following:

Let $\theta: R \to R$ be strictly increasing with $\theta(0) = 0$. Define $h(a, b) := \theta(a) + \theta(b) - \theta(|a - b|)$. Then $h(a, b) = 0$ iff $a, b \geqq 0$ and $ab = 0$. Hence, defining F, as in (2), by

$$F_i(x) := h(x_i, f_i(x)), \tag{A.2}$$

again, $F(x) = 0$ iff x is a solution to CP. Also, since

$$(a + b) - |a - b| = 2 \min(a, b), \tag{A.3}$$

the device (A.1) corresponds to the case $\theta(t) := t$, and then F is continuous when f is. Mangasarian notes that even for simple θ's (e.g., $\theta(t) = t \cdot |t|$), F is differentiable when f is.

(iv) With regard to the differentiable case (6)–(8) of NLP, and an equivalence of (5) and (4), consider the *convex* case of NLP, namely, when h_0 and h_i are convex (so that D is then a convex region), a necessary and sufficient condition that $\bar{x}$ solve (6) is that $z = [\bar{x}, u] \geqq 0$ satisfies [8] for some u. The KKT conditions are "geometrical" and also show that a solution $\bar{x}$ to (6) also solves min $x^{\mathrm{T}} Dh_0(\bar{x})$, in which the functional to be minimized is linear, which further shows that $\bar{x}$ solves the VIP (5) with $g(x) := Dh_0(x)$. In this sense, VIP (5) for the region D is more general since a "general" g is not the derivative (gradient) of a scalar function.

On the other hand, as observed by Rockafellar (see his paper in the recent [G.10]), consider (5) for the case of D which has the representation as in (6). If $\bar{x}$ is a stationary point, of (g, D) upon writing $c := g(\bar{x})$, $\bar{x}$ also solves the NLP $\min_D c^{\mathrm{T}}x$, whose KKT conditions are (8) after replacing $h_0(x)$ by $c^{\mathrm{T}}x = x^{\mathrm{T}}g(\bar{x})$, so that, for some $u \geqq 0$, $[\bar{x}, u]$ is a solution to CP:

$$g(x) + Dh(x) \cdot u \geqq 0; \qquad -h(x) \geqq 0$$

and (A.4)

$$x^{\mathrm{T}}[g(x) + Dh(x) \cdot u] = 0 = u^{\mathrm{T}}h(x)$$

(where $D_h := [Dh_1, \ldots, Dh_m]$ is the derivative (Jacobian) of h).

Further, therefore, for such VIP (5) where the h_i are convex, $\bar{x}$ solves VIP iff for some $u \geqq 0$, $[x, u]$ solves (A.4), establishing an equivalence of (5) with (4).

APPENDIX 2: PIVOTING ON A LINEAR SYSTEM

Consider a linear system in the form

$$L: y := b + Ax \qquad (A, m \times n. \quad \text{Write:} \quad z := [y, x]). \tag{A.5}$$

$L \subset R^{m+n}$ is the set of all *solutions*, a linear set of dimension n in z-space. There are as many such forms in the *system variables* z_k as there are nonsingular $m \times m$ submatrices of the matrix $(I, -A)$ (equivalently—one plus the number of nonsingular submatrices of all orders of A), each such form equally well expressing L explicitly. A typical such form we may write as

$$y^k := b^k + A_k x^k \qquad (A_k,\ m \times n;\ \ z^k := [y^k,\ x^k]), \tag{A.6}$$

where therefore z^k is some permutation of z.

[Note: If $F(z) := (I, -A)z - b$, then $DF(z) = (I, -A)$, and the above remarks relate to the implicit function theorem, and "solving $F(z) = 0$ for some m of the $m + n$ variables."]

For the typical form (A.6), on the left are the m *dependent* (also called basic) variables; on the right the n (current) *independent* (nonbasic) variables, and associated with the form is the *basic solution* (b.s.), namely $\bar{z}^k = [\bar{y}^k, \bar{x}^k] = [b^k, 0]$, obtained by setting all independent x_j^k to 0.

A *pivot* (according to Gauss) from a typical form such as (A.6) entails an exchange of (the roles of) an independent and a dependent variable: the "(r, s) pivot" from (A.6) involves the exchange of the pair (y_r^k, x_s^k) and the "standard updating" to the equivalent form.

In z-space the (r, s)-pivot is seen as generating a line segment of solutions (points of L) from the basic solution of the first to that of the second.

In this way, a sequence of pivots from a given form [such as (A.5)] may be viewed as generating a PL path in L. In particular, since the number of forms is finite, a sequence generated so that *no form repeats* must terminate.

A point in L *may* possibly be the basic solution for more than one form. This is precluded if the system is *nondegenerate*. A linear system L [such as represented in (A.5)] is called nondegenerate iff no point z of L has *more than* n 0's [equivalently, iff for every form, such as (A.6), for each i, $b_i^k \neq 0$]. In the case of nondegenerates, the basic solutions are then all points of L having exactly n 0's, and there is then a one–one correspondence with basic solutions and forms for L.

The SM applied to LP (11) may be said to generate a sequence of pivots from the form $y = b + Ax$, with a functional such as $c^{\mathrm{T}}x$ "guiding the pivoting." When the linear set L is nondegenerate, no form may repeat, and hence the sequence terminates and in a resolution of LP.

Similarly for the CPS: When the linear set $L\colon w = q + Mz$ is nondegenerate, and the "artificial" column d is introduced, proper variation of z_0, and retaining the conditions $(q + z_0 d) \geqq 0$, $z^{\mathrm{T}}w = 0$ while pivoting (i.e., complementary pivoting), "guides the pivoting" on L so that no form repeats.

APPENDIX 3

A "fundamental theorem of linear inequalities" is Ville's theorem:

For any matrix A, either

$$\textit{for some} \quad x \geqq 0, \quad x \neq 0, \qquad Ax \geqq 0,$$

or

$$\textit{for some} \quad u \geqq 0, \qquad A^{\mathrm{T}}u < 0. \tag{A.7}$$

This fact motivates the "largest," most basic classes (S_0) and (S) of matrices M considered below.

Following the symbolic definitions of the classes that have been studied, tables of class inclusions are given together with other results.

In "definitions" below, the line headed (P), for example, is to be read as:

$$(P) := \{M : \text{each principal minor of } M \text{ is positive}\}.$$

Each class is supposed to contain matrices of all orders.

A3.1. Summary of Definitions

- (S_0): For some $0 \neq x \geqq 0$, $Mx \geqq 0$
- (S): For some $x > 0$, $Mx > 0$
- (E_0): For each principal submatrix $\overline{M}$ of M, $\overline{M} \in (S_0)$
- (E): for each principal submatrix $\overline{M}$ of M, $\overline{M} \in (S)$
- (P_0): Each principal minor of M is nonnegative
- (P): Each principal minor of M is positive
- (D_0): For all x: $x^{\mathrm{T}}Mx \geqq 0$ (M is psd)
- (D): $M \in (D_0)$ and $x^{\mathrm{T}}Mx = 0$ iff $x = 0$ (M is pd)
- (C_0): For all $x \geqq 0$: $x^{\mathrm{T}}Mx \geqq 0$ (M is "copositive")
- (C): $M \in (C_0)$ and $0 \neq x \geqq 0$ implies $x^{\mathrm{T}}Mx > 0$ (M is strictly copositive).
- (U_0^R): M is "row principally dominant"—the rank of any submatrix of rows of M is the rank of the principal submatrix corresponding to that set of rows
- (U_0^C): Replace "rows" by "columns" in the above
- (U_0): $:= U_0^R \cap U_0^C$
- (U) Each principal minor of M is nonzero ("nondegenerate" M)
- (Z): M such that for $i \neq j$, $M_{ij} \leqq 0$ (nonpositive off-diagonal elements)
- $E_0(0)$: If $w = Mz$, $w, z \geqq 0$, and $w^{\mathrm{T}}z = 0$, then for some $y = -M^{\mathrm{T}}x$, $0 \leqq [y, x] \leqq [w, z]$, and if $z \neq 0$, then $x \neq 0$
- $E(0)$: If $w = Mz$, $w, z \geqq 0$, and $w^{\mathrm{T}}z = 0$, then $z = 0$
- $E(p)$: For $p > 0$, if $w = p + Mz$, $w, z \geqq 0$, and $w^{\mathrm{T}}z = 0$, then $z = 0$
- (C_0^+): $M \in (C_0)$ and $[x \geqq 0, x^{\mathrm{T}}Mx = 0]$ imply $[(M + M^{\mathrm{T}})x = 0]$
- (P_0^+): $:= (P_0) \cap (U_0^R)$ ("row-adequate" matrices)

In Table I a × denotes inclusion; e.g., $(C) \subset (E_0^+)$; a ○ denotes definite noninclusion. Note that the subscript ○ on a class, such as (C_0), denotes classes weaker, larger than those, such as (C), without the subscript. Note in particular the class relationships with Q and Q_0.

In Table I under "some special or composed classes," "d" denotes "inclusion by definition." For example,

$$(K_0) := (S_0) \cap (Z).$$

Whereas some of the facts in the table follow easily, or are by definition, some are by no means easy.

It should be noted that the classes are not limited to *symmetric* matrices M.

TABLE I

Table of Class Inclusions[a]

	S_0												Q_0	Q
E_0	×	E_0												
C_0	×	×	C_0											
P_0	×	×	○	P_0										
E_0^+	×	×	○	○	E_0^+									
C_0^+	×	×	×	○	×	C_0^+							×	
P_0^+	×	×	○	×	×	○	P_0^+						×	
D_0	×	×	×	×	×	×	○	D_0					×	
S	×	○	○	○	○	○	○	○	S				○	
S	×	○	○	○	○	○	○	○	S					
E	×	×	○	○	×	○	○○	○	×	E				
C	×	×	×	○	×	×	○	○	×	×	C		×	×
P	×	×	○	×	×	○	×	○	×	×	○	P	×	×
G	×	×	×	×	×	×	×	×	×	×	×	×	×	×

Some Special or Composed Classes

	S_0	S	E_0	P_0	Z	U	$E_0(0)$	$E(0)$	$E(p)$	Q_0	Q
Z					d					×	
K_0	d		×	×	d					×	
K	×	d	×	×	d	×				×	×
E_0^+			d				d			×	
			d			d				×	×
				d			d			×	
				d				d	×	×	×
$R(p)$								d	d	×	×

[a] All inclusions noted are proper.

A3.2. Some Special Relations

(If A denotes a class: $A^{\mathrm{T}} := \{M^{\mathrm{T}} : M \in A\}$, $A^s := \{M \in A : M = M^{\mathrm{T}}\}$, and $A^{-1} := \{M \in A : M$ is nonsingular$\}$.)

Transpose: $E_0 = E_0^{\mathrm{T}}$, $E = E^{\mathrm{T}}$ (although $S_0 \neq S_0^{\mathrm{T}}$, $S \neq S^{\mathrm{T}}$). Hence, for example, $E_0 \cap U = (E_0 \cap U)^{\mathrm{T}}$.

Inverse: $(U_0^R)^{-1} = U = (U_0^C)^{-1}$. Hence, $(P_0^+)^{-1} = P$ (although $P_0^{-1} \neq P$, $D_0^{-1} \neq D$).

Symmetry: $P_0^s = D_0^s$, $P^s = D^s \subset U_0$ (classical, although $P_0 \neq D_0$, $D_0 \not\subset U_0$) $C_0^s = E_0^s$, $C^s = E^s$.

SELECTED REFERENCES

THREE RECENT VOLUMES:

A. M. L. BALINSKI AND R. W. COTTLE, eds., "Complementarity and Fixed Point Problems," Mathematical Programming Study, No. 7, North–Holland Publ., Amsterdam, 1978.

B. R. W. COTTLE AND C. E. LEMKE, eds., *Nonlinear Programming: Proc. Symp. Appl. Math., Amer. Math. Soc., Soc. Indust. Appl. Math.*, Amer. Math. Soc., Providence, Rhode Island, 1976.

C S. KARAMARDIAN, ed., in collaboration with C. B. GARCIA, "Fixed-Points: Algorithms and Applications," Academic Press, New York, 1977.

L.1. R. W. COTTLE, "Complementarity and Variational Problems," Symposia Mathematica, Vol. XIX, Inst. Naz. Alta Math., Bologna, 1976.

L.2. R. W. COTTLE AND C. E. LEMKE, Classes of Matrices and the Linear Complementarity Problem, Tech. Rep. Dept. Oper. Res., Stanford Univ., Stanford, California (forthcoming).

L.3. G. B. DANTZIG AND A. S. MANNE, A complementarity algorithm for an optimal capital path with invariant proportions, *J. Econom. Theory* **9** (1974).

L.4. I. KANEKO, LCP with an $n \times 2n$ P-matrix, *in* "Complementarity and Fixed Point Problems" (M. L. Balinski and R. W. Cottle, eds.), Mathematical Programming Study No. 7, North–Holland Publ., Amsterdam, 1978.

L.5. J.-S. PANG, I. KANEKO, AND W. P. HALLMAN, On the Solution of Some (Parametric) LCP's with Applications to Portfolio Analysis, Structural Engineering, and Graduation, MRC Tech. Summary Rep., WP 77–27, Univ. of Wisconsin, Madison, 1977.

L.6. R. W. COTTLE, Fundamentals of quadratic programming and linear complementarity, *NATO–ASI Proc. Confer. Engrg. Plasticity Math. Programming, Univ. of Waterloo, 1977.*

GENERAL REFERENCES

G.1. J. C. ALEXANDER AND J. A. YORKE, The homotopy continuation method: Numerically implementable topological procedures, *Trans. Amer. Math. Soc.* (to appear).

G.2. S.-N. CHOW, J. MALLET-PARET, AND J. A. YORKE, Finding zeroes of maps: Homotopy methods that are constructive with probability one (to appear).

G.3. B. C. EAVES, A short course in solving equations with piecewise-linear homotopies, *Nonlinear Programming, SIAM–AMS Proc.* **9** (1976).

G.4. C. B. GARCIA AND W. I. ZANGWILL, On a New Approach to Homotopy and Degree Theory, Report, Univ. of Chicago, Chicago, Illinois, 1978 (to appear).

G.5. H. E. SCARF, Introduction, in reference C.

G.6. R. B. KELLOGG, T. Y. LI, AND J. YORKE, A constructive proof of the Brouwer fixed-point theorem and computational results, *SIAM J. Numer. Anal.* **13** (1976).

G.7. L. S. PONTRYAGIN, Smooth manifolds and their applications in homotopy theory, *Trans. Amer. Math. Soc.* **11** (1959), 1–114.

G.8. R. SAIGAL, "Fixed Point Computing Methods," Encyclopedia of Computer Science and Technology, Vol. 8, Decker, New York, 1977.

G.9. M. HIRSCH, A proof of the nonretractibility of a cell onto its boundary, *Proc. Amer. Math. Soc.* **14** (1963), 364–365.

G.10. R. W. COTTLE AND F. GIANESSI, eds., Variational Problems and Complementarity Problems in Mathematical Physics and Economics, *Proc. Confer., 2nd Course*, Internat. Sch. Math., Center Sci. Culture, Erice, Trapani, Sicily, 1978 (to appear).

Work partially supported by The National Science Foundation Grant NSF MCS78-02096, and Institut für Operations Research, Eidgenössische Technische Hochschule, Zürich. An earlier version of this survey appears in [G.10].

DEPARTMENT OF MATHEMATICAL SCIENCES
RENSSELAER POLYTECHNIC INSTITUTE
TROY, NEW YORK

On Discovering Hidden Z-Matrices

JONG-SHI PANG

In this paper, we (1) present a finite procedure for determining if a given matrix is hidden Minkowski, (2) establish a necessary and sufficient condition for a nondegenerate matrix M to be hidden Z, and (3) use this latter characterization to show that the class of hidden Minkowski matrices is properly contained in the class of P-matrices. We also discuss a few other consequences of the characterization.

1. INTRODUCTION

A *Z-matrix* is a real square matrix whose off-diagonal entries are nonpositive. The importance of a Z-matrix in the study of the linear complementarity problem has been documented in numerous places (see [1, 3, 5, 17–19]). Recently, an extension of the class of Z-matrices has been introduced by Mangasarian [12, 13] in his study of solving linear complementarity problems as linear programs. In [15], we proposed to call matrices belonging to this extended class *hidden* Z. Specifically, a real square matrix M is hidden Z if there exist Z-matrices X and Y satisfying the two defining conditions below:

(M1) $$MX = Y,$$

(M2) $$r^{\mathrm{T}}X + s^{\mathrm{T}}Y > 0 \qquad \text{for some} \quad r, s \geqq 0.$$

Recalling that a *Minkowski matrix* is a Z-matrix which is a *P-matrix* as well (i.e., which has positive principal minors), we shall call a hidden Z-matrix *hidden Minkowski* if it is also a P-matrix.

Numerous properties of a hidden Z-matrix have been obtained (see [4, 14–16]). In particular, such a matrix bears very much the same relationship to a certain *hidden Leontief matrix* as a Z-matrix to a *Leontief matrix* (see [14]). Incidentally, it is precisely this fact which has motivated us to call such matrices hidden Z. Recall that a real matrix A is Leontief if it has at most one positive entry in each column and there is a vector $x \geqq 0$ such that $Ax > 0$ and that a real matrix A is hidden Leontief if there is a nonsingular matrix D such that DA is Leontief. Properties and applications of Leontief matrices are well recognized in the literature (see [7, 20]). Hidden Leontief matrices were introduced by Saigal [18]. An application exploiting the hidden Leontief property can be found in [11].

ISBN 0-12-178150-X

Despite the fact that various interesting classes of matrices have been shown to belong to the class of hidden Z-matrices, due to the nonlinearity of condition (M2), it is in general, not easy to determine if an arbitrary matrix is hidden Z. The purpose of this paper is to provide at least a partial resolution to this problem by (1) developing a finite procedure for determining if a given matrix is hidden Minkowski and (2) establishing a useful characterization for a *nondegenerate matrix* (i.e., one whose principal submatrices are all nonsingular) to be hidden Z. Among the consequences of this latter characterization, we show that the class of hidden Minkowski matrices is properly contained in that of P-matrices.

2. DISCOVERING HIDDEN MINKOWSKI MATRICES

In this section, we develop a finite procedure which can be used to determine if a given matrix is hidden Minkowski. The procedure is motivated by the one described in Dantzig and Veinott [8] for discovering a *hidden totally Leontief matrix* and involves solving two linear programs. We recall that a Leontief matrix A is *totally Leontief* if there is a vector $y \geqq 0$ such that $y^{\mathrm{T}}A > 0$ and that a hidden Leontief matrix A is hidden totally Leontief if the matrix DA is in fact totally Leontief for some nonsingular matrix D. As in the case of a hidden Z-matrix, a hidden Minkowski matrix also bears the same relationship to a certain hidden totally Leontief matrix as a Minkowski matrix to a totally Leontief matrix (see [15]).

The cornerstone of our procedure is a result which shows how the nonlinear condition (M2) can be replaced by a linear one if M is hidden Minkowski. We state this key result in the theorem below.

Theorem 1. *Let M be a real square matrix. Then M is hidden Minkowski if and only if the following conditions are satisfied*:

(1) $MX = Y$ *for some Z-matrices X and Y*;
(2) $Xe > 0$ *where e is the vector of* 1's;
(3) *M is an S-matrix*; *i.e., there is a vector $x \geqq 0$ such that $Mx > 0$.*

The proof of the theorem depends on the lemma below whose proof can be found in [15].

Lemma 1. *Let M be a hidden Z-matrix. Then it is hidden Minkowski if and only if it is an S-matrix. Moreover, if M is indeed hidden Minkowski and if X and Y are any two Z-matrices satisfying the defining conditions* (M1) *and* (M2), *then X is itself Minkowski.*

Proof of Theorem 1. (Sufficiency) Conditions (1) and (2) imply that M is hidden Z. Hence, by the above lemma, M is hidden Minkowski.

(Necessity) Again by Lemma 1, there exist Z-matrices X' and Y' such that $MX' = Y'$ and that X' is Minkowski. By [9, Theorem 4.3], there is a diagonal matrix D with positive diagonal entries such that $X'De > 0$. Let $X = X'D$ and $Y = Y'D$. Clearly, conditions (1) and (2) are then satisfied. Condition (3) is a well-known consequence of a P-matrix [10, Theorem 2.6]. This completes the proof of the theorem. ■

Using Theorem 1, we may now state a finite procedure for determining if a given matrix is hidden Minkowski.

2.1. The Finite Procedure

Step 1. Determine whether M is an S-matrix by solving the linear program

$$\begin{array}{ll} \text{minimize} & e^{\mathrm{T}}x \\ \text{subject to} & x \geqq 0 \quad \text{and} \quad Mx \geqq e. \end{array}$$

If this program is infeasible, stop. The matrix M cannot be a P-matrix, therefore it cannot be hidden Minkowski. Otherwise, go to Step 2.

Step 2. Determine whether there is a matrix $X = (x_{ij})$ satisfying conditions (1) and (2) by checking the consistency of the *linear* inequality system below:

(LIS)
$$\begin{array}{ll} x_{ij} \leqq 0 & \text{for all} \quad i \neq j, \\ \sum_{k=1}^{n} m_{ik} x_{kj} \leqq 0 & \text{for all} \quad i \neq j, \\ \sum_{j=1}^{n} x_{ij} > 0 & \text{for all } i. \end{array}$$

If the system is inconsistent, stop. The matrix M is not hidden Minkowski. Otherwise, terminate with the desired conditions (1)–(3) satisfied.

2.2. A Computational Remark

The consistency of the linear inequality system (LIS) can be checked by solving the linear program

$$\begin{array}{lll} \text{maximize} & \xi & \\ \text{subject to} & x_{ij} \leqq 0 & \text{for all} \quad i \neq j, \\ & \sum_{k=1}^{n} m_{ik} x_{kj} \leqq 0 & \text{for all} \quad i \neq j, \\ & \sum_{j=1}^{n} x_{ij} \geqq \xi & \text{for all } i. \\ & \xi \geqq 0. & \end{array}$$

It is obvious that the system (LIS) is consistent if and only if the above linear program has a positive optimum objective value.

3. CHARACTERIZING NONDEGENERATE HIDDEN Z-MATRICES

In this section, we establish a necessary and sufficient condition for a nondegenerate matrix to be hidden Z and describe some of its consequences. We recall some terminology. If M is a square matrix, the matrix $P^{\mathrm{T}}MP$, where P is a permutation matrix, is called a *principal rearrangement* of M. Let $M_{\alpha\alpha}$ be a nonsingular principal submatrix of the $n \times n$ matrix M with $\alpha \subseteq \{1, \ldots, n\}$. Let P be a permutation matrix such that

$$P^{\mathrm{T}}MP = \begin{bmatrix} M_{\alpha\alpha} & M_{\alpha\beta} \\ M_{\beta\alpha} & M_{\beta\beta} \end{bmatrix},$$

where $\beta = \{1, \ldots, n\} \setminus \alpha$. Then the *principal pivot transform* of M (corresponding to the principal submatrix $M_{\alpha\alpha}$) is defined as the matrix PM^*P^{T} where M^* is given by

$$M^* = \begin{bmatrix} M_{\alpha\alpha}^{-1} & -M_{\alpha\alpha}^{-1}M_{\alpha\beta} \\ M_{\beta\alpha}M_{\alpha\alpha}^{-1} & M_{\beta\beta} - M_{\beta\alpha}M_{\alpha\alpha}^{-1}M_{\alpha\beta} \end{bmatrix}.$$

Lemma 2. *Let M be a hidden Z-matrix. Then*

(4) *so is each principal rearrangement of M,*
(5) *so is each principal pivot transform of M.*

Proof. The proof of (4) is trivial. To prove (5), it suffices to show that the matrix M^* given above is hidden Z. The proof of this latter fact can be found in [15, Lemma 2] and is thus omitted.

Theorem 2. *Let M be a nondegenerate square matrix of order n. Then M is hidden Z if and only if there is a principal pivot transform of M, which after a suitable rearrangement, has the form*

$$M^* = \begin{bmatrix} F & a \\ b^{\mathrm{T}} & c \end{bmatrix},$$

where F is $(n-1) \times (n-1)$ satisfying the conditions:

(6) $F\tilde{X} = \tilde{Y}$ *for some Z-matrices $\tilde{X}$ and $\tilde{Y}$ with $\tilde{X}$ Minkowski.*

(7) *There exist an $(n-1)$-vector $\tilde{x}$ and a positive scalar $\tilde{z}$ such that for $i = 1, \ldots, n-1$ we have*

$$0 \leqq \tilde{x}_i \leqq \min\{(-\tilde{X}_{ij})/b^{\mathrm{T}}\tilde{X}_{\cdot j} : j \in J \text{ and } j \neq i\}$$

$$0 \leqq (F\tilde{x} - a\tilde{z})_i \leqq \min\{(-\tilde{Y}_{ij})/b^{\mathrm{T}}\tilde{X}_{\cdot j} : j \in J \text{ and } j \neq i\}$$

and that

$$(*) \qquad -b^{\mathrm{T}}\tilde{x} + c\tilde{z} \geqq 1 \qquad \text{if } J \neq \varnothing.$$

Here $J = \{j : b^{\mathrm{T}}\tilde{X}_{\cdot j} > 0\}$ *with* $\tilde{X}_{\cdot j}$ *denoting the* jth *column of* $\tilde{X}$. *The above two minimums are taken to be* $+\infty$ *if* $J = \varnothing$.

Proof. (Sufficiency) By Lemma 2, it suffices to show that M^* is hidden Z. Define a vector

$$\tilde{y} = \begin{cases} \max\{0, b^{\mathrm{T}}\tilde{X}/(-b^{\mathrm{T}}\tilde{x} + c\tilde{z})\} & \text{if } \ J \neq \varnothing \\ 0 & \text{if } \ J = \varnothing. \end{cases}$$

and two matrices

$$X = \begin{bmatrix} \tilde{X} + \tilde{x}\tilde{y}^{\mathrm{T}} & -\tilde{x} \\ -\tilde{y}^{\mathrm{T}}\tilde{z} & \tilde{z} \end{bmatrix}$$

and

$$Y = \begin{bmatrix} \tilde{Y} + (F\tilde{x} - a\tilde{z})\tilde{y}^{\mathrm{T}} & -(F\tilde{x} - a\tilde{z}) \\ b^{\mathrm{T}}\tilde{X} - (-b^{\mathrm{T}}\tilde{x} + c\tilde{z})\tilde{y}^{\mathrm{T}} & -b^{\mathrm{T}}\tilde{x} + c\tilde{z} \end{bmatrix}.$$

Then clearly, $M^*X = Y$. We show that $\tilde{X} + \tilde{x}\tilde{y}^{\mathrm{T}}$ is a Z-matrix. Indeed if $i \neq j$ and $\tilde{y}_j > 0$ then condition (7) together with the definition of the vector $\tilde{y}$ implies that

$$\tilde{X}_{ij} + \tilde{x}_i\tilde{y}_j \leqq 0.$$

On the other hand, this inequality clearly holds if $i \neq j$ and $\tilde{y}_j = 0$. Hence $\tilde{X} + \tilde{x}\tilde{y}^{\mathrm{T}}$ is a Z-matrix. Similarly, we may deduce that $\tilde{Y} + (F\tilde{x} - a\tilde{z})\tilde{y}^{\mathrm{T}}$ is a Z-matrix. Therefore, so are X and Y. We claim that X is Minkowski. In fact since the matrix $\tilde{X} + \tilde{x}\tilde{y}^{\mathrm{T}}$ is Minkowski [9, Theorem 4.2], all leading principal minors of X which are of order not exceeding $n - 1$ are positive. Moreover, by Schur's determinantal formula (see [2], e.g.), we have

$$\det X = \tilde{z} \det \tilde{X} > 0.$$

Hence by [9, Theorem 4.3], X is Minkowski. Consequently, M^* is hidden Z.

(Necessity). Suppose now that M is hidden Z. Let X and Y be the Z-matrices satisfying the defining conditions (M1) and (M2). Then according to [4, Theorem 3.3], there exist a permutation matrix P and partitioning of M, X and Y such that

$$P^{\mathrm{T}}MP = \begin{bmatrix} M_{\alpha\alpha} & M_{\alpha\beta} \\ M_{\beta\alpha} & M_{\beta\beta} \end{bmatrix}, \qquad P^{\mathrm{T}}XP = \begin{bmatrix} X_{\alpha\alpha} & X_{\alpha\beta} \\ X_{\beta\alpha} & X_{\beta\beta} \end{bmatrix}, \qquad P^{\mathrm{T}}YP = \begin{bmatrix} Y_{\alpha\alpha} & Y_{\alpha\beta} \\ Y_{\beta\beta} & Y_{\beta\beta} \end{bmatrix}$$

and that the matrix

$$\begin{bmatrix} Y_{\alpha\alpha} & Y_{\alpha\beta} \\ X_{\beta\alpha} & X_{\beta\beta} \end{bmatrix}$$

is Minkowski. Here $\alpha \subseteq \{1, \dots, n\}$ and β is the complement of α.

It is easy to deduce that (see [15, Lemma 2])

$$\begin{bmatrix} M_{\alpha\alpha}^{-1} & -M_{\alpha\alpha}^{-1}M_{\alpha\beta} \\ M_{\beta\alpha}M_{\alpha\alpha}^{-1} & M_{\beta\beta} - M_{\beta\alpha}M_{\alpha\alpha}^{-1}M_{\alpha\beta} \end{bmatrix} \begin{bmatrix} Y_{\alpha\alpha} & Y_{\alpha\beta} \\ X_{\beta\alpha} & X_{\beta\beta} \end{bmatrix} = \begin{bmatrix} X_{\alpha\alpha} & X_{\alpha\beta} \\ Y_{\beta\alpha} & Y_{\beta\beta} \end{bmatrix}.$$

Consider now the matrix

$$M^* = \begin{bmatrix} M_{\alpha\alpha}^{-1} & -M_{\alpha\alpha}^{-1}M_{\alpha\beta} \\ M_{\beta\alpha}M_{\alpha\alpha}^{-1} & M_{\beta\beta} - M_{\beta\alpha}M_{\alpha\alpha}^{-1}M_{\alpha\beta} \end{bmatrix},$$

which is a principal rearrangement of a principal pivot transform of M. We may write

$$M^* = \begin{bmatrix} F & a \\ b^{\mathrm{T}} & c \end{bmatrix},$$

where F is the leading principal submatrix of M^* which is of order $n - 1$. We may partition the matrices

$$\begin{bmatrix} Y_{\alpha\alpha} & Y_{\alpha\beta} \\ X_{\beta\alpha} & X_{\beta\beta} \end{bmatrix} \quad \text{and} \quad \begin{bmatrix} X_{\alpha\alpha} & X_{\alpha\beta} \\ Y_{\beta\alpha} & Y_{\beta\beta} \end{bmatrix}$$

accordingly and write

$$\begin{bmatrix} Y_{\alpha\alpha} & Y_{\alpha\beta} \\ X_{\beta\alpha} & X_{\beta\beta} \end{bmatrix} = \begin{bmatrix} X_{11} & X_{12} \\ X_{21} & X_{22} \end{bmatrix} \quad \text{and} \quad \begin{bmatrix} X_{\alpha\alpha} & X_{\alpha\beta} \\ Y_{\beta\alpha} & Y_{\beta\beta} \end{bmatrix} = \begin{bmatrix} Y_{11} & Y_{12} \\ Y_{21} & Y_{22} \end{bmatrix},$$

where X_{11} and Y_{11} are of order $(n - 1) \times (n - 1)$. Again, it is easy to deduce that

$$F(X_{11} - X_{12}X_{22}^{-1}X_{21}) = Y_{11} - Y_{12}X_{22}^{-1}X_{21}.$$

It is obvious that both $X_{11} - X_{12}X_{22}^{-1}X_{21}$ and $Y_{11} - Y_{12}X_{22}^{-1}X_{21}$ are Z-matrices. Moreover, it has been shown in [6] that $X_{11} - X_{12}X_{22}^{-1}X_{21}$ is in fact Minkowski. Hence condition (6) is satisfied with

$$\tilde{X} = X_{11} - X_{12}X_{22}^{-1}X_{21} \quad \text{and} \quad \tilde{Y} = Y_{11} - Y_{12}X_{22}^{-1}X_{21}.$$

Now let

$$\tilde{x} = -X_{12}, \quad \tilde{y}^{\mathrm{T}} = -X_{22}^{-1}X_{21}, \quad \text{and} \quad \tilde{z} = X_{22}.$$

Then we have $\tilde{x} \geqq 0$ and $\tilde{z} > 0$. Moreover,

$$\begin{bmatrix} Y_{11} & Y_{12} \\ Y_{21} & Y_{22} \end{bmatrix} = \begin{bmatrix} F & a \\ b^{\mathrm{T}} & c \end{bmatrix} \begin{bmatrix} X_{11} & X_{12} \\ X_{21} & X_{22} \end{bmatrix} = \begin{bmatrix} F & a \\ b^{\mathrm{T}} & c \end{bmatrix} \begin{bmatrix} \tilde{X} + \tilde{x}\tilde{y}^{\mathrm{T}} & -\tilde{x} \\ -\tilde{y}^{\mathrm{T}}\tilde{z} & \tilde{z} \end{bmatrix}$$

$$= \begin{bmatrix} \tilde{Y} + (F\tilde{x} - a\tilde{z})\tilde{y}^{\mathrm{T}} & -(F\tilde{x} - a\tilde{z}) \\ b^{\mathrm{T}}\tilde{X} - (-b^{\mathrm{T}}\tilde{x} + c\tilde{z})\tilde{y}^{\mathrm{T}} & -b^{\mathrm{T}}\tilde{x} + c\tilde{z} \end{bmatrix}.$$

The Z-property of the matrix

$$\begin{bmatrix} Y_{11} & Y_{12} \\ Y_{21} & Y_{22} \end{bmatrix}$$

implies that

$$F\tilde{x} - a\tilde{z} \geqq 0, \qquad b^{\mathrm{T}}\tilde{X} \leqq (-b^{\mathrm{T}}\tilde{x} + c\tilde{z})\tilde{y}^{\mathrm{T}},$$

and for all $i \neq j$,

$$\tilde{Y}_{ij} + (F\tilde{x} - a\tilde{z})_i \tilde{y}_j \leqq 0, \qquad \tilde{X}_{ij} + \tilde{x}_i \tilde{y}_j \leqq 0.$$

Hence condition (7) follows immediately if $J = \varnothing$. On the other hand, if $J \neq \varnothing$ and if j is an index such that $b^{\mathrm{T}}\tilde{X}_{\cdot j} > 0$, then it follows that

$$-b^{\mathrm{T}}\tilde{x} + c\tilde{z} > 0 \qquad \text{and} \qquad \tilde{y}_j > (b^{\mathrm{T}}\tilde{X}_{\cdot j})/(-b^{\mathrm{T}}\tilde{x} + c\tilde{z}).$$

Consequently, if $i \neq j$, we must have

$$0 \geqq \tilde{X}_{ij} + \tilde{x}_i \tilde{y}_j \geqq \tilde{X}_{ij} + \tilde{x}_i (b^{\mathrm{T}}\tilde{X}_{\cdot j})/(-b^{\mathrm{T}}\tilde{x} + c\tilde{z})$$

and

$$0 \geqq \tilde{Y}_{ij} + (F\tilde{x} - a\tilde{z})_i \tilde{y}_j \geqq \tilde{Y}_{ij} + (F\tilde{x} - a\tilde{z})_i (b^{\mathrm{T}}\tilde{X}_{\cdot j})/(-b^{\mathrm{T}}\tilde{x} + c\tilde{z}).$$

By redefining

$$x = \tilde{x}/(-b^{\mathrm{T}}\tilde{x} + c\tilde{z}) \qquad \text{and} \qquad z = \tilde{z}/(-b^{\mathrm{T}}\tilde{x} + c\tilde{z}),$$

condition (7) is satisfied readily. This completes the proof of the theorem. ■

Remark 1. The proof above shows that the theorem remains valid if $(*)$ is replaced by

$$(*') \qquad -b^{\mathrm{T}}\tilde{x} + c\tilde{z} = 1 \qquad \text{if} \quad J \neq \varnothing.$$

Remark 2. The nondegeneracy assumption is used only in proving the necessity part of the theorem.

The corollary below shows that a stronger necessary condition must hold if M is hidden Minkowski.

Corollary 1. *Suppose that M is hidden Minkowski. Then conditions* (6) *and* (7) *must be satisfied for* all *principal submatrices F of M which are of order $n - 1$.*

Proof. It suffices to show this for the leading principal submatrix of order $n - 1$. But this follows immediately from the above proof and from the fact that if X and Y are any Z-matrices satisfying the defining conditions (M1) and (M2) for M, then X must necessarily be a Minkowski matrix (by Lemma 1). This completes the proof of the corollary. ■

The next corollary identifies two sufficient conditions for condition (7) to hold.

Corollary 2. *Condition* (7) *holds if either one of the following two conditions is satisfied:*

(8) $c > 0$ *and for each* i,

$$0 \leqq -a_i \leqq c \min\{-\tilde{Y}_{ij}/b^{\mathrm{T}}\tilde{X}_{\cdot j} : j \in J \text{ and } j \neq i\};$$

(9) $b^{\mathrm{T}}\tilde{X} \leqq 0$ *and there exist a nonnegative vector $\tilde{x}$ and a positive scalar $\tilde{z}$ such that*

$$F\tilde{x} - a\tilde{z} \geqq 0.$$

Proof. If condition (8) is satisfied, then by choosing $\tilde{x} = 0$ and $\tilde{z} = 1/c$, condition (7) is satisfied readily. On the other hand, if $b^{\mathrm{T}}\tilde{X} \leqq 0$, then condition (7) asserts the existence of a nonnegative vector $\tilde{x}$ and of a positive scalar $\tilde{z}$ such that $F\tilde{x} - a\tilde{z} \geqq 0$. But this assertion is precisely that contained in condition (9). This establishes the corollary. ■

Corollary 3. *Conditions* (6) *and* (7) *hold if either one of the following two conditions is satisfied:*

(10) *F is a Z-matrix, $b \leqq 0$ and there is a nonnegative vector $\tilde{x}$ such that $F\tilde{x} \geqq a$.*

(11) *F^{-1} exists and is a Minkowski matrix and $b^{\mathrm{T}}F^{-1} \leqq 0$.*

Proof. If (10) holds, choose $\tilde{X} = I$ and $\tilde{Y} = F$, then condition (6) obviously holds. The conclusion follows immediately from Corollary 2. On the other hand, if (11) holds, choose $\tilde{X} = F^{-1}$ and $\tilde{Y} = I$. Since F^{-1} is Minkowski, F is in fact a P-matrix, thus an S-matrix. Consequently, condition (9) is satisfied. Therefore, the conclusion follows as before. This completes the proof of the corollary. ■

Corollary 4. *A P-matrix which is either upper triangular or lower triangular is hidden Z.*

Proof. The case of a lower (upper) triangular P-matrix follows immediately from an inductive argument and the fact that condition (8) [(9)] is satisfied. The conclusion therefore is a direct consequence of Theorem 2 and Corollary 2. ■

Finally, we show how Theorem 2 can sometimes be used to show that a given matrix is *not* hidden Z.

Theorem 3. *The class of hidden Minkowski matrices is properly contained in that of P-matrices.*

Proof. It suffices to exhibit a P-matrix which is not hidden Z. Consider the matrix

$$M = \begin{bmatrix} 2 & -1 & 2 \\ -3 & 2 & -2 \\ -4 & 3 & 5 \end{bmatrix}.$$

It is easy to verify that this is indeed a P-matrix. According to Corollary 1 and the remark immediately following Theorem 2, to show that M is not hidden Z, it suffices to show that the leading principal submatrix

$$F = \begin{bmatrix} 2 & -1 \\ -3 & 2 \end{bmatrix}$$

does not satisfy conditions (6) and (7) where $(*)$ is replaced by $(*)'$. In fact, let

$$\tilde{X} = \begin{bmatrix} x_{11} & -x_{12} \\ -x_{21} & x_{22} \end{bmatrix}$$

be any Z-matrix satisfying condition (6). Then the corresponding $\tilde{Y}$-matrix is given by

$$\tilde{Y} = \begin{bmatrix} 2 & -1 \\ -3 & 2 \end{bmatrix} \begin{bmatrix} x_{11} & -x_{12} \\ -x_{21} & x_{22} \end{bmatrix} = \begin{bmatrix} 2x_{11} + x_{21} & -2x_{21} - x_{22} \\ -3x_{11} - 2x_{21} & 3x_{12} + 2x_{22} \end{bmatrix}.$$

We have

$$b^{\mathrm{T}}\tilde{X} = [-4 \quad 3] \begin{bmatrix} x_{11} & -x_{12} \\ -x_{21} & x_{22} \end{bmatrix} = [-4x_{11} - 3x_{21} \quad 4x_{12} + 3x_{22}].$$

The Z-property of the matrix $\tilde{X}$ implies that $J = \{2\}$. Therefore, condition (7) would assert the existence of a nonnegative vector

$$\tilde{x} = \begin{bmatrix} x_1 \\ x_2 \end{bmatrix}$$

and a positive scalar $\tilde{z}$ such that

$$(**)\qquad x_1 \leqq x_{12}/(4x_{12} + 3x_{22}),$$

$$(**')\qquad 0 \leqq 2x_1 - x_2 - 2\tilde{z} \leqq (2x_{11} + x_{22})/(4x_{12} + 3x_{22}),$$

$$(**'')\qquad 0 \leqq -3x_1 + 2x_2 + 2\tilde{z},$$

and

$$4x_1 - 3x_2 + 5\tilde{z} = 1.$$

The last equation implies that

$$\tilde{z} = (1 - 4x_1 + 3x_2)/5.$$

Substituting into $(**')$ and $(**'')$, we obtain

$$0 \leqq 18x_1 - 11x_2 - 2,$$

$$0 \leqq -23x_1 + 16x_2 + 2.$$

Adding these two inequalities, we deduce that $x_2 \geqq x_1$. Hence, we have

$$0 \leqq 18x_1 - 11x_2 - 2 \leqq 7x_1 - 2,$$

which implies

$$x_1 \geqq \tfrac{2}{7}.$$

However, $(**)$ implies that $x_1 < \frac{1}{4}$ which is a contradiction. This establishes the theorem. ■

We conclude this paper by establishing

Theorem 4. *All 2×2 P-matrices are hidden Z.*

Proof. Let $M = \left[\begin{smallmatrix} a & b \\ c & d \end{smallmatrix}\right]$ be a 2×2 P-matrix. There are three cases.

Case 1: Both b and c are nonpositive. In this case, M is itself a Z-matrix.

Case 2: Either b or c is positive but not both. Suppose, say, that $b > 0$ and $c \leqq 0$. Then we have

$$\begin{bmatrix} a & b \\ c & d \end{bmatrix}\begin{bmatrix} 1 & -1 \\ 0 & \alpha \end{bmatrix} = \begin{bmatrix} a & -a + \alpha b \\ c & -c + \alpha d \end{bmatrix}.$$

If $\alpha > 0$ is chosen small enough so that $a \geqq \alpha b$, it then follows that the defining conditions (M1) and (M2) are satisfied for M. Hence M is hidden Z.

Case 3: Both b and c are positive. We have

$$\begin{bmatrix} a & b \\ c & d \end{bmatrix}^{-1} = \frac{1}{ad - bc}\begin{bmatrix} d & -b \\ -c & a \end{bmatrix}$$

which is a Z-matrix. Hence, M is hidden Z. This establishes the theorem. ■

REFERENCES

1. R. CHANDRASEKARAN, A special case of the complementary pivot problem, *Opsearch* **7** (1970), 263–268.
2. R. W. COTTLE, Manifestations of the Schur complement, *Linear Algebra and Appl.* **8** (1974), 189–211.
3. R. W. COTTLE, G. H. GOLUB, AND R. S. SACHER, On the Solution of Large, Structured, Linear Complementarity Problems: III, Tech. Rep. 74–7, Dept. Oper. Res., Stanford Univ., Stanford, California, 1974.
4. R. W. COTTLE AND J. S. PANG, On solving linear complementarity problems as linear programs, *Math. Programming Stud.* **7** (1978), 88–107.
5. R. W. COTTLE AND A. F. VEINOTT, JR., Polyhedral sets having a least element, *Math. Programming* **3** (1969), 238–249.
6. D. E. CRABTREE, Applications of M-matrices to nonnegative matrices, *Duke Math. J.* **33** (1966), 197–208.
7. G. B. DANTZIG, Optimal solution of a dynamic Leontief model with substitution, *Econometrica* **23** (1955), 295–302.
8. G. B. DANTZIG AND A. F. VEINOTT, JR., Discovering hidden totally Leontief substitution systems, *Math. Oper. Res.* **3** (1978), 102–103.
9. M. FIEDLER AND V. PTAK, On matrices with non-positive off-diagonal elements and positive principal minors, *Czechoslovak Math. J.* **12** (1962), 382–400.
10. M. FIEDLER AND V. PTAK, Some generalizations of positive definiteness and monotonicity, *Numer. Math.* **9** (1966), 163–172.
11. G. J. KOEHLER, A. B. WHINSTON, AND G. P. WRIGHT, "Optimization Over Leontief Substitution Systems," North–Holland Publ., Amsterdam, 1975.
12. O. L. MANGASARIAN, Linear complementarity problems solvable by a single linear program, *Math. Programming* **10** (1976), 263–270.
13. O. L. MANGASARIAN, Solution of linear complementarity problems by linear programming, *in* "Numerical Analysis, Dundee 1975" (G. W. Watson, ed.), Lecture Notes in Mathematics, No. 506, Springer-Verlag, Berlin and New York, 1976, pp. 166–175.
14. J. S. PANG, On Cone Orderings and the Linear Complementarity Problem, MRC Tech. Rep. No. 1757, Math. Res. Center, Univ. of Wisconsin, Madison, 1977. [*Linear Algebra and Appl.* (to appear).]
15. J. S. PANG, Hidden Z-Matrices with Positive Principal Minors, MRC. Tech. Rep. No. 1776, Math. Res. Center, Univ. of Wisconsin, Madison, 1977. [*Linear Algebra and Appl.* (to appear).]
16. J. S. PANG, A new characterization of real H-matrices with positive diagonals, *Linear Algebra and Appl.* (to appear).
17. R. S. SACHER, On the Solution of Large, Structured Linear Complementarity Problems, Ph.D. Thesis, Dep. Oper. Res., Stanford Univ., Stanford, California, 1974.
18. R. SAIGAL, On a Generalization of Leontief Substitution Systems, Working Paper No. CP-325, Center Res. Management Sci., Univ. of California, Berkeley, 1971.
19. A. TAMIR, The complementarity Problem of Mathematical Programming, Ph.D. Thesis, Dept. Oper. Res., Case Western Reserve Univ., Cleveland, Ohio, 1973.
20. A. F. VEINOTT, JR., Extreme points of Leontief substitution systems, *Linear Algebra and Appl.* **1** (1968), 181–194.

GRADUATE SCHOOL OF INDUSTRIAL ADMINISTRATION
CARNEGIE–MELLON UNIVERSITY
PITTSBURGH, PENNSYLVANIA

Hidden Optima in Engineering Design

DOUGLASS J. WILDE

1. INTRODUCTION

It is becoming uncomfortably apparent that there are important engineering systems having several local optima and for which existing optimization schemes consequently may not work. If started near the global optimum, the locally convergent techniques of modern mathematical programming and numerical optimization can be successful even when there are multiple local optima. In defense of such procedures it has long been argued that any engineer reasonably well acquainted with the system studied would know where to make such a start. This article will, however, demonstrate the possibility of *hidden* optima, which are global optima so far from a reasonable start that short step optimization procedures will usually converge to the wrong local optimum.

The next section briefly describes three engineering systems in the literature that have hidden optima, one of them costing $4.2(10^6) annually. Following this is a section showing the mathematical methods used to find these hidden optima and not only identify what caused them, but also show how they were overlooked in the first place.

2. THREE HIDDEN OPTIMA

This section describes three engineering systems with optima that are hidden in the sense that conventional optimization procedures started at a reasonable place can converge to the wrong local optimum. The first problem involves design of an ammonia storage system; the second, operation of a steel plant; the third, design of an irrigation reservoir. In the first two, the global optimum is hidden, whereas in the third the hidden local optimum will only become globally optimal as land cost increases sufficiently.

ISBN 0-12-178150-X

The first example was formulated by Stoecker [4] to illustrate the steepest descent (gradient) direct search method. It is proposed to attach a vapor recondensation refrigeration system to lower the temperature, and consequently vapor pressure, of liquid ammonia stored in a steel pressure vessel, for this would permit thinner vessel walls. The tank cost saving must be traded off against the refrigeration and thermal insulation cost to find the temperature and insulation thickness minimizing the total annual cost. Stoecker showed the total cost to be the sum of insulation cost $i \equiv 400x^{0.9}$ (x is the insulation thickness, in.), the vessel cost $v \equiv 1000 + 22(p - 14.7)^{1.2}$ (p is the absolute pressure, psia), and the recondensation cost $r \equiv 144(80 - t)/x$ (t is the temperature, °F). The pressure is related to the temperature by

$$\ln p = -3950(t - 460)^{-1} + 11.86.$$

By direct gradient search, iterated 16 times from a starting temperature of 50°F, the total annual cost is found to be locally minimum at $x = 5.94$ in. and $t = 6.29$°F, where the cost is \$53,400/yr. The reader can verify, however, that an ambient system (80°F) without any recondensation only costs \$52,000/yr, a saving of 3%.

The second example comes from Ray and Szekely [2]. The problem is to find the minimum cost operating conditions for producing crude steel at a given production rate in an integrated steel plant with two blast furnaces, a basic oxygen furnace, and an open hearth shop. There are 20 variables, including feed rates for coke, silicon carbide, steel scrap, and both sintered and pelleted iron ore, as well as production rates for all four units. Cost, and all constraints both equalities and inequalities, are linear except for the ratio of coke to hot metal produced in each blast furnace, which in the first blast furnace for example takes the form

$$R_1 = 0.4[1 + 0.5\exp(-0.71x_2/x_1)] + 0.3(1 - x_4)^2,$$

where R_1 is the mass ratio of coke to hot metal produced, x_1 is the sintered ore input rate (M tons/yr), x_2 is the pelleted ore input rate (M tons/yr), and x_4 is the hot metal (iron) output (M tons/yr). The coke input rate (M tons/yr) x_3 equals $R_1 x_4$, and a mass balance gives the linear relation

$$x_4 = 0.715x_1 + 0.91x_2.$$

The variable annual cost in U.S. dollars is $21x_1 + 30x_2 + 25x_3$, also linear. Similar relations with different parameters only in the coke/metal ratio hold for the second blast furnace.

Using the well-known optimization procedures of linear programming (see [2, Section 4.4]) and gradient projection [3], Ray and Szekely generated an operating plan using pure sintered ore unmixed with pellets ($x_1 = 1.70$, $x_2 = 0$), giving a hot metal output of $x_4 = 1.215$. The reader can verify that this requires a coke rate of $x_3 = 0.744$ for an annual variable cost of \$54.3(10^6)

in the first blast furnace. Since only the hot metal connects this furnace to the rest of the plant, ore feed changes affect only the furnace cost as long as x_4 is unchanged. So it is of great interest that different feed ($x_1 = 0.21, x_2 = 1.17, x_3 = 0.418, x_4 = 1.215$) gives an annual variable cost in the first blast furnace of only \$52.1(10^6)—4% less than when pure sinter is fed. Pellets instead of sinter in the second blast furnace lowers cost comparably. Even more can be saved by permitting the hot metal rates to respond to the economic changes generated by the altered ore composition. In this example, finding the hidden optimum is worth over \$4 million a year.

The third example concerns an irrigation pipeline with irrigation reservoir described by Wilde and McNeill [7]. The designer is to choose diameter d, number of pumping stations n, and maximum flow rate q to minimize the total annual cost of piping, pumps, stations, energy, and most pertinent to this discussion, the reservoir. Conventional practice calls for the flow to be held constant at the annual average demand rate A; the reservoir being made large enough to absorb seasonal fluctuations. In principle, however, the system could operate without a reservoir (or with a very small one) if the pipeline could at all times meet the varying demand rate. Thus the maximum flow rate q is bounded below by A and above by the maximum irrigation demand rate M:

$$A \leqq q \leqq M.$$

The particular form of the reservoir cost r was found empirically to have the form $r = R(M - q)^\alpha$ in the case of the world's largest irrigation project near Setif, Algeria. Here R is a known parameter, and for the Kebir reservoir the exponent α turns out to be less than unity (0.84). What is interesting is that since the first derivative $\partial r/\partial q$ decreases without limit as q approaches M, the total cost must be locally minimum when $q = M$, where there is no reservoir at all. This local minimum, being as far as possible from the usual design $q = A$, is certainly hidden, for no short step optimization procedure starting near A, also a local minimum, would ever find the one at M. In the case studied, and perhaps for all water conveyance-reservoir systems ever built, the global minimum was fortunately in the usual place. But although for Kebir the optimal cost was only half the maximum possible, it was only 10% below the cost of a system with no reservoir. Shifting economics in future projects, especially as ambitious as the one now being studied for irrigating the Sahara Desert, might well cause the global optimum to jump to large pipes and small reservoirs.

3. DIAGNOSIS

This section deals with what caused these optima to be hidden and how to prevent similar ones from being overlooked in the future. In the first example the ammonia storage system, after partial optimization (see [6,

Section 8-03]) with respect to insulation thickness using partial condensation [1], the cost depends only on temperature, the optimized sum of recondensation and insulation being $492(80\text{-}t)^{0.474}$. Since the exponent is less than unity, the derivative of this sum with respect to temperature decreases without limit as 80-t approaches zero, indicating a local minimum at the ambient temperature 80°F. This local optimum happens also to be globally optimal in this case, although a 10% increase in material cost relative to energy cost would shift the global optimum to the interior local one found by Stoecker. The global minimum is hidden by a sharp local *maximum* very near by, and one would hardly expect a designer of auxiliary equipment, without being forewarned, to guess that the best system would have no auxiliary equipment at all.

Viewed mathematically, the third example with the irrigation reservoir resembles the first with the ammonia system, since both have a term which can vanish and whose exponent is less than unity. In both situations there is a local minimum where a major component vanishes—the reservoir in the third example. Optimization theorists would find interesting the fact that if this problem is formulated as a geometric program with a reversed inequality, the only interior solution is where the cost is *maximum*, that is, as bad as possible. This is because no variable in a geometric program can vanish, although here such a solution has engineering significance and is a local minimum. Wilde [5] shows how to construct lower bounding functions for proving the global optimality of the designs given here.

Consider now the global optimization of a typical blast furnace (e.g., the first) for a given iron production rate x_4. Let $u \equiv x_1/x_4$ and use the constraints to express the cost $c(x_4, v(u))$ as a function only of the parameter x_4 and the variable cost rate

$$v(u) \equiv -2.57u + 8.75 \exp(-0.78u^{-1}).$$

Then

$$c(x_4, v(u)) = x_4(u) + 43x_4 + 7.5x_4(1 - x_4)^2.$$

For fixed x_4, this is minimum with respect to u $(0 \leqq u \leqq 1.40)$ where $v(u)$ is minimum.

The first derivative is

$$dv/du = -2.57 + 6.83u^{-2} \exp(-0.78u^{-1}).$$

At $x_1 = 1.70$, the schedule given by Ray and Szekeley, $u = 1.40$ and $dv/du = -1.38 < 0$, indicating a local minimum there because u is at its maximum value. At the other extreme $(u = 0)$, the variable part of the derivative vanishes in the limit, so $dv/du = -2.57 < 0$, indicating a local maximum because u

cannot decrease. However, the value of v at this local maximum is less than at the local minimum.

$$v(0) = 0 < 1.41 = v(1.40).$$

Hence the global minimum must be in the interior of the interval.

To see that this global minimum is at the unique interior local minimum, consider the second derivative

$$d^2v/du^2 = 6.83u^{-3} \exp(-0.78u^{-1})[0.78u^{-1} - 2].$$

This vanishes where $u = 0.39$, which is consequently the unique inflection point for v. For $0 \leqq u < 0.39$, $d^2v/du^2 > 0$, so there can be no more than one stationary point (a local minimum) in that interval. For $0.39 < u \leqq 1.40$, $d^2v/du^2 < 0$, so this interval has no more than one stationary point—this time a local maximum. Therefore the global minimum must be at the unique interior local minimum, found to be at $u_* = 0.179$, for which $v = v(u_*) = -0.348$. Here the globally minimum cost is

$$c(x_4, v(u_*)) = c_*(x_4) = 42.64x_4 + 7.5x_4(1 - x_4)^2.$$

For $x_4 = 1.21$, this is \$2.2(10^6) per year less than for pure sinter, itself only 0.1(10^6) per year below the global maximum value at $u^* = 1.15$. The saving is almost as great in the second blast furnace, and when the iron production rates are adjusted to exploit the true optima, a total savings of \$4.2(10^6) annually is realized. Incidentally, the steel plant downstream of the furnaces is not affected by these revised feed schedules.

The research of Ray and Szekeley that produced the nonlinear coke rate relation has therefore great economic potential. It shows that an ore mixture that is largely, but not entirely, pellets reduces the coke requirement drastically. This seems to be worth several million dollars a year in each of the example blast furnaces. This global optimum appears to have been hidden from the local optimum found (pure sinter) by a long, flat maximum in the middle of the range of interest. This misdirected the short step algorithm toward the wrong local optimum.

REFERENCES

1. R. J. Duffin, Linearizing geometric programs, *SIAM Rev.* **12** (1970), 211–227.
2. W. H. Ray and J. Szekely, *in* "Process Optimization—With Applications in Metallurgy and Chemical Engineering," Wiley (Interscience), New York, 1973, pp. 299–310.
3. J. B. Rosen, The Gradient Projection Method for Nonlinear Programming, Part I, Linear Constraints, *SIAM J.* **8** (1960), 181.
4. W. F. Stoecker, *in* "Design of Thermal Systems," McGraw-Hill, New York, 1971, pp. 152–155.

5. D. J. WILDE, "Globally Optimal Design," Wiley (Interscience), New York, 1978.
6. D. J. WILDE AND C. S. BEIGHTLER, "Foundations of Optimization," Prentice-Hall, Englewood Cliffs, New Jersey, 1967.
7. D. J. WILDE AND B. W. MCNEILL, Economic Design of a Pipeline with Discharge Reservoir, *Eng. Optim.* **4** (1978), 1.

MECHANICAL ENGINEERING DEPARTMENT
STANFORD UNIVERSITY
STANFORD, CALIFORNIA

Part IV

DIFFERENTIAL EQUATIONS

Analysis of Finite Element Methods Using Mesh Dependent Norms

I. BABUŠKA

J. OSBORN

This paper presents a new approach to the analysis of finite element methods for the approximate solution of second order boundary value problems in which error estimates are derived directly in terms of two mesh dependent norms that are closely related to the L_2 norm and to the second order Sobolev norm, respectively, and in which there is no assumption of quasi-uniformity on the mesh family. This is in contrast to the usual analysis in which error estimates are first derived in the first order Sobolev norm and subsequently are derived in the L_2 norm and in the second order Sobolev norm—the second order Sobolev norm estimates being obtained under the assumption that the functions in the underlying approximating subspaces lie in the second order Sobolev space and that the mesh family is quasi-uniform.

1. INTRODUCTION

During the last several years there has been an extensive development of finite element methods for the approximate solution of boundary value problems for differential equations. In the most standard approach to the analysis of such methods, as they pertain to second order elliptic equations, error estimates are first derived in the first order Sobolev norm and subsequently are derived in the L_2 norm and in the second order Sobolev norm—the second order Sobolev norm error estimates being obtained under the assumptions that the functions in the underlying approximating subspaces lie in the second order Sobolev space and that the mesh family is quasi-uniform.

It is the purpose of this paper to present a new approach to the analysis of finite element methods in which error estimates are derived directly in terms of two mesh dependent norms that are closely related to the L_2 norm and to the second order Sobolev norm, respectively, and in which there is no assumption of quasi-uniformity of the mesh family.

In Section 2 we review results on the approximate solution of variationally formulated boundary value problems. In Section 3 we describe the (mesh dependent) norms and spaces used in the paper. Section 4 contains the main

ISBN 0-12-178150-X

results and applications of the paper. These involve L_2 estimates for the error and its second derivative (considered on each subinterval of the mesh). In Section 5 we state results on L_1 and L_∞ estimates of the error and its second derivatives. Section 6 contains the proof of one of the theorems of the paper. In this paper we discuss only boundary value problems for ordinary differential equations. Most of the results are presented without proofs. A complete development of this material, as well as results for partial differential equations, will be presented elsewhere.

Throughout the paper $H_p^k = H_p^k(\mathscr{I})$, $k = 0, 1, \ldots$, $1 \leqq p \leqq \infty$, will denote the kth Sobolev space on an interval $\mathscr{I}$ in R^1 consisting of functions with k derivatives in $L_p(\mathscr{I})$. On this space we have the usual norm given by

$$\|u\|_{k,p,\mathscr{I}} = \begin{cases} \left(\sum_{j=0}^k \int_{\mathscr{I}} |u^{(j)}|^p \, dx\right)^{1/p}, & 1 \leqq p < \infty \\ \sum_{j=0}^k \operatorname{ess\,sup} |u^{(j)}|, & p = \infty. \end{cases}$$

$\overset{0}{H}{}_p^1(\mathscr{I})$ denotes the subspace of $H_p^1(\mathscr{I})$ of functions that vanish at the endpoints of $\mathscr{I}$. In the special case $p = 2$ we use the notation $H^k = H_2^k$ and $\|u\|_{k,\mathscr{I}} = \|u\|_{k,2,\mathscr{I}}$. Note that $H_p^0 = L_p$.

2. ABSTRACT CONVERGENCE RESULTS

In this section we review certain results on the approximate solution of variationally formulated boundary value problems.

Let $\mathscr{H}_{1,\Delta}$ and $\mathscr{H}_{2,\Delta}$ be two Hilbert spaces (indexed by the parameter Δ for Δ in some index set) with norms $\|\cdot\|_{1,\Delta}$ and $\|\cdot\|_{2,\Delta}$, respectively, and let B_Δ be a bilinear form on $\mathscr{H}_{1,\Delta} \times \mathscr{H}_{2,\Delta}$. We suppose the following are satisfied:

$$|B_\Delta(u, v)| \leqq C_1 \|u\|_{1,\Delta} \|v\|_{2,\Delta} \qquad \text{for all} \quad u \in \mathscr{H}_{1,\Delta}, v \in \mathscr{H}_{2,\Delta}, \tag{2.1}$$

$$\inf_{\substack{u \in \mathscr{H}_{1,\Delta} \\ \|u\|_{1,\Delta} = 1}} \sup_{\substack{v \in \mathscr{H}_{2,\Delta} \\ \|v\|_{2,\Delta} = 1}} |B_\Delta(u, v)| \geqq C_2 > 0, \tag{2.2a}$$

$$\sup_{u \in \mathscr{H}_{1,\Delta}} |B_\Delta(u, v)| > 0, \qquad \text{for each} \quad 0 \neq v \in \mathscr{H}_{2,\Delta}, \tag{2.2b}$$

where C_1 and C_2 are constants that do not depend on Δ. As a consequence of (2.1) and (2.2) we have

Theorem 1 [1, 2, 5]. *If $f \in (\mathscr{H}_{2,\Delta})'$, then there is a unique solution u to the problem*

$$\begin{aligned} &u \in \mathscr{H}_{1,\Delta}, \\ &B_\Delta(u, v) = f(v) \qquad \textit{for all} \quad v \in \mathscr{H}_{2,\Delta}. \end{aligned} \tag{2.3}$$

We will refer to (2.3) as an abstract, variationally formulated boundary value problem. In this paper we study the approximate solution of (2.3).

Toward this end we suppose $S_{1,\Delta}$ and $S_{2,\Delta}$ are finite dimensional subspaces of $\mathscr{H}_{1,\Delta}$ and $\mathscr{H}_{2,\Delta}$, respectively, and we assume

$$\inf_{\substack{u \in S_{1,\Delta} \\ \|u\|_{1,\Delta}=1}} \sup_{\substack{v \in S_{2,\Delta} \\ \|v\|_{2,\Delta}=1}} |B_\Delta(u, v)| \geqq C_2' > 0, \tag{2.4a}$$

$$\sup_{u \in S_{1,\Delta}} |B_\Delta(u, v)| > 0 \qquad \text{for each} \quad 0 \neq v \in S_{2,\Delta}, \tag{2.4b}$$

where C_2' is a constant independent of Δ. As a consequence of (2.4) (cf. Theorem 1) there is a unique solution u_Δ to the problem

$$\begin{aligned} &u_\Delta \in S_{1,\Delta}, \\ &B_\Delta(u_\Delta, v) = f(v) \qquad \text{for} \quad v \in S_{2,\Delta}. \end{aligned} \tag{2.5}$$

Since $S_{1,\Delta}$ and $S_{2,\Delta}$ are finite dimensional, (2.5) is computationally resolvable (in terms of the solution of a linear system of equations). u_Δ is the Ritz–Galerkin approximation to the exact solution u of (2.3).

u_Δ can also be characterized by

$$\begin{aligned} &u_\Delta \in S_{1,\Delta}, \\ &B_\Delta(u_\Delta, v) = B_\Delta(u, v) \qquad \text{for} \quad v \in S_{2,\Delta}. \end{aligned}$$

Thus we refer to u_Δ as the B_Δ projection of u onto $S_{1,\Delta}$ with respect to $S_{2,\Delta}$, or, more briefly, as the left B_Δ projection of u. This projection is defined for each $u \in \mathscr{H}_{1,\Delta}$. Also, for each $v \in \mathscr{H}_{2,\Delta}$ we can define the B_Δ projection of v onto $S_{2,\Delta}$ with respect to $S_{1,\Delta}$ (the right B_Δ projection of v) by

$$\begin{aligned} &v_\Delta \in S_{2,\Delta}, \\ &B_\Delta(u, v_\Delta) = B_\Delta(u, v) \qquad \text{for} \quad u \in S_{1,\Delta}. \end{aligned}$$

We now state the fundamental estimates for the errors $u - u_\Delta$ and $v - v_\Delta$.

Theorem 2. [1, 2].

$$\|u - u_\Delta\|_{1,\Delta} \leqq (1 + C_1/C_2') \inf_{\chi \in S_{1,\Delta}} \|u - \chi\|_{1,\Delta} \tag{2.6a}$$

and

$$\|v - v_\Delta\|_{2,\Delta} \leqq (1 + C_1/C_2') \inf_{\eta \in S_{2,\Delta}} \|v - \eta\|_{2,\Delta}, \tag{2.6b}$$

with C_1 and C_2' as in (2.1) and (2.4a).

These inequalities are called quasi-optimal error estimates.

In many applications of the results (cf. [2]) in this section the spaces $\mathscr{H}_{1,\Delta}$ and $\mathscr{H}_{2,\Delta}$ and the form B_Δ do not in fact depend on the parameter Δ, i.e., $\mathscr{H}_{1,\Delta} = \mathscr{H}_1$ and $\mathscr{H}_{2,\Delta} = \mathscr{H}_2$ are fixed Hilbert spaces and $B_\Delta = B$ is a fixed form on $\mathscr{H}_1 \times \mathscr{H}_2$. The spaces $S_{i,\Delta}$ typically are spaces of piecewise polynomials with respect to a mesh Δ of some domain and, of course, depend on Δ. In the applications we consider, both the spaces $\mathscr{H}_{i,\Delta}$ and $S_{i,\Delta}$ will depend on Δ; the constants C_1, C_2 and C_2', however, will be independent of Δ (cf. [2, Chapter 7]). In these applications the solution u of (2.3) will lie in $\mathscr{H}_{1,\Delta}$ for all Δ. Thus the estimate (2.6a) provides a convergence estimate for $u - u_\Delta$, provided the family $S_{1,\Delta}$ satisfies an approximability assumption. For typical finite element applications, this would involve the assumption that $\inf_{\chi \in S_{1,\Delta}} \|u - \chi\|_{1,\Delta}$ tends to zero as the maximum mesh length of Δ tends to zero. Finally, we remark that for most applications (2.4a) is the major assumption. (2.4a) is called the stability condition.

For a complete treatment of the material in this section we refer to [1, 2, 5].

3. MESH DEPENDENT NORMS AND SPACES

In this section we define the (mesh dependent) norms, spaces, and forms that we will use in this paper. Let $\Delta = \{0 = x_0 < x_1 < \cdots < x_n = 1\}$, $n = n(\Delta)$ = a positive integer, be an arbitrary mesh on the interval $I = [0, 1]$ and set $h_j = x_j - x_{j-1}$ and $I_j = (x_{j-1}, x_j)$ for $j = 1, \ldots, n$, $\delta_j = (h_j + h_{j+1})/2$ for $j = 1, \ldots, n-1$, and $h = h(\Delta) = \max_j h_j$.

We now define two new spaces H_Δ^2 and H_Δ^0 and a bilinear form B_Δ, all of which depend on the mesh Δ. We let

$$H_\Delta^2 = \{u \in \overset{0}{H}{}^1(I) : u|_{I_j} \in H^2(I_j), \quad j = 1, \ldots, n\}$$

and

$$\|u\|_{2,\Delta}^2 = \sum_{j=1}^{n} \|u\|_{2,I_j}^2 + \sum_{j=1}^{n-1} \delta_j^{-1} |Ju'(x_j)|^2,$$

where $Ju'(x_j) = u'(x_j^+) - u'(x_j^-)$. H_Δ^2 is complete with respect to $\|\cdot\|_{2,\Delta}$. H_Δ^0 is defined to be the completion of $\overset{0}{H}{}^1(0, 1)$ with respect to

$$\|u\|_{0,\Delta}^2 = \int_0^1 u^2\, dx + \sum_{j=1}^{n-1} \delta_j |u(x_j)|^2.$$

H_Δ^0 can be identified with $L_2 \oplus R^{n-1}$. H_Δ^2 and H_Δ^0 are Hilbert spaces. For $u \in H_\Delta^0$ and $v \in H_\Delta^1$ we define

$$B_\Delta(u, v) = \sum_{j=1}^{n} \int_{I_j} u[-(av')' + cv]\, dx - \sum_{j=1}^{n-1} a(x_j) u(x_j) Jv'(x_j), \qquad (3.1)$$

where $a \in C^1(I)$, $c \in C(I)$, and $a(x) \geqq a_0 > 0$, $c(x) \geqq 0$ for $x \in I$. Note that $B_\Delta(u, v) = \int_0^1 (au'v' + cuv)\, dx$ for $u \in H^1$ and $v \in H_\Delta^2$.

For $r = 2, 3, \ldots$ let

$$S_\Delta = \{u \in C^0(I) : u|_{I_j} = \text{polynomial of degree } r - 1,\ u(0) = u(1) = 0\}. \quad (3.2)$$

Clearly S_Δ lies in H_Δ^0 and H_Δ^2.

In Section 4 we apply the results of Section 2 with $\mathscr{H}_{1,\Delta} = H_\Delta^0$, $\mathscr{H}_{2,\Delta} = H_\Delta^2$, B_Δ as defined in (3.1), and $S_{1,\Delta} = S_{2,\Delta} = S_\Delta$. We conclude this section with a statement of theorems corresponding to assumptions (2.1), (2.2), and (2.4).

Theorem 3. *With $\mathscr{H}_{1,\Delta} = H_\Delta^0$, $\mathscr{H}_{2,\Delta} = H_\Delta^2$, and B_Δ as in (3.1), (2.1) and (2.2) hold with C_1 and C_2 constants which do not depend on the mesh Δ.*

Theorem 4. *With $S_{1,\Delta} = S_{2,\Delta} = S_\Delta$, (2.4) holds with C_2' a constant which does not depend on Δ.*

4. APPLICATIONS

In this section we use the mesh dependent spaces H_Δ^0 and H_Δ^2 to analyze the finite element method for two different problems.

(a) Consider the two point boundary value problem

$$Lu \equiv -(au')' + cu = f, \qquad x \in I,$$
$$u(0) = u(1) = 0,$$

where $a \in C^1(I)$, $c \in C(I)$, $a(x) \geqq a_0 > 0$, and $c(x) \geqq 0$ for $x \in I$, and $f \in L_2(I)$. The usual variational characterization of the solution u is given by

$$u \in \overset{0}{H}{}^1,$$
$$\int_0^1 (au'v' + cuv)\,dx = \int_0^1 fv\,dx \qquad \text{for all} \quad v \in \overset{0}{H}{}^1. \quad (4.1)$$

However, u can also be characterized in the following two ways in terms of the form B_Δ defined in (3.1):

$$u \in H_\Delta^0,$$
$$B_\Delta(u, v) = \int_0^1 fv\,dx \qquad \text{for all} \quad v \in H_\Delta^2, \quad (4.2)$$

and

$$u \in H_\Delta^2,$$
$$B_\Delta(v, u) = \int_0^1 fv\,dx \qquad \text{for all} \quad v \in H_\Delta^0. \quad (4.3)$$

We note that the solution u lies in the spaces H_Δ^0 and H_Δ^2 for all Δ.

The usual Ritz approximation u_Δ of u is defined by

$$u_\Delta \in S_\Delta,$$
$$\int_0^1 (au'_\Delta v' + cu_\Delta v)\,dx = \int_0^1 fv\,dx \qquad \text{for all} \quad v \in S_\Delta,$$

with S_Δ as defined in (3.2), and it is easily seen that u_Δ also satisfies

$$\begin{aligned} &u_\Delta \in S_\Delta, \\ &B_\Delta(u_\Delta, v) = \int_0^1 fv\,dx \qquad \text{for all} \quad v \in S_\Delta \end{aligned} \tag{4.4}$$

and

$$\begin{aligned} &u_\Delta \in S_\Delta, \\ &B_\Delta(v, u_\Delta) = \int_0^1 fv\,dx \qquad \text{for all} \quad v \in S_\Delta. \end{aligned} \tag{4.5}$$

Using (4.2)–(4.5) we see that

$$B_\Delta(u_\Delta, v) = B(u, v) \qquad \text{for all} \quad v \in S_\Delta$$

and

$$B_\Delta(v, u_\Delta) = B(v, u) \qquad \text{for all} \quad v \in S_\Delta,$$

i.e., that u_Δ is simultaneously the left and the right B_Δ projection of u onto S_Δ. Since assumptions (2.1), (2.2), and (2.4) hold (Theorems 3 and 4) we can thus apply the estimates (2.6) in Theorem 2.

From (2.6a) we get

$$\|u - u_\Delta\|_{0,\Delta} \leqq (1 + C_1/C'_2) \inf_{\chi \in S_\Delta} \|u - \chi\|_{0,\Delta}. \tag{4.6}$$

Thus we have obtained a quasi-optimal error estimate in the norm $\|\cdot\|_{0,\Delta}$. We can estimate the right side of (4.6) in terms of the mesh parameter h as follows. Let $v = \mathcal{I}u$ be that S_Δ interpolant of u which satisfies

$$v(x_j) = u(x_j), \qquad j = 0, 1, \ldots, n,$$
$$\int_{I_j} v(x - x_{j-1})^i\,dx = \int_{I_j} u(x - x_{j-1})^i\,dx, \qquad j = 1, \ldots, n, \quad i = 0, \ldots, r-3.$$

Using standard results in approximation theory we then have

$$\begin{aligned} \|u - u_\Delta\|_{0,I} \leqq \|u - u_\Delta\|_{0,\Delta} &\leqq (1 + C_1/C'_2)\|u - \mathcal{I}u\|_{0,\Delta} \\ &= (1 + C_1/C'_2)\|u - \mathcal{I}u\|_0 \\ &\leqq Ch^k\|u\|_k, \qquad 0 \leqq k \leqq r, \end{aligned} \tag{4.7}$$

provided $u \in H^k(I)$. (4.7) contains the standard L_2 error estimate for the problem we are considering. (The usual proof of this estimate consists in first obtaining an estimate for $\|u - u_\Delta\|_{1,I}$ and then obtaining an estimate for $\|u - u_\Delta\|_{0,I}$ by a duality argument.) (4.5) is closely related to an estimate obtained by Eisenstat *et al.* [4]. They showed that

$$\|u - u_\Delta\|_{0,I} \leqq C \inf \{\|u - \chi\|_{0,I} : \chi \quad \text{an} \quad S_\Delta \text{ interpolant of } u\}$$

under the assumption that the mesh family is quasi-uniform if $r \geqq 3$ and for the case $r = 2$ (piecewise linear approximating functions) for arbitrary mesh families.

We have thus seen that, on the one hand, the norm $\|\cdot\|_{0,\Delta}$ is closely related to the L_2 norm, and, on the other, a quasi-optimal error estimate holds with respect to $\|\cdot\|_{0,\Delta}$. We remark that the estimate

$$\|u - u_\Delta\|_{0,I} \leqq C \inf_{\chi \in S_\Delta} \|u - \chi\|_{0,I}$$

is false, i.e., the Ritz approximation is not quasi-optimal with respect to the L_2 norm.

Since u_Δ is also the right B_Δ projection of u we can apply (2.6b) to get

$$\begin{aligned}\|u - u_\Delta\|_{2,\Delta} &\leqq (1 + C_1/C_2') \inf_{\eta \in S_\Delta} \|u - \eta\|_{2,\Delta} \\ &\leqq (1 + C_1/C_2') \inf_{\eta \in S_\Delta \cap C^1(I)} \|u - \eta\|_2 \\ &\leqq Ch^{k-2}\|u\|_k, \qquad 2 \leqq k \leqq r.\end{aligned}$$

From the definition of $\|\cdot\|_{2,\Delta}$ we see that

$$\left(\sum_{j=1}^{n} \|u - u_\Delta\|_{2,I_j}^2\right)^{1/2} \leqq Ch^{k-2}\|u\|_k \tag{4.8a}$$

and

$$|u_\Delta'(x_j^+) - u_\Delta'(x_j^-)| \leqq Ch^{k-3/2}\|u\|_k, \qquad j = 1, \ldots, n. \tag{4.8b}$$

(4.8a) provides an estimate on the L_2 norm of the second derivatives of the error on each I_j (since we merely assume the continuity of the functions in S_Δ, u_Δ will not in general possess second derivatives on I). We note that the estimate $\|u - u_\Delta\|_{2,I} \leqq Ch^{k-2}\|u\|_{k,I}$ is known to hold provided $S_\Delta \subset H^2(I)$ (this requires C^1 finite elements) and the mesh family is quasi-uniform [i.e., $h_i/h_j \leqq K =$ constant, for all $1 \leqq i, j \leqq n = n(\Delta)$ and for all Δ in the mesh family $\{\Delta\}$ under consideration]. (4.8b) gives an estimate on the convergence to zero of the jumps in u_Δ' at the nodes x_j.

(b) Consider now the two point boundary value problem

$$Lu = -(au')' + cu = \delta'_p,$$
$$u(0) = u(1) = 0$$

with $f = \delta'_p$ equal to the derivative of the Dirac distribution δ_p at $p, 0 < p < 1$. $u \notin \overset{0}{H}{}^1$ and hence we cannot characterize u as in (4.1). u can, however, be characterized by

$$u \in H^0 = L_2,$$
$$\int_0^1 uLv\,dx = -v'(p) \qquad \text{for all} \quad v \in H^2 \cap \overset{0}{H}{}^1, \tag{4.9}$$

and also (more importantly for our purpose) by

$$u \in H^0_\Delta,$$
$$B_\Delta(u, v) = -v'(p) \qquad \text{for all} \quad v \in H^2_\Delta, \tag{4.10}$$

provided p is not a node of Δ. Note that $\delta'_p \in (H^2_\Delta)'$ and thus that existence and uniqueness follows from Theorem 1. If an approximation procedure is based on (4.9) one would have to assume that the finite dimensional spaces of approximating functions lie in $H^2(I)$. However, if we base our approximation procedure on (4.10) we can use S_Δ [as defined in (3.2)].

The Ritz approximation u_Δ to u can now be defined by

$$u \in S_\Delta,$$
$$B_\Delta(u_\Delta, v) = \int_0^1 (au_\Delta' v' + cu_\Delta v)\,dx = v'(p) \qquad \text{for all} \quad v \in S_\Delta.$$

Thus u_Δ is the left B_Δ projection of u and from (2.6) we have

$$\|u - u_\Delta\|_{0,I} \leqq \|u - u_\Delta\|_{0,\Delta} \leqq C \inf_{\chi \in S_\Delta} \|u - \chi\|_{0,\Delta} \leqq C\sqrt{h},$$

if $r = 2$ (piecewise linear approximating functions).

5. L_1 AND L_∞ ESTIMATES

In Section 4 we obtained quasi-optimal error estimates in the norms $\|\cdot\|_{0,\Delta}$ and $\|\cdot\|_{2,\Delta}$. Both of these norms are based on the L_2 norm. We turn now to similar results for mesh dependent norms based on the L_1 and L_∞ norms. Prior to defining the mesh dependent norms and spaces we will use, we briefly sketch the relevant abstract convergence results (cf. Section 2).

Let $X_{1,\Delta}$ and $X_{2,\Delta}$ be two Banach spaces (indexed by the parameter Δ)

with norms $\|\cdot\|_{1,\Delta}$ and $\|\cdot\|_{2,\Delta}$, respectively, and let B_Δ be a bilinear form on $X_{1,\Delta} \times X_{2,\Delta}$. We assume

$$|B_\Delta(u, v)| \leqq C_1 \|u\|_{1,\Delta} \|v\|_{2,\Delta} \qquad \text{for all} \quad u \in X_{1,\Delta}, \quad v \in X_{2,\Delta}, \tag{5.1}$$

where C_1 is independent of Δ.

Let $S_{1,\Delta}$ and $S_{2,\Delta}$ be two finite dimensional subspaces of $X_{1,\Delta}$ and $X_{2,\Delta}$, respectively, and assume

$$\inf_{\substack{u \in S_{1,\Delta} \\ \|u\|_{1,\Delta}=1}} \quad \sup_{\substack{v \in S_{2,\Delta} \\ \|v\|_{2,\Delta}=1}} |B(u, v)| \geqq C_2' > 0, \tag{5.2a}$$

$$\sup_{u \in S_{1,\Delta}} |B_\Delta(u, v)| > 0 \qquad \text{for each} \quad 0 \neq v \in S_{2,\Delta}, \tag{5.2b}$$

where C_2' is independent of Δ.

If $X_{1,\Delta}$ and $X_{2,\Delta}$ are reflexive and we make assumptions analogous to (2.1), (2.2), and (2.4), the results of Section 2 [the existence and uniqueness result for (2.3) and the error estimates (2.6)] are valid without change. In the nonreflexive case (which is of most interest to us) the analog of (2.2) cannot in general hold. Thus we assume only the analog of (2.1) and (2.4), namely, (5.1) and (5.2). The analogs of the estimates (2.6) can be shown to hold under just the assumptions (5.1) and (5.2), and we can thus derive the desired error estimates. This is related to the fact that only the constants C_1 and C_2' appear in (2.6). Note that (2.2) is most directly associated with the existence and uniqueness of solutions to (2.3) for all $f \in (\mathscr{H}_{2,\Delta})'$. Since we will be working with a subclass of problems for which existence and uniqueness is known (from other principles) the fact that we have not assumed the analog of (2.2) will not affect our analysis.

For $u \in X_{1,\Delta}$ we define the left B_Δ projection u_Δ of u by

$$u_\Delta \in S_{1,\Delta},$$

$$B(u_\Delta, v) = B(u, v) \qquad \text{for all} \quad v \in S_{2,\Delta},$$

and for $v \in X_{2,\Delta}$ we define the right B_Δ projection v_Δ of v by

$$v_\Delta \in S_{2,\Delta},$$

$$B(u, v_\Delta) = B(u, v) \qquad \text{for all} \quad u \in S_{2,\Delta}.$$

Then we have

$$\|u - u_\Delta\|_{1,\Delta} \leqq (1 + C_1/C_2') \inf_{\chi \in S_{1,\Delta}} \|u - \chi\|_{1,\Delta} \tag{5.3a}$$

and

$$\|v - v_\Delta\|_{2,\Delta} \leqq (1 + C_1/C_2') \inf_{\eta \in S_{2,\Delta}} \|v - \eta\|_{2,\Delta}. \tag{5.3b}$$

These estimates are analogous to the estimates (2.6) in Theorem 2.

Let

$$H^2_{1,\Delta} = \{u \in \overset{0}{H}{}^1_1(I) : u|_{I_j} \in H^2_1(I_j)\},$$

$$\|u\|_{H^2_{1,\Delta}} = \sum_{j=1}^{n} \|u\|_{2,1,I_j} + \sum_{j=1}^{n-1} |(Ju')(x_j)|$$

and

$$H^2_{\infty,\Delta} = \{u \in \overset{0}{H}{}^1_\infty(I) : u|_{I_j} \in H^2_\infty(I_j)\},$$

$$\|u\|_{H^2_{\infty,\Delta}} = \max_j \|u\|_{2,\infty,I_j} + \max_j \delta_j^{-1} |Ju'(x_j)|.$$

For $u \in \overset{0}{H}{}^1_\infty(I)$ let

$$\|u\|_{H^0_{\infty,\Delta}} = \|u\|_{L_\infty(I)}$$

and for $u \in \overset{0}{H}{}^1_1(I)$ let

$$\|u\|_{H^0_{1,\Delta}} = \int_0^1 |u|\,dx + \sum_{j=1}^{n-1} \delta_j |u(x_j)|,$$

and then define $H^0_{\infty,\Delta}$ to be the completion of $\overset{0}{H}{}^1_\infty(I)$ with respect to $\|\cdot\|_{H^0_{\infty,\Delta}}$ and $H^0_{1,\Delta}$ to be the completion of $\overset{0}{H}_1(I)$ with respect to $\|\cdot\|_{H^0_{1,\Delta}}$. $H^0_{\infty,\Delta}$ and $H^2_{1,\Delta}$ are Banach spaces for each Δ. It is easily seen that $H^0_{\infty,\Delta} = \{u : u \in C(I), u(0) = u(1) = 0\}$.

Consider once again the two point boundary value problem

$$Lu = f, \qquad x \in I,$$
$$u(0) = u(1) = 0.$$

We now analyze the error $u - u_\Delta$ between the solution u and the Ritz approximation u_Δ in two different ways.

First we observe that

$$B_\Delta(u, v) = \sum_j \int_{I_j} uLv\,dx + \sum_{j=1}^{n-1} a(x_j)u(x_j)Jv'(x_j) \tag{5.4}$$

can be considered as a bilinear form on $H^0_{1,\Delta} (\equiv X_{1,\Delta}) \times H^2_{\infty,\Delta} (\equiv X_{2,\Delta})$. It is clear that

$$|B_\Delta(u, v)| \leqq C_1 \|u\|_{H^0_{1,\Delta}} \|v\|_{H^2_{\infty,\Delta}}, \tag{5.5}$$

and it can be shown that the stability condition holds:

$$\inf_{\substack{u \in H^0_{1,\Delta} \\ \|u\|_{H^0_{1,\Delta}} = 1}} \sup_{\substack{v \in H^2_{\infty,\Delta} \\ \|v\|_{H^2_{\infty,\Delta}} = 1}} |B(u, v)| \geqq C_2' > 0; \tag{5.6}$$

here C_1 and C_2' are independent of Δ. (5.5) and (5.6) correspond to (5.1) and (5.2), respectively. The solution u is easily seen to simultaneously satisfy

$$u \in H^0_{1,\Delta},$$

$$B_\Delta(u, v) = \int_0^1 fv\,dx \qquad \text{for all} \quad v \in H^2_{\infty,\Delta}$$

and

$$u \in H^2_{\infty,\Delta},$$

$$B_\Delta(v, u) = \int_0^1 fv\,dx \qquad \text{for all} \quad v \in H^0_{1,\Delta},$$

and the usual Ritz approximation u_Δ is seen to simultaneously satisfy

$$u_\Delta \in S_\Delta,$$

$$B_\Delta(u, v) = \int_0^1 fv\,dx \qquad \text{for all} \quad v \in S_\Delta$$

and

$$u_\Delta \in S_\Delta,$$

$$B_\Delta(v, u) = \int_0^1 fv\,dx \qquad \text{for all} \quad v \in S_\Delta.$$

Thus u_Δ is both the left and right B_Δ projection of u, with B_Δ defined in (5.4). Hence we can apply (5.3) to obtain

$$\|u - u_\Delta\|_{H^0_{1,\Delta}} \leqq C \inf_{\chi \in S_\Delta} \|u - \chi\|_{H^0_{1,\Delta}} \tag{5.7}$$

and

$$\|u - u_\Delta\|_{H^2_{\infty,\Delta}} \leqq C \inf_{\chi \in S_\Delta} \|u - \chi\|_{H^2_{\infty,\Delta}}. \tag{5.8}$$

The second estimate here shows that the second derivatives of $u - u_\Delta$, considered on each I_j, converge to zero uniformly.

Next we note that

$$B_\Delta(u, v) = \sum_j \int_{I_j} uLv\,dx + \sum_{j=1}^{n-1} a(x_j)u(x_j)Jv'(x_j)$$

can be considered as a bilinear form on $H^0_{\infty,\Delta}$ $(\equiv X_{1,\Delta}) \times H^2_{1,\Delta}$ $(\equiv X_{2,\Delta})$. With this choice of spaces the form B_Δ satisfies (5.1) and (5.2). As above we are led to

$$\|u - u_\Delta\|_{H^0_{\infty,\Delta}} \leqq C \inf_{\chi \in S_\Delta} \|u - \chi\|_{H^0_{\infty,\Delta}} \tag{5.9}$$

and

$$\|u - u_\Delta\|_{H^0_{\infty,\Delta}} \leqq C \inf_{\chi \in S_\Delta} \|u - \chi\|_{H^2_{1,\Delta}}. \tag{5.10}$$

(5.7)–(5.10) provide quasi-optimal error estimates in the indicated norms. From these estimates we can derive estimates in terms of the mesh parameter h. For example, from (5.9) we get

$$\|u - u_\Delta\|_{L_\infty(I)} \leqq Ch^k \|u\|_{H^k_\infty(I)}, \qquad 0 \leqq k \leqq r. \tag{5.11}$$

(5.11) was proved by Wheeler ([6]; cf. also [3]).

6. PROOF OF THEOREM 4 IN A SPECIAL CASE

In order to apply the results of Section 2 (Theorems 1 and 2) to a problem, one must check assumptions (2.1), (2.2), and (2.4). For most problems the most difficult assumption to verify is (2.4a). In this section we sketch the proof of (2.4a) in the special case where $r = 3$ (piecewise quadratic approximating functions) and $Lu = -u''$. This corresponds to Theorem 4 for this special case.

We need three preliminary results. The first is

$$K_1^{-1}\|u\|_{2,\Delta}^2 \leqq \sum_{j=1}^{n} \|u''\|_{L_2(I_j)}^2 + \sum_{j=1}^{n-1} \delta_j^{-1}|Ju'(x_j)|^2$$

$$\leqq K_1\|u\|_{2,\Delta} \qquad \text{for all} \quad u \in H_{2,\Delta}, \tag{6.1}$$

where K_1 is independent of Δ. The right side of (6.1) is obvious. To prove the left we proceed as follows. For $u \in H_\Delta^2$ we have

$$\begin{aligned}
\|u\|_{1,I}^2 &= \int_0^1 (|u'|^2 + |u|^2)\, dx \\
&\leqq 2\int_0^1 |u'|^2\, dx \\
&= 2\left\{\sum_{j=1}^{n} \int_{I_j} (-u'')u\, dx - \sum_{j=1}^{n-1} u(x_j)Ju'(x_j)\right\} \\
&\leqq 2\left(\sum_{j=1}^{n} \int_{I_j} |u''|^2\, dx\right)^{1/2} \left(\int_0^1 |u|^2\, dx\right)^{1/2} \\
&\quad + 2\left(\sum_{j=1}^{n-1} \delta_j|u(x_j)|^2\right)^{1/2} \left(\sum_{j=1}^{n-1} \delta_j^{-1}|Ju'(x_j)|^2\right)^{1/2} \\
&\leqq 2\left(\sum_{j=1}^{n} \int_{I_j} |u''|^2\, dx\right)^{1/2} \|u\|_{1,I} \\
&\quad + 2\left(\sum_{j=1}^{n-1} \delta_j\|u\|_{1,I}^2\right)^{1/2} \left(\sum_{j=1}^{n-1} \delta_j^{-1}|Ju'(x_j)|^2\right)^{1/2} \\
&\leqq 2\|u\|_{1,I}\left\{\left(\sum_{j=1}^{n} \int_{I_j} |u''|^2\, dx\right)^{1/2} \right. \\
&\quad \left. + \left(\sum_{j=1}^{n-1} \delta_j^{-1}|Ju'(x_j)|^2\right)^{1/2}\right\},
\end{aligned}$$

and hence

$$\|u\|_{1,I}^2 \leqq 4\left\{\sum_{j=1}^{n} \int_{Ij} |u''|^2\, dx + \sum_{j=1}^{n-1} \delta_j^{-1}|Ju'(x_j)|^2\right\}.$$

(6.1) follows immediately from this inequality.

Second, we note that

$$\int_{I_j} u^2\, dx \leqq K_2\left\{(|u(x_{j-1})|^2 + |u(x_j)|^2)h_j + \left[\left(\int_{I_j} u\, dx\right)^2 \Big/ h_j\right]\right\} \tag{6.2}$$

for any quadratic polynomial u, where K_2 is independent of u and j. This is proved by first noting that there is a constant K_2' such that

$$\int_0^1 u^2\, dx \leqq K_2'\left[(|u(0)|^2 + |u(1)|^2 + \left(\int_0^1 u\, dx\right)^2\right] \tag{6.3}$$

for all quadratic polynomials u. Since both sides of (6.3) define norms on the space of quadratic polynomials, this inequality follows from the fact that all norms on a finite dimensional space are equivalent. Now we derive (6.2) from (6.3) by a change of scale argument.

The third result is

$$\inf_{\substack{u\in S_\Delta\\ \|u\|_{0,\Delta}=1}} \sup_{\substack{v\in S_\Delta\\ \|v\|_{2,\Delta}=1}} |B_\Delta(u,v)| = \inf_{\substack{v\in S_\Delta\\ \|v\|_{2,\Delta}=1}} \sup_{\substack{u\in S_\Delta\\ \|u\|_{0,\Delta}=1}} |B_\Delta(u,v)|. \tag{6.4}$$

This follows from the fact that the norm of a matrix and its transpose are equal.

Now let $v \in S_\Delta$. Since v is piecewise quadratic, $b_j \equiv v''|_{I_j}$ is a constant. Choose $u \in S_\Delta$ so that

$$u(x_j) = -(Jv')(x_j)\delta_j^{-1}, \qquad 1 \leqq j \leqq n-1,$$

$$u(0) = u(1) = 0,$$

$$\int_{I_j} u\, dx = b_j h_j, \qquad 1 \leqq j \leqq n.$$

Then, using (6.1), we have

$$\begin{aligned} B_\Delta(u,v) &= \sum_{j=1}^{n} \int_{I_j} u(-v'')\, dx - \sum_{j=1}^{n-1} u(x)(Jv')(x_j) \\ &= \sum_{j=1}^{n} b_j^2 h_j + \sum_{j=1}^{n-1} \delta_j^{-1}|Jv'(x_j)|^2 \\ &= \sum_{j=1}^{n} \int_{I_j} |v''|^2\, dx + \sum_{j=1}^{n-1} \delta_j^{-1}|Jv'(x_j)|^2 \\ &\geqq C\|v\|_{2,\Delta}^2, \end{aligned}$$

where $C > 0$ is independent of Δ, and, using (6.1) and (6.2),

$$\begin{aligned}
\|u\|_{0,\Delta}^2 &= \int_0^1 u^2\,dx + \sum_{j=1}^{n-1} \delta_j |u(x_j)|^2 \\
&\leqq C'\left\{\sum_{j=1}^{n} (|u(x_{j-1})|^2 + |u(x_j|^2)h_j + \sum_{j=1}^{n} \left[\left(\int_{I_j} u\,dx\right)^2 \Big/ h_j\right]\right\} \\
&\quad + \sum_{j=1}^{n-1} \delta_j |u(x_j)|^2 \\
&= C' \sum_{j=1}^{n} \left[\left(\int_{I_j} u\,dx\right)^2 \Big/ h_j\right] + 3C' \sum_{j=1}^{n-1} \delta_j |u(x_j)|^2 \\
&= C' \sum_{j=1}^{n} b_j^2 h_j + 3C' \sum_{j=1}^{n-1} \delta_j^{-1} |Jv'(x_j)|^2 \\
&= C' \sum_{j=1}^{n} \int_{I_j} |v''|^2\,dx + 3C' \sum_{j=1}^{n-1} \delta_j^{-1} |Jv'(x_j)|^2 \\
&\leqq C''\|v\|_{2,\Delta}^2,
\end{aligned}$$

where C'' is independent of Δ. From these two inequalities we immediately get

$$\inf_{\substack{v \in S_\Delta \\ \|v\|_{2,\Delta}=1}} \quad \sup_{\substack{v \in S_\Delta \\ \|u\|_{0,\Delta}=1}} |B_\Delta(u, v)| \geqq \frac{C}{C''} \equiv C_2' > 0, \tag{6.5}$$

where C_2' is independent of Δ. (6.5) and (6.4) yield (2.4a).

REFERENCES

1. I. Babuška, Error-bounds for finite element method, *Numer. Math.* **16** (1971), 322–333.
2. I. Babuška and A. K. Aziz, Survey lectures on the mathematical foundation of the finite element method, *in* "The Mathematical Foundations of the Finite Element Method with Applications to Partial Differential Equations" (A. K. Aziz, ed.), Academic Press, New York, 1973, pp. 5–359.
3. J. Douglas, Jr., T. Dupont, and L. Wahlbin, Optimal L_∞ error estimates for Galerkin approximations to solutions of two-point boundary value problems, *Math. Comp.* **29** (1975), 475–483.
4. S. Eisenstat, R. Schreiber, and M. Schultz, On the Optimality of the Rayleigh–Ritz Approximation, Res. Rep. No. 83, Dept. Comput. Sci., Yale Univ. New Haven.
5. J. Nečas, "Les Méthodes Directes en Théorie des Équations Elliptiques," Masson, Paris, 1967.
6. M. Wheeler, An optimal L_∞ error estimate for Galerkin approximations to solutions of two point boundary value problems, *SIAM J. Numer. Anal.* **10** (1973), 914–917.

The work of the first author was partially supported by the Department of Energy under Contract E(40–1)3443 and that of the second author by NSF Grants MCS76–06963 and MCS78–02851.

I. Babuška
INSTITUTE FOR PHYSICAL SCIENCE AND TECHNOLOGY

and

DEPARTMENT OF MATHEMATICS
UNIVERSITY OF MARYLAND
COLLEGE PARK, MARYLAND

J. Osborn
DEPARTMENT OF MATHEMATICS
UNIVERSITY OF MARYLAND
COLLEGE PARK, MARYLAND

The Fundamental Mode of Vibration of a Clamped Annular Plate Is Not of One Sign

C. V. COFFMAN

R. J. DUFFIN

D. H. SHAFFER

1. INTRODUCTION

We are concerned here with the modes of vibration of a clamped annular plate; that is to say, with the eigenvalue problem

$$\Delta^2\varphi = \lambda\varphi \quad \text{in } D_\varepsilon, \qquad \varphi = \frac{\partial\varphi}{\partial n} = 0 \quad \text{on } \partial D_\varepsilon, \tag{1.1}$$

where

$$D_\varepsilon = \{z = x + iy : \varepsilon < |z| < 1\}, \qquad \varepsilon > 0,$$

and $\partial/\partial n$ denotes differentiation with respect to the outer normal.

Apparently it was A. Weinstein who first raised the question of whether the fundamental mode of vibration of a clamped plate must in general be of one sign. Interest in this question increased when Szegö showed [10] that an affirmative answer would imply that among all plates of a given area the fundamental eigenvalue is smallest for a circular plate.

In 1952 Duffin and Shaffer [5], [6] showed that the response to Weinstein's question is negative. More specifically they showed that there exists a critical value $\varepsilon_c \approx 1/715$ such that for $0 < \varepsilon < \varepsilon_c$ the fundamental eigenvalue has multiplicity two and the corresponding eigenfunctions have a diametric nodal line. At least when ε is not too close to one, the fundamental eigenvalue is simple and the corresponding eigenfunction is of one sign for

ISBN 0-12-178150-X

$\varepsilon > \varepsilon_c$. When $\varepsilon = \varepsilon_c$ there is a fundamental eigenfunction which is of one sign but the fundamental eigenvalue has multiplicity three.

There are other regions for which numerical evidence or theoretical considerations strongly suggest that the fundamental eigenfunction is not of one sign. Among these are simply connected and even convex regions [1, 4]. However, to our knowledge the only case where this can be considered as established beyond doubt is that of the annulus with small ratio of inner to outer radius. We note that Hadamard had shown, [7] that the biharmonic Green's function for such a region is not positive.

For the performance of the work reported on in [6] the only computational aids were a table of Bessel functions (namely, [2]) and a desk computer. With such limited facilities the task was made feasible only by the introduction of certain simplifying approximations. While the qualitative results were established beyond doubt we thought it would be of interest to rework the computations with a view to obtaining greater accuracy and in particular to obtain the critical value of ε_c with more precision. The computations are quite within the capability of the hand-held programmable calculators now available. A general description of the procedure and some numerical results are given below.

As was done in [6] we also consider here the "punctured disk," i.e., (1.1) with $\varepsilon = 0$. One encounters some difficulty in formulating the problem correctly due to the degeneracy of the boundary point $z = 0$, but when this is resolved the analysis of the resulting problem is substantially simpler than when $\varepsilon > 0$. Indeed to verify that in this case the fundamental eigenfunctions have nodal lines, no computations are required, only a consultation of the tables. Continuity considerations then enable one to deduce the same result for small positive ε.

2. THE EIGENVALUES OF THE ANNULUS

From the factorization

$$\Delta^2 - \mu^4 = (\Delta - \mu^2)(\Delta + \mu^2)$$

one sees that the equation

$$\Delta^2 u - \mu^4 u = 0$$

admits as solution any function of the form

$$u(re^{i\theta}) = [c_1 J_n(\mu r) + c_2 Y_n(\mu r) + c_3 I_n(\mu r) + c_4 K_n(\mu r)] \cos n\theta; \quad (2.1)$$

here we employ the standard notation for the Bessel functions and the modified Bessel functions. In order that the problem (1.1) admit an eigenfunction of the form (2.1) it is necessary and sufficient that μ satisfy

$$\begin{vmatrix} J_n(\mu\varepsilon) & J_n'(\mu\varepsilon) & J_n(\mu) & J_n'(\mu) \\ Y_n(\mu\varepsilon) & Y_n'(\mu\varepsilon) & Y_n(\mu) & Y_n'(\mu) \\ I_n(\mu\varepsilon) & I_n'(\mu\varepsilon) & I_n(\mu) & I_n'(\mu) \\ K_n(\mu\varepsilon) & K_n'(\mu\varepsilon) & K_n(\mu) & K_n'(\mu) \end{vmatrix} = 0. \tag{2.2}$$

Let $\mu_n(\varepsilon)$ denote the least positive root of (2.2). We shall show that *if* ε *is sufficiently small, then*

$$\mu_1(\varepsilon) < \mu_0(\varepsilon). \tag{2.3}$$

To see that this suffices to prove the contention of the title, observe that if $u(re^{i\theta})$ is an eigenfunction of (1.1), then so is the radially symmetric function

$$\bar{u}(r) = \frac{1}{2\pi} \int_0^{2\pi} u(re^{i\theta})\, d\theta;$$

thus the latter function is an eigenfunction of the problem

$$\frac{1}{r}\left\{\frac{d}{dr}\left[\frac{1}{r}\frac{d}{dr}\left(r\frac{du}{dr}\right)\right]\right\} = \mu^4 u, \qquad \varepsilon < r < 1, \tag{2.4}$$

$$u(\varepsilon) = \frac{du(\varepsilon)}{dr} = u(1) = \frac{du(1)}{dr} = 0. \tag{2.5}$$

In particular $\bar{u}$ has the form (2.1) with $n = 0$, consequently the eigenvalue that corresponds to a positive eigenfunction cannot be less than $[\mu_0(\varepsilon)]^4$. In fact it was shown in [6] that the fundamental eigenfunction of (2.4), (2.5) is of one sign. General results implying this can also be found in [8] or [9].

3. VERIFICATION OF INEQUALITY (2.3)

It will simplify our computations if we put

$$\rho = \mu\varepsilon$$

in (2.2) and consider the resulting implicit relationship between μ and ρ. Accordingly we henceforth let

$$\mu_n = \mu_n(\rho)$$

denote the least positive root of (2.2) corresponding to the fixed value

$$\rho = \mu\varepsilon.$$

We make use of the following standard relations:

$$J_0' = -J_1, \qquad Y_0' = -Y_1, \qquad I_0' = I_1, \qquad K_0' = -K_1 \tag{3.1}$$

and

$$J_1' = -(1/x)J_1 + J_0, \qquad Y_1' = -(1/x)Y_1 + Y_0,$$
$$I_1' = -(1/x)I_1 + I_0, \qquad K_1' = -(1/x)K_1 - K_0.$$

Using these relations and the fundamental rules for the manipulation of determinants we find that the Eqs. (2.2) with $n = 0$ and $n = 1$ are equivalent, respectively, to

$$D_0(\mu\varepsilon, \mu) = 0 \qquad \text{and} \qquad D_1(\mu\varepsilon, \mu) = 0,$$

where

$$D_i(\rho, \mu) = \begin{vmatrix} J_0(\rho) & (\rho/2)J_1(\rho) & J_0(\mu) & (\mu/2)J_1(\mu) \\ \pi Y_0(\rho) & (\pi/2)\rho Y_1(\rho) & \pi Y_0(\mu) & (\pi/2)\mu Y_1(\mu) \\ I_0(\rho) & (-1)^{i+1}(\rho/2)I_1(\rho) & I_0(\mu) & (-1)^{i+1}(\mu/2)I_1(\mu) \\ (-1)^i K_0(\rho) & (\rho/2)I_1(\rho) & (-1)^i K_0(\mu) & (\mu/2)K_1(\mu) \end{vmatrix}. \tag{3.2}$$

To evaluate the Bessel functions the following series representations were used.

$$J_0(x) = 1 - \sum_{k=1}^{\infty} \frac{(-1)^{k-1}}{(k!)^2}\left(\frac{x}{2}\right)^{2k}, \qquad \frac{x}{2}J_1(x) = \sum_{k=1}^{\infty} \frac{(-1)^{k-1}}{k!(k-1)!}\left(\frac{x}{2}\right)^{2k},$$

$$\pi Y_0(x) = \sum_{k=0}^{\infty} \frac{(-1)^k}{(k!)^2}\left(\frac{x}{2}\right)^{2k}\left[2\log\left(\frac{x}{2}\right) - 2\psi(k+1)\right],$$

$$\frac{\pi}{2}xY_1(x) = -1 - \sum_{k=1}^{\infty} \frac{(-1)^k}{k!(k-1)!}\left(\frac{x}{2}\right)^{2k}\left[2\log\left(\frac{x}{2}\right) - \psi(k) - \psi(k+1)\right],$$

$$I_0(x) = 1 + \sum_{k=1}^{\infty} \frac{1}{(k!)^2}\left(\frac{x}{2}\right)^{2k}, \qquad \frac{x}{2}I_1(x) = \sum_{k=1}^{\infty} \frac{1}{k!(k-1)!}\left(\frac{x}{2}\right)^{2k},$$

$$K_0(x) = -\sum_{k=0}^{\infty} \frac{1}{(k!)^2}\left(\frac{x}{2}\right)^{2k}\left[\log\left(\frac{x}{2}\right) - \psi(k+1)\right],$$

$$xK_1(x) = 1 + \sum_{k=1}^{\infty} \frac{1}{k!(k-1)!}\left(\frac{x}{2}\right)^{2k}\left[2\log\left(\frac{x}{2}\right) - \psi(k) - \psi(k+1)\right],$$

where

$$\psi(0) = \psi(1) = -\gamma, \qquad \psi(k) = \sum_{j=1}^{k-1}\frac{1}{j} - \gamma, \qquad k > 1, \qquad \gamma = 0.5772156649.$$

The calculator was programmed to evaluate these series, truncating after the last term exceeding 10^{-10}. With this program we obtained complete agreement, for all eight functions, with the values given in the tables [2] as long as the argument x did not exceed 8.

Using the above mentioned program as a subroutine we next wrote a main program to locate the roots of $D_i(\rho, \mu)$ by successive approximation. By allowing for the appropriate variation of signs [see (3.2)] this could be written simultaneously for $i = 0$ and $i = 1$. Note also that $D_i(\rho, \mu)$ is symmetric in ρ, μ so that the same program could be used either to determine the value of μ corresponding to a given value of ρ or vice versa. Of course the program was written so that the values of the functions corresponding to the fixed argument were computed only once during the approximation procedure; a considerable saving of computation time resulted from choosing μ as the fixed argument. In beginning the successive approximation we used as first approximation values from the graph in [6] of the data obtained in that earlier work. Results of our computations are tabulated in Tables I and II.

TABLE I

Roots of $D_0(\rho, \mu) = 0$

μ	ρ	ε
4.76830995	0.0001	0.0000209718
4.768319895	0.0005	0.0001048587
4.768344645	0.001	0.0002097164
4.768425778	0.002	0.0004194256
4.768685966	0.004	0.0008388055
4.769102418	0.0062557144	0.0013117174
4.77	0.0099299122	0.0020817426
4.78	0.0339204678	0.0070963322

TABLE II

Roots of $D_1(\rho, \mu) = 0$

μ	ρ	ε
4.70	0.0001041288	0.000022155
4.72	0.0005795765	0.0001227916
4.74	0.0018980407	0.0004004305
4.75	0.0030239677	0.0006366248
4.76	0.0045275591	0.0009511679
4.769102418	0.0062557144	0.0013117174
4.77	0.0064455384	0.0013512659
4.78	0.00883428	0.0018417213

4. DETERMINATION OF THE CRITICAL VALUE

The determination of ε_c is equivalent to the determination of the least positive value of ρ for which the equations

$$D_0(\rho, \mu) = 0 \qquad \text{and} \qquad D_1(\rho, \mu) = 0$$

have their least positive root μ in common.

Let ρ_1 be chosen so that

$$\mu_1(\rho_1) < \mu_0(\rho_1) \tag{4.1}$$

and let $\rho_2, \rho_3, \ldots, \mu^1, \mu^2, \ldots$ be defined inductively in such a way that

$$\begin{aligned} \mu_1 &= \mu_0(\rho_1) \\ \mu_1(\rho_2) &= \mu^1 \\ &\vdots \\ \mu^n &= \mu_0(\rho_n) \\ \mu_1(\rho_{n+1}) &= \mu^n. \\ &\vdots \end{aligned}$$

Since $\mu_0(\rho)$ and $\mu_1(\rho)$ are increasing functions it follows by induction that

$$\rho_{n+1} > \rho_n \qquad \text{and} \qquad \mu_1(\rho_n) < \mu_0(\rho_n)$$

for all n. It is clear from the remarks in the preceding section that the program described there can be conveniently used to compute successively the values of ρ_n, μ^n. When this was implemented we obtained sequences converging to the values

$$\rho_c = 0.0062557144, \qquad \mu_c = 4.769102418.$$

From these values we obtain the critical value

$$\varepsilon_c = 0.0013117174$$

or

$$R_c = 762.359356 = 1/\varepsilon_c.$$

The disk D_{ε_c} has the triply degenerate fundamental eigenvalue

$$\lambda_0(\varepsilon_c) = 517.3049053$$

For $i = 0$, $\rho = \rho_c$, $\mu = \mu_c$ the entries of the matrix that appears on the right in (3.2) are listed below. By changing the signs of the entries marked by an asterisk one obtains the matrix corresponding to $i = 1$, $\rho = \rho_c$, $\mu = \mu_c$. To enable the reader to verify that both of these matrices are singular we also give solutions $x_1^0, x_2^0, x_3^0, x_4^0$ and $x_1^1, x_2^1, x_3^1, x_4^1$ of the homogeneous systems with these coefficient matrices; for example, $(x_1^0, x_2^0, x_3^0, x_4^0)$ is orthogonal to all of the rows of the first of the abovementioned matrices.

0.9999902165	0.0000097834	−0.2495615904	−0.6982901238
−10.38026177	−1.000111339	−0.8342644509	1.672698802
1.000009783	−0.0000097835*	22.17669622	−46.94488161*
5.190252006*	0.4999443298	0.0047561682*	0.0124780646
x_1^0	x_2^0	x_3^0	x_4^0
1.213059213	−12.63814517	2.062150576	1
x_1^1	x_2^1	x_3^1	x_4^1
0.1680972254	1.699958712	−2.124437141	1

5. THE PUNCTURED DISK

We now consider the problem (1.1) with $\varepsilon = 0$. Because of singularities of the Bessel functions Y_0, Y_1, K_0, K_1, at 0 this leads to a loss of two of the four degrees of freedom that appear in (2.1), and this simplifies the problem. However, as was found in [6], in order to obtain an eigenfunction with exactly one diametric nodal line one must give up the requirement that the gradient vanish at 0. One way in which one can explain this apparent anomaly is by considering the problem from a variational point of view. This will also provide the basis for the continuity argument by which we deduce the nodal properties of the fundamental eigenfunction of (1.1) when $\varepsilon > 0$ is small from the results for the punctured disk.

For $\varepsilon > 0$ the eigenvalues and eigenfunctions are found as indicated in Section 2. As is well known, the fundamental (i.e., least) eigenvalue can then be characterized variationally as the infimum of the Rayleigh quotient

$$J_\varepsilon(u) = \iint_{D_\varepsilon} (\Delta u)^2\, dx\, dy \Big/ \iint_{D_\varepsilon} u^2\, dx\, dy$$

over the class of "test functions" u such that

$$u \in C^2(D_\varepsilon) \cap C^1(\bar{D}_\varepsilon), \tag{5.1}$$

$$u = |\text{grad } u| = 0 \qquad \text{on } \partial D_\varepsilon, \tag{5.2}$$

$$u \not\equiv 0 \qquad \text{in } D_\varepsilon \tag{5.3}$$

(in fact the test functions can be taken to vanish identically in a neighborhood of ∂D_ε). The higher eigenvalues can be characterized by means of the Poincaré principle or the mini–max principle in the same space of test functions. For

$\varepsilon = 0$ we *define* the eigenvalues of (1.1) by this variational principle. Now for $0 \leqq \varepsilon < 1$ let $\lambda_0(\varepsilon), \lambda_1(\varepsilon), \ldots$ be the eigenvalues of (1.1) *indexed in nondecreasing order*; $\lambda_0, \lambda_1, \ldots$ will denote the eigenvalues of the same problem for the unit disk D. It is easy to see from the variational characterization that for $n = 0, 1, 2, \ldots,\quad 0 < \varepsilon < \varepsilon' < 1$,

$$\lambda_n \leqq \lambda_n(0) \leqq \lambda_n(\varepsilon) \leqq \lambda_n(\varepsilon')$$

and

$$\lambda_n(0) = \lim_{\varepsilon \downarrow 0} \lambda_n(\varepsilon).$$

Since, as it turns out, J_{D_0} does not assume its infimum in the space of test functions for D_0 it is useful to consider the completion $H(D_0)$ of the test functions in the norm

$$\|u\| = \left(\iint_{D_0} (\Delta u)^2 \, dx \, dy \right)^{1/2}.$$

Similarly we form the function spaces $H(D)$, $H(D_\varepsilon)$, $0 < \varepsilon < 1$. We then have the continuous inclusions

$$H(D) \supset H(D_0) \supset H(D_\varepsilon),$$

moreover [3], $H(D_0)$ is of codimension 1 in $H(D)$ and $u \in H(D)$ belongs to $H(D_0)$ whenever $u(0) = 0$. The eigenfunctions of D that correspond to the second eigenvalue λ_1 possess a diametric nodal line and thus belong to $H(D_0)$. It follows that

$$\lambda_0(0) \leqq \lambda_1.$$

We shall show that *the equality holds* and that *the positive eigenfunction of D_0 corresponds to an eigenvalue strictly greater than* $\lambda_0(0)$. It is easily seen that as $\varepsilon \to 0$ the minimizing functions for J_{D_ε} tend in $H(D_0)$ to a minimizing function for J_{D_0}. Thus the above result will imply that the fundamental eigenfunctions of D_ε are not of one sign provided ε is sufficiently small.

The positive eigenfunction for D_0 has the form

$$\varphi(re^{i\theta}) = C_1[Y_0(\mu r) + \tilde{K}_0(\mu r)] + C_2[I_0(\mu r) - J_0(\mu r)]$$

where

$$\tilde{K}_0(s) = (2/\pi)K_0(s),$$

and μ is the least positive root of

$$[Y_0(\mu) + \tilde{K}_0(\mu)][I_0'(\mu) - J_0'(\mu)] - [Y_0'(\mu) + \tilde{K}_0'(\mu)][I_0(\mu) - J_0(\mu)] = 0. \tag{5.4}$$

Among the eigenfunctions of D with a single nodal line is one of the form

$$\psi(re^{i\theta}) = [d_1 I_1(vr) + d_2 J_1(vr)] \sin \theta$$

where v is the least positive root of

$$I_0(v)J_0'(v) + I_0'(v)J_0(v) = 0. \tag{5.5}$$

We first write down some of the properties of the Bessel functions. The modified functions I_0 and K_0 satisfy

$$0 < I_0'(s) < I_0(s) \tag{5.6}$$

and

$$0 < K_0(s) < -K_0'(s). \tag{5.7}$$

As is customary, the positive roots of

$$J_i(s) = 0, \qquad Y_i(s) = 0$$

will be denoted by $j_{i,n}$, $y_{i,n}$, respectively, and indexed in increasing order. From the definitions, the relations (3.1), and elementary Sturm comparison (or the tables [2]) there follows the interlacing

$$0 < y_{0,1} < y_{1,1} < j_{0,1} < j_{1,1} < y_{0,2} < y_{1,2} < j_{0,2}.$$

The signs of J_0, J_0', Y_0, Y_0' are as tabulated in Table III.

TABLE III

	$J_0(s)$	$J_0'(s)$	$Y_0(s)$	$Y_0'(s)$
$(0, y_{0,1})$	+	−	−	+
$(y_{0,1}, y_{1,1})$	+	−	+	+
$(y_{1,1}, j_{0,1})$	+	−	+	−
$(j_{0,1}, j_{1,1})$	−	−	+	−
$(j_{1,1}, y_{0,2})$	−	+	+	−
$(y_{0,2}, y_{1,2})$	−	+	−	−
$(y_{1,2}, j_{0,2})$	−	+	−	+

The positive eigenfunction of the disk has the form

$$c_1 J_0(\tau r) + c_2 I_0(\tau r),$$

where τ is the least positive root of

$$I_0(\tau)J_0'(t) - I_0'(\tau)J_0(\tau) = 0. \tag{5.8}$$

From (5.6) and Table III it is clear that this root τ lies in the interval $(j_{0,1}, j_{1,1})$. Since the positive eigenfunction is the dominant one for the disk it is clear that the positive roots of (5.4) and (5.5) must lie to the right of that of (5.8), hence to the right of $j_{0,1}$.

A consultation of Table III shows that $J_0'(s)/J_0(s)$ is positive in $(j_{0,1}, j_{1,1})$ and decreases from 0 to $-\infty$ as s increases from $j_{1,1}$ to $j_{0,2}$. Thus, if α is the root of

$$J_0'(\alpha) = -J_0(\alpha)$$

in $(j_{1,1}, j_{0,2})$, then it follows from (5.6) that the root ν of (5.5) lies in the interval $(j_{1,1}, \alpha)$. From the tables [2], we have

$$j_{1,1} = 3.8317, \qquad 4.68 < \alpha < 4.681.$$

We next attempt to locate the least positive root μ of (5.4); as we have seen, $\mu > j_{0,1}$. We henceforth use without mention (5.6), (5.7), and Table III. First we have

$$\frac{I_0'(s) - J_0'(s)}{I_0(s) - J_0(s)} < \frac{I_0'(s)}{I_0(s)} < 1$$

on $(y_{0,2}, j_{0,2})$. Next we note that if (5.4) has a root μ in $(j_{0,1}, y_{1,2})$, then $Y_0(\mu) + \tilde{K}_0(\mu) < 0$; thus

$$Y_0(\beta) + \tilde{K}_0(\beta) = 0$$

must have a root β with

$$y_{0,2} < \beta < \mu < y_{1,2}.$$

However,

$$\frac{Y_0'(s) + \tilde{K}_0'(s)}{Y_0(s) + \tilde{K}_0(s)} > \frac{Y_0'(s)}{Y_0(s)}$$

on $(\beta, y_{1,2})$. Thus, if $\mu < y_{1,2}$, then μ must lie to the right of the root γ of

$$Y_0'(\gamma) = Y_0(\gamma)$$

that belongs to $(y_{0,2}, y_{1,2})$. In any case therefore $\mu > \gamma$. From the tables we have

$$4.68 < \alpha < 4.681, \qquad 4.69 < \gamma < 4.70,$$

and thus

$$\mu > \gamma > \alpha > \nu.$$

The values μ, ν are in fact

$$\mu = 4.768309396, \qquad \nu = 4.610899880.$$

It would be interesting if one could obtain a proof of the inequality

$$\gamma > \alpha \tag{5.9}$$

without appealing to the tables. We remark that (5.9) would follow from the inequality

$$\pi s Y_0(s) J_0(s) < 1,$$

which we believe is valid.

REFERENCES

1. L. BAUER AND E. REISS, Block five diagonal metrics and the fast numerical computation of the biharmonic equation, *Math. Comp.* **26** (1972), 311–326.
2. British Association for the Advancement of Science, "Mathematical Tables VI, Bessel Functions," Part I, Cambridge Univ. Press, London, 1937.
3. C. V. COFFMAN AND C. L. GROVER, Obtuse cones in Hilbert spaces and applications to partial differential equations, *J. Funct. Anal.* (to appear).
4. R. J. DUFFIN, Nodal lines of a vibrating plate, *J. Math. Phys.* **31** (1953), 294–299.
5. R. J. DUFFIN AND D. H. SHAFFER, On the modes of vibration of a ring-shaped plate, *Bull. Amer. Math. Soc.* **58** (1952), 652.
6. R. J. DUFFIN AND D. H. SHAFFER, On the Vibration of a Ring-Shaped Plate, Dept. Air Force Tech. Rep., AF8-TRIG, Carnegie Inst. of Tech., Pittsburgh, Pennsylvania, 1952.
7. J. HADAMARD, Sur certains cas intéressants du problème biharmonique, *Atti Congr. Internaz. Mat., Rome I*, Part III (1909), 12–14. ("Oeuvres de Jacques Hadamard," Vol. III, CNRS, Paris, 1968, pp. 1297–1299.)
8. S. KARLIN, "Total Positivity," Vol. I, Stanford Univ. Press, Stanford, California, 1968.
9. W. LEIGHTON AND Z. NEHARI, On the oscillation of solutions of self-adjoint linear differential equations of the fourth order, *Trans. Amer. Math. Soc.* **89** (1958), 325–377.
10. G. SZEGÖ, On membranes and plates, *Proc. Nat. Acad. Sci. U.S.A.* **36** (1950), 210–216.

C. V. Coffman and *R. J. Duffin*
DEPARTMENT OF MATHEMATICS
CARNEGIE–MELLON UNIVERSITY
PITTSBURGH, PENNSYLVANIA

D. H. Shaffer
WESTINGHOUSE RESEARCH LABS.
PITTSBURGH, PENNSYLVANIA

On Explicit Norm Inequalities and Their Applications in Partial Differential Equations

L. E. PAYNE

1. INTRODUCTION

Many of the important properties of solutions of initial and/or boundary value problems for partial differential equations follow from the existence of a priori inequalities relating suitable norms of the unknown solution to norms of the data. If one is interested only in qualitative properties of the solution then the a priori inequalities need not be explicit. However, if one wishes to derive quantitative properties he usually requires explicit inequalities. Questions of existence, uniqueness, regularity, and order of convergence of numerical schemes are in the first category, while error bounds, limits of existence and stability, decay rates, etc., normally fall into the second category. We are concerned in this paper only with explicit norm inequalities, and since sharp quantitative results require sharp inequalities we shall attempt to derive explicit inequalities that are in some sense "best possible," e.g., isoperimetric.

In this paper we consider three different classes of problems which lead to different types of explicit norm inequalities—classes which are in the same general area and hopefully in the same spirit as some of the interesting work of R. J. Duffin. We first discuss some recently derived consequences of the Hopf maximum principles [4, 5] for second order elliptic equations. Although the specific results are due to Payne and Philippin [16], this latter work is actually a generalization and extension of earlier work of Payne [12], Payne and Stakgold [19, 30], Stakgold [29], Schaefer and Sperb [24, 25], and others [13–15, 18, 23, 26, 32]. The Hopf principles are used to derive maximum principles for linear combinations of a solution of a second order nonlinear elliptic equation and its first derivatives. Several examples are given.

ISBN 0-12-178150-X

The second problem to be dealt with arises in the derivation of explicit a priori inequalities appropriate to the Dirichlet problem for a doubly connected domain in $\mathbb{R}^2$. The optimal constants in the inequality turn out to be eigenvalues of interesting related problems. An isoperimetric bound for one of these eigenvalues was recently obtained by Payne and Sperb [17] and is given here.

Finally, we shall remark briefly on some recent results of Khosrovshahi *et al.* [8] on the positive spectrum of the Schrödinger operator.

We reiterate that in so far as possible we are interested in deriving "best possible" maximum principles and isoperimetric bounds for eigenvalues. There exist in the literature numerous results on maximum principles for gradients of solutions of nonlinear elliptic equations (see, e.g., Protter and Weinberger [21, 22], Serrin [27], Ladyzhenskaya and Uraltseva [9], and many others). However, since these authors were for the most part studying qualitative properties of solutions they were not concerned with finding best possible inequalities.

For our purposes we shall say that a maximum principle for some function of the solution of an elliptic boundary value problem and its derivatives is "best possible" if there is some region of definition of the problem, D, for which this function is constant throughout D. We illustrate with a simple example.

Consider the problem

$$\begin{aligned} \Delta u + 1 &= 0 \quad \text{in} \quad D, \\ u &= 0 \quad \text{on} \quad \partial D, \end{aligned} \tag{1.1}$$

where D is a bounded simply connected region in $\mathbb{R}^2$ with smooth boundary ∂D, and Δ is the Laplace operator.

Now define

$$\Phi(x; \alpha) = |\text{grad}\, u|^2 + \alpha u, \qquad \alpha = \text{const.} \tag{1.2}$$

It turns out that for any $\alpha \leqq 1$, $\Phi(x; \alpha)$ assume its maximum value on ∂D, i.e.,

$$|\text{grad}\, u|^2 + \alpha u \leqq \max_{\partial D} |\text{grad}\, u|^2, \qquad \alpha \leqq 1. \tag{1.3}$$

For fixed α the maximum principle will then be called "best possible" if for that value of α the equality sign holds throughout D for some geometry. For $\alpha = 1$ it is easily checked that the equality sign holds in (1.3) if D is the interior of a circle. For $\alpha < 1$ there is no region D on which $\Phi(x; \alpha) \equiv \text{const.}$

Our goal then in the first part of this paper will be to find "best possible" maximum principles for combinations of the solution of a second order nonlinear elliptic equation and its gradient. If the maximum principle is "best possible" then it may be integrated over D or subdomains of D and the result will be isoperimetric. We remark in passing that if this combination of function and derivatives is not constant throughout D, but for some domain D the combination is constant along a surface γ in D, then it may be integrated along γ and the result will be isoperimetric.

As we know from classical mathematical physics the gradient of the solution of an elliptic equation often represents an interesting physical quantity, and a bound for its absolute value is often of considerable value. It is of course helpful if we can show that the gradient assumes its maximum value on the boundary of the region, but we would also usually like to know the point on the boundary at which the maximum value is assumed. Even for the simple problem (1.1) this is likely to be an impossible task. The type of inequality we shall discuss does, however, give some information in this direction.

We assume throughout that D is a bounded region in $\mathbb{R}^N$ with a boundary ∂D which is a $C^{2+\varepsilon}$ surface. In order not to become bogged down in details we assume throughout that at each point of ∂D the average curvature, K, is nonnegative. Both of these assumptions can of course be relaxed. We shall adopt the summation convention (summation from 1 to N on repeated indices), use the comma to denote partial differentiation, and employ the symbol $\partial/\partial n$ to denote the outward directed normal derivative on ∂D.

2. MAXIMUM PRINCIPLES INVOLVING FIRST DERIVATIVES

We begin this section by pointing out that one can frequently obtain information on the maximum value of certain combinations of first derivatives of a function, quite irrespective of any equation that the function might satisfy. We list here two specific examples, one of which was previously given in [11].

Property 1. *Let $u(x)$ be defined in a bounded region $D \subset \mathbb{R}^3$, and suppose $\partial u/\partial n$ vanishes on a C^2 portion Σ of ∂D which at each point has positive Gaussian curvature; if $u \in C^2(D \cup \Sigma)$, then $|\text{grad } u|^2$ can take its maximum value on Σ iff $|\text{grad } u| \equiv 0$ in D.*

Property 2. *Let $u(x)$ be defined in a bounded region $D \subset \mathbb{R}^N$ and suppose $u = \text{const}$ and $\Delta u - \alpha(x)(\partial u/\partial n) = 0$ on a C^2 portion Σ of ∂D, at each point of which the following condition is satisfied*

$$(N-1)K - \alpha > 0. \tag{2.1}$$

If $u \in C^2(D \cup \Sigma)$, then $|\text{grad } u|^2$ can take its maximum value on ∂D iff $|\text{grad } u| \equiv 0$ in D; i.e., $u \equiv \text{const}$ in D.

Note that in neither of the two above cases is $u(x)$ required to satisfy a specific equation. By applying Property 1 to the free membrane equation

$$\Delta v + \nu v = 0 \quad \text{in} \quad D \subset \mathbb{R}^2, \quad \nu \neq 0,$$
$$\frac{\partial v}{\partial n} = 0 \quad \text{on} \quad \partial D. \tag{2.2}$$

one concludes that if D is convex, $|\text{grad } v|^2$ cannot take its maximum value on ∂D. On the other hand, if Property 2 is applied to the classical elastic plate problem

$$\Delta^2 \varphi = p(x) \quad \text{in} \quad D \subset \mathbb{R}^2,$$
$$\varphi = 0 \quad \text{on} \quad \partial D, \tag{2.3}$$
$$\Delta \varphi - \sigma K \frac{\partial \varphi}{\partial n} = 0 \quad \text{on} \quad \partial D,$$

since Poisson's ratio σ is always less than unity it follows that $|\text{grad } \varphi|$ cannot take its maximum value on ∂D. In (2.3), $p(x)$ is the prescribed loading, which may be arbitrary.

We turn now to some recent results of G. Philippin and the author [16] on maximum principles involving the solutions of nonlinear second order elliptic equations and their derivatives. We shall give several maximum principles involving solutions of the following nonlinear equation

$$[g(q^2)u_{,i}]_{,i} + \rho(q^2)f(u) = 0 \quad \text{in} \quad D \subset \mathbb{R}^N, \quad N \geqq 2, \tag{2.4}$$

where

$$q^2 = |\text{grad } u|^2. \tag{2.5}$$

It is assumed throughout that g is a positive C^2 function of its argument and that $\rho\,(>0)$ and f are C^1 functions of their respective arguments. We assume further the strong ellipticity condition

$$g(\xi) + 2\xi g'(\xi) > 0, \qquad \forall \xi \geqq 0. \tag{2.6}$$

We now introduce the function

$$\Phi(x; \alpha) \equiv \int_0^{q^2} \frac{[g(\xi) + 2\xi g'(\xi)]}{\rho(\xi)} \, d\xi + \alpha \int_0^u f(\eta)\, d\eta \tag{2.7}$$

defined on solutions of (2.4) and depending on a parameter α. It turns out that if $N = 1$ and $\alpha = 2$, $\Phi(x; \alpha) \equiv$ const for any solution of (2.4). It is to be expected then that if $\Phi(x; 2)$ can be shown to satisfy a maximum principle then it will be a best possible principle in the sense that $\Phi(x; 2)$ will tend to a constant in the limit as the domain tends to an infinite slab. On the other hand it is sometimes the case that if $\alpha = 2/N$, $\Phi(x; 2/N)$ will be a constant if D is the interior of the N-sphere. Thus if $\Phi(x; 2/N)$ can be shown to satisfy a maximum principle, this principle will frequently be best possible. In [16] the following theorems were established.

Theorem 1. *Let $u(x)$ be a solution of (2.4)–(2.6) in some bounded domain $D \subset \mathbb{R}^N$. For $N \geqq 2$, the function $\Phi(x; 2)$ defined by (2.7) takes its maximum value on ∂D or at a critical point of u (i.e., a point in D at which* grad $u = 0$). *For $N = 2$ the same statement is true for the minimum value of $\Phi(x; 2)$.*

Theorem 2. *If $u(x)$ is a solution of (2.4)–(2.6) in D and satisfies $u \equiv$* const *on ∂D, then since $K \geqq 0$ on ∂D it follows that*

$$\frac{\partial \Phi}{\partial n}(x; 2) \leqq 0 \qquad on \quad \partial D. \tag{2.8}$$

Thus $\Phi(x; 2)$ cannot take its maximum value on ∂D unless $q \equiv 0$ on ∂D.

Theorem 3. *Let $u(x)$ be a solution of (2.4)–(2.6) in some domain $D \subset \mathbb{R}^N$. If the following condition is satisfied,*

$$(2/N)\rho' f^2 - (g + 2q^2 g')f' \geqq 0 \qquad in \quad D, \tag{2.9}$$

then the function $\Phi(x; 2/N)$ defined by (2.7) takes its maximum value on ∂D.

Theorem 4. *If $u(x)$ is a solution of (2.4)–(2.6) in D, (2.9) holds throughout D and $u \equiv$* const *on ∂D then at the point $Q \in \partial D$ at which $\Phi(x, 2/N)$ assumes its maximum value the following inequality holds:*

$$qg(q^2)K(Q) \leqq |f|\rho/N. \tag{2.10}$$

In the proofs of Theorems 1 and 2 it is shown that $\Phi(x, 2)$ satisfies an inequality of the type

$$\Delta\Phi + (2g'/g)u_{,k}u_{,j}\Phi_{,kj} + W_k\Phi_{,k} \geqq 0 \qquad \text{in} \quad D, \tag{2.11}$$

where W_k is a vector function which is singular at the critical points of u. Theorem 1 is then a consequence of Hopf's first principle, and Theorem 2 a consequence of Hopf's second principle and the fact that $K \geqq 0$ on ∂D. The proofs of Theorems 3 and 4 are arrived at in a similar manner except that in this case $\Phi(x; 2/N)$ can be shown to satisfy an inequality similar to (2.11) where now W_k is bounded throughout D. Theorems 3 and 4 are thus consequences of Hopf's first and second principles, respectively. It should be remarked that on ∂D, K need not be everywhere nonnegative in order for Theorem 3 to hold.

Theorems 1 and 2 imply, for $f(u) \geqq 0$,

$$\int_0^{q^2} \frac{[g(s) + 2sg'(s)]}{\rho(s)}\, ds \leqq 2 \int_u^{u_M} f(\eta)\, d\eta, \qquad u_M = \max_D u. \tag{2.12}$$

If this inequality is reducible to an inequality for q^2 of the form

$$q^2 \leqq G(u, u_M), \tag{2.13}$$

then this inequality may be integrated from the point P at which u assumes its maximum value to the nearest boundary point Q, thus yielding the inequality

$$\int_0^{u_M} \frac{dy}{\sqrt{G(y, u_M)}} \leqq \delta(P, Q), \tag{2.14}$$

where δ denotes the distance from P to Q. Clearly $\delta(P, Q) \leqq d$, the radius of the largest inscribed sphere in D. Thus we arrive at the result

$$\int_0^{u_M} \frac{dy}{\sqrt{G(y, u_M)}} \leqq d. \tag{2.15}$$

On the other hand (2.10) often leads to useful information on the location of the point at which q takes its maximum value. It was shown in [16] that if $f' \leqq 0$ and $\rho' \geqq 0$, then $\Phi(x; \alpha)$ takes its maximum value on ∂D for any $\alpha \in [0, 2/N]$. The case $\alpha = 0$ implies q takes its maximum value q_0 on ∂D. Thus $\Phi(x; 2/N)$ takes its maximum value (assuming $u \equiv a$, a constant, on ∂D) at the point on ∂D at which q takes its maximum value. Inequality (2.10) then implies

$$\frac{q_0 g(q_0^2)}{\rho(q_0^2)} \leqq \frac{|f(a)|}{K_{\min} N}, \tag{2.16}$$

where $K_{\min}$ is the minimum value of K on ∂D. Alternatively we have

$$\left| \int_D f(u)\rho(q^2)\, dx \right| = \left| \oint_{\partial D} \frac{\partial u}{\partial n} g(q^2)\, ds \right| \leqq L q_0 g(q_0^2) \leqq \frac{|f(a)|\rho(q_0^2) L}{K(Q) N} \tag{2.17}$$

or

$$K(Q) \leqq \frac{|f(a)|\rho(q_0^2) L}{N |\int_D f(u)\rho(q^2)\, dx|}, \qquad L \equiv \text{surface area of } \partial D. \tag{2.18}$$

Thus, q cannot assume its maximum value in D, nor can it take its maximum value at a point on ∂D at which (2.18) is violated. In the particular case in which f and ρ are constants (2.18) reduces to the purely geometric criterion

$$K(Q) \leqq L/(NV), \qquad V \equiv \text{volume of } D. \tag{2.19}$$

2.1. Applications

2.1.1. Elastic Torsion

The mathematical model of the elastic torsion problem for a simply connected region $D \subset \mathbb{R}^2$ asks for the solution of

$$\begin{aligned} \Delta u &= -2 \quad \text{in} \quad D, \\ u &= 0 \quad \text{in} \quad \partial D. \end{aligned} \tag{2.20}$$

The stress at any point is proportional to the gradient of u and the torsional rigidity is proportional to S, the Dirichlet integral of u over D.

Theorems 1 and 2 imply

(i) $|\text{grad } u|^2 \leqq 4(u_M - u)$,
(ii) $q_0^2 = \max_{\partial D} |\text{grad } u|^2 \leqq 4u_M \leqq 4d^2$,

where d is the radius of the largest inscribed circle in D, and $u_M = \max_D u$;

(iii) $A^{-1}S \leqq (4/3)u_M \leqq d^2/3$,

where A is the area of D. In (i), (ii), and (iii) the equality sign holds in the limit for a rectangular section as the length tends to infinity. Somewhat sharper results have been obtained by different methods by Fu and Wheeler [3].

Theorems 3 and 4 yield for $\Phi(x; 1)$

(i) $q_0 \leqq K_{\min}^{-1}$,

where $K_{\min} \equiv \min_{\partial D} K$, and if Q is the point on ∂D at which q assumes its maximum value

(ii) $K(Q) \leqq L/2A$,

where L is the length of ∂D. Also

(iii) $S \leqq Aq_0^2/2 \leqq \frac{1}{2}AK_{\min}^{-2}$.

The equality sign holds in the above three inequalities if D is the interior of a circle. Note that the second inequality states that $|\text{grad } u|$ cannot take its maximum value at a point on ∂D at which (ii) is violated.

2.1.2. Clamped Membrane

Here the governing equation for the first eigenfunction is

$$\begin{aligned} \Delta u + \lambda u &= 0 \quad \text{in} \quad D \subset \mathbb{R}^N, \\ u &= 0 \quad \text{on} \quad \partial D, \\ u &> 0 \quad \text{in} \quad D. \end{aligned} \tag{2.21}$$

Theorems 1 and 2 now lead to (see [19])

(i) $|\text{grad}\, u|^2 \leqq \lambda[u_M^2 - u^2]$,
(ii) $\lambda \geqq \pi^2/(4d^2)$, $E = (\int_D u\, dx)/(Vu_M)$, $V =$ volume of D,
(iii) $E \equiv \int_D u\, dx/Vu_M \leqq 2/\pi$.

The equality sign holds in the preceding inequalities in the limit for an infinite slab. Note that on the one hand (ii) yields a lower bound for the first eigenvalue of (2.21). On the other hand the parenthetical expression indicates that the point at which u assumes its maximum value must lie at least as far as $\pi/2\lambda^{1/2}$ from the boundary of D.

For this problem Theorem 3 is inapplicable.

2.1.3. Torsional Creep

Here the governing equation is

$$\frac{\partial}{\partial x_i}\left[g(q^2)\frac{\partial u}{\partial x_i}\right] = -2 \qquad \text{in} \quad D \subset \mathbb{R}^2,$$
$$u = 0 \qquad \text{on} \quad \partial D. \tag{2.22}$$

The function $g(x)$ is assumed to be a C^2 function of its argument for $x > 0$ and to further satisfy

$$g(x) > 0, \qquad x \geqq 0, \tag{2.23a}$$

$$g(x) + 2xg'(x) > 0, \qquad x \geqq 0. \tag{2.23b}$$

The maximum shear strain intensity E_0 is given by

$$E_0 = \max_{\partial D}\{qg(q^2)\}. \tag{2.24}$$

In this case Theorems 1 and 2 imply

(i) $\displaystyle\int_0^{q^2} [g(s) + 2sg'(s)]\, ds \leqq 4(u_M - u)$,

while Theorems 3 and 4 imply

(ii) $E_0 \leqq K_{\min}^{-1}$,
(iii) $K(Q) \leqq L/2A$.

Note that (iii) is independent of the form of $g(x)$. (Here L, A, and Q have the same meaning as in the case of elastic torsion; i.e., $g \equiv 1$.) Again the equality sign holds in the limit in (i) as D tends to the infinite strip while equality holds in (ii) and (iii) if D is the interior of a circle.

In [16] a number of other applications of the theorem were given for problems connected with surfaces of constant mean curvature, for capillary tube problems, etc. We have listed here only three applications and have given only a sample of the types of results that follow from the theorems in each case.

An interesting maximum principle involving harmonic functions and their gradients was recently obtained by Payne and Philippin [15], i.e.:

Theorem 5. *Let H and h be two harmonic functions in D with $H \in C^1(\bar{D})$ and $h \in C^0(\bar{D})$. If $h > 0$ in $\bar{D}$, then for $N \geqq 3$ the function*

$$\psi = |\text{grad } H|^2/h^\alpha$$

takes its maximum value on ∂D for any $\alpha \in [0, 2(N-1)/(N-2)]$. For $N = 2$ the function

$$\psi = |\text{grad } H|^2/f(h)$$

assumes its maximum value on ∂D for any positive function f which satisfies the condition

$$(\log f)'' \leqq 0.$$

Here K is not required to be nonnegative on ∂D. Some applications to Green's functions and to electrostatic capacity were given in [15].

We end this section with the following remarks.

(1) Numerous nonisoperimetric results follow from the five theorems.
(2) The principles can be extended to equations of the form

$$[g(u, q^2)u_{,i}]_{,i} + h(u, q^2) = 0.$$

This will be done in a forthcoming paper of Payne and Philippin.
(3) Theorems 3 and 4 are extendable to some special elliptic systems.

3. THE DIRICHLET PROBLEM FOR A DOUBLY CONNECTED REGION IN $\mathbb{R}^2$

The results of this section were obtained by Payne and Sperb [17]. Here we are concerned with the following Dirichlet problem for a doubly connected region D with outer boundary Γ and inner boundary γ. The inner boundary γ is assumed to bound a simply connected region H.

$$\begin{aligned} \Delta V &= F \quad \text{in} \quad D, \\ V &= f \quad \text{on} \quad \Gamma, \\ \frac{\partial V}{\partial s} &= g \quad \text{on} \quad \gamma, \\ \oint_\gamma \frac{\partial V}{\partial n}\, ds &= \alpha \quad \text{(a prescribed constant).} \end{aligned} \tag{3.1}$$

Here F and g are prescribed bounded functions on their respective domains of definition while f is assumed to have bounded first derivatives. For simplicity both γ and Γ are assumed to be C^1 surfaces. One physical interesting problem of type (3.1) is the problem of elastic torsion of a hollow beam.

One method for approximating the solution of (3.1) is through explicit a priori inequalities, which for arbitrary smooth functions φ are as follows:

$$\left[\int_D |\mathrm{grad}(V - \varphi)|^2 \, dx\right]^{1/2} \leqq C_1 \left\{ \frac{1}{A(H)} \left(\alpha - \oint_\gamma \frac{\partial \varphi}{\partial n} \, ds\right)^2 + \int_D (F - \Delta\varphi)^2 \, dx \right\}^{1/2} + C_2 \left\{ \oint_\Gamma \left[\frac{\partial}{\partial s}(f - \varphi)\right]^2 ds + \oint_\gamma \left[g - \frac{\partial \varphi}{\partial s}\right]^2 ds \right\}^{1/2}. \tag{3.2}$$

Here $A(H)$ is the area of H and C_1 and C_2 are constants. After squaring, using the arithmetic–geometric mean inequality on the right and using the Rayleigh-Ritz procedure one can compute upper and lower bounds for $\int_D |\mathrm{grad}\, u|^2 \, dx$. Clearly, one would like to have optimal constants C_1 and C_2 which can be shown to be related to the following eigenvalue problems:

(a) Let u be the first eigenfunction of

$$\begin{aligned} \Delta u + \lambda u &= 0 \quad \text{in} \quad D, \\ u &= 0 \quad \text{on} \quad \Gamma, \\ u &= \beta \quad \text{on} \quad \gamma, \end{aligned} \tag{3.3}$$

where β is not specified a priori but determined by the condition

$$\oint_\gamma \frac{\partial u}{\partial n} \, ds = \lambda \beta A(H). \tag{3.4}$$

Then the optimal C_1 is

$$C_1 = \lambda^{-1/2}. \tag{3.5}$$

Now the problem (3.3), (3.4) is just the mathematical model of the fundamental mode of a vibrating elastic membrane (covering D) with a rigid inclusion (covering H) which is clamped along Γ and constrained in such a way that the rigid portion remains parallel to the xy plane. What we seek then is an isoperimetric lower bound for λ.

(b) The optimal C_2 is given by

$$C_2 = p_2^{-1/2}, \tag{3.6}$$

where p_2 is the first nonzero eigenvalue in a certain Stekloff problem. No isoperimetric lower bound for p_2 is known when D is doubly connected, though a nonisoperimetric bound may be easily obtained.

Guided by the analogous result obtained by Polya and Weinstein [20] for the problem of elastic torsion one is tempted to conjecture that of all regions with fixed $A(D)$ and fixed $A(H)$, λ is a minimum if D is a circular annulus (i.e., H is centrally located). However, this is not always the case; it depends on the ratio of $A(H)$ to $A(D \cup H)$. In [17] the following theorem was proved.

Theorem. *For given values of $A(D)$ and $A(H)$ the eigenvalue λ defined by* (3.3), (3.4) *is a minimum either* (1) *for a circular annulus with the rigid portion at the center or* (2) *for the circular membrane of area $A(D)$.*

It turns out that if the ratio $A(H)/A(D \cup H)$ is sufficiently small (1) will yield a minimum, but as the ratio of $A(H)/A(D \cup H)$ approaches 1 the circular membrane of area D will yield the minimum value.

4. RESULTS ON THE POSITIVE SPECTRUM OF THE SCHRÖDINGER OPERATOR

In this section we describe some recent results of Khosrovshahi *et al.* [8], which extend earlier results of Kato [6], Odeh [10], Weidmann [31], Agmon [1], Simon [28], and others [2, 7]. We are concerned in particular with nontrivial solutions of the equation

$$\Delta u + [\lambda - V(x)]u = 0 \qquad \text{in} \quad \Omega \subset \mathbb{R}^N, \tag{4.1}$$

where $\Omega \supset \{x, \|x\| \geqq R_0\}$. The object is to determine how the asymptotic behavior of $V(x)$ affects the admissible range of positive eigenvalues, i.e., positive values of λ corresponding to L_2 solutions. We assume throughout that for $r \equiv \|x\| \geqq R^* > R_0$, where R^* is sufficiently large, the potential $V(x)$ may be decomposed as

$$V(x) = V_0(x) + V_1(x) + V_2(x), \tag{4.2}$$

where $V_0(x)$ is a real continuous function in $r \geqq R^*$ with a continuous radial derivative which satisfies

$$\text{(i)} \quad V_0(x) = 0(1) \quad \text{as} \quad r \to \infty, \limsup_{r\to\infty} r(\partial V_0/\partial r) = \Lambda_0,$$

$V_1(x)$ satisfies

$$\text{(ii)} \quad \limsup_{r\to\infty} |rV_1(x)| \leqq k,$$

and

$$\text{(iii)} \quad \sup_{\|\omega\|=1} |\textstyle\int_r^\rho \sigma V_2(\sigma\omega)\, d\sigma| \leqq M < \frac{1}{4}, \forall r, \rho \geqq R^*.$$

Under these hypotheses and additional hypotheses on $V(x)$ sufficient to guarantee that solutions of (4.1) possess a unique continuation property it can be shown that there are no possible eigenvalues in the range (α, ∞) where

$$\alpha = \max\left\{\frac{[k + \sqrt{k^2 + 2\Lambda_0(1 - 2M)^2}]^2}{4(1 - 2M)^2}, \frac{2k^2 + \Lambda_0(1 - 4M)}{2(1 - 4M)^2}\right\}.$$

The proof uses convexity arguments similar to those used by Agmon [1] and Khosrovshahi [7].

Example. $V(x) = Ar^\delta \sin r^\beta$. Choosing $V \equiv V_2$, the above result implies that thre exists no positive eigenvalue provided $\delta + 2 < \beta$.

In a forthcoming paper of Khosrovshahi and Payne these results will be sharpened somewhat for a special class of potentials.

It should be pointed out that the results of this section are not sharp. For special classes of potentials sharp results have been obtained by Atkinson [2], and the author has been informed that sharp results have recently been obtained for a wider class of potentials by Dollard and Friedman.

REFERENCES

1. S. Agmon, Lower bounds for solutions of Schrödinger equations, *J. Analyse Math.* **23** (1970), 1–25.
2. F. V. Atkinson, The asymptotic solution of second order differential equations, *Ann. Mat. Pura Appl.* **37** (1954), 347–378.
3. S. Fu, and L. Wheeler, Stress bounds for bars in torsion, *J. Elasticity* **3** (1973), 1–13.
4. E. Hopf, Elementare Bemerkung über die Lösungen partieller Differentialgleichungen zweiter Ordnung vom elliptischen Typus, *Ber. Sitzungsber. Preuss. Akad. Wiss.* **19** (1927), 147–152.
5. E. Hopf, A remark on elliptic differential equations of the second order, *Proc. Amer. Math. Soc.* **3** (1952), 791–793.
6. T. Kato, Growth properties of the reduced wave equation with variable coefficients, *Comm. Pure Appl. Math.* **12** (1959), 403–425.

7. G. B. KHOSROVSHAHI, Nonexistence of nontrivial solutions of Schrödinger type systems, *SIAM J. Math. Anal.* **8** (1977), 998–1013.
8. G. B. KHOSROVSHAHI, H. A. LEVINE, AND L. E. PAYNE, On the positive spectrum of Schrödinger operators with long range potentials, *Trans. Amer. Math. Soc.* (to appear).
9. O. A. LADYZHENSKAYA AND N. N. URALTSEVA, "Équations aux derivees partielles de type elliptique," Dunod, Paris, 1968.
10. F. ODEH, Notes on differential operators with purely continuous spectrum, *Proc. Amer. Math. Soc.* **16** (1965), 363–366.
11. L. E. PAYNE, Some remarks on maximum principles, *J. Analyse Math.* **30** (1976), 421–433.
12. L. E. PAYNE, Bounds for the maximum stress in the Saint Venant torsion problem, *Indian J. Mech. Math.* Special Issue (1968), 51–59.
13. L. E. PAYNE AND G. A. PHILIPPIN, Some applications of the maximum principle in the problem of torsional creep, *SIAM J. Appl. Math.* **33** (1977), 446–455.
14. L. PAYNE AND G. A. PHILIPPIN, Some remarks on the problems of elastic torsion and torsional creep," Some Aspects of the Mechanics of Continua," Sen Memorial Committee, 1977, pp. 32–40.
15. L. E. PAYNE AND G. A. PHILIPPIN, On some maximum principles involving harmonic functions and their derivatives, *SIAM J. Math. Anal.* **10** (1979), 96–104.
16. L. E. PAYNE AND G. A. PHILIPPIN, Some maximum principles for nonlinear elliptic equations in divergence form with applications to capillary surfaces and to surfaces of constant mean curvature, *Nonlinear Anal.* **3** (1979), 193–211.
17. L. E. PAYNE AND R. P. SPERB, Some a priori and isoperimetric inequalities associated with a class of boundary value problems for doubly connected domains; *SIAM J. Math. Anal.* **7** (1976), 451–460.
18. L. E. PAYNE, R. P. SPERB, AND I. STAKGOLD, On Hopf type maximum principles for convex domains, *Nonlinear Anal.* **1** (1977), 547–559.
19. L. E. PAYNE AND I. STAKGOLD, On the mean value of the fundamental mode in the fixed membrane problem, *Applicable Anal.* **3** (1973), 295–303.
20. G. POLYA AND A. WEINSTEIN, On the torsional rigidity of multiply connected cross sections, *Ann. of Math.* **52** (1950), 154–163.
21. M. H. PROTTER AND H. F. WEINBERGER, "Maximum Principles In Differential Equations," Prentice–Hall, Englewood Cliffs, New Jersey, 1967.
22. M. H. PROTTER AND H. F. WEINBERGER, A maximum principle and gradient bounds for linear elliptic equations, *Indiana Univ. Math. J.* Math **23** (1973–1974), 239–249.
23. P. W. SCHAEFER, Some maximum principles for nonlinear elliptic boundary value problems, *Quart. Appl. Math.* **35** (1978), 517–523.
24. P. W. SCHAEFER AND R. P. SPERB, Maximum principles for some functionals associated with the solution of elliptic boundary value problems, *Arch. Rational Mech. Anal.* **61** (1976), 65–76.
25. P. W. SCHAEFER AND R. P. SPERB, Maximum principles and bounds in some inhomogeneous boundary value problems, *SIAM J. Math. Anal.* **8** (1977), 871–878.
26. P. W. SCHAEFER AND R. P. SPERB, A maximum principle for a class of functionals in nonlinear Dirichlet problems, Lecture Notes in Mathematics No. 564, Springer-Verlag, Berlin and New York, 1976, pp. 400–406.
27. J. B. SERRIN, Gradient estimates for solutions of nonlinear elliptic and parabolic equations, *in* "Contributions to Nonlinear Functional Analysis," Univ. of Wisconsin Press, Madison, 1971, pp. 565–601.
28. B. SIMON, On positive eigenvalues of one-body Schrödinger operators, *Comm. Pure Appl. Math.* **22** (1967), 531–538.
29. I. STAKGOLD, Global estimates for nonlinear reaction and diffusion, Lecture Notes in Mathematics, No. 415, Springer-Verlag, Berlin and New York, 1974, pp. 252–266.

30. I. Stakgold and L. E. Payne, Nonlinear problems in nuclear reactor analysis, Lecture Notes in Mathematics, No. 322, Springer-Verlag, Berlin and New York, 1973, pp. 298–307.
31. J. Weidmann, The virial theorem and its applications to the spectral theory of Schrödinger operators, *Bull. Amer. Math. Soc.* **73** (1967), 452–459.
32. L. T. Wheeler, M. J. Turtletaub, and C. O. Horgan, A Saint Venant principle for the gradient in the Neumann problem, *Z. Angew. Math. Phys.* **26** (1975), 141–153.

This material is based upon work supported by the National Science Foundation under Grant No. MCS77-01273 AO1.

DEPARTMENT OF MATHEMATICS
CORNELL UNIVERSITY
ITHACA, NEW YORK

Genetic Wave Propagation, Convex Sets, and Semiinfinite Programming

H. F. WEINBERGER

1. INTRODUCTION

Fisher [10] constructed a model for the evolution in time of a spatial distribution of genotypes. More specifically, he studied a migrating population of diploid individuals whose ability to survive is determined by the genotype with respect to a single gene locus. If one supposes that the gene in question occurs in two allelic forms, say A and a, then, because the gene locus in an individual contains two genes, there are three possible genotypes, designated AA, Aa, and aa.

The variable of interest is the gene fraction $u(x, t)$, which is the ratio of the number of alleles A to the total number of alleles A and a in the individuals which are in a small neighborhood of the point x at the time t. Fisher's model for the time evolution of the gene fraction is the semilinear diffusion equation

$$\frac{\partial u}{\partial t} = D\,\nabla^2 u + f(u). \tag{1.1}$$

The function $f(u)$ is determined by the relative survival properties of the three genotypes and their birth and death rates. Since one does not consider the effect of mutation, the states $u \equiv 0$ and $u \equiv 1$ which correspond to the absence of one of the allelic types must be solutions of this equation. Consequently, $f(0) = f(1) = 0$. We shall deal here with the case in which the heterozygote individuals Aa are more likely to survive than the aa homozygotes but less likely to survive than the AA homozygotes, which is called the *heterozygote intermediate* case. Then $f(u)$ is positive for $0 < u < 1$. Because of its definition,

$$0 \leqq u(x, t) \leqq 1.$$

ISBN 0-12-178150-X

The term $D \nabla^2 u$ on the right of (1.1) is designed to model the effect of random migration.

The model (1.1) has some interesting properties. In the one-dimensional case with a special class of functions $f(u)$ it was shown by Fisher that there are traveling wave solutions of the form $u(x, t) = \varphi(x - ct)$ for all values of c which are at least as large as a certain critical value c^*. Fisher conjectured that if at time zero a population of *aa* homozygotes ($u = 0$) living on the right half $x > 0$ of the real line encountered a population of *AA* homozygotes ($u = 1$) living on the half-line $x < 0$, then the advantageous allele *A* would advance to the right with the minimal wave speed c^*. Kolmogoroff *et al.* [15] proved this conjecture in an asymptotic sense and showed that the shape of $u(x, t)$ approaches that of the wave φ which travels at this speed. This result for more general functions f was obtained by Kanel' [13].

Aronson and the author [1, 2] showed that c^* also gives the asymptotic speed at which a local mutant *A* advances into an *aa* homozygote population, and that this result is also valid in two or more dimensions.

While (1.1) can be expected to model the qualitative features of the genetic system under consideration, it cannot stand up to close scrutiny. The processes being modeled, mating, birth, and death, are by their very nature statistical, but the model is deterministic. One can, of course, justify a deterministic model for a statistical process by invoking the law of large numbers. However, the presence of the partial derivative with respect to t means that we should deal with the limit of small time intervals. The number of rabbits born in a microsecond would appear to be too small to apply the law of large numbers,

To overcome this difficulty, we have suggested [19, 20] replacing the model (1.1) by a more general kind of model. We discretize the time into multiples of a unit which is sufficiently large to permit us to apply the law of large numbers to the processes which occur during this time interval. We only look at the functions $u(x, n)$, $n = 0, 1, 2, \ldots$, and denote these functions by $u_n(x)$.

We construct a deterministic model by supposing that the function $u_{n+1}(x)$ is uniquely defined by the function $u_n(x)$:

$$u_{n+1} = Q[u_n], \tag{1.2}$$

where Q is an operator on the space of continuous functions with values on the interval [0, 1]. The particular model is specified by prescribing a particular operator Q, which is to be determined experimentally.

The fact that the operator Q does not depend upon n means that we are neglecting long-term changes in the environment. This does not preclude, for instance, seasonal variations, provided the unit of time is an integral multiple of one year.

One example of an operator Q is the solution operator of (1.1), which takes the solution with initial function $u(x, 0) = \varphi(x)$ into the function $u(x, 1)$, which we denote by $Q[\varphi]$. Then if $u(x, t)$ is a solution of (1.1) and $u_n(x) \equiv u(x, n)$, the equation (1.2) is satisfied.

The same objection which applies to the use of a continuous time variable also applies to the use of continuous space variables. For instance, one would like to define $u(x, t)$ as the limit of the gene fraction of all the individuals who are located in a neighborhood of the point x at time t as the neighborhood shrinks to x, but one cannot make a significant statistical analysis of all the rabbits in a one-centimeter square.

This difficulty can be overcome by discretizing the habitat. That is, the space, say R^2, is broken into congruent rectangles, each of which is sufficiently large that the law of large numbers can be applied to the various processes which occur within each rectangle in a unit of time. We identify each rectangle with its midpoint, and define $u_n(x)$ only when x is such a midpoint. Then Q is an operator on the space of functions defined on a discrete mesh. Models of the form (1.2) when x varies on a discrete mesh are called stepping-stone models (see [14, 17]).

It was shown in [20] that under the rather natural hypotheses (2.1) one can associate with each unit vector ξ a propagation speed $c^*(\xi)$ with the property that if $u_0(x) = 0$ for $x \cdot \xi \geqq 0$, if $u_0 \leqq \alpha < 1$, and if $c > c^*(\xi)$, then $u_n(x)$ is uniformly small for $x \cdot \xi \geqq nc$ when n is large.

Such a result clearly says something about how a disturbance u_0 which is confined to a bounded set spreads out for large n. We wish to construct a set $\mathscr{S}'$ such that the disturbance u_n is small outside the dilation $n\mathscr{S}'$ of $\mathscr{S}'$.

Any reader of Courant and Hilbert [5, Chapter VI] knows that even in the simple case of a linear partial differential equation with constant coefficients the problem of how the wave speeds in the various directions determine the domain of influence of a bounded set, or even a point, leads to a rather sophisticated use of the theory of convex sets. Because our wave velocity is only asymptotic, the problem becomes more complicated.

One would expect that in any realistic model for the genetic process under discussion, the migration velocity is bounded above. It is then easy to show that $c^*(\xi)$ cannot exceed this bound, so that it is also bounded. For this simple and interesting case the problem of constructing $\mathscr{S}'$ has been solved in [20].

The hypotheses (2.1) do not imply that $c^*(\xi)$ is bounded above and, in fact, permit c^* to take on the value $+\infty$ for some unit vectors ξ. This paper is devoted to constructing the set $\mathscr{S}'$ in the presence of these added complications.

This is certainly a problem of mathematical interest. It is not clear whether there are any biologically interesting models in which c^* is unbounded. While we have motivated our model (1.2) in terms of population genetics, similar models occur in the theory of population growth [12] and the theory of epidemics [7, 16].

In Section 2 we state and motivate the hypotheses on the operator Q and sketch the construction of the wave speeds $c^*(\xi)$, which was carried out in more detail in [20].

The solution to the problem of finding a set $\mathscr{S}'$ with the desired properties is carried through in Section 3 by means of the theory of convex sets.

In Section 4 we show that the perfect duality theorem of Duffin [8] can be used to reduce our problem to that of determining conditions under which there is no duality gap between a certain semiinfinite linear program and its dual. We then demonstrate that one of the principal lemmas of Section 3 is equivalent to a well-known theorem of Duffin and Karlovitz [9]. The principal idea here is that the lower semicontinuity of the function c^* is essentially equivalent to the property that the corresponding set of linear inequalities is canonically closed.

I am very pleased to dedicate this work to my teacher and friend Dick Duffin. I find it particularly appropriate to do so, not just because of its connection with Dick's work, but because it makes use of the close connection between apparently diverse fields of mathematics and between mathematics and the sciences, which is the thread that runs through all of Dick's work, and which constitutes Dick's most important legacy to his students.

I am indebted to Robert Gulliver for valuable discussion of some of the geometric questions which occur in Section 3. I am most grateful to Robert Jeroslow for guiding me to the pertinent linear programming literature.

I wish to say a few words about Joe Diaz, whose untimely death prevented him from giving one of the plenary lectures at this symposium. I was but one of a multitude of beginning mathematicians whom Joe befriended and assisted. My first paper was joint work with him [6]. It was Joe who taught me that the mathematician's job does not end when the answer is found, but includes the laborious and less inspiring task of writing up the result for publication.

2. THE WAVE SPEED

We begin with some hypotheses about the recursion operator Q, which defines the model (1.2).

The habitat $\mathscr{H}$ is a given subset of a Euclidean N-space $\mathscr{R}^N$ which is closed under vector addition and subtraction. That is, if x and y lie in $\mathscr{H}$, so do $x + y$ and $x - y$. For instance, $\mathscr{H}$ may consists of all of $\mathscr{R}^N$ or of the set of points in $\mathscr{R}^N$ whose coordinates are integers.

Let $\mathscr{B}$ be the set of continuous functions $u(x)$ on $\mathscr{H}$ with values on the interval $[0, 1]$. (Of course, if $\mathscr{H}$ is discrete, all functions are continuous.) For any y in $\mathscr{H}$ we define the translation operator

$$T_y[u](x) = u(x - y)$$

on $\mathscr{B}$.

Let Q be an operator from $\mathscr{B}$ to $\mathscr{B}$ which has the following properties:

$$\begin{array}{rl} \text{(i)} & Q[0] = 0, Q[1] = 1. \\ \text{(ii)} & QT_y = T_yQ, \forall y \in \mathscr{H}. \\ \text{(iii)} & u \leq v \Rightarrow Q[u] \leq Q[v]. \\ \text{(iv)} & \text{For any constant } \alpha \in (0, 1), Q[\alpha] > \alpha. \\ \text{(v)} & u_k \to u \text{ uniformly on each bounded subset of } \mathscr{H} \Rightarrow Q[u_k] \to Q[u] \\ & \text{at each point of } \mathscr{H}. \end{array} \tag{2.1}$$

Property (i) requires the states $u \equiv 0$ and $u \equiv 1$ to be fixed points of Q. In the case of population genetics, it says that the model does not permit mutation into either of the two allelic forms under consideration. That is, if at one time one of the alleles is nowhere to be found, it will not appear later.

Property (ii) states that the habitat is homogeneous.

Property (iii) states that the higher the gene fraction is in one generation, the higher it will be in the next generation. This is a hypothesis which need not be true in all situations. When Q is the solution operator for the problem (1.1), property (iii) is a consequence of the maximum principle for parabolic equations.

Hypothesis (iv) is a severe restriction of the model. In the case of population genetics it states that, in the absence of migration, the ability of the heterozyte individuals with genotype Aa to survive is greater than that of the hyomozygotes aa but less than that of the homozygotes AA. For this reason we refer to the case where this model applies as the *heterozygote intermediate* case.

The last hypothesis (v) is a mathematical one which can be expected to be satisfied by any reasonable model.

Our basic tool in establishing properties of solutions of the recursion

$$u_{n+1} = Q[u_n] \tag{2.2}$$

is the following simple comparison lemma.

Lemma 2.1. *Let R be an operator on a set $\tilde{\mathscr{B}}$ of real-valued functions which is order-preserving in the sense that*

$$w \leqq v \Rightarrow R[w] \leqq R[v]. \tag{2.3}$$

If v_n is a sequence in $\tilde{\mathscr{B}}$ which satisfies the inequalities

$$v_{n+1} \geqq R[v_n], \qquad n = 0, 1, \ldots,$$

if the sequence w_n satisfies

$$w_{n+1} \leqq R[w_n], \qquad n = 0, 1, \ldots,$$

and if

$$w_0 \leqq v_0,$$

then

$$w_n \leqq v_n$$

for all n.

Proof. The proof is by induction. Suppose that for some n, $w_n \leqq v_n$. Then by (2.3)

$$w_{n+1} \leqq R[w_n] \leqq R[v_n] \leqq v_{n+1}. \quad \blacksquare$$

We define the wave speeds of the operator Q by means of the following construction.

We first choose a continuous nonincreasing function $\varphi(s)$ of one variable with the properties

$$\varphi(s) = \begin{cases} \alpha \in (0, 1) & \text{for} \quad s \leqq -1 \\ 0 & \text{for} \quad s \geqq 0. \end{cases} \tag{2.4}$$

For any unit vector ξ and any real number c we define the operator $R_{c,\xi}$ on the space $\tilde{\mathscr{B}}$ of continuous functions on the real line with values on the interval $[0, 1]$:

$$R_{c,\xi}[w](s) = \max\{\varphi(s), Q[w(x \cdot \xi + s + c)](0)\}. \tag{2.5}$$

Here $w(x \cdot \xi + s + c)$ is to be regarded as a function of x in $\mathscr{H}$, and the maximum means the larger of the two numbers for each fixed s.

We now define the sequence $a_n(c, \xi; s)$ of functions of s by means of the recursion

$$a_{n+1} = R_{c,\xi}[a_n], \qquad a_0(s) = \varphi(s). \tag{2.6}$$

Because of the hypothesis (2.1, iii) the operator $R_{c,\xi}$ has the order preserving property (2.3). We see from the definition of $R_{c,\xi}$ that $a_1 \geqq \varphi = a_0$. Therefore Lemma 2.1 with $v_n = a_{n+1}$ and $w_n = a_n$ shows that

$$a_{n+1}(c, \xi; s) \geqq a_n(c, \xi; s), \qquad n = 0, 1, 2, \ldots.$$

That is, a_n is nondecreasing in n. Since $a_n \leqq 1$, we conclude that the limit function

$$a(c, \xi; s) = \lim_{n\to\infty} a_n(c, \xi; s) \tag{2.7}$$

exists

It is easily seen that because of the translation invariance (2.1, ii) of Q and because $\varphi(s)$ is nonincreasing, each $a_n(c, \xi; s)$ is nonincreasing in s and in c (see [20]). Consequently the same is true of $a(c, \xi; s)$.

The translation invariance (2.1, ii) and the continuity (2.1, v) show that

$$\lim_{s\to-\infty} a_n(c, \xi; s) = \lambda_n,$$

where the sequence of constants λ_n is the solution of the recursion

$$\lambda_{n+1} = Q[\lambda_n], \qquad \lambda_0 = \alpha.$$

We see from (2.1, iv) that λ_n increases to 1 as n tends to ∞. Therefore

$$a(c, \xi; -\infty) = 1.$$

We can show that when c is sufficiently small, $a(c, \xi; s) \equiv 1$, and that if $a(c, \xi; s)$ is not identically 1, then

$$\lim_{s\to\infty} a(c, \xi; s) = 0.$$

We now define the *wave speed* $c^*(\xi)$ in the direction ξ by the formula

$$c^*(\xi) = \sup\{c \mid a(c, \xi; s) \equiv 1\}. \tag{2.8}$$

If $a(c, \xi; s) \equiv 1$ for all c, we set $c^*(\xi) = +\infty$. Thus the function c^* has values in $(-\infty, +\infty]$.

Because $a(c, \xi; s)$ is the limit of a nondecreasing sequence of continuous functions a_n of c, ξ, and s, it is lower semicontinuous. Because a is nonincreasing in c and s, it is right continuous in these variables.

It can also be shown [20, Lemma 1] that $a(c, \xi; 0) \leqq \alpha$ for $c > c^*(\xi)$. This and the right continuity of a prove that $c^*(\xi)$ *is a lower semicontinuous function of* ξ and that, if $c^*(\xi)$ is finite,

$$\lim_{s\to\infty} a(c^*(\xi), \xi; s) = 0. \tag{2.9}$$

It is easily seen that the function $c^*(\xi)$ does not depend upon the particular function $\varphi(s)$ which is used in this construction or upon α.

The following lemma shows that, at least in an asymptotic sense, $c^*(\xi)$ provides a limit for the speed of propagation in the direction ξ.

Lemma 2.2. *Let u_n be a solution of the recursion $u_{n+1} = Q[u_n]$. Suppose that*

$$0 \leqq u_0(x) \leqq \alpha < 1,$$

that for some unit vector ξ there is a constant ρ such that

$$u_0(x) = 0 \qquad \text{for} \quad x \cdot \xi \geqq \rho,$$

and that $c^*(\xi) < \infty$.

Then for any $c > c^*(\xi)$

$$\lim_{n\to\infty} \sup_{x \cdot \xi \geqq nc} u_n(x) = 0. \tag{2.10}$$

Proof. Choose a continuous nonincreasing function $\varphi(s)$ with the properties (2.4). Then

$$u_0(x) \leqq \varphi(x \cdot \xi - \rho - 1) = a_0(c^*(\xi), \xi; x \cdot \xi - \rho - 1).$$

We now define the sequence

$$b_n(x) = a_n(c^*(\xi), \xi; x \cdot \xi - \rho - 1 - nc^*(\xi)).$$

We see from the recursion (2.6) and the definition (2.5) of $R_{c,\xi}$ that

$$b_{n+1} \geqq Q[b_n].$$

Since $u_0 \leqq b_0$, Lemma 2.1 shows that

$$\begin{aligned} u_n(x) \leqq b_n(x) &= a_n(c^*(\xi), \xi; x \cdot \xi - \rho - 1 - nc^*(\xi)) \\ &\leqq a(c^*(\xi), \xi; x \cdot \xi - \rho - 1 - nc^*(\xi)) \end{aligned} \tag{2.11}$$

for all n. Since a is nonincreasing in s, we find that if $c > c^*(\xi)$

$$\sup_{x \cdot \xi \geqq nc} u_n(x) \leqq a(c^*(\xi), \xi; n(c - c^*(\xi)) - \rho - 1), \tag{2.12}$$

so that (2.10) follows from (2.9). ■

If no signal could travel in any direction ξ with a speed faster than $c^*(\xi)$, then a disturbance concentrated at the origin at time 0 would, at time n, be confined to the set $n\mathscr{S}$, where

$$\mathscr{S} = \{x \mid x \cdot \xi \leqq c^*(\xi)\ \forall \xi \in S^{N-1}\} \tag{2.13}$$

and $n\mathscr{S}$ is the dilatation

$$n\mathscr{S} = \{nx \mid x \in \mathscr{S}\}.$$

S^{N-1} is the unit sphere in $\mathscr{R}^N$.

The results for the recursion (2.2) can be expected to be less sharp because the requirement $c > c^*(\xi)$ in Lemma 2.2 means that $c^*(\xi)$ is only an asymptotic propagation speed.

Let $\xi_1, \ldots, \xi_K$ be unit vectors and ε a positive number, and define the set

$$\mathcal{M} = \{x \mid x \cdot \xi_i \leqq c^*(\xi_i) + \varepsilon, i = 1, \ldots, K\}. \tag{2.14}$$

Clearly the set $\mathcal{M}$ contains the set $\mathcal{S}$. Our basic theorem on propagation of disturbances is the following.

Theorem 2.1. *Let $\mathcal{S}'$ be an open set which contains a set $\mathcal{M}$ of the form* (2.14) *with $\varepsilon > 0$.*

If u_n is a solution of the recursion $u_{n+1} = Q[u_n]$, if $0 \leqq u_0 < 1$, and if $u_0 = 0$ outside a bounded set, then

$$\lim_{n\to\infty} \max_{x \notin n\mathcal{S}'} u_n(x) = 0. \tag{2.15}$$

Proof. Suppose that $u_0(x) = 0$ for $|x| \geqq \rho$. Then $u_0 = 0$ for $x \cdot \xi \geqq \rho$, and hence we can derive the inequality (2.11) for all ξ. Since a is nonincreasing in s, we find that

$$\max_{x \notin n\mathcal{S}'} u_n \leqq \sup_{x \notin n\mathcal{M}} u_n(x) \leqq \max_{1 \leqq i \leqq K} a(c^*(\xi_i), \xi_i; n\varepsilon - \rho - 1),$$

and (2.15) follows from (2.9). [It follows from (2.11) that $u_n(x)$ approaches zero at infinity so that its maximum outside the open set $n\mathcal{S}'$ is attained.] ■

We state an obvious but interesting corollary of this theorem.

Corollary 2.1. *If there is a set $\mathcal{M}$ of the form* (2.14) *with $\varepsilon > 0$ which is empty, $u_0(x) = 0$ for $|x| \geqq \rho$, and $0 \leqq u_0(x) < 1$, then*

$$\lim_{n\to\infty} \max_{x \in \mathcal{H}} u_n(x) = 0. \tag{2.16}$$

That is, u_n approaches zero uniformly, so that no propagation occurs.

Proof. Let $\mathcal{S}'$ be the empty set and apply Theorem 2.1. ■

Theorem 2.1 only shows that $c^*(\xi)$ is an upper bound for the asymptotic propagation speed in the direction ξ. The fact that it is actually a propagation speed, at least when $\mathcal{S}$ has interior points, is shown by the following complementary theorem, whose proof will be given elsewhere.

Theorem 2.2. *Let $\mathcal{S}''$ be any closed bounded subset of the interior of $\mathcal{S}$. For any positive γ there exists a radius r_γ such that if $u_0 \geqq \gamma$ on a ball of radius r_γ and $u_{n+1} = Q[u_n]$, then*

$$\lim_{n\to\infty} \min_{x \in n\mathcal{S}''} u_n(x) = 1.$$

3. GEOMETRIC PROPERTIES OF THE PROPAGATION

Theorem 2.1 gives a sufficient (and not too far from necessary) condition on a set $\mathscr{S}'$ for the property (2.15) to be valid. This condition is that $\mathscr{S}'$ contain a set $\mathscr{M}$ of the form (2.14) with $\varepsilon > 0$. This condition depends not only on the set $\mathscr{S}$ but also on the function $c^*(\xi)$ which is used to define it.

The function $c^*(\xi)$ is by no means determined by the set $\mathscr{S}$. For example, it is easily seen that the set

$$\mathscr{A} = \{(x, y, z) | x \sin\theta \cos\tfrac{1}{2}\theta + y \sin\theta \sin\tfrac{1}{2}\theta + z\cos\theta \leqq 0 \;\forall\, \theta \in [0, \pi]\} \quad (3.1)$$

in $\mathscr{R}^3$ is of the form (2.13) with

$$c^*(\sin\theta\cos\varphi, \sin\theta\sin\varphi, \cos\theta) = \begin{cases} 0 & \text{for} \quad \varphi = \tfrac{1}{2}\theta, \quad 0 \leqq \theta \leqq \pi, \\ +\infty & \text{otherwise.} \end{cases}$$

The same set can be written more simply as

$$\mathscr{A} = \{(x, y, z) | x \leqq 0, y \leqq 0, z = 0\}. \quad (3.2)$$

Thus, it can also be obtained from (2.13) by setting

$$c^*(\xi, \eta, \zeta) = \begin{cases} 0 & \text{for} \quad (\xi, \eta, \zeta) = (1, 0, 0), (0, 1, 0), (0, 0, 1), (0, 0, -1), \\ +\infty & \text{otherwise,} \end{cases}$$

or by setting

$$c^*(\xi, \eta, \zeta) = \begin{cases} 0 & \text{for} \quad \xi \geqq 0, \quad \eta \geqq 0, \\ +\infty & \text{otherwise.} \end{cases}$$

All these functions c^* are lower semicontinuous.

In this section we shall endeavor to obtain some conditions on $\mathscr{S}'$ which depend only on the set $\mathscr{S}$ and which imply that $\mathscr{S}'$ satisfies the hypotheses of Theorem 3.1.

We shall need some elementary concepts and results from the theory of convex sets.

Let $\mathscr{J}$ be the convex set in $\mathscr{R}^N$ which is defined by a set of linear inequalities:

$$\mathscr{J} = \{x \in \mathscr{R}^N | x \cdot \xi \leqq \tilde{c}(\xi) \;\forall \xi \in S^{N-1}\}. \quad (3.3)$$

Here $\tilde{c}$ is a given function on the unit sphere S^{N-1} with values in $(-\infty, +\infty]$.

Associated with the function $\tilde{c}(\xi)$ is what Rockafellar [18] calls the *recession cone*, which we prefer to call the *cone of recession*,

$$\mathscr{K} = \{z \in \mathscr{R}^N | z \cdot \xi \leqq 0 \;\forall \xi \in S^{N-1} \ni \tilde{c}(\xi) < \infty\}. \quad (3.4)$$

It is clearly a closed cone. This set has the following properties [18, Theorems 8.1, 8.4].

Lemma 3.1. *If $\mathscr{J}$ is not empty, then*

$$\mathscr{K} = \{z \in \mathscr{R}^N \mid x + z \in \mathscr{J} \ \forall x \in \mathscr{J}\}. \tag{3.5}$$

$\mathscr{J}$ is bounded if and only if either $\mathscr{K} = \{0\}$ or $\mathscr{J}$ is empty.

Proof. If $z \in \mathscr{K}$ and if $x \in \mathscr{J}$, then for all ξ with $\tilde{c}(\xi) < \infty$

$$(x + z) \cdot \xi \leqq \tilde{c}(\xi) + 0$$

so that $x + z \in \mathscr{J}$. Thus $\mathscr{J} + \mathscr{K} \subset \mathscr{J}$.

If $\mathscr{J} + z \subset \mathscr{J}$ and $\mathscr{J}$ is not empty, then for any $x \in \mathscr{J}$ the sequence $x + nz$ lies in $\mathscr{J}$. Hence whenever $\tilde{c}(\xi) < \infty$,

$$(x + nz) \cdot \xi \leqq \tilde{c}(\xi).$$

We divide this inequality by n and let $n \to \infty$ to see that $z \cdot \xi \leqq 0$. Thus if $\mathscr{J} + z \subset \mathscr{J}$ and $\mathscr{J}$ is not empty, $z \in \mathscr{K}$, and we have proved (3.5).

If $\mathscr{J}$ is unbounded, there is a sequence $x_n \in \mathscr{J}$ with $|x_n| \to \infty$ as $n \to \infty$. By the Weierstrass theorem there is a sequence $n_i \to \infty$ such that the unit vectors $x_{n_i}/|x_{n_i}|$ converge to a unit vector z. Because $x_{n_i} \in \mathscr{J}$ we have

$$x_{n_i} \cdot \xi \leqq \tilde{c}(\xi)$$

whenever $\tilde{c}(\xi) < \infty$. Hence

$$(x_{n_i}/|x_{n_i}|) \cdot \xi \leqq \tilde{c}(\xi)/|x_{n_i}|.$$

Letting $i \to \infty$, we see that

$$z \cdot \xi \leqq 0,$$

whenever $\tilde{c}(\xi) < \infty$. Thus if $\mathscr{J}$ is unbounded, $\mathscr{K}$ contains the unit vector z, so that $\mathscr{K} = \{0\}$ implies that $\mathscr{J}$ is bounded.

Clearly, if $\mathscr{J}$ is empty it is also bounded.

Suppose, finally, that $\mathscr{J}$ is not empty and that $\mathscr{K}$ contains a nonzero element z. Then by (3.5) $\mathscr{J}$ contains the unbounded sequence $x + nz$, $n = 0, 1, 2, \ldots$, so that it is unbounded. ■

We see from (3.5) that if $\mathscr{J}$ is not empty, the cone of recession $\mathscr{K}$ depends on $\mathscr{J}$ and not on the particular representation (3.3). Of course, when $\mathscr{J}$ is empty $\mathscr{K}$ depends upon the function $\tilde{c}(\xi)$.

The definition (3.4) shows that $\mathscr{K}$ is the dual cone of the convex cone

$$C = \left\{ \sum_{i=1}^{L} \alpha_i \eta_i \,|\, \alpha_i \geqq 0, |\eta_i| = 1, \tilde{c}(\eta_i) < \infty, i = 1, \ldots, L, L < \infty \right\}. \tag{3.6}$$

In particular, $\mathscr{K} = \{0\}$ if and only if $C = \mathscr{R}^N$ or, equivalently, if and only if the origin is an interior point of C. That is, $\mathscr{K} = \{0\}$ if and only if there are unit vectors $\eta_1, \ldots, \eta_L$ with $\tilde{c}(\eta_i) < \infty$ such that the set of vectors $\{\sum \alpha_i \eta_i \,|\, \alpha_i \geqq 0\}$ contains a neighborhood of 0.

Our simplest results are based on the following elementary lemma.

Lemma 3.2. *Let $\mathscr{J}$, a given set of the form* (3.3), *be empty. Suppose that there are unit vectors $\eta_1, \ldots, \eta_L$ such that $\tilde{c}(\eta_i) < \infty$ and that the set*

$$\left\{ \sum_{i=1}^{L} \alpha_i \eta_i \,|\, \alpha_i \geqq 0 \right\}$$

contains a neighborhood of the origin. Then there exist a positive ε and a finite set of unit vectors $\zeta_1, \ldots, \zeta_M$ such that the set

$$\{x \,|\, x \cdot \zeta_\nu \leqq \tilde{c}(\zeta_\nu) + \varepsilon, \nu = 1, \ldots, M\}$$

is empty.

Proof. Define the family of sets

$$\mathscr{J}_{\xi,\delta} = \{x \,|\, x \cdot \xi \leqq \tilde{c}(\xi) + \delta, x \cdot \eta_i \leqq \tilde{c}(\eta_i) + \delta, i = 1, \ldots, L\}$$

for $\xi \in S^{N-1}$, $\delta > 0$.

By hypothesis the cone of recession of each $\mathscr{J}_{\xi,\delta}$ is $\{0\}$, so that by Lemma 3.1 each $\mathscr{J}_{\xi,\delta}$ is closed and bounded.

If the intersection of any finite collection of members of this family is nonempty, the finite intersection property (Helly's theorem) implies that the intersection of the whole family is nonempty. This intersection is $\mathscr{J}$ which is, by hypothesis, empty. Therefore the intersection of some finite collection of $\mathscr{J}_{\xi,\delta}$ is empty. We let ε be the smallest value of δ in this collection and let the ζ_ν be the η_i together with the ξ in the collection to obtain the statement of the lemma. ■

We apply this lemma to show that under some conditions a half-space $\{x \,|\, x \cdot \xi_0 < t_0\}$ which contains $\mathscr{S}$ also contains a set $\mathscr{M}$ of the form (2.14).

Lemma 3.3. *Let the unit vector ξ_0 be an interior point of the cone*

$$C^* = \left\{ \sum_{i=1}^{L} \alpha_i \eta_i \,|\, \alpha_i \geqq 0, |\eta_i| = 1, c^*(\eta_i) < \infty, i = 1, \ldots, L, L < \infty \right\}.$$

If $\mathscr{S}$ is not empty and lies in the open half-space $\{x \,|\, x \cdot \xi_0 < t_0\}$, then there exists a set $\mathscr{M}$ of the form (2.14) *with $\varepsilon > 0$ which also lies in this half-space.*

Proof. Define the function

$$\tilde{c}(\xi) = \begin{cases} -t_0 & \text{for} \quad \xi = -\xi_0 \\ c^*(\xi) & \text{for} \quad \xi \neq -\xi_0. \end{cases}$$

Because the additional condition $-x \cdot \xi_0 \leqq -t_0$ makes the set $\mathscr{S}$ empty, we must have $-t_0 < c^*(-\xi_0)$. Then the condition $x \cdot \xi_0 \geqq t_0$ implies that $-x \cdot \xi_0 \leqq c(-\xi_0)$. Therefore the set $\mathscr{I}$ defined by (3.3) is just

$$\mathscr{S} \cap \{x \mid x \cdot \xi \geqq t_0\},$$

which is empty.

Because ξ_0 is an interior point of C^*, one can find unit vectors $\eta_1, \ldots, \eta_L$ with $c^*(\eta_i) < \infty$ such that the set $\{\sum \alpha_i \eta_i \mid \alpha_i \geqq 0\}$ contains a neighborhood of ξ_0. Therefore the convex hull of $-\xi_0, \eta_1, \ldots, \eta_L$ contains a neighborhood of the origin, and $\tilde{c}(-\xi_0)$ and $\tilde{c}(\eta_i)$, $i = 1, \ldots, L$, are finite.

By Lemma 3.2 there exist a positive ε and unit vectors $\zeta_1, \ldots, \zeta_M$ such that the set

$$\{x \mid x \cdot \zeta_\nu \leqq \tilde{c}(\zeta_\nu) + \varepsilon, \nu = 1, \ldots, M\}$$

is empty.

If none of the ζ_ν were equal to $-\xi_0$, then this set would contain the non-empty set $\mathscr{S}$, which is impossible. Thus one of the ζ_ν, say ζ_M, equals $-\xi_0$ while the others are different from $-\xi_0$. Therefore the intersection of the set

$$\mathscr{M} = \{x \mid x \cdot \zeta_\nu \leqq c^*(\zeta_\nu) + \varepsilon, \nu = 1, \ldots, M-1\}$$

with the set $\{x \mid x \cdot \xi_0 \geqq t_0 - \varepsilon\}$ is empty, which implies the statement of the lemma. ■

The cone C^* depends upon the function $c^*(\xi)$, and so it appears that the same is true of the hypotheses of Lemma 3.3. We recall, however, that the dual cone of C^* is the cone of recession

$$\mathscr{K}^* = \{z \mid z \cdot \xi \leqq 0 \; \forall \xi \in S^{N-1} \ni c^*(\xi) < \infty\}$$

of $\mathscr{S}$. If $\mathscr{S}$ is not empty, we see from Lemma 3.1 that

$$\mathscr{K}^* = \{z \mid x + z \in \mathscr{S} \; \forall x \in \mathscr{S}\},$$

so that $\mathscr{K}^*$ depends on $\mathscr{S}$ but not on $c^*(\xi)$. The interior of the convex cone C^* can now be written as

$$\text{int } C^* = \{\xi \in R^N \mid z \cdot \xi < 0 \; \forall z \in \mathscr{K}^* \cap S^{N-1}\}$$

so that this set is also independent of $c^*(\xi)$.

We now apply this remark and Lemma 3.3 to obtain a geometric sufficient condition for $\mathscr{S}'$ to contain a set $\mathscr{M}$ of the form (2.14) with $\varepsilon > 0$.

Theorem 3.1. *Suppose that the set $\mathscr{S}$ is not empty, and that the open set $\mathscr{S}'$ contains an open convex polyhedron P with finitely many faces which, in turn, contains $\mathscr{S}$. If the outward unit normal ξ_j to each of the faces of P satisfies the condition*

$$z \cdot \xi_j < 0 \quad \forall z \in \mathscr{K}^* \cap S^{N-1}, \tag{3.7}$$

then $\mathscr{S}'$ satisfies the hypothesis of Theorem 2.1.

Proof. The polyhedron P can be written in the form

$$P = \{x \mid x \cdot \xi_j < t_j, j = 1, \ldots, R\} = \bigcap_{j=1}^{R} \{x \mid x \cdot \xi_j < t_j\}.$$

Since P contains $\mathscr{S}$, $\mathscr{S}$ lies in each of the half-spaces $\{x \mid x \cdot \xi_j < t_j\}$. Lemma 3.3 shows that $\mathscr{S}'$ contains the intersection of R sets $\mathscr{M}_j$ of the form (2.14) with each ε positive. This intersection contains a set of the form (2.14) with $\varepsilon > 0$, which proves the theorem. ■

The simplest and most important case for biological applications is that in which $\mathscr{S}$ is bounded (see also [20, Theorem 1]).

Theorem 3.2. *Let the set $\mathscr{S}$ be nonempty and bounded. Then any open set $\mathscr{S}'$ which contains $\mathscr{S}$ satisfies the hypothesis of Theorem* 2.1.

Proof. Lemma 3.1 shows that in the present case $\mathscr{K}^* = \{0\}$. Therefore $\mathscr{K}^* \cap S^{N-1}$ is empty, and the condition (3.7) is vacuous. Thus one only needs to recall the well-known fact (see, e.g., [3, Section 27]) that any neighborhood of a closed bounded convex set $\mathscr{S}$ contains a convex polyhedron which contains $\mathscr{S}$, and to apply Theorem 3.1. ■

We observe that if ξ_0 is outside the closure of C^*, then there is a z in the cone of recession $\mathscr{K}^*$ of $\mathscr{S}$ such that $z \cdot \xi_0 > 0$. If x_0 is any element of $\mathscr{S}$, the sequence $x_0 + nz$ lies in $\mathscr{S}$, and consequently, the function $x \cdot \xi_0$ is not bounded above for x in $\mathscr{S}$. Thus ξ_0 cannot be an outward normal to any face of a convex polyhedron which contains $\mathscr{S}$. Thus what Theorem 3.1 misses is the case in which $\mathscr{S}$ lies in a convex polyhedron one or more of whose outward normals lies on the boundary of C^*.

Geometrically, the hypotheses of Theorem 3.1 require that no face of the enclosing polyhedron be parallel to a "supporting plane at infinity" of $\mathscr{S}$. In other words, the distance between $\mathscr{S}$ and the part of the exterior of the polyhedron outside the ball $\{x \mid |x| \leqq R\}$ must approach infinity with R.

For example, if $\mathscr{S}$ is the two-dimensional set

$$\{(x, y) | y \geqq -\sqrt{1 - x^2}, -1 \leqq x \leqq 1\},$$

a polygon with a vertical edge is not permitted in Theorem 3.1.

In order to treat such cases we shall use the following lemma.

Lemma 3.4. *Suppose that the set $\mathscr{J}$ defined by* (3.3) *has no interior points, and that the function $\tilde{c}(\xi)$ is lower semicontinuous. Then there exists a linear subspace $\mathscr{L}$ of $\mathscr{R}^N$ of dimension at least* 1 *such that the subset*

$$\hat{\mathscr{J}} = \{x \in \mathscr{L} | x \cdot \xi \leqq \tilde{c}(\xi)\ \forall \xi \in \mathscr{L} \cap S^{N-1}\}$$

of $\mathscr{L}$ also has no interior points and that the cone of recession of the restriction of $\tilde{c}(\xi)$ to $\mathscr{L} \cap S^{N-1}$ consists only of 0.

Proof. The proof is by induction. If the cone of recession of $\tilde{c}(\xi)$ is $\{0\}$, the lemma is valid with $\mathscr{L} = \mathscr{R}^N$.

Suppose instead that there is a $z_1 \neq 0$ in the cone of recession of $\tilde{c}$, and define

$$\mathscr{L}^{(1)} = \{\xi \in \mathscr{R}^N | \xi \cdot z_1 = 0\}.$$

Because $\mathscr{J}$ contains no interior points, there is for any x in $\mathscr{R}^N$ and any positive integer n a unit vector ξ_n such that

$$(x + nz_1) \cdot \xi_n \geqq \tilde{c}(\xi_n). \tag{3.8}$$

By the Weierstrass theorem there is a subsequence $\{\xi_{n_i}\}$ of $\{\xi_n\}$ which converges to a unit vector $\bar{\xi}$. We divide (3.8) by n, set $n = n_i$, and let $i \to \infty$. Because $\tilde{c}$ is lower semicontinuous on the compact set S^{N-1}, it is bounded below, and we find that

$$z_1 \cdot \bar{\xi} \geqq 0. \tag{3.9}$$

On the other hand (3.8) shows that $\tilde{c}(\xi_n)$ is finite, so that $z_1 \cdot \xi_{n_i} \leqq 0$ because z_1 is in the cone of recession of $\tilde{c}$. Letting $i \to \infty$, we see that $z_1 \cdot \bar{\xi} \leqq 0$, which, together with (3.9), shows that $z_1 \cdot \bar{\xi} = 0$. That is, $\bar{\xi} \in \mathscr{L}^{(1)}$.

We now return to (3.8) and note that because $z_1 \cdot \xi_{n_i} \leqq 0$, we must have

$$x \cdot \xi_{n_i} \geqq \tilde{c}(\xi_{n_i}).$$

We let $i \to \infty$ and use the fact that $\tilde{c}$ is lower semicontinuous to see that

$$x \cdot \bar{\xi} \geqq \tilde{c}(\bar{\xi}). \tag{3.10}$$

Thus for each x in $\mathscr{R}^N$ there is a $\bar{\xi}$ in $\mathscr{L}^{(1)} \cap S^{N-1}$ such that the inequality (3.10) is valid. This means that the set

$$\mathscr{J}^{(1)} = \{x \in \mathscr{L}^{(1)} | x \cdot \xi \leqq \tilde{c}(\xi)\ \forall \xi \in \mathscr{L}^{(1)} \cap S^{N-1}\}$$

has no interior points.

If the cone of recession of the restriction of $\tilde{c}$ to $\mathscr{L}^{(1)} \cap S^{N-1}$ is $\{0\}$, the lemma is valid with $\mathscr{L} = \mathscr{L}^{(1)}$ and $\mathscr{J} = \mathscr{J}^{(1)}$. If not, we choose a nonzero element z_2 in the new cone of recession, and define

$$\mathscr{L}^{(2)} = \{x \in \mathscr{L}^{(1)} | x \cdot z_2 = 0\}.$$

We show as above that the set

$$\mathscr{J}^{(2)} = \{x \in \mathscr{L}^{(2)} | x \cdot \xi \leqq \tilde{c}(\xi)\ \forall \xi \in \mathscr{L}^{(2)} \cap S^{N-1}\}$$

has no interior points.

We pass from $\mathscr{L}^{(j)}$ and $\mathscr{J}^{(j)}$ to $\mathscr{L}^{(j+1)}$ and $\mathscr{J}^{(j+1)}$ if the cone of recession of the restriction of $\tilde{c}$ to $\mathscr{L}^{(j)} \cap \mathscr{S}^{N-1}$ is not $\{0\}$.

The dimension of $\mathscr{L}^{(j)}$ is clearly $N - j$, so that, if the process has not terminated before, we eventually obtain the one-dimensional set $\mathscr{J}^{(N-1)}$ with no interior points. Since a one-space only contains two unit vectors and since it takes at least two inequalities to force a set to have no interior points, we conclude that the cone of recession of $\tilde{c}$ on $\mathscr{L}^{(N-1)} \cap S^{N-1}$ must be $\{0\}$. Thus there is always some j, $0 \leqq j \leqq N - 1$ such that $\mathscr{J}^{(j)}$ has no interior points and the cone of recession of $\tilde{c}$ on $\mathscr{L}^{(j)} \cap S^{N-1}$ is $\{0\}$. This proves the lemma. ■

We shall use this lemma to prove a result which can be used in place of Lemma 3.3.

Lemma 3.5. *Let the set $\mathscr{S}$ have interior points and let $c^*(\xi)$ be lower semicontinuous. Suppose that $\mathscr{S}$ lies in the half-space $\{x | x \cdot \xi_0 \leqq t_0\}$. Then for any positive δ there is a set $\mathscr{M}$ of the form* (2.14) *with* $\varepsilon > 0$ *which lies in the half-space $\{x | x \cdot \xi_0 \leqq t_0 + \delta\}$.*

Proof. We define the lower semicontinuous function

$$\tilde{c}(\xi) = \begin{cases} -t_0 & \text{for} \quad \xi = -\xi_0 \\ c^*(\xi) & \text{for} \quad \xi \neq -\xi_0. \end{cases} \tag{3.11}$$

Because $\mathscr{S}$ is not empty while its intersection with the half-space $\{x | x \cdot \xi > t_0\}$ is empty, we must have $-t_0 \leqq c^*(-\xi_0)$, so that $\tilde{c}$ is again lower semicontinuous.

By hypothesis the set $\mathscr{J}$ defined by (3.3) has no interior points. Lemma 3.4 shows that there is a linear subspace $\mathscr{L}$ of $\mathscr{R}^N$ such that the set

$$\hat{\mathscr{J}} = \{x \in \mathscr{L} | x \cdot \xi \leqq \tilde{c}(\xi)\ \forall \xi \in \mathscr{L} \cap S^{N-1}\}$$

has no interior points and the cone of recession of the restriction of $\tilde{c}$ to $\mathscr{L} \cap S^{N-1}$ is $\{0\}$.

If $-\xi_0$ did not lie in $\mathscr{L}$, then the restriction of $\tilde{c}$ to $\mathscr{L} \cap S^{N-1}$ would agree with that of c^*, so that $\hat{\mathscr{J}}$ would contain the projection of $\mathscr{S}$ on $\mathscr{L}$. Because this projection has interior points but $\hat{\mathscr{J}}$ does not, we conclude that $-\xi_0 \in \mathscr{L}$. Thus $\hat{\mathscr{J}}$ is the intersection of a convex set with interior and the half-space $\{x | -x \cdot \xi_0 \leqq -t_0\}$.

Since $\hat{\mathscr{J}}$ contains no interior points, the set

$$\hat{\mathscr{J}} \cap \{x \mid -x \cdot \xi_0 \leqq -t_0 - \delta\}$$

is empty whenever δ is positive. The cone of recession of this set is, of course, still $\{0\}$. We apply Lemma 3.2 to this set to find that there exist a positive ε and unit vectors $\zeta_1, \ldots, \zeta_M$ in $\mathscr{L}$ such that the set

$$(x \in \mathscr{L} \mid x \cdot \zeta_\nu \leqq \tilde{\tilde{c}}(\zeta_\nu) + \varepsilon, \nu = 1, \ldots, M\} \tag{3.12}$$

is empty. Here

$$\tilde{\tilde{c}}(\xi) = \begin{cases} -t_0 - \delta & \text{for} \quad \xi = -\xi_0 \\ c^*(\xi) & \text{for} \quad \xi \neq -\xi_0. \end{cases}$$

If none of the ζ_ν were equal to $-\xi_0$, this set would contain the projection of $\mathscr{S}$ on $\mathscr{L}$, and hence could not be empty. We conclude that one of the ζ_ν, say ζ_M, is $-\xi_0$ while $\zeta_1, \ldots, \zeta_{M-1}$ are different from $-\xi_0$. Thus the fact that the set (3.12) is empty means that the set

$$(x \in \mathscr{L} \mid x \cdot \zeta_\nu \leqq c^*(\zeta_\nu) + \varepsilon, \nu = 1, \ldots, M - 1\}$$

lies in the half-space

$$\{x \in \mathscr{L} \mid x \cdot \xi_0 < t_0 + \delta - \varepsilon\}.$$

Since all the ζ_ν lie in $\mathscr{L}$, it follows that the set

$$\{x \in \mathscr{R}^N \mid x \cdot \zeta_\nu \leqq c^*(\zeta_\nu) + \varepsilon, \nu = 1, \ldots, M - 1\}$$

lies in the half-space

$$\{x \in \mathscr{R}^N \mid x \cdot \xi_0 < t_0 + \delta - \varepsilon\}.$$

Thus our lemma is proved. ∎

When $N = 2$ the hypothesis that $\mathscr{S}$ contains interior points can be dropped.

Lemma 3.6. *Let the set $\mathscr{S}$ be nonempty, let the dimension N be 2, and let $c^*(\xi)$ be lower semicontinuous. Suppose that $\mathscr{S}$ lies in the half-space $\{x \mid x \cdot \xi_0 \leqq t_0\}$. Then for any positive δ there is a set $\mathscr{M}$ of the form* (2.14) *with $\varepsilon > 0$ which lies in the half-space $\{x \mid x \cdot \xi \leqq t_0 + \delta\}$.*

Proof. If $\mathscr{S}$ contains interior points, the result is a consequence of Lemma 3.5. If $\mathscr{S}$ is bounded, it follows from Lemma 3.3. Suppose, then, that $\mathscr{S}$ is unbounded and has no interior points. Since $\mathscr{S}$ is closed and convex, it must be either a half-line or a line. Therefore its cone of recession consists of either the ray $\{\alpha z_1 \mid \alpha \geqq 0\}$ or the line $\{\alpha z_1 \mid -\infty < \alpha < \infty\}$ where z_1 is a unit vector tangent to $\mathscr{S}$.

We now define the function $\tilde{c}(\xi)$ by means of (3.11) and the set $\mathscr{J}$ by means of (3.3). By hypothesis the set $\mathscr{J}$ has no interior points.

If ξ_0 is not orthogonal to z_1, then the cone of recession of $\mathscr{J}$ is $\{0\}$. Therefore the subspace $\mathscr{L}$ provided by Lemma 3.4 is just $\mathscr{R}^2$.

If, on the other hand, ξ_0 is orthogonal to z_1, then the subspace $\mathscr{L}$ given by Lemma 3.4 is the orthogonal complement of z_1. In either case, $\mathscr{L}$ contains ξ_0.

The property that $\mathscr{S}$ contains interior points was used in the proof of Lemma 3.5 only to show that ξ_0 lies in $\mathscr{L}$. Since we have established this fact, we may now use the remainder of the proof of Lemma 3.5 to prove the present lemma. ■

These two lemmas immediately yield the following theorem.

Theorem 3.3. *Suppose that $c^*(\xi)$ is lower semicontinuous and that either $\mathscr{S}$ contains interior points or $\mathscr{S}$ is nonempty and the dimension N is 2. If the open set $\mathscr{S}'$ contains an open convex polyhedron P with finitely many faces whose exterior lies at a positive distance from $\mathscr{S}$, then $\mathscr{S}'$ satisfies the hypothesis of Theorem* 2.1.

Proof. Let the polyhedron P be given in the form

$$P = \bigcap_{j=1}^{R} \{x \mid x \cdot \xi_j < t_j\}.$$

If δ is the distance from $\mathscr{S}$ to the exterior of P, then

$$\mathscr{S} \subset \{x \mid x \cdot \xi_j \leqq t_j - \delta\}$$

for $j = 1, \ldots, R$. According to Lemma 3.5 or Lemma 3.6 each half-space $\{x \mid x \cdot \xi_j \leqq t_j - \frac{1}{2}\delta\}$ contains a set $\mathscr{M}_j$ of the form (2.14) with a positive ε. Then $\mathscr{S}'$ contains the intersection of the $\mathscr{M}_j$, which contains a set $\mathscr{M}$ of the form (2.14) with $\varepsilon > 0$. ■

The example of the set $\mathscr{A}$ defined by (3.1) shows that when $N > 2$ the condition that $\mathscr{S}$ contain interior points is needed for this result as well as for Lemma 3.5. In fact, for this $c^*(\xi)$ the intersection with the plane $z = 0$ of any set $\mathscr{M}$ of the form (2.14) with $\varepsilon > 0$ contains a sector in whose interior the quarter-space $x \leqq 0$, $y \leqq 0$ lies. Thus the half-space $\{x \mid x \leqq 1\}$, which is also a polyhedron and which certainly contains $\mathscr{A}$, does not contain a set $\mathscr{M}$ of the form (2.14) with $\varepsilon > 0$ or, for that matter, with $\varepsilon = 0$.

If, however, the same set $\mathscr{A}$ is represented in the form (3.2), then the sets $\mathscr{M}$ are of the form $x \leqq \varepsilon$, $y \leqq \varepsilon$, $-\varepsilon \leqq z \leqq \varepsilon$. Thus for the set $\mathscr{A}$ the question of whether or not $\mathscr{S}'$ satisfies the condition of Theorem 2.1 depends upon the function $c^*(\xi)$ which is used to represent it.

The function

$$c^*(\cos\theta, \sin\theta) = \begin{cases} 0 & \text{for} \quad 0 \leqq \theta < \pi/2 \\ +\infty & \text{for} \quad \pi/2 \leqq \theta < 2\pi \end{cases}$$

represents the two-dimensional set $\{(x, y) | x \leqq 0,\ y \leqq 0\}$. However, any set $\mathscr{M}$ of the form (2.14) with $\varepsilon > 0$ contains an open sector which contains the negative x-axis. Thus the statements of Lemmas 3.5 and 3.6 and Theorem 3.3 are not true for this function c^*, which is lower semicontinuous except at $\theta = \pi/2$. This shows that the condition of lower semicontinuity is essential.

As might be expected, the fact that $\mathscr{S}$ is empty is not sufficient to conclude that a set $\mathscr{M}$ of the form (2.14) with $\varepsilon > 0$ is empty. For instance, the two-dimensional set

$$\{(x, y) | x\cos\theta + y\sin\theta \leqq -\sqrt{\sin\theta} \qquad \text{for} \quad 0 \leqq \theta \leqq \pi\}$$

for which

$$c^*(\cos\theta, \sin\theta) = \begin{cases} -\sqrt{\sin\theta} & \text{for} \quad 0 \leqq \theta \leqq \pi \\ +\infty & \text{for} \quad \pi < \theta < 2\pi \end{cases}$$

is easily seen to be empty, while any set of the form (2.14) with $\varepsilon \geqq 0$ contains a semiinfinite part of the negative y-axis. Thus any criterion which permits $\mathscr{S}$ to be empty must involve the function $c^*(\xi)$. We shall give two such criteria.

Theorem 3.4. *Let the set $\mathscr{S}$ be empty and suppose that the convex hull of the set $\{\xi \in S^{N-1} | c^*(\xi) < \infty\}$ contains a neighborhood of the origin. Then there is a set $\mathscr{M}$ of the form* (2.14) *with $\varepsilon > 0$ which is empty, so that Corollary* 2.1 *can be applied.*

Proof. Lemma 3.2 with $\mathscr{J} = \mathscr{S}$ gives this result.

The simplest and most important case in which this theorem can be used is that in which $c^*(\xi)$ is known to be bounded.

The following theorem gives another sufficient condition for Corollary 3.1 to be applicable.

Theorem 3.5. *Suppose that $c^*(\xi)$ is lower semicontinuous and that for some positive δ the set*

$$\mathscr{S}_\delta = \{x | x \cdot \xi \leqq c^*(\xi) + \delta\ \forall \xi \in S^{N-1}\}$$

is empty. Then there is an empty set $\mathscr{M}$ of the form (2.14) *with $\varepsilon > 0$, so that Corollary* 2.1 *can be applied.*

Proof. Apply Lemma 3.4 to the set $\mathscr{J} = \mathscr{S}_\delta$ to find a linear subspace $\mathscr{L}$ such that the set

$$\hat{\mathscr{J}} = \{x \in \mathscr{L} \mid x \cdot \xi \leqq c^*(\xi) + \delta \ \forall \xi \in \mathscr{L} \cap S^{N-1}\}$$

has no interior points and the cone of recession $\{0\}$. Therefore the set

$$\mathscr{J}' = \{x \in \mathscr{L} \mid x \cdot \xi \leqq c^*(\xi) \ \forall \xi \in \mathscr{L} \cap S^{N-1}\}$$

is empty and has the cone of recession $\{0\}$. Lemma 3.2 now shows that there are $\varepsilon > 0$ and $\zeta_1, \ldots, \zeta_M$ in $\mathscr{L}$ such that the set

$$\{x \in \mathscr{L} \mid x \cdot \zeta_\nu \leqq c^*(\zeta_\nu) + \varepsilon, \nu = 1, \ldots, M\}$$

is empty. Because $\zeta_1, \ldots, \zeta_M$ are in $\mathscr{L}$, the set $\mathscr{M}$ defined by (2.14) with $K = M$ and $\xi_\nu = \zeta_\nu$ is then also empty, which proves the theorem. ∎

We end this section with a theorem which is rather general, although not necessarily very useful.

Theorem 3.6. *A necessary and sufficient condition for the open set $\mathscr{S}'$ to contain a set $\mathscr{M}$ of the form* (2.14) *with $\varepsilon > 0$ is that $\mathscr{S}'$ contain a convex polyhedron P of finitely many faces which, in turn, contains a set of the form*

$$\mathscr{S}_\delta = \{x \mid x \cdot \xi \leqq c^*(\xi) + \delta \ \forall \xi \in S^{N-1}\}$$

with $\delta > 0$.

Proof. It is clear that $\mathscr{M}$ is a convex polyhedron of finitely many faces and that it contains $\mathscr{S}_\varepsilon$, so the condition is certainly necessary.

To prove its sufficiency, we again write the polyhedron P in the form

$$P = \bigcap_{i=1}^{R} \{x \mid x \cdot \xi_j \leqq t_j\}.$$

If for some j

$$y \in \mathscr{S}_{\frac{1}{2}\delta} \cap \{x \mid x \cdot \xi_j > t_j - \tfrac{1}{2}\delta\},$$

then $x_1 = y + \frac{1}{2}\delta\xi_j$ satisfies

$$x_1 \in \mathscr{S}_\delta \cap \{x \mid x \cdot \xi_j > t_j\}$$

which is, by hypothesis, empty. Therefore the set $\mathscr{S}_{\frac{1}{2}\delta} \cap \{x \mid x \cdot \xi_j > t_j - \frac{1}{2}\delta\}$ is empty. Thus the set $\mathscr{S}_{\frac{1}{2}\delta}$ is at a distance at least $\frac{1}{2}\delta$ from the exterior of P.

If the set $\mathscr{S}_{\frac{1}{4}\delta}$ is not empty, its members are interior points of $\mathscr{S}_{\frac{1}{2}\delta}$, and Theorem 3.3 shows that P contains a set $\mathscr{M}$ of the form (2.14) with $\varepsilon > 0$.

If, on the other hand, $\mathscr{S}_{\frac{1}{4}\delta}$ is empty (and this is certainly the case if $\mathscr{S}_\delta$ is empty), then Theorem 3.5 shows that there is an empty set $\mathscr{M}$ of the form

(2.14) with $\varepsilon > 0$. Thus our theorem is proved whether $\mathscr{S}_{\frac{1}{4}\delta}$ is empty or not. ■

We remark that if $\mathscr{S}_\delta$ is empty, it lies in an empty polyhedron such as

$$\{x \mid x \cdot \xi_0 \leqq -1\} \cap \{x \mid x \cdot \xi_0 \geqq 1\}$$

so that Theorem 3.6 implies Theorem 3.5.

4. CONNECTION WITH SEMIINFINITE PROGRAMMING

As is clear from the proofs, Theorem 3.1 is equivalent to Lemma 3.3 and Theorem 3.3 is equivalent to Lemmas 3.5 and 3.6. The problem we have treated is thus reduced to the question: Under what conditions is it true that if the set $\mathscr{S}$ lies at a positive distance from a half-space $\{x \mid x \cdot \xi_0 \geqq t_0\}$, then there is also a set $\mathscr{M}$ of the form (2.14) with $\varepsilon > 0$ which lies outside this half-space?

The infimum of t_0 for which $\mathscr{S}$ is outside such a half-space is, by definition, the support function

$$s(\xi_0) = \sup\{x \cdot \xi_0 \mid x \cdot \xi \leqq c^*(\xi)\ \forall \xi \in S^{N-1}\} \tag{4.1}$$

of $\mathscr{S}$.

The question is whether one can obtain this $s(\xi_0)$ as the limit of the support functions of sets of the form $\mathscr{M}$. That is, is

$$\hat{s}(\xi_0) \equiv \lim_{\varepsilon \searrow 0} \inf\{\sup\{x \cdot \xi_0 \mid x \cdot \xi_i \leqq c^*(\xi_i) + \varepsilon, i = 1, \ldots, K\} \mid \xi_1, \ldots, \xi_K \in S^{N-1}, K < \infty\} \tag{4.2}$$

equal to $s(\xi_0)$?

For fixed $\xi_1, \ldots, \xi_K$, and ε the supremum on the right is a finite linear program. By the Gale–Kuhn–Tucker theorem [11] its value is equal to that of the dual program. Therefore

$$\begin{aligned}
&\inf\{\sup\{x \cdot \xi_0 \mid x \cdot \xi_i \leqq c^*(\xi_i) + \varepsilon, i = 1, \ldots, K\} \mid \xi_1, \ldots, \xi_K \in S^{N-1}, K < \infty\} \\
&\quad = \inf\left\{ \sum_{i=1}^{K} \lambda_i [c^*(\xi_i) + \varepsilon] \mid \lambda_i \geqq 0, \sum_{i=1}^{K} \lambda_i \xi_i = \xi_0, \right. \\
&\qquad\qquad \left. \xi_1, \ldots, \xi_K \in S^{N-1}, K < \infty \right\}.
\end{aligned} \tag{4.3}$$

The program on the right is the dual of the semiinfinite program

$$s_\varepsilon(\xi_0) = \sup\{x \cdot \xi_0 \mid x \cdot \xi \leqq c^*(\xi) + \varepsilon, \xi \in S^{N-1}\} \tag{4.4}$$

which gives the support function s_ε of the set $\mathscr{S}_\varepsilon$. This program and its dual need not have the same value. That is, there may be a duality gap. However, the perfect duality theorem of Duffin [8, Theorem 1] states that the value of the program (4.3) is equal to the negative of what Duffin called the subvalue

$$\lim_{\delta \searrow 0} [-s_{\varepsilon+\delta}(\xi_0)]$$

of the dual program (4.4). We therefore see that

$$\hat{s}(\xi_0) = \lim_{\varepsilon \searrow 0} \lim_{\delta \searrow 0} s_{\varepsilon+\delta}(\xi_0) = \lim_{\varepsilon \searrow 0} s_\varepsilon(\xi_0). \tag{4.5}$$

Our question is thus reduced to the question of whether the support function of $\mathscr{S}_\varepsilon$ converges to that of $\mathscr{S}$ as ε decreases to zero.

In the language of Duffin, $-\hat{s}(\xi_0)$ is the subvalue of the program for $-s(\xi_0)$. Duffin's perfect duality theorem [8] now states that $\hat{s}$ is equal to the value of the dual program. Thus

$$\hat{s}(\xi_0) = \inf\left\{ \sum_{i=1}^{K} \lambda_i c^*(\xi_i) | \xi_1, \ldots, \xi_K \in S^{N-1}, K < \infty \right\}, \tag{4.6}$$

and our question is reduced to whether or not the support function $s(\xi_0)$ can be computed from this standard geometric formula. In the language of semiinfinite programming, we ask whether the programs (4.1) and (4.6) are in perfect duality. Duffin and Karlovitz [9, p. 127] have shown that this is the case, provided the following two conditions are satisfied:

(i) $\mathscr{S}$ has interior points.

(ii) The set of points $\{(\xi, c^*(\xi))\}$ in $\mathscr{R}^{N+1}$ is canonically closed in the sense of Charnes *et al.* [4]; that is, the set

$$\left\{\left(\frac{1}{1 + |c^*(\xi)|}\xi, \frac{c^*(\xi)}{1 + |c^*(\xi)|}\right)\right\}$$

is closed in $\mathscr{R}^{N+1}$.

Under these conditions, then, we have $s(\xi) = \hat{s}(\xi)$ for all ξ.

We wish to relate the theorem of Duffin and Karlovitz to our Lemma 3.5. We first note that any nonempty set $\mathscr{S}$ of the form

$$\mathscr{S} = \{x | x \cdot a_h \leqq \beta_h, h \in H\}$$

for an arbitrary index set H can be put into the form (2.13) by defining

$$c^*(\xi) = \inf\{\beta_h/|a_h| \, | a_h/|a_h| = \xi\}. \tag{4.7}$$

If there is no a_h with $a_h/|a_h| = \xi$, we set $c^*(\xi) = +\infty$. Because $\mathscr{S}$ is nonempty, $\beta_h/|a_h|$ is bounded below, and there is no h with $a_h = 0$ and $\beta_h < 0$. The trivial inequalities where $a_h = 0$ and $\beta_h \geqq 0$ do not contribute to the definition of $\mathscr{S}$.

The following two lemmas relate the canonical closure of a set of inequalities with the lower semicontinuity of the corresponding function c^*.

Lemma 4.1. *If the set* $\{(a_h, \beta_h)\}$ *in* R^{N+1} *is canonically closed, and contains no point of the form* $(0, \beta)$ *with* $\beta < 0$, *then* $c^*(\xi)$ *is lower semicontinuous.*

Proof. The set

$$\mathscr{I} \equiv \left\{\left(\frac{1}{|a_h| + |\beta_h|}\, a_h, \frac{1}{|a_h| + |\beta_h|}\, \beta_h\right) \middle| h \in H\right\}$$

is closed. Therefore if the quantity $\beta_h/|a_h|$ were not bounded below, $\mathscr{I}$ would contain the point $(0, -1)$, contrary to hypothesis. Thus c^* is bounded below.

Again because $\mathscr{I}$ is closed, the infimum in the definition (4.7) of c^* is attained. That is, there exist (a_h, β_h) in $\mathscr{I}$ such that

$$\frac{a_h}{|a_h|} = \xi, \qquad \frac{\beta_h}{|a_h|} = c^*(\xi).$$

Let ξ_ν be a sequence of unit vectors with the limit ξ_0 and suppose that $c^*(\xi_\nu)$ has the finite limit c. Then the vector

$$\left\{\frac{1}{1 + |c|}\, \xi_0, \frac{c}{1 + |c|}\right\}$$

lies in $\mathscr{I}$ and consequently

$$c^*(\xi_0) \leqq c,$$

which shows that c^* is lower semicontinuous. ■

Lemma 4.2. *Let* $c^*(\xi)$ *be defined by* (4.7) *and let* $\bar{c}(\xi)$ *be the corresponding function for the canonical closure of the set* $\{(a_h, \beta_h)\}$. *Then*

$$\bar{c}(\xi) = \min\left\{c^*(\xi), \liminf_{\eta \to \xi} c^*(\eta)\right\}.$$

In particular, $\bar{c}(\xi) \equiv c^*(\xi)$ *if and only if* c^* *is lower semicontinuous.*

Proof. Clearly $\bar{c}(\xi) \leqq c^*(\xi)$. Suppose that

$$\bar{c}(\xi_0) < c^*(\xi_0).$$

Then there must be a sequence $\{\alpha_\nu \xi_0, B_\nu\}$ with $\alpha_\nu > 0$ in the closure of the set $\mathscr{I}$ such that B_ν/α_ν converges to $\bar{c}(\xi_0)$. Each point $\{\alpha_\nu \xi_0, B_\nu\}$ is approximated by a member of $\mathscr{I}$. Therefore there is a sequence $\{a_{h_\nu}, \beta_{h_\nu}\}$ such that

$$\lim_{\nu \to \infty} \frac{a_{h_\nu}}{|a_{h_\nu}|} = \xi_0, \qquad \lim_{\nu \to \infty} \frac{\beta_{h_\nu}}{|a_{h_\nu}|} = \bar{c}(\xi_0).$$

By definition, then,

$$\limsup_{\nu \to \infty} c^*\left(\frac{a_{h_\nu}}{|a_{h_\nu}|}\right) \leqq \bar{c}(\xi_0).$$

Thus, we see that

$$\bar{c}(\xi_0) \geqq \min\left\{c^*(\xi_0), \liminf_{\xi \to \xi_0} c^*(\xi)\right\}.$$

Because $\bar{c}$ is lower semicontinuous by Lemma 4.1, and because $\bar{c}(\xi) \leqq c^*(\xi)$, we see that

$$\bar{c}(\xi_0) \leqq \min\left\{c^*(\xi_0), \liminf_{\xi \to \xi_0} c^*(\xi)\right\}.$$

This proves the lemma. ∎

Lemmas 4.1 and 4.2 show that the canonical closedness of $\{(a_h, \beta_h)\}$ implies the lower semicontinuity of c^* and that the semicontinuity of c^* implies that the value of the dual program (4.6) is not changed when the set of inequalities $x \cdot a_h \leqq \beta_h$ is replaced by its canonical closure. (The original program

$$\sup\{x \cdot \xi_0 \,|\, x \cdot a_h \leqq \beta_h, h \in H\}$$

is clearly not changed when the set of inequalities is replaced by its canonical closure, because the set $\mathscr{S}$ is not altered.) Thus we see that Lemma 3.5 is equivalent to the perfect duality theorem of Duffin and Karlovitz. Lemma 3.6 shows that the condition that $\mathscr{S}$ have interior points can be dropped if $N = 2$. Lemma 3.3 shows that the additional conditions are only needed if ξ_0 is on the boundary of the convex cone C^*.

REFERENCES

1. D. G. Aronson and H. F. Weinberger, Nonlinear diffusion in population genetics, combustion, and nerve propagation, *in* "Partial Differential Equations and Related Topics" (J. Goldstein, ed.), Lecture Notes in Mathematics, No. 446, Springer-Verlag, Berlin and New York, 1975, pp. 5–49.
2. D. G. Aronson and H. F. Weinberger, Multidimensional nonlinear diffusion arising in population genetics, *Adv. in Math.* **30** (1978), 33–76.
3. T. Bonnesen and W. Frenchel, Theorie der Konvexen Körper, *in* "Ergeb. d. Mathematik und ihrer Grenzgeb.," No. 3, Chelsea, New York, 1948.
4. A. Charnes, W. W. Cooper, and K. Kortanek, Duality in semi-infinite programs and some works of Haar and Carathéodory, *Management Sci.* **9** (1963), 209–228.
5. R. Courant and D. Hilbert, "Methods of Mathematical Physics," Vol. 2, Wiley (Interscience), New York, 1962.
6. J. B. Diaz and H. F. Weinberger, A solution of the singular initial value problem for the Euler–Poisson–Darboux equation, *Proc. Amer. Math. Soc.* **4** (1953), 703–718.

7. O. DIEKMANN, Threshhold and Travelling Waves for the Geographical Spread of Infection, *J. Math. Biol.* **6** (1978), 109–130.
8. R. J. DUFFIN, Infinite programs, *in* "Linear Inequalities and Related Systems" (H. W. Kuhn and A. W. Tucker, eds.), Annals of Mathematics Studies, Vol. 38, Princeton Univ. Press, Princeton, New Jersey, 1956, pp. 157–170.
9. R. J. DUFFIN AND L. A. KARLOVITZ, An infinite linear program with a duality gap, *Management Sci.* **12** (1965), 122–134.
10. R. A. FISHER, The advance of advantageous genes, *Ann. Eugenics* **7** (1937), 355–369.
11. D. GALE, H. W. KUHN, AND A. W. TUCKER, Linear programming and the theory of games, in "Activity Analysis of Production and Allocation" (T. C. Koopmans, ed.), Wiley, New York, 1951, pp. 317–329.
12. F. HOPPENSTEADT, "Mathematical Theories of Populations: Demographics, Genetics, and Epidemics," C.B.M.S. Regional Conferences in Applied Mathematics, No. 20, SIAM, Philadelphia, Pennsylvania, 1975.
13. JA. I. KANEL', Certain problems on equations in the theory of burning, *Dokl. Akad. Nauk SSSR* **136** (1961), 277–280; *Soviet Math. Dokl.* **2** (1961), 48–51.
14. S. KARLIN, Population subdivision and selection migration interaction, *in* "Population Genetics and Ecology" (S. Karlin and E. Nevo, eds.), Academic Press, New York, 1976, pp. 617–657.
15. A. KOLMOGOROFF, I. PETROVSKY, AND N. PISCOUNOFF, Étude de l'équations de la diffusion avec croissance de la quantité de matière et son application a un problème biologique, *Bull. Univ. Moscow, Ser. Internat., Sect. A* **1**(6) (1937), 1–25.
16. D. MOLLISON, The rate of spatial propagation of simple epidemics, *Proc. 6th Berkeley Symp. Math. Statist. Probability* **3** (1972), 579–614.
17. T. NAGYLACKI, "Selection in One-and Two-Locus Systems," Lecture Notes in Biomathematics, No. 15, Springer-Verlag, Berlin and New York, 1977.
18. R. T. ROCKAFELLAR, "Convex Analysis," Princeton Univ. Press, Princeton, New Jersey, 1970.
19. H. F. WEINBERGER, Asymptotic behavior of a model in population genetics, *in* "Nonlinear Partial Differential Equations" (J. Chadam, ed.), Lecture Notes in Mathematics, No. 648, Springer-Verlag, Berlin and New York, 1978, pp. 47–96.
20. H. F. WEINBERGER, Asymptotic behavior of a class of discrete-time models in population genetics, *in* "Applied Nonlinear Analysis" (V. Lakshmikantham, ed.), Academic Press, New York, 1979, pp. 407–422.

This work was supported by the National Science Foundation through Grants MCS76-06128 A01 and MCS78-02182.

SCHOOL OF MATHEMATICS
UNIVERSITY OF MINNESOTA
MINNEAPOLIS, MINNESOTA

Part V

MATHEMATICAL MODELS

Analysis of a Stochastic Reynolds Equation and Related Problems

P. L. CHOW

E. A. SAIBEL

This work is mainly concerned with the analysis of a stochastic Reynolds equation in the hydrodynamic theory of lubrication. A differential equation for the mean pressure distribution is derived rigorously in the asymptotic limit by invoking an ergodic theorem. Also two upper bounds to the absolute mean and the root-mean-square deviations of a normalized load carrying capacity from the smooth case is obtained for a general one-dimensional problem. Finally some related problems are discussed briefly.

1. INTRODUCTION

In the present paper, we shall first summarize two main results contained in our recent paper [3], to which the reader is referred for a detailed presentation. The first result is concerned with the derivation of a valid, averaged Reynolds equation under appropriate conditions. Broadly stated, the sufficient conditions require that the roughness parameters for the two surfaces be characterized by weakly dependent random functions of the longitudinal variable, and that the length of the bearing be large relative to the correlation length of the roughness parameters. As our second result, two upper bounds for the absolute mean and the root-mean-square deviations of a normalized load carrying capacity from a smooth case are obtained for a general one-dimensional problem, valid for an arbitrary probability distribution. The result shows a critical dependence of the upper bounds on the correlation of the roughness. These results are presented in Section 2.

In the last section, we shall give a general discussion of some related questions, the two-dimensional problem and its connection with other random boundary problems in a different physical context.

ISBN 0-12-178150-X

2. ANALYSIS OF A STOCHASTIC REYNOLDS EQUATION

For a wide slider bearing of length L, the Reynolds equation in one dimension reads

$$\frac{d}{dx}\left(H^3 \frac{dp}{dx}\right) = \Lambda \frac{d}{dx}[h + v(\delta_1 - \delta_2)], \tag{1}$$

$$p(0) = p(L) = 0. \tag{2}$$

where

$p = p(x)$ is the pressure,
$H = H(x) = h(x) + \delta_1(x) + \delta_2(x)$ is the film thickness,
$h(x) = \langle H(x) \rangle$ is the average (or mean) film thickness,
$\delta_1(x)$, $\delta_2(x)$ are the roughness profiles of the lower and upper surfaces, respectively,
$\Lambda = 6\mu(u_1 + u_2)$ and μ is the viscosity of the lubricant,
$v = (u_2 - u_1)/(u_1 + u_2)$,
u_1 and u_2 are the rolling or sliding speeds of the lower and upper surfaces.

In the stochastic model, the roughness parameters $\delta_1 = \delta_1(x, \omega)$ and $\delta_2 = \delta_2(x, \omega)$ are random functions (or processes) of x for which the probability distributions are given. The goal is to compute the mean pressure $\langle p \rangle$ from (1) subject to the boundary conditions (2). It is a common practice to derive an equation for the mean pressure $\langle p \rangle$.

Consider the boundary-value problem (1) and (2). For each realization of δ_1 and δ_2, an integration of (1) yields

$$p(x) = M \int_0^x \frac{dy}{H^3(y)} + \Lambda \int_0^x \frac{h(y) + v[\delta_1(y) - \delta_2(y)]}{H^3(y)}\, dy, \tag{3}$$

where by applying the boundary condition (2) at $x = L$,

$$M = -\Lambda \int_0^L \frac{h(y) + v[\delta_1(y) - \delta_2(y)]}{H^3(y)}\, dy \Bigg/ \int_0^L \frac{dy}{H^3(y)}. \tag{4}$$

For convenience, let

$$\xi = \delta_1 + \delta_2, \qquad \eta = \delta_1 - \delta_2. \tag{5}$$

Then (4) becomes

$$M = -\Lambda \int_0^L \frac{h(y) + v\eta(y)}{H^3(y)}\, dy \Bigg/ \int_0^L \frac{dy}{H^3(y)}. \tag{6}$$

Suppose that the roughness profiles $\delta_1(x, \omega)$, $\delta_2(x, \omega)$, $x \geqq 0$, are random processes satisfying

(a) δ_1, δ_2 are continuous stochastic processes, defined for $x \geqq 0$,

(b) the maximum $\max_{0 \leqq x \leqq L}\{|\delta_1(x)| + |\delta_2(x)|\}$ is less than the minimum h_m of the smooth film thickness, $h_m = \min_{0 \leqq x \leqq L} h(x)$, for almost every realization,

(c) the deviation $D(x) = |H^{-3}(x) - \langle H^{-3}(x)\rangle|$ satisfies the following condition of asymptotic weak dependence:

$$\lim_{L \to \infty} \frac{1}{L} \int_0^L \langle D(L)D(x)\rangle \, dx = 0.$$

Since the random functions involved are bounded and positive, the ergodic theorem stated on [3, p. 177] becomes applicable to (6) under conditions (a)–(c). By rewriting the expression (6) and invoking the stated theorem, we have, with almost certainty,

$$-M = \Lambda \frac{1}{L} \int_0^L \frac{h(y) + v\eta(y)}{H^3(y)} \, dy \Bigg/ \frac{1}{L} \int_0^L \frac{1}{H^3(y)} \, dy$$

$$\to \Lambda \int_0^L \left\langle \frac{h(y) + v\eta(y)}{H^3(y)} \right\rangle dy \Bigg/ \int_0^L \left\langle \frac{1}{H^3(y)} \right\rangle dy, \qquad \text{as} \quad L \to \infty. \tag{7}$$

Noting (7) and averaging the equation (3), we differentiate the resulting equation twice to get

$$\frac{d}{dx}\left[\left\langle \frac{1}{H^3} \right\rangle^{-1} \frac{d\langle p\rangle}{dx}\right] = \Lambda \frac{d}{dx}\left[h + v \frac{\langle \delta_1/H^3\rangle - \langle \delta_2/H^3\rangle}{\langle 1/H^3\rangle}\right], \tag{8}$$

which is our first announced result.

For a smooth bearing, the pressure p_0 satisfies the equation

$$\frac{d}{dx}\left(h^3 \frac{dp_0}{dx}\right) = \Lambda \frac{dh}{dx}, \qquad p_0(0) = p_0(L) = 0. \tag{9}$$

Subtracting (9) from (1), one obtains, noting (5),

$$\frac{d}{dx}\left(H^3 \frac{dp}{dx} - h^3 \frac{dp_0}{dx}\right) = \Lambda v \frac{d\eta}{dx}. \tag{10}$$

After adding and subtracting the term $H^3(dp_0/dx)$ in the above parentheses, the equation (10) can be written as

$$\frac{d}{dx}\left(H^3 \frac{d\delta p}{dx}\right) = \frac{dQ}{dx}, \qquad \delta p(0) = \delta p(L) = 0, \tag{11}$$

where $\delta p = p - p_0$ is the deviation of p from the smooth case, and

$$Q = \Lambda v\eta - (H^3 - h^3) \frac{dp_0}{dx}. \tag{12}$$

A simple integration of (11) yields

$$\delta p(x) = \int_0^x Q(y)H^{-3}(y)\,dy - \int_0^x H^{-3}(y)\,dy \frac{\int_0^L Q(y)H^{-3}(y)\,dy}{\int_0^L H^{-3}(y)\,dy}. \tag{13}$$

Splitting each of the integrals from 0 to L into two parts, from 0 to x and x to L, the equation (13) is reduced to

$$\delta p(x) = \frac{-\int_0^x H^{-3}\,dy \int_x^L QH^{-3}\,dy + \int_0^x QH^{-3}\,dy \int_x^L H^{-3}\,dy}{\int_0^L H^{-3}\,dy}. \tag{14}$$

It follows that

$$|\delta p(x)| \leqq \frac{\int_0^x H^{-3}\,dy \int_x^L |Q| H^{-3}\,dy + \int_x^L H^{-3}\,dy \int_0^x |Q| H^{-3}\,dy}{\int_0^L H^{-3}\,dy}$$

$$\leqq \int_0^L |Q| H^{-3}\,dy \tag{15}$$

By applying Schwarz's inequality, one gets the mean (absolute) deviation

$$\sigma_1 = \max_{0 \leqq x \leqq L} \langle |\delta p(x)| \rangle$$

$$\leqq \left\{ \int_0^L \langle Q^2(x) \rangle\,dx \int_0^L \langle H^{-6}(x) \rangle\,dx \right\}^{1/2} \tag{16}$$

Similarly, squaring the expression (15), a bound for the mean-square deviation σ_2^2 is obtained

$$\sigma_2^2 = \max_{0 \leqq x \leqq L} \langle |\delta p|^2 \rangle$$

$$\leqq \left\{ \int_0^L \int_0^L \langle Q^2(x) Q^2(y) \rangle\,dx\,dy \int_0^L \int_0^L \langle H^{-3}(x) H^{-3}(y) \rangle\,dx\,dy \right\}^{1/2}. \tag{17}$$

The above bounds can be made explicit by estimating the correlation function $\langle Q^2 \rangle$ and $\langle Q^2(x) Q^2(y) \rangle$ in terms of the roughness correlations of ξ and η. For brevity, only the mean deviation σ_1 will be treated. Let $R_\xi = \langle \xi^2 \rangle$ and $R_\eta = \langle \eta^2 \rangle$. Then, after a sequence of algebraic inequalities, it can be shown that

$$\sigma_1 \leqq \sqrt{2}A \left\{ v^2 \int_0^L R_\eta(x)\,dx + 49 \int_0^L R_\xi(x) \left[\frac{\Delta h(x)}{h(x)} \right]^2 dx \right\}^{1/2},$$

$$\text{with} \quad A \equiv \Lambda \left\{ \int_0^L \langle H^{-6}(x) \rangle\,dx \right\}^{1/2}. \tag{18}$$

Let $|\delta W|$ denote the mean deviation of the load-carrying capacity defined as

$$|\delta W| = \int_0^L \langle |\delta p(x)| \rangle \, dx. \tag{19}$$

Then, noting (26), the unit mean deviation ρ is given by

$$\rho = \left| \frac{\partial W}{W_0} \right| \leq \frac{\sigma_1 L}{\int_0^L p_0 \, dx}. \tag{20}$$

A similar bound for the unit mean-square deviation can be derived in terms of the fourth order correlations. For details and a concrete example, one is referred to [3].

3. RELATED PROBLEMS

The problem of a slider bearing of finite width is governed by the Reynolds equation in two dimensions. This partial differential equation was derived from the Stokes equations in hydrodynamics in three dimensions, based on the main assumption that the film thickness is small [1]. This is an example of many random boundary problems of "thin domain," that is, one component of the domain is small compared with others. In general, the boundary-value problem with an irregular domain is difficult to deal with. However, for a thin domain, the problem can often be reduced to one of regular domain, as exemplified by the Reynolds theory of hydrodynamic lubrication. To illustrate this principle, we consider, symbolically, the boundary-value problem

$$Lu = f \qquad \text{in } D, \tag{21}$$

$$u|_{\partial D} = g, \tag{22}$$

where L is an elliptic operator, such as a Laplacian, D is a random domain of two or higher dimensions with ∂D as its boundary, and f, g are given functions. Suppose that D is decomposable as $D = D_1 \times D_2$, such that the diameter $d(D_1)$ of the random set D_1 is much smaller than a characteristic length of the problem, but the component D_2 is regular. Then, for each realization, one can take a spatial average of (21) over the set D_1 and make use of the divergence, mean-value theorems or others in the integral calculus to incorporate the boundary condition on ∂D_1. This procedure often yields an approximate boundary-value problem in a reduced, regular domain D_2:

$$\tilde{L}(\partial D_1)\tilde{u} = \tilde{f} \qquad \text{in } D_2, \tag{23}$$

$$\tilde{u}|_{\partial D_2} = \tilde{g}, \tag{24}$$

Here $\tilde{L}(\partial D_1)$ is an elliptic or ordinary differential operator with random coefficients resulting from the random part of the boundary ∂D_1.

For example, in the problem of the optical fiber as a wave guide, D_1 is the thin cross section of the fiber with random imperfections, and L designates the reduced wave operator. In this case, the system (23) and (24) constitute a two-point boundary-value problem for a random ordinary differential equation. As another example, the system (21) and (22) may describe the motion of long water waves over the ocean bed of a random topography. Since the wave length is much greater than the water depth, a reduced equation (23) similar to the Reynolds equation in two dimensions can be derived.

The reduced problem (23)–(24) is simpler than the original problem (21)–(22) in the sense that differential equations with random coefficients are better understood, in contrast with a random boundary problem. For instance there are more reliable methods of approximation in solving the random system (23)–(24) [2]. However, applications of such methods to specific problems will not be discussed here.

REFERENCES

1. A. Cameron, "Principles of Lubrication," Longmans, London, 1966.
2. P. L. Chow, Perturbation methods in stochastic wave propagation, *SIAM Rev.* **17** (1975), 57–81.
3. P. L. Chow and E. A. Saibel, On the roughness effect in hydrodynamic lubrication, *J. Lubrication Tech.* **100** (1978), 176–180.

The work of P. L. Chow was supported by the U.S. Army Research Office Grant DAAG29-78-G-0042.

P. L. Chow
DEPARTMENT OF MATHEMATICS
WAYNE STATE UNIVERSITY
DETROIT, MICHIGAN

E. A. Saibel
ENGINEERING SCIENCES DIVISION
U.S. ARMY RESEARCH OFFICE
RESEARCH TRIANGLE PARK, NORTH CAROLINA

Optimization Methods for Solving Nonoptimization Problems

DAVID GALE

1. INTRODUCTION

One of the basic techniques in all of mathematics is the so-called variational method. One wishes to prove the existence of an object with certain properties. One defines some maximum or minimum problem which one knows has a solution and then shows that any solution to this problem must have the desired property. A famous example is Riemann's proof of the existence of a harmonic function with prescribed boundary values. The appropriate optimization problem in this case is to find a function which takes on the given boundary values and minimizes the Dirichlet integral. Variational techniques, however, are at least two thousand years older than this. Recall Euclid's proof, that every pair of nonzero integers a and b has a *greatest common divisor*. That is, there exists a positive integer d which is a common divisor of a and b and such that any other common divisor also divides d. The proof of existence is achieved by setting up the following

Minimum Problems. *Let*

$$\Phi(x, y) = xa + yb, \qquad x, y \in Z.$$

Find the smallest positive value of Φ.

One knows this problem has a solution because every set of positive integers has a smallest member. One then proves that this smallest member is the desired d by a "variational" argument. Namely if d did not divide a or b one could produce (by the Euclidean algorithm) a smaller positive value of Φ.

This example is typical of variational problems. The function Φ does not seem to bear any obvious relation to the original question. Notice that

ISBN 0-12-178150-X

although we seek to find the *greatest* common divisor our method involves finding the *smallest* value of Φ. The more natural approach would be to try the *largest* positive integer which divides both a and b. This turns out to be the right number but this approach does not lead to a proof of its properties.

Modern optimization theory which has been developed essentially over the past 30 years has produced new techniques for finding global optima. Combinatorial optimization has emerged as an important branch of discrete mathematics while the parallel development of convex optimization has extended the classical differentiable theory. It is not surprising that along with these new techniques one has discovered new existence theorems, typically for situations with an economic flavor. In the rest of this paper I will present three such examples. The first is solved by the combinatorial theory, the last two by convex optimization. The first two examples are 20 years old but the last and probably most significant is very recent. In each case the existence problem will be described and then the appropriate maximum problem will be given along with an indication of how to establish the equivalence of the two problems.

2. COMPETITIVE ALLOCATION OF INDIVISIBLE COMMODITIES

The problem is due to Shapley and is treated in [2, pp. 160–162].

There are n goods, say houses, $H_1, \ldots, H_n$ and m buyers $B_1, \ldots, B_m$, $m \leqq n$. Buyer B_i places a value v_{ij} on house H_j. One may imagine, for example, that B_i is a real estate speculator who believes he can realize v_{ij} dollars from selling H_j. It is assumed that each buyer can be assigned at most one house and we are interested in determining how a "free market economy" will allocate the houses among the buyers. The formulation is the following:

Let $p = (p_1, \ldots, p_n)$ be nonnegative *prices* on $H_1, \ldots, H_n$. We say that H_j is *acceptable* to B_i at prices p if

$$v_{ij} - p_j \geqq 0, \tag{2.1}$$

$$v_{ij} - p_j \geqq v_{ik} - p_k \qquad \text{for all} \quad k. \tag{2.2}$$

The condition states that B_i will accept H_j if his *profit* is nonnegative and at least as great as it would be for any other house. Thus, each B_i behaves *competitively* in that he acts to maximize his profit. An assignment of houses to buyers is then called competitive if there exist prices p such that each buyer is assigned a house which is acceptable to him. We now have

Existence Theorem. *For any* value matrix $V = (v_{ij})$ *there exists a competitive assignment.*

The corresponding maximum problem is the classical optimal assignment problem. Let Σ be the set of all one–one mappings (assignments) from $N = \{1, \ldots, n\}$ to $M = \{1, \ldots, m\}$. Define Φ from Σ to R by

$$\Phi(\sigma) = \sum_{j=1}^{n} v_{\sigma(j)j}.$$

Maximum Problem. *Find $\sigma \in \Sigma$ such that $\Phi(\sigma)$ is a maximum.*

Since σ is a finite set the maximum clearly exists. The assertion is that this assignment is in fact competitive. This follows from the duality theorem for the optimal assignment problem which states that there exist nonnegative numbers $q_1, \ldots, q_m, p_1, \ldots, p_n$ such that

$$q_i + p_j \begin{cases} \geqq v_{ij} & \text{for all} \quad i, j \\ = v_{ij} & \text{if} \quad i = \sigma(j). \end{cases} \tag{2.3}$$

One easily verifies that the p_j are the desired competitive prices (also, q_i represents the profit of B_i under the assignment).

In the above example the function Φ has a rather natural interpretation. Namely, $\Phi(\sigma)$ is the total value to the set of buyers of the assignment σ, and the optimal assignment is then the one which achieves the "greatest good for the greatest number." For the next two examples, however, we have not been able to find any natural economic interpretation for the maximum problem.

3. THE PARIMUTUEL PROBLEM

This problem is treated in [1]. The problem is to determine how under the system of parimutuel betting the track's odds on a horse race are determined by the subjective probabilities of the bettors.

We assume there are n horses $H_1, \ldots, H_n$ and m bettors $B_1, \ldots, B_m$. After studying the racing form each B_i arrives at a *subjective probability vector* $p^i = (p_{i1}, \ldots, p_{in})$ where p_{ij} represents the probability in the opinion of B_i that H_j will win the race. It is assumed that for each j there is an i such that p_{ij} is positive (if not one would simply eliminate H_j from consideration since no one would bet on H_j under any circumstances). We suppose B_i is prepared to bet the amount β_i. Let $x^i = (x_{i1}, \ldots, x_{in})$ be B_i's *bet vector* meaning that B_i bets the amount x_{ij} on H_j. The *budget condition* then takes the form

$$\sum_{j=1}^{n} x_{ij} = \beta_i \qquad \text{for all} \quad i. \tag{3.1}$$

The *track's odds* are defined to be a vector $p = (p_1, \ldots, p_n)$ where $0 \leqq p_j \leqq 1$ for all j. The meaning is that a bet of x dollars placed on H_j wins x/p_j dollars if H_j wins the race. We now assume that, given the odds p, each B_i will act so as to maximize his expected winnings. One easily sees that he will therefore bet only on that horse or those horses for which p_{ij}/p_j is a maximum; that is, for all i and j

$$x_{ij} > 0 \quad \text{only if} \quad p_{ij}/p_j \geqq p_{ik}/p_k \quad \text{for all} \quad k. \tag{3.2}$$

Finally, the parimutuel condition requires that the track's odds on each horse be proportioned to the amount bet on it. That is

$$p_j = \lambda \sum_{i=1}^{m} x_{ij} \qquad \text{for some} \quad \lambda > 0 \quad \text{and all} \quad j. \tag{3.3}$$

Existence Problem. *Given any vectors p^i and numbers β_i find vectors p and x^i which satisfy* (3.1)–(3.3).

The appropriate optimization problem turns out to be the following

Maximum Problem. *Find nonnegative n-vectors $(y^1, \ldots, y^n)$ which maximize*

$$\Phi(y^1, \ldots, y^n) = \sum_i \beta_i \log(p^i y^i) \tag{3.4}$$

subject to

$$\sum_i y^i = e^n = (1, 1, \ldots, 1). \tag{3.5}$$

The maximum clearly exists since Φ is continuous and its domain is compact. In fact this is the nicest possible concave programming problem with Φ strictly concave and separable in the y^i. The solution of the existence problem is now obtained by applying the Kuhn–Tucker theorem to this problem. The multipliers p corresponding to the constraints (3.5) turn out to be precisely the desired parimutuel odds and the bet vectors x^i are obtained from the maximizing y^i by setting $x_{ij} = p_j y_{ij}$. As to the interpretation of the function Φ one notes that the vectors y^i represent the winnings of B_i. Namely y_{ij} is the amount of money B_i wins if H_j wins the race, and hence the quantity $p^i y^i$ represents B_i's expected winnings. The function Φ thus represents a weighted sum of the logarithms of the expected winnings, where the weights are the budgets of the bettors. The occurrence of the logarithm here is essential as is seen if one goes through the manipulations for deriving the conditions (3.1)–(3.3).

4. A MODEL OF EQUITABLE INSURANCE

We introduce the model by means of an example. There are two grain farmers one of whom raises wheat, the other rye. In a normal year each earns an income of say \$20,000 from selling his harvest. However, in a very wet year the wheat harvest is cut in half while the rye harvest is unaffected. In a drought year the situation is just the reverse. In such a situation the farmers might decide to enter into an "insurance agreement" where, for example, each would agree to pay the other \$3,000 of their own income in the event of a bad year for the other. Such an arrangement would be considered beneficial by both participants. Suppose, however, it should turn out that, drought years occurred twice as often as wet years. Say, one year in five was dry while only one in 10 was wet. In this case there would be an *expected net* payment of \$300 from the wheat to the rye farmer, which would seem inequitable. There are various ways in which the "balance of expected payments" could be reestablished. For example the wheat farmer might argue that he should only have to pay \$1,500 as against \$3,000 for the rye farmer. A more satisfactory arrangement might be for the rye farmer to pay the wheat farmer \$430 if the year turned out to be normal.

There are clearly an infinity of ways in which the balance of transfers could be established. We are concerned with finding among them one which is "best" in the sense that no other arrangement is preferred by both farmers. It turns out, rather remarkably, that there exists essentially only one such arrangement.

There is an easy and natural way to formalize the preceding to very general situations. Let S be a probability space representing the possible states of the world. We assume there are n individuals $A_1, \ldots, A_n$. The *income* I_i of A_i is a random variable defined on S. Further, each individual has a *utility function* u_i which measures his utility of income as a function of the state of nature. Thus u_i is defined on $S \times R^+$ and $u_i(s, x)$ represents the value of x dollars to A_i when the state of the world is s.

An *insurance scheme* d is a function from S to R^n_+ where $d_i(s)$ is the income distributed to A_i in world state s. The scheme d is called *feasible* if it satisfies

$$\sum_i d_i(s) = \sum_i I_i(s). \tag{4.1}$$

Thus, a feasible scheme is just a redistribution of income among agents. Given a scheme d the *expected utility* of d to A_i is given by

$$U_i(d) = \int_S u_i(s, d_i(s)).$$

Definition 1. The feasible scheme d is called (Pareto) *optimal* if there is no other scheme d' such that

$$U_i(d') > U_i(d) \qquad \text{for all} \quad i. \tag{4.2}$$

Clearly any acceptable scheme should be optimal since otherwise there would be another scheme which everyone would prefer, but, of course, there are infinitely many optimal schemes and among those we seek one which is "fair" to all the A_i.

Definition 2. The scheme d is called *equitable* if it is feasible and

$$\int_S d_i(s) = \int_S I_i(s). \tag{4.3}$$

This is the balance condition expressing the requirement that no agent be asked to subsidize any of the others. Thus, each agent's expected income is the same as it would be if he were to refrain from participating in the insurance scheme.

Our main result now states that under a set of rather standard assumptions there exists essentially only one insurance scheme which is both optimal and equitable. The significant feature here is the uniqueness which shows that if people agree that schemes should be optimal and equitable then a unique scheme is determined. Such situations are rather rare in economics so that results of this type are of considerable interest.

The standard assumptions are boundedness of the I_i, and

The function u_i is integrable in s and twice differentiable in x, and $u_i'(s, x)$ is positive, bounded in s, and decreasing in x. (4.4)

Theorem *There exists an optimal, equitable insurance scheme d which is unique up to changes on sets of measure zero.*

We will now describe the maximum problem whose solution gives the desired scheme.

We define the function ψ_i by

$$\psi_i(s, x) = \int_1^x \log u_i'(s, t)\, dt.$$

One verifies by differentiation that these functions are strictly concave because of the concavity of the u_i. Now define Φ by

$$\Phi(d) = \sum_{i=1}^{m} \int_S \psi_i(s, d_i(s)). \tag{4.5}$$

Notice that Φ is defined on the (infinite dimensional) space of all insurance schemes.

Maximum Problem. *Find a scheme d which maximizes Φ among all equitable schemes.*

The assertion is now that a scheme is fair and optimal if and only if it solves the maximum problem. The proof is somewhat more involved than those for the previous examples and proceeds by stages. (The details in a somewhat different context can be found in [4].) First one must show that the maximum problem actually has a solution. This is not a triviality since Φ is now defined on an infinite dimensional space, and the proof makes use of some elementary functional analysis.

The crucial tool is again the Kuhn–Tucker theorem, but here it is applied to a maximum problem on an infinite dimensional space. This is the only application I know of which actually requires the infinite dimensional version. The constraints in this problem are the n equitability conditions (4.3). For the case of finitely many linear constraints one of the standard proofs [3] of the Kuhn–Tucker theorem carries over so this part of the argument presents no difficulty.

The Kuhn–Tucker theorem reduces the original problem to one of maximizing the Lagrangian

$$\int_S \left(\sum_{i=1}^{m} (\psi_i(s, d_i(s)) - v_i d_i(s) \right) \tag{4.6}$$

over the set of all feasible distributions, where the v_i are the set of Kuhn–Tucker multipliers. Notice that the infinite dimensional constraints have been removed and the problem is now one with infinitely many finite constraints, the feasibility condition (4.1). Clearly the integral (4.6) will be maximized exactly when the integrand is a maximum, so the last part of the proof is concerned with the elementary finite dimensional problem of maximizing for each s

$$\sum_i (\psi_i(s, d_i(s)) - v_i d_i(s)) \tag{4.7}$$

subject to

$$\sum_i d_i(s) = \sum_i I_i(s). \tag{4.1}$$

The simple derivative condition for optimality for the problem asserts the existence of a function g such that

$$\begin{aligned} \psi_i'(s, d_i(s)) - v_i &\leqq g(s), \\ &= g(s) \qquad \text{if} \quad d_i(s) > 0 \end{aligned}$$

for almost all s. The rest of the proof is a matter of juggling inequalities, and of course it makes use of the specific form of the functions ψ_i.

There is one point in which this last example is quite different from the previous two. In the latter the dual variables or Kuhn–Tucker multipliers give the required solution. In this last example on the other hand the multipliers play an essential role in the proof but are not themselves a part of the solution which only involves the "primal variables," the functions d_i.

We should remark that it is possible to prove the existence of equitable schemes using fixed point theorems and one can also give a proof of uniqueness by a combinatorial type of argument. It was this latter discovery that led us to suspect that something was being optimized and eventually to the functions ψ_i. At this time we still have no interpretation of their economic significance.

REFERENCES

1. E. Eisenberg and D. Gale, Consensus of subjective probabilities: The pari-mutuel method, *Annals of Math. Statist.* **30** (1959).
2. D. Gale, "The Theory of Linear Economic Models," McGraw–Hill, New York, 1960.
3. D. Gale, A mathematical theory of optimal economic development, *Bull. Amer. Math. Soc.* **74** (1968), 207–223.
4. D. Gale and J. Sobel, Fair division of a random harvest (to appear).

DEPARTMENT OF INDUSTRIAL ENGINEERING AND OPERATIONS RESEARCH
UNIVERSITY OF CALIFORNIA, BERKELEY
BERKELEY, CALIFORNIA

The Influence of Dick Duffin on an Engineer

CLARENCE ZENER

History is replete with examples of older men having a profound influence upon young men. Today I shall give a converse example, of how a young man completely changed the life of a colleague many years his senior.

At the beginning of the 1960s Dick Duffin and I were both at the Westinghouse Research Laboratories, he as a part time consultant, I as a full time employee. All of you here know that by this time Dick had already established himself as an eminent mathematician with a passionate hobby for formulating real world problems in a form amenable to rigorous mathematical logic. At that time I had become fascinated by the possibility of applying my background in the physical sciences to develop economic processes for tampering with nature on a large scale, such as desalination of sea water, obtaining power from the reversible mixing of fresh river water with the ocean, etc. In such processes the only unknown is the cost per unit production rate. But how is one to estimate this cost for untried systems? The only way that I know is to first settle on a feasible system, then write the cost of the major components of the system in terms of the system parameters. These parameters include the physical dimensions of the component parts, as well as the various fluxes, such as heat flows, material flows, electrical currents, etc. One must then find that combination of parameters which minimizes the cost. The joker is of course that these parameters are related to one another. One must therefore minimize the cost function subject to the constraints which express these interrelations.

Whereas in the early sixties experts in computers could no doubt have solved such problems, my generation has not acquired this facility. I was, however, greatly encouraged by a comment of Dick that one's fingers are frequently smarter than one's mind. Taking this comment seriously, one night I sat down with a fresh pad of paper, a sharp pencil, and a large waste-basket, and started to doodle.

ISBN 0-12-178150-X

The next week I proudly showed Dick a manuscript which completely solved the problem of minimizing a polynomial containing n terms in $n - 1$ variables. Each term in the polynomial represented the cost of a component part of the system. The remarkable feature of the solution was that the minimum value of the polynomial was determined before finding the minimizing values of the variables. Furthermore, at the optimized design the relative contribution of a given term was independent of the value of the various coefficients. The following week Dick proudly showed me a manuscript giving the solution to the generalization of my problem. Rather than a simple minimization, he introduced a minimization subject to constraints. Further, he had developed a methodology for handling the case where the number of terms was arbitrarily greater than the number of variables. During this week he had laid the foundation for what is now called geometric programming. During this one week he had evolved a new and elegant approach to engineering.

This contact with Dick Duffin marked a turning point in my career. Previously I had been a physical scientist interested primarily in how nature works. Dick had given me the intellectual tools to find economically viable ways to induce nature to perform desirable tasks. Henceforth, I became an engineer.

Whereas Elmor Peterson has transformed geometric programming into a tool used by schools of business management, I always think of it as an approach in developing an engineering system to perform a given function at the minimum possible cost. In my discussion of Dick's contribution, I shall therefore confine my remarks to the insight he has given to engineering design.

The first insight I shall give is the relative insensitivity of the ratio of the cost of the various components at optimum design. This insensitivity is absolute in the particular case where the total number of terms is one greater than the number of parameters. As an example, I shall speak of a problem solved by Lord Kelvin more than 100 years ago. Suppose a factory connects a wire between its plant and a power station. Further suppose its cost of transmitting power over this wire has only two components, the cost of the power lost in the wire, and the cost of the wire itself. This power lost is inversely proportional to the wire cross section, the cost of the wire is directly proportional to this cross section. Lord Kelvin showed that when the factory selects that wire cross section which minimizes the total cost—the cost of the power dissipated and the cost of the wire itself, properly amortized—the cost of the two are equal. At first sight this conclusion seems ridiculous. If we used gold wire rather than copper wire, surely the cost of the wire relative to the cost of the power loss would increase. Not so, provided the change from copper to gold is accompanied by a reoptimization of cross section. The cross section of the gold wire will be so small that its power loss will be very high, so high in fact that the cost of its power loss will be identical to the cost of the wire itself.

This principle of insensitivity of the ratio of the various cost components at optimum design can be extended. When constraints are present, the relative values of each term in the constraints are also invariant. This principle has been extremely useful to those of us at CMU working on the design of power plants operating on the temperature gradients of the tropical oceans, including the Gulf of Mexico. We conclude, e.g., that the power required to push the warm and cold water through the heat exchanger is not more than one-third of the gross power output. When we first publicized our concept of designing an economically viable plant to extract power from the ocean, many scientists and engineers ridiculed us with the comment that because of our low efficiency of $\sim 2.5\%$, the power to push the water through the heat exchangers would be greater than the total gross power generated. They were evidently not aware of the general relationships which are so easily seen by geometric programming, but which are completely obscured by computer computation.

The second insight which Dick has given to engineering design is in the area of formulation of constraints. It has been customary to formulate constraints as equality constraints. In the formulation of design as a problem in geometric programming, Dick showed that constraints must be in the form of inequalities rather than equalities. To me this seemed a very odd way to write constraints. It soon became apparent, however, that for real engineering problems inequality rather than equality constraints were the more natural. As an example, I shall discuss the design of a power plant operated by the tropical ocean's thermal gradient. Since the major cost of these plants is the heat exchanger, and since the cost of a heat exchanger is proportional to the heat exchanger area, we choose our design objective to minimize this area. However, the area must be such that the gross power generated is *at least equal* to the specified power output plus the power required to push the water through the heat exchanger pipes. This constraint is easier to satisfy, the larger the temperature drop across the heat exchangers, and the larger the temperature drop across the gas turbine. But the sum of these two temperature drops *cannot exceed* the total available temperature drop in the ocean.

So far I have discussed only how Dick has influenced my professional life. I now turn to how he has influenced my personal life. Because of the constant application of Dick's optimization principles in my professional work, I have acquired a special sensitivity to recognizing the presence of optimization problems. My primary recreation is working with trees on our farm in the Appalachian hills east of Pittsburgh. This pleasure has been greatly enhanced by understanding certain geometrical ratios in leaves of different types of trees, and of explaining deviations from the normal ratio.

This almost constant geometrical ratio is a consequence of nature's strategy to reduce the impedance of sap flow to the various parts of the leaf. As an introduction to this problem I shall describe a manifolding problem we solved for an ocean power plant. Our condensers will consist of banks of vertical tubes through which cold ocean water flows, vertically, and with ammonia vapor to be condensed on the outer surface of the tubes. The vapor is fed into the condenser bank from one side, as illustrated in Fig. 1. Our design objective was to minimize the maximum pressure drop of the ammonia vapor for a given ammonia input flux. Suppose we had settled on a design where the tubes formed a regular square array. The pressure drop across a given gap is

$$c\frac{\Gamma}{w^3},$$

where Γ is the volume flux of vapor across the gap, w is the width of the gap, and c is a numerical constant. The total pressure drop from the front to the back of the condenser is

$$\frac{1}{2}cq\frac{N^2}{w^3}, \qquad N \gg 1. \tag{1}$$

Here q is the volume of vapor condensed in unit time by each unit length of tube, N is the number of tube rows which are crossed by the vapor which reaches the last row. Thus

$$\Gamma = Nq.$$

We now attempt to reduce this pressure drop without changing either the tube bank overall dimensions, or the number of tubes within the bank. The design change we make is to group the bank into subgroups. Each subgroup has the same original number of N rows deep with the original gap width. Within each subgroup the gaps are however eliminated in each of the N rows. We now utilize this space saved by separating adjacent sub-banks by a large gap of width

$$\sigma w,$$

where σ is the number of gaps eliminated within each sub-bank.

In the new design the vapor now flows normal to the bank face in a few wide channels, and flows therefrom through the tube bank in many narrow channels, as illustrated in Fig. 2. The greatest pressure drop is now

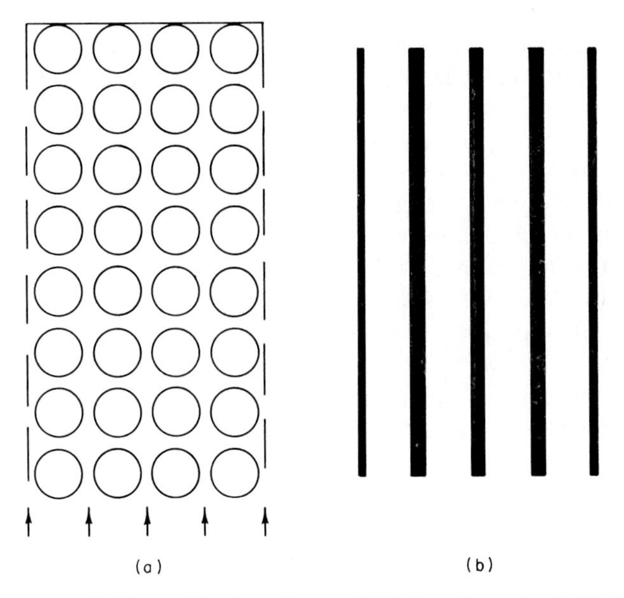

Fig. 1. No manifolding. (a) Section of unmanifolded condenser bank. (b) Flow lanes of vapor in this condenser bank.

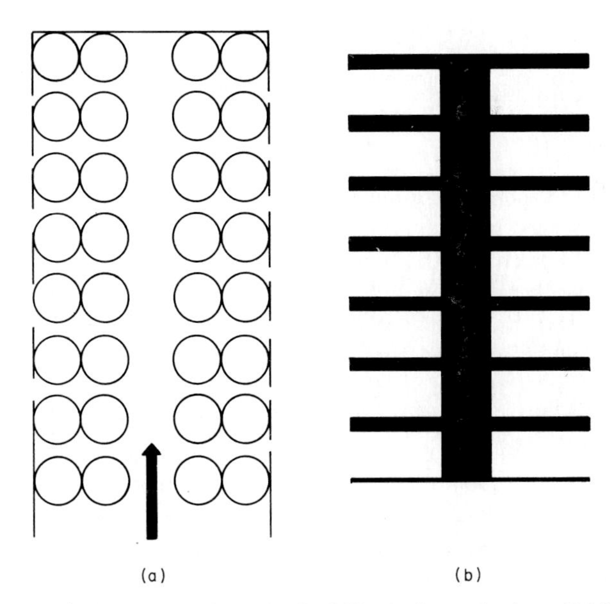

Fig. 2. Manifolding. (a) Condenser bank of Fig. 1 after optimal manifolding with same average tube density. (b) Flow lanes in manifolded condenser fed by one main channel.

the sum of the pressure drops along these two types of channels. To obtain the pressure drop along the main channels we change (1) by replacing

$$q \quad \text{by} \quad \sigma q$$

and

$$w \quad \text{by} \quad \sigma w.$$

To obtain the maximum pressure drop within each tributary channel, we replace in (1)

$$N \quad \text{by} \quad \tfrac{1}{2}\sigma.$$

We now observe that the maximum pressure drop along the main channels is proportional to σ^{-2} and the drop along each tributary is proportional to σ^2. We now choose the previous unspecified number σ such as to minimize the sum of these two pressures. We obtain for the maximum pressure

$$\frac{1}{2} cq \frac{N}{w^3},$$

a value less than for the unmanifolded regular square by the factor $1/N$. Since banks of hundreds of tubes deep are contemplated, a truly tremendous value in the pressure drop is obtained by introducing this single parameter.

In our improved design, each main channel nourishes with vapor a rectangular area having a length L equal to the length of the main channel, and a breadth B equal to

$$B = \sqrt{2SL},$$

where S is the spacing between the tributary channels. Thus the breadth of the region served by each main channel is essentially the geometric mean of the channel length and the tributary spacing.

The next weekend following solving this manifolding problem, I looked upon my trees with astonishment. They too have a manifolding problem. The water which evaporates from every part of their leaves must be supplied with a minimum pressure drop. During their long evolutionary history Darwinian selection had gradually developed an efficient manifolding system, having a minimum pressure drop. Table I gives the ratio of the width of my tree leaves to the theoretical value calculated for heat exchanger manifolding. The bulk of my trees are well behaved, having widths within 20% of the theoretical value.

How can we explain away the unusually smaller ratio of the willows and the walnuts, as well as the unusually large breadth of the tulip poplars. So far I had considered only osmotic pressure as the driving force for the

TABLE I

Tree	$B/\sqrt{2SL}$[a]
Willow	0.5
Walnut	0.8
Butternut	0.9
Ash	1.0
Choke cherry	1.1
Black birch	1.15
Beech	1.2
Cottonwood	1.45
Magnolia	1.45
Tulip poplar	1.5

[a] B, leaf breadth; L, leaf length; S, tributary spacing.

movement of sap through the veins. However, in drooping leaves gravity aids the flow along the main channel, only to a lesser extent along the tributaries. Thus my drooping willows have anomalously narrow leaves.

I have always been able to recognize my tulip poplars from a long distance provided a slight breeze is blowing. Particularly in the early spring the growing leaves stand upright. Only a gentle breeze thus sets the leaves in continual oscillation, thus exposing first the shiny so-called upper surface, then the dull lower surface. Gravity is now acting against osmotic pressure, in contrast to the willows. The leaves are therefore unusually broad.

Evolution aids of course only those species which take advantage of the possibilities of evolution. Thus those trees which have not evolved a manifolding vein system have been crowded out by those faster growing trees which did develop manifolding. Fossil records reveal that before 200,000,000 years ago, trees did not have a manifolding system. Ginko trees forested all land between the North and South Poles. Their leaves consisted solely of a close-packed array of parallel veins. I have none of these on my farm. In fact, no place in the world can they survive the competition from fast growing trees with manifolded systems. They have survived final extinction only by grace of the thousands of years old custom of the Chinese emperors to plant these trees in their courtyards. As you leave the Mellon Institute you can see an avenue of these trees along the street between the Mellon Institute and the University of Pittsburgh. As you look at these spindly ginko trees, you can readily understand why they have become living fossils.

CARNEGIE–MELLON UNIVERSITY
PITTSBURGH, PENNSYLVANIA

Part VI

RELATED AREAS

The Algebra of Multivariate Interpolation

GARRETT BIRKHOFF

1. INTRODUCTION

Over the past 20 years, many papers and several books have been written about so-called *finite element methods* (FEM) and *spline functions*. Most of these have been written from the standpoint of *numerical mathematics*—i.e., of their applications or of approximation theory.

However, they are also interesting from a purely algebraic standpoint, and I will consider them from this standpoint here. I shall try to avoid overlap with Philip Davis' excellent book [9], which deals mainly with schemes of *global* interpolation and approximation. Instead I will concentrate on *piecewise polynomial* interpolation in a general *polyhedral complex*, as defined by Poincaré [14].

Not only will I confine my attention on (piecewise) *interpolation* schemes, as contrasted with more general approximation and extrapolation methods, but I will also consider only interpolation schemes I which are *linear*, in the sense that $I(c_1 f_1 + c_2 f_2) = c_1 I(f_1) + c_2 I(f_2)$. Most (piecewise) polynomial and some rational interpolation schemes are linear, and I will concentrate on these.

1.1. Univariate Interpolation

Multivariate interpolation is most easily understood as a generalization of the more familiar and more highly developed theory of interpolation by functions of *one* (real or complex) variable, and so I shall begin by recalling some facts about this. As Lagrange showed, exactly one polynomial $f(x)$ of degree n or less can be found which assumes given values $y_j = f(x_j)$, $j = 0, \ldots, n$, at $n + 1$ specified points $x_0, \ldots, x_n$. Partial similar results were proved by Cauchy for rational functions of the form (see [18a])

$$f(x) = (a_0 + a_1 x + \cdots + a_l x^l)/(b_0 + b_1 x + \cdots + b_m x^m), \qquad (1.1)$$

ISBN 0-12-178150-X

$l + m + 1 = n$. By letting the x_j coalesce, we get as limiting cases Hermite and (generalized) Padé approximations, in which $n + 1$ equations of the form

$$f^{(k)}(x_j) = y_j^k, \qquad k = 0, \ldots, \kappa(j), \quad j = 1, \ldots, m, \tag{1.2}$$

are given, where $\kappa(1) + \cdots + \kappa(m) = n + 1$.

Intrigued by the possibilities of generalized Hermite interpolation, George D. Birkhoff [7] propounded in 1906 the following question. For what sets of pairs (j, k) of nonnegative integers is the interpolation problem $f^{(k)}(x_j) = y_j^k$ well set, for *any* choice of y_j^k and distinct x_j? Pólya solved this "Hermite–G. D. Birkhoff interpolation problem" for the case of two points (but any number of equations) in 1931 [14a]. The problem then lay dormant until the last decade, during which it has been studied by Atkinson and Sharma, D. Ferguson, Schoenberg, Karlin, G. G. Lorentz, and others; a forthcoming monograph by G. G. Lorentz will survey what is now known about this problem.

In [3, pp. 38–45], I pointed out that this interpolation problem could be generalized in two directions: (a) by replacing polynomial by *piecewise polynomial* functions, and (b) by considering functions of more than one variable. I suggested that it would be interesting to know when and how one could interpolate generalized "splines" of given degree through given values and (partial) derivatives, possessing general "smoothness properties," on the vertices of a general topological complex. I referred there to [16, pp. 185–348] for surveys of what was known about this problem in 1970, with special emphasis on multivariate splines.

2. "IDEAL" INTERPOLATION SCHEMES

In this paper, I shall concentrate on *interpolation* schemes, and especially on "ideal" interpolation schemes for which functions having zero interpolants form an ideal. Accordingly, let F^X be a space of functions with domain X and values in a field F, and let Φ be a set of *functionals* $\varphi: F^X \to F$ on F^X. These data define a *mapping*

$$\alpha: F^X \to F^\Phi \tag{2.1}$$

which assigns to each function $g \in F^X$ a *vector* $\boldsymbol{\varphi}[g] \in F^\Phi$ having the $\varphi[g]$ as components. (*Caution*: $\varphi[g(x)]$ has no meaning for individual $x \in X$.)

A function $I: \Phi \to F^X$ will be called an *interpolation* scheme on Φ when

$$\alpha[I[\Phi]] = \varphi \qquad \text{for all} \quad \varphi \in \Phi, \tag{2.1'}$$

i.e., when I is a *right inverse* of α. This trivially implies that $I\alpha I = I$ and $I\alpha I\alpha = I\alpha$, whence $\mathscr{P} = I\alpha: F^X \to F^X$ is *idempotent*. The simplest case occurs when $\Phi = F^Y$ for some (perhaps finite) set $Y \subset X$; in this case, α is a natural endomorphism of linear algebras,† and $I[\alpha[g]](y) = g(y)$.

Definition. An *ideal* interpolation scheme is a pair consisting of a *linear* mapping (2.1) and a *linear* right inverse I as in (2.1′) for which the set of all functions with $I[\alpha[g]] = 0$ an *ideal* of F^X: $I[\alpha[g]] = 0$ implies $I[\alpha[fg]] = 0$ for all $f \in F^X$. I will call the set of all functions g with $I[\alpha[g]] = 0$ the *kernel* of the given ideal interpolation scheme.

Assumption. From now on, we will assume that α and I are *linear*, as is almost always the case in practice. This assumption makes $\mathscr{P}$ a "projector"—i.e., an idempotent linear mapping $\mathscr{P}: V \to V$, V a linear space. Following de Boor, we can associate with each such projector $\mathscr{P}$ the "approximation" of neglecting the component of f in the null space of $\mathscr{P}$ (the "error"). Hence any ideal interpolation scheme I as defined above is also associated with an "approximation" in the sense of de Boor; $f - \mathscr{P}[f]$ is its "remainder (error)" in his sense.

Example 1. Lagrange polynomial interpolation of degree n through $p(a_0), p(a_1), \ldots, p(a_n)$ assigns as interpolant to each polynomial function $p(x)$ its remainder $\alpha(x)$ in the (Euclidean) polynomial ring $F[x]$ after division by $a(x) = \prod_{j=0}^{n} (x - a_j)$:

$$p(x) = q(x)a(x) + r(x). \tag{2.2}$$

Caution. Note that in number theory $r(x)$ above is called (by algebraists) the "remainder" of $p(x)$ [under division by $a(x)$], whereas in interpolation theory it is $q(x)a(x)$ that is called the "remainder" (by numerical analysts): it is the difference between the "true" function $p(x)$ and its interpolant. Note also that for $F = \mathbb{R}$ (or $F = \mathbb{C}$), $r(x)$ is the polynomial of least degree (i.e., having the lowest order of growth at ∞) among all polynomials $\varphi(x)$ satisfying $\varphi(a_i) = p(a_i)$ for $i = 0, 1, \ldots, n$.

For a given support set $A = \{a_0, a_1, \ldots, a_n\}$, Lagrangian interpolation has the cardinal interpolation functions given by the usual Lagrangian formula

$$\beta_i(x) = [\prod(x - a_j)]/[\prod(a_i - a_j)]: \tag{2.3}$$

if $J_{\mathbf{a}}$ is the principal ideal of multiples of $a(x) = \prod(x - a_j)$, the mappings $p \mapsto p(a_i)$ are the summands (over any field F) of the decomposition of the *commutative ring* $F[x]/J_{\mathbf{a}}$ into its direct summands.

† A linear algebra is a ring which is also a vector space; an endomorphism of linear algebras is a mapping that preserves products and linear combinations. I am assuming that F^X is closed under multiplication as well as linear combination of functions.

Example 2. Hermite interpolation is the limiting case of Lagrange interpolation with multiple ("coalescent") points of interpolation. For given a_j with multiplicity $\kappa(j)$ $(j = 0, \ldots, r)$, it interpolates to any function $f \in \mathscr{C}^{\kappa}(-\infty, \infty)$ $[\kappa = \max \kappa(j)]$ that polynomial $u(x)$ which satisfies $u^{(k)}(x_j) = f^{(k)}(x_j)$ for $j = 0, \ldots, r$ and $k = 0, \ldots, \kappa(j)$. If $f(x) = p(x)$ is a *polynomial*, then

$$p(x) = q(x)m(x) + u(x), \qquad m(x) = \prod_{j=0}^{r} (x - x_j)^{\kappa(j)};$$

i.e., the interpolant $u(x)$ is the "remainder" of $p(x)$ after division by $m(x)$. This shows that any Hermite interpolation scheme is the *ideal* interpolation scheme defined in the polynomial ring $\mathbb{R}[x]$ by the principal polynomial ideal, usually written $(m(x))$, of all multiples of $m(x)$.

A similar results holds (with respect to formal derivatives) over any field of sufficiently large characteristic; moreover in $\mathbb{C}[z]$, it is true that conversely every ideal interpolation scheme is such a scheme of Hermite interpolation.

Example 3. We next consider the G. D. Birkhoff interpolation scheme which assigns to any $f(x) \in \mathscr{C}^2(-\infty, \infty)$ the cubic polynomial $p(x)$ satisfying

$$p(0) = f(0), \qquad p(1) = f(1), \qquad p''(0) = f''(0), \qquad p''(1) = f''(1). \quad (2.4)$$

This scheme assigns to $f(x) = x - 2x^3 + x^4$ the interpolant 0, yet to $xf(x) = x^2(1 - 2x^2 + x^3)$ the nonzero interpolant $2x^3 + x^2 - 3x$. Hence it is *not* an ideal interpolation scheme.

Example 4. Cubic spline interpolation to given $f(x_0), \ldots, f(x_r)$ and $f'(x_0)$, $f'(x_r)$ is "ideal." However, as Example 3 shows, cubic spline interpolation ceases to be ideal if the endpoint conditions $u'(x_0) = f'(x_0)$ and $u'(x_r) = f'(x_r)$ are replaced by the "minimum strain energy" conditions $u''(x_0) = u''(x_r) = 0$.

Example 5. In the space $F[x, y]$ of all polynomial functions in two variables over any field F, linear interpolation to the values of $p(0, 0)$, $p(1, 0)$, $p(0, 1)$ has for its kernel $(xy, x^2 - x, y^2 - y)$.† The three-dimensional subspace of all (linear) interpolants is *complementary* to this ideal, which thus has linear codimension three.

The interpolation scheme defined by this ideal and the complementary subspace of all linear functions $a + bx + cy$ is the simplest finite element method (FEM), a subject to which I shall now turn my attention.

† Although the ideal $(xy, x^2 - x, y^2 - y)$ of all linear combinations $xy\, a(x, y) + (y^2 - x)b(x, y) + (y^2 - y)c(x, y)$ is not principal, one can easily compute the "remainder" of any polynomial by systematically replacing xy by zero, x^n by x, and y^n by y, in any expression containing one of these as a factor.

3. FEM PRECURSORS

In their masterful survey [10] of the FEM, Felippa and Clough (who coined the name "finite element" in 1960) state that "the Ritz method for minimizing the governing energy functional . . . is entirely equivalent to the FEM . . . ," and attribute the recognition of this fact (in 1963–1964) to Melosh, Besseling, and de Veubeke. Actually, in the 45 years between the appearance of Ritz's classic paper [15] and the coining of the name "finite element," there were many other precursors of the FEM.

Thus a notable early application (perhaps the earliest!) of *piecewise linear* multivariate interpolation was to topology. In 1912, Brouwer [8] considered the displacement function $\mathbf{y} = f(\mathbf{x}) - \mathbf{x}$ associated with a general continuous map $f: K \to K$ of the q-dimensional unit cube $[0, 1]^q$ into itself. As Brouwer realized, his techniques apply equally well to any map $g: K \to L \subset V$ of K onto any subset L of any real vector space V.

Brouwer first divided K into n^q congruent subcubes

$$[S_{\mathbf{r}} : a_j = (r_j - 1)h \leqq x_j \leqq r_j h = b_j, \qquad h = 1/n]. \tag{3.1}$$

He then subdivided each of these into $q!$ congruent simplices, each having for vertices $q + 1$ of the 2^q vertices of $S_{\mathbf{r}}$. This can be done most simply by associating with each permutation β of $1, \ldots, q$ the simplex $S_{\mathbf{r},\beta}$ of points whose coordinates satisfy

$$0 \leqq \mathbf{x}_{\beta(1)} - \mathbf{a}_{\beta(1)} \leqq \cdots \leqq \mathbf{x}_{\beta(q)} - \mathbf{a}_{\beta(q)} \leqq h. \tag{3.2}$$

These simplices all have the diagonal $\lambda\mathbf{b} + (1 - \lambda)\mathbf{a}$ $(0 \leqq \lambda \leqq 1)$ of $S_{\mathbf{r}}$ as common edge; the $q + 1$ vertices of $S_{\mathbf{r},\beta}$ are

$$\mathbf{x}_q = \mathbf{b}, \qquad \mathbf{x}_{q-1} = \mathbf{b} - h\boldsymbol{\epsilon}_{\beta(q)}, \ldots, \mathbf{x}_0 = \mathbf{b} - h(\boldsymbol{\epsilon}_{\beta(q)} + \cdots + \boldsymbol{\epsilon}_{\beta(1)}) = \mathbf{a}. \tag{3.3}$$

He then defined an approximating map $g_h(\mathbf{x})$ piecewise, by assigning to each point $\mathbf{x} = \sum_{j=0}^{q} \alpha_j \mathbf{x}_j$ $(0 \leqq \alpha_j \leqq 1, \sum a_j = 1)$ in $S_{\mathbf{r},\beta}$ the image $\mathbf{y} = \sum \alpha_j g(\mathbf{x}_j)$ —i.e., by mapping $S_{\mathbf{r},\beta}$ barycentrically onto the simplex $T_{\mathbf{r},\beta}$ spanned by the $g(\mathbf{x}_j)$.

I will not try to review here the way in which Brouwer applied his piecewise linear (or "barycentric") interpolation scheme to prove (i) the topological invariance of dimension, and (ii) the fact that any continuous map $f: K \to K$ must have a fixpoint. But it seems relevant that effective schemes for *computing* fixpoints of any such f have been developed recently by Scarf and others on the basis of Brouwer's ideas. Also, that Brouwer's mapping g_h has $O(h^2)$ accuracy, in the sense that for any $g \in \mathscr{C}^2(K \to L)$, we have the inequality $|g_h(\mathbf{x}) - g(\mathbf{x})| \leqq Mh^2$ for some finite M independent of $h = 1/n$ or $\mathbf{x}$.

3.1. Courant's Observation

In a well-known paper, Courant [8a, see esp. p.15) proposed the use of "polyhedral functions" as trial functions for the Rayleigh–Ritz method (as substitutes for the global functions proposed by Ritz), and observed that piecewise linear trial functions in the isosceles right triangles formed by slicing squares by parallel diagonals, as in Fig. 1a, give rise to the standard 5-point approximation for the Laplace equation.

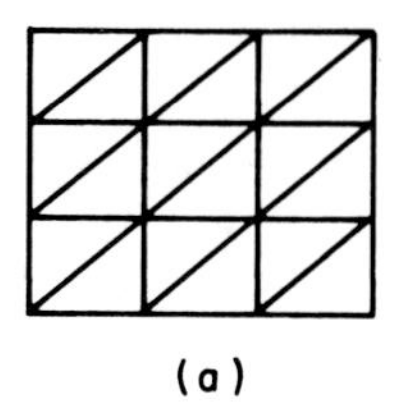
(a)

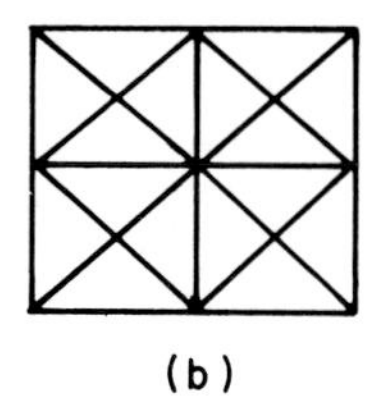
(b)

Fig. 1.

Courant's observation can be generalized to q dimensions, for the following reason. The q-simplices incident on any interior node $\mathbf{a}$ are those contained in the cube of edge length $2h$ and principal diagonal extending from $\mathbf{a} - \mathbf{1}$ to $\mathbf{a} + \mathbf{1}$, where $\mathbf{1} = (1, 1, \ldots, 1)$. There are $(q + 1)(q!)$ of them, since the $S_{\mathbf{r},\beta}$ are bijective to the *isotone graph paths* that contain $\mathbf{a}$ whose nodes increase successively by $\boldsymbol{\epsilon}_{\beta(1)}, \ldots, \boldsymbol{\epsilon}_{\beta(q)}$. Each such path can contain $\mathbf{a}$ in any of $q + 1$ places $0, 1, \ldots, q$.

On the other hand, the Dirichlet integral of a linear interpolant to the nodal values over any such simplex is the simplex volume $h^q/(q!)$ times half the gradient squared, and the latter is the sum $\sum \Delta\varphi_j^2$ of the squares of the jumps in φ over the graph segments. Those jumps which actually depend on $\varphi(\mathbf{a})$ are the $|\varphi(\mathbf{a} \pm \boldsymbol{\epsilon}_j) - \varphi(\mathbf{a})|^2$. The sum of these is clearly minimized (by symmetry) when $\varphi(\mathbf{a})$ is the arithmetic mean of the $2q$ adjacent nodal $\varphi(\mathbf{a} \pm \boldsymbol{\epsilon}_j)$, as was observed by Courant for $q = 2$.

Courant suggested much more: that difference equations (ΔEs) could be derived in general by analogous variational considerations. However, it is now known that analogous calculations for the Poisson and Helmholtz equations do *not* give the usual difference approximations to these DEs [5a].

3.2. Pólya

George Pólya was another pre-1960 FEM user [146]. He pointed out in 1954 that on a *rectangular* mesh, one could use *piecewise bilinear interpolants* to the nodal values at the corners of each mesh rectangle, to construct an approximating subspace of (automatically continuous) functions. By minimizing the Dirichlet integral $D(\varphi, \varphi) = \frac{1}{2} \iint (\nabla\varphi \cdot \nabla\varphi)\, dx\, dy$ on this subspace, one could get approximate solutions to the Dirichlet problem. The same procedure can be followed for many other second-order self-adjoint elliptic problems, because these minimize some quadratic variational integral involving at most first derivatives.

3.3. Prager and Synge

In 1957, Synge published a notable book on *The Hypercircle Method in Mathematical Physics* [17a], which showed how the function-theoretic "hypercircle method," which he and Prager had invented a decade earlier [17b], could be implemented to solve a wide variety of variational problems. He used piecewise linear "pyramidal" functions like those proposed by Courant, but on a more isotropic mesh (see Fig. 1b), to solve numerically several specific torsion problems.

The nodal interpolation formulas used by Pólya and Synge are well known to have $O(h^2)$ accuracy in φ and $O(h)$ accuracy in grad φ. It occurred to me in reviewing Synge's book, that he and Pólya could have achieved *higher order accuracy by using smoother approximating functions*, since elliptic DEs with analytic coefficients are well known to have "smooth" (analytic) solutions. Although this is true, as will become clearer in Sections 4–5, the interesting phenomenon of *superconvergence* makes the nodal values obtained by such primitive methods quite accurate in some cases.

4. HIGHER-ORDER FINITE ELEMENTS

In the early 1960s, ingenious structural engineers such as Clough, Irons, and Argyris developed "finite element" interpolation schemes to nodal values that approximated smooth functions much more accurately than the piecewise linear and bilinear schemes of Courant, Pólya, and Synge. These are described succinctly in [10], and at length in [19]; they involve mostly triangular and quadrilateral "elements."

From an algebraic standpoint, the most interesting triangular elements are *Lagrangian polynomial interpolants of degree* r to nodal values on a *uniformly subdivided* q*-simplex*. By affine similarity, it suffices to consider the *standard* q*-simplex* Σ_q with nodes at $\mathbf{0}, \boldsymbol{\epsilon}_1, \ldots, \boldsymbol{\epsilon}_q$. The mapping

$$y = h\left\{\sum_{j=1}^{q} r_j \boldsymbol{\epsilon}_j + A \sum_{j=1}^{q} x_j \boldsymbol{\epsilon}_{\beta(j)}\right\}, \qquad 0 \leqq x_j \leqq 1, \qquad \sum x_j = 1,$$

where A is the lower triangular matrix of 0's and 1's, with entries $a_{ij} = 0$ if $i < j$ and $a_{ij} = 1$ otherwise, maps Σ_q into the $S_{\mathbf{r},\beta}$ of Section 3.

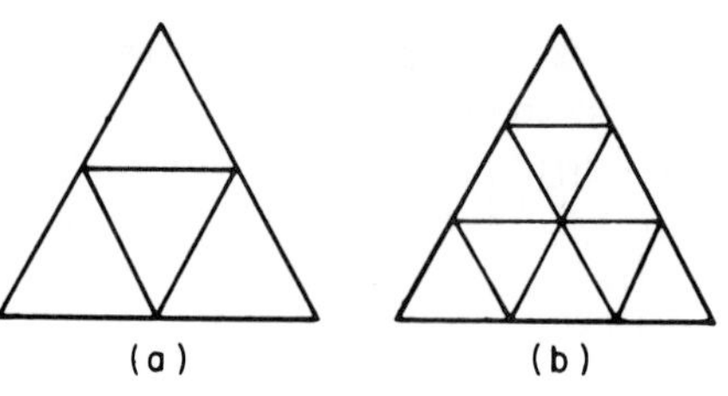

Fig. 2.

Geometrically, the $\frac{1}{2}(q+r)!/(r!)$ nodes $\sum_{j=1}^{q} x_j \boldsymbol{\epsilon}_j$ with $x_j = k_j/r$ ($k_j = 0, 1, \ldots, r$ and $\Sigma\, k_j \leqq r$) subdivide Σ_q into r^q congruent simplices. The cases of quadratic and cubic bivariate polynomials correspond to the subdivisions sketched in Figs. 2a and 2b; see [1a, p. 369] for more details.

The resulting piecewise polynomial function is *continuous* across edges, faces, and other k-cells for any $k = 1, 2, \ldots, q-1$. Moreover, it approximates any function $\varphi(\mathbf{x}) \in \mathscr{C}^{r+1}$ with $O(h^{r+1})$ accuracy, as well as approximating each lth derivative ($0 < l \leqq r$) with $O(h^{r+1-l})$ accuracy.

4.1. Multivariate Splines

Also in the early 1960s, bicubic, biquintic, and other odd degree multivariate spline interpolation schemes were developed at the General Motors Research Laboratories for representing smooth sheet metal surfaces. The principles involved were explained in [11, pp. 152–190]; the key idea (due to de Boor) was to use *tensor products* of univariate (cubic and quintic) spline interpolation schemes, of which cubic splines go back to St. Venant (1840) if not to the Euler–Bernoulli theory of thin beams (1740).

Since 1960, an enormous amount has been learned about spline interpolation and approximation; I shall only extract from the relevant literature a few salient algebraic facts. (For applications, see [16a, 17]. For the underlying theory, see [1] and penetrating books by S. Karlin.).

For any odd integer $2m-1$, and for any partition $x_0 < x_1 < \cdots < x_n$ of $[x_0, x_n]$ and set of values y_j at these x_j there exists a unique function $\varphi(x) \in \mathscr{C}^{2m-2}$ that equals a polynomial function of degree $2m-1$ on each subinterval $[x_{j-1}, x_j]$ *and* has specified derivatives $y_0', \ldots, y_0^{(m-1)}$ and $y_n', \ldots, y_n^{(m-1)}$ at the two endpoints. Finally, the interpolation problem of satisfying these conditions is not only *algebraically* but *analytically* well set, in the sense that small changes in the data have a small effect on the interpolant, and that the approximation error in interpolating to smooth functions is $O(h^{2m})$ [1a].

4.2. B-Spline Basis

The "smooth splines of degree $2m-1$" just characterized have a very convenient and well-conditioned basis of so-called B-splines $\varphi_j(x)$, with support confined to $[x_{j-m}, x_{j+m}]$—i.e., to $2m$ subintervals. In the case of a *uniform* mesh, these B-splines have an elegant algebraic construction: (i) $\varphi_j(x) = \psi_{2m}(x - x_j)$, where $\psi_{2m}(\xi)$ is the *convolution*

$$\overbrace{\psi_1 * \cdots * \psi_1}^{2m \text{ Factors}},$$

ψ_1 being the step function with value $1/h$ on $(-h/2, h/2)$ and vanishing elsewhere (see [9a]).

4.3. Tensor Products

In any rectangularly subdivided rectangle $[x_0, x_n] \times [y_0, y_n]$, the $\varphi_j(x)\tilde{\varphi}_k(y)$, where the φ_j and $\tilde{\varphi}_k$ are defined as above, form a "tensor product basis" of functions for piecewise polynomial interpolation by functions of the piecewise polynomial form $\sum_{j=0}^{2m-1} \sum_{k=0}^{2m-1} x^j y^k$ in each subrectangle, of class $\mathscr{C}^{2m-1,2m-1}$, with accuracy of order $O(h^{2m})$.†

5. INTERPOLATION TO EDGE VALUES

In 1932, Mangeron [12b] developed an ingenious method for interpolating to *edge* ("boundary") values in rectangles, so as to obtain a unique solution to the hyperbolic DE $u_{xxyy} = 0$. His method, which is today called *bilinear blending*, can be described as follows. (For a more recent analysis, see Birkhoff and Gordon [5b].)

By an affine transformation, we can reduce to the case of the unit square $S = [0, 1]^2$, with corners at (0, 0), (1, 0), (0, 1), and (1, 1). Bilinear interpolation to corner values is an "ideal" interpolation scheme having for kernel the ideal $K = (x^2 - x, y^2 - y)$; this is because, by repeatedly substituting $x \equiv x^2 - (x^2 - x) \bmod K$ for x^2, and y for y^2, we can reduce every polynomial $p(x, y)$ to the form $a + bx + cy + dxy$.

Likewise, modulo $K_2 = (y^2 - y)$ alone, $p(x, y)$ is congruent to $p_0(x) + p_1(x)y$ where $p_0(x) = p(x, 0)$ and $p_1(x) = p(x, 1) - p(x, 0)$. Likewise, modulo $K_1 = (x^2 - x)$, $p(x, y)$ is congruent to $p(0, y) + x[p(1, y) - p(0, y)]$. Finally, by the unique factorization theorem for polynomials,

$$K_1 \cap K_2 = ((x^2 - x)(y^2 - y)) = J$$

is the set of polynomial functions vanishing identically on ∂S, the boundary of the unit square.

Bilinear blending and its generalizations to higher dimensions are also associated with an interesting Boolean algebra (hence ring) of projectors, as

† By $f \in \mathscr{C}^{k,l}$, we mean that all derivatives of $\partial^{i+j}\varphi/\partial x^i \partial y^j$ with $i \leq k$ and $j \leq l$ exist and are continuous. For the sharpest results on spline interpolation, see Hall and Meyer [11a].

was first pointed out by W. J. Gordon in a highly original paper (see Schoenberg [16, pp. 223–277]).† Namely, the four projectors defined by K, K_1, K_2, and J *commute* under composition; hence, they form a distributive lattice with respect to composition (product) and the (Boolean) union $\mathscr{P}_i \vee \mathscr{P}_j = \mathscr{P}_i + \mathscr{P}_j - \mathscr{P}_i\mathscr{P}_j = \mathscr{P}_i + \mathscr{P}_j - \mathscr{P}_j\mathscr{P}_i$. And, as was implied above, a polynomial is in the range of the projector $\mathscr{P}_4$ defined by J if and only if it is in the null space of the differential operator $\partial^4/\partial x^2\, \partial y^2$. Moreover, similar results may be derived for "boxes" (products of intervals) in $\mathbb{R}^q$, i.e., in q dimensions, and show that such interpolation schemes have $O(h^{2q})$ order of accuracy in a q-dimensional box. Much of the theory generalizes from $\mathbb{R}$ to any ordered field F.

Example 6. More generally, consider the quadrangle‡ with corners (0, 0), $(a, 0)$, $(0, b)$, and (c, c'). Then modulo the ideal

$$K = (x^2 - \mu xy + ax, y^2 - \lambda xy - by), \qquad \mu = \frac{c - a}{c'}, \qquad \lambda = \frac{c' - b}{c},$$

of all polynomial functions which *vanish* at these vertices, every polynomial $p(x, y)$ is congruent to a bilinear function $a + bx + cy + dxy$, because $x^2 \equiv \mu xy + ax$ and $y^2 \equiv \lambda xy + by \mod K$.

Moreover if J is the principal ideal $(xy(x - \mu y - a)(y - \lambda x + b))$, then we have $x^2y^2 \equiv q(x) \mod J$, where q is a polynomial whose terms are all linear in x or y. Hence every polynomial function $p(x, y)$ is congruent to $a + xp_1(y) + yp_2(x) \mod J$, for suitable polynomials $p_2(x)$ and $p_1(y)$. In particular, bilinear interpolation to corner values and interpolation to edge values by polynomials of the form $a + xp_1(y) + yp_2(x)$ are both "ideal" interpolation schemes.

Unfortunately, the function xy is not *linear* on the edges; hence *piecewise* bilinear interpolation in a general complex whose 2-cells are quadrilaterals does not define a globally continuous function. A scheme for constructing a globally continuous interpolant to edge values in any polygon that has been subdivided into quadrilaterals will be described in the Appendix.

Finally, I call your attention to the ingenious extensions by Wachspress [18] of some of the preceding constructions to *rational* interpolation in arbitrary quadrangles. [Since the vertices (3, 1), (1, 3), (−3, −1), and (−1, −3) of the rectangle $(x + y + 4)(x + y - 4)(x - y + 4)(x - y - 4) = 0$ all satisfy $xy - 3 = 0$, it is evident that bilinear interpolants to corner values cannot be used generally.]

† In this paper, Gordon cites important (and relevant) earlier work of Stancu and Coons.

‡ If Q is any quadrilateral with edges $L_j(x, y) = 0$, all L_j linear ($j = 1, 2, 3, 4$), one can construct analogous ideal interpolation schemes $\mathrm{mod}(L_1L_3, L_2L_4)$ and $(L_1L_2L_3L_4)$.

6. TRIANGLES AND SIMPLICES

Interpolation to edge values in triangles, to face values in tetrahedra, and more generally to values on the "skeleton" of all values on the k-cells of a q-simplex is more subtle. I will begin with the case of triangles, first treated in [1a, pp. 363–386; 2, 4].

The algebra of interpolation in triangles becomes most symmetrical if one uses barycentric (or "areal") coordinates x, y, and $z = 1 - x - y$. The ideal $K = (xy, yz, zx)$ consists of those functions that vanish at all three vertices; it is complementary to the subspace of all linear functions $a + bx + cy + dz$. This shows again (see Example 5) that linear interpolation in a triangle to the corner values is "ideal."

Although the scheme of interpolation to edge values described in [2] requires division, and might *seem* to involve rational functions, it is in fact an "ideal" interpolation scheme carrying polynomials into polynomials, and associated with the principal ideal (xyz) as null space, and the complementary subspace of polynomials satisfying the analog $\partial^3 u/\partial x\, \partial y\, \partial z = 0$ of the Mangeron DE $u_{xxyy} = 0$ as range. To the remainder obtained after subtracting the linear interpolant to corner values, it interpolates half the sum of the linear interpolants along parallels to the three edges to the appropriate pairs of edge values; see Fig. 3. Thus it interpolates $f(x, y, z) = 2xyz$ to the function $f(x, y) = (x + y) - (x^3 + y^3)$.

The interpolation scheme of [2] yields an interesting *algebra* of projectors, including the linear interpolation projector $\mathscr{L}$ to corner values and the three interpolants $\mathscr{P}_i$ $(i = 1, 2, 3)$ along parallels to the ith edge, interpolating linearly between values on the other two sides.

Another interesting algebra of projectors is generated by the three interpolation schemes $\mathscr{R}_i$, which interpolate linearly along straight line segments ("rays") from the value at the ith vertex to values on the opposite side. It is plausible to use $(\mathscr{R}_0 + \mathscr{R}_1 + \mathscr{R}_2)/3$ as an interpolation scheme, as has been suggested by D. H. Thomas and J. Wixom (personal communication). This scheme also has the ideal (xyz) for null space; unfortunately, it interpolates to the *polynomial* function $(x + y) - (x^3 + y^3)$ a *rational*

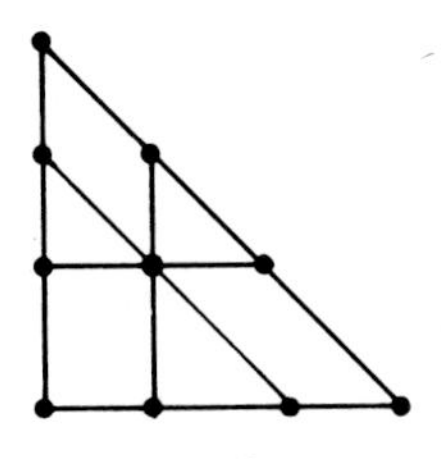

Fig. 3.

function having singularities at the vertices. The scheme is most conveniently analyzed using various projectors, such as the scheme $\mathscr{L}$, which interpolate linearly to corner values.

One can proceed similarly in a q-simplex, using again barycentric coordinates $x_1, \ldots, x_q$ and $x_0 = 1 - x_1 - \cdots - x_q$. The faces are associated with the principal ideals $K_i = (x_i)$; the $(n-2)$-cells with the ideals $K_{ij} = K_i + K_j = (x_i, x_j)$ and finally the vertices with the ideals of the form $P_k = \bigvee_{i \neq k}(x_i) = \sum_{i \neq k}(x_i)$.

The case $n = 3$ of tetrahedra has been treated in full by Mansfield in [13], where some aspects of the noncommutative algebra of projectors are also discussed. More important from the standpoint of finite element methods, she has used this procedure to construct an optimal FEM for interpolating by piecewise polynomial functions of class $C^1(D)$ in a simplicial 3-complex. Such a procedure had been derived previously by Zlámál [20] using other methods (see also Zenisek [18b]).

7. THE DIRICHLET PROBLEM

Interpolation to edge values has an interesting application to the two-dimensional *Dirichlet problem*: the problem of finding the harmonic function assuming specified values on the boundary ∂R of a given region R. One knows that the *exact* solution $\varphi(x, y)$ minimizes the Dirichlet integral

$$D(\varphi, \varphi) = \iint_R (\nabla\varphi \cdot \nabla\varphi)\, dx\, dy, \qquad (7.1)$$

in the space $C \cap H_1$ of all functions continuous in $R \cup \partial R$ for which $D(\varphi, \varphi)$ exists (as a Lebesgue integral) and is finite.

Now suppose that R is a *rectangular polygon* that has been subdivided into rectangles $R_{i,j}$ by a partition π, and that approximate values $u_{i,j} \cong \varphi(x_i, y_j)$ have been computed for the values of the solution at the corners of these rectangles (the "mesh points"), by some high-order difference approximation such as HODIE [12a]. How should one then estimate φ at other points, the gradient field $\nabla\varphi$ anywhere, or $D(\varphi, \varphi)$? To do this well is important, not only because $\nabla\varphi$ and $D(\varphi, \varphi)$ may be of *physical* interest, but also because the smallness of $D(u, u) - D(\varphi, \varphi)$ is a reliable measure of the *accuracy* with which $u(x, y)$ approximates φ.

Scheme 1. Much better than Pólya's use of piecewise bilinear interpolation is the following scheme. We first interpolate *continuously* along *mesh lines* by piecewise cubic polynomials, as follows. Along *boundary* segments, φ and all its derivatives are known. On mesh segments *abutting* on the boundary, therefore, one knows φ and its second normal derivative

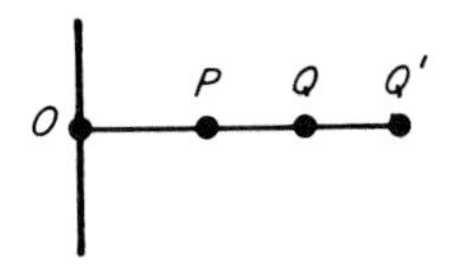

Fig. 4.

where it meets ∂R (at 0 in Fig. 4). Hermite–Birkhoff interpolation determines from these and the values at the next two mesh points (P and Q) a unique *cubic* interpolant on $\overline{OP}$. Finally, along any mesh segment PQ interior to $\overline{OPQQ'}$ in R (see Fig. 4), one can copy Bessel and use the (cubic) Lagrangian interpolant to $\varphi(O)$, $\varphi(P)$, $\varphi(Q)$, and $\varphi(Q')$.

One can then interpolate to the preceding continuous, piecewise cubic values along mesh lines by *bilinear blending* [5] in each rectangle to the edge values, thus obtaining a *continuous, piecewise* bicubic interpolant of a 12-parameter family ascribed to Adini [20, Section 10.4].

Scheme 2. There are seven linearly independent harmonic polynomials of degree three or less: $1, x, y, x^2 - y^2, xy, x^3 - 3xy^2$, and $3x^2y - y^3$. On the other hand, there are 12 linearly independent choices of u, u_x, u_y at the four corners of a rectangle, and each such choice defines a linearly independent continuous function on $\partial R_{i,j}$, cubic on each side. Since the four classes (even–even, even–odd, odd–even, and odd–odd) (see Birkhoff and Garabedian [5c]) of polynomials in x and y are orthogonal in H_0 and in H_1, it would be easy to compute numerically the Dirichlet integral of the *harmonic* interpolant (in $R_{i,j}$) to any continuous function on $\partial R_{i,j}$ which is cubic on each side. This would minimize $D(u, u)$ subject to the interpolated values on mesh lines. I have not yet computed the coefficients of the quadratic function determining $D_{i,j} = \iint_{R_{ij}} \nabla u \cdot \nabla u \, dx\, dy$, even for a square.

8. DISCUSSION

In the case of the boundary values shown in Fig. 5a, the exact harmonic interpolant is $\varphi(x, y) = x^2 - y^2$, and this minimizes $D(u, u)$ in the class of functions $u(x, y) \in C(S) \cap H_1(S)$ assuming the given values on ∂S, S the unit square $[0, 1] \times [0, 1]$.

Surprisingly, the (Pólya) piecewise bilinear interpolant to the specified boundary values at mesh *points* has an even smaller Dirichlet integral! This fact is easily verified as follows. Obviously, $\nabla\varphi \cdot \nabla\varphi = 4(x^2 + y^2)$. Letting $i' = i - \frac{1}{2}$ and $j' = j - \frac{1}{2}$, we get over the $h \times h$ mesh square with center $(i'h, j'h)$

$$\iint_S \nabla\varphi \cdot \nabla\varphi \, dx\, dy = 4h^4(i'^2 + j'^2) + 2h^4/3$$

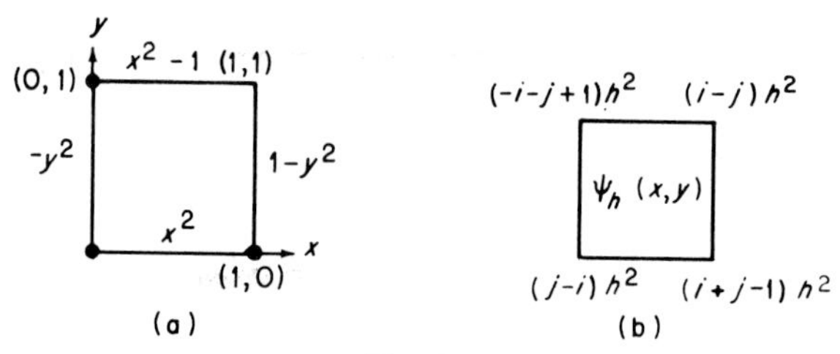

Fig. 5.

as the *exact* value of the Dirichlet integral. This is shown in Fig. 5b. On the other hand, the Dirichlet integral of the bilinear interpolant is $4h^4(i'^2 + j'^2)$, which is a clearly less. Summing over i and j, we see that the Pólya interpolant to $x^2 - y^2$ has a smaller Dirichlet integral than $x^2 - y^2$ itself!

Since the function $x^2 - y^2$ *minimizes* $D(\varphi, \varphi)$ for the given boundary values, this might seem paradoxical. However, it only shows that to flatten the graph of the boundary values between adjacent mesh points reduces $D(u, u)$ more than piecewise bilinear interpolation increases it.

Indeed, one can show easily that for *any* finite set of nodal values on the boundary, an interpolant can be found whose Dirichlet integral is arbitrarily small: the problem of minimizing the Dirichlet integral for given *boundary* nodal values is not well set. We shall now prove this for the unit disk $r = 1$, where the proof is simpler.

In polar coordinates, the Dirichlet integral of

$$u(r, \theta) = a_0/2 + \sum_{k=1}^{\infty} r^j(a_j \cos j\theta + b_j \sin j\theta)$$

is $\sum_{j=1}^{\infty} j(a_j^2 + b_j^2)$. On the other hand, consider $u(1, \theta) = \exp(-\lambda|\theta|)$, $-\pi < \theta \leqq \pi$; every $b_j = 0$, while $a_j = 2/\pi(\lambda^2 + j^2)$. Hence for the harmonic interpolant to $u(1, \theta)$,

$$\sum_{j=1}^{\infty} j(a_j^2 + b_j^2) = (4/\pi^2\lambda^4) \sum_{j=1}^{\infty} j/[1 + (j^2/\lambda^2)]^2. \tag{8.1}$$

This sum is clearly bounded (the integral test) by

$$(4/\pi^2) \int_0^{\infty} dx/(\lambda^2 + x^2)^2 = (4/\pi^2\lambda^3) \int_0^{\infty} d\xi/(\xi^2 + 1)^2, \qquad \xi = x/\lambda. \tag{8.2}$$

The result claimed is now obvious, letting $\lambda \uparrow \infty$.

Likewise, in the unit cube $[0, 1]^3$, functions can be found assuming any finite set of nodal values, whose Dirichlet integral is arbitrarily small. This follows similarly for $u = e^{-\lambda r}$, letting $\lambda \uparrow \infty$, since

$$\int_0^{\infty} (\nabla u \cdot \nabla u)^2 r^2 \, dr = (1/\lambda) \int_0^{\infty} e^{-2\lambda r}(\lambda r)^2 \, d(\lambda r) = 1/4\lambda. \tag{8.3}$$

Hence it is also impossible to minimize the Dirichlet integral in the set of functions assuming given nodal values: the Dirichlet problem is ill posed for values given on a uniform *interior* mesh in $n \geqq 3$ dimensions.

APPENDIX: COMPLEXES OF QUADRILATERALS

The purpose of this Appendix is to extend to *general* polygons decomposed into arbitrary convex quadrilateral cells, the method described in [5] for approximating smooth functions in *rectangular* polygons that have been decomposed into rectangular cells. The key idea is that of *interpolation to edge values*, already discussed in Sections 5 and 6. As in the case of rectangular polygons and quadrilaterals, the method can be applied to higher dimensional domains, such as polyhedra that have been subdivided into hexahedral cells. The inclusion of boundary (and interior) triangles (in three dimensions, triangular prisms, and tetrahedra) does not lead to any serious algorithmic complications, but does result in mild loss of order of accuracy.

Consider for example Fig. 5, which can be imagined as representing a polygonal channel through which water is flowing. If the velocity field peaks sharply near reentrant corners and is mildly irregular near convex corners (where it is nearly stagnant), then the most desirable subdivision might be of the type indicated in the figure. Here the polygonal domain has been subdivided into 69 quadrilaterals.

The first step towards approximating a given function $u(x, y)$ in this subdivided polygon consists in applying *univariate interpolation* to its restriction to mesh lines, subdivided by meshpoints where two or more mesh lines meet. For brevity, we shall refer to this configuration of mesh lines and mesh points at their intersections as the *graph* of the subdivision. In Fig. 6, this graph has 100 nodes (mesh points) and about 165 edges (mesh segments).

The simplest reasonably accurate scheme of univariate interpolation on mesh lines is by piecewise linear interpolation between nodal values; this gives $O(h^2)$ accuracy for functions $f \in C^2$. In Fig. 6, this leads to a 59-parameter

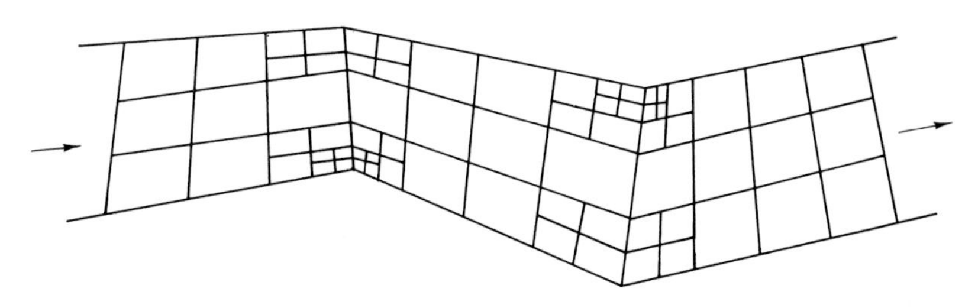

Fig. 6.

family for Dirichlet-type boundary conditions, but a 100-parameter family of approximating functions in general. As was shown for the rectangular case in [5], and is true for the same reasons in the general quadrilateral case under slightly weaker hypotheses, one can however achieve a higher $O(h^3)$ accuracy by blending continuous piecewise quadratic univariate interpolants and $O(h^4)$ accuracy by blending univariate piecewise cubic Hermite or spline interpolants, when $f \in C^4$.

Namely, let P be any polygon, partitioned into convex quadrilateral 2-cells by two families of mesh lines, L_i and M_j, in such a way that:

(i) the boundary of each quadrilateral 2-cell consists of segments of two L_i and two M_j, and
(ii) all four corners of each 2-cell are at intersections $L_i \cap M_j$.

It will *not* be assumed that P itself is convex or even simply connected.

Then, just as in [5], there exist a unique compatible† univariate *piecewise cubic* Hermite and spline interpolants on each mesh line to the appropriate nodal values and derivatives (gradients). Even more simply, there exist *piecewise linear* interpolants to nodal values, given at all meshpoints.

Now parametrize the four edges B_0, B_1 and C_0, C_1 of any quadrilateral 2-cell Q by arc lengths along them (with t increasing in the direction indicated by the arrows in Fig. 7). Then each of the preceding interpolation schemes defines *edge interpolants* $b_0(t)$, $b_1(t)$, and $c_0(t)$, $c_1(t)$ along these edges which

(i) are polynomial functions of t (cubic or linear depending on the interpolation scheme), and
(ii) differ from the given function $u = u(x(t), y(t))$ by univariate *edge errors* $e_j(t)$ satisfying inequalities specified by the following lemma.

Lemma. *If* $u \in C^4(Q)$ *and (univariate) piecewise* cubic *(Hermite or spline) interpolation is used along edges, then there exist universal constants* M_j *such that*

$$|e_j(t)| \leqq M_0 h^4, \qquad e_j'(t) \leqq M_1 h^3, \qquad |e_j''(t)| \leqq M_2 h^2, \qquad \text{(A.1)}$$

where h *is the diameter of* Q. *If* $u \in C^2(Q)$ *and piecewise* linear *interpolation is used, then*

$$|e_j(t)| \leqq M_0 h^2, \qquad |e_j'(t)| \leqq M_1 h. \qquad \text{(A.1')}$$

We now describe a simple *linear blending* technique for passing from compatible edge functions defined on the mesh lines of the partition to

† For the purpose of this note, edge functions will be called "compatible" when they have the same limiting values at corners.

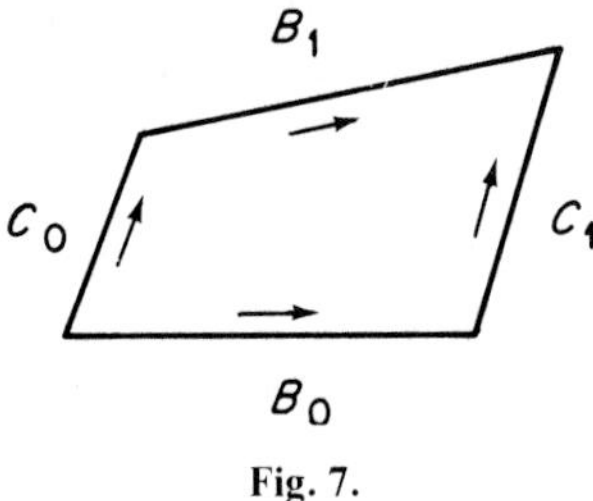

Fig. 7.

functions defined in all of P. If the edge functions are piecewise analytic, then the same will be true of the resulting function on P. This linear blending technique is based on a combination of ordinary local linear blending with the well-known *bilinear parametrization* of an arbitrary convex quadrilateral Q with vertices (x_j, y_j) $(j = 0, 1, 2, 3)$; namely,

$$\begin{aligned} x(r, s) &= (1 - r)(1 - s)x_0 + r(1 - s)x_1 + (1 - r)sx_2 + rsx_3, \\ y(r, s) &= (1 - r)(1 - s)y_0 + r(1 - s)y_1 + (1 - r)sy_2 + rsy_3. \end{aligned} \tag{A.2}$$

Given this parametrization, which maps the unit square $S = [0, 1] \times [0, 1]$ in the (r, s)-plane onto Q, one can interpolate to arbitrary edge values $f_j(r)$ and $g_i(s)$ by the formulas

$$v(r, s) = v_0(r, s) + v_1(r, s), \tag{A.3}$$

where v_0 is the bilinear interpolant to the four corner values, and v_1 blends the remainders $\varphi_j(r) = f_j(r) - v_0(r, j)$ and $\psi_i(s) = g_i(s) - v_0(i, s)$ according to the formula

$$v_1(r, s) = (1 - s)f_0(r) + sf_1(r) + (1 - r)g_0(s) + rg_1(s). \tag{A.3$'$}$$

The rest of this Appendix explains how to use the preceding scheme as a FEM for solving second-order elliptic boundary value problems, and establishes its main properties. Any three vertices of Q can be mapped into (0, 0), (0, 1), and (1, 0), by a unique *affine* transformation. The parametrization (A.2) then reduces to the case

$$x = r - \gamma rs, \qquad y = s - \delta rs, \tag{A.4}$$

making the fourth vertex at

$$\mathbf{x}_3(c, d) = (1 - \gamma, 1 - \delta). \tag{A.4$'$}$$

Moreover, we can always choose the vertices in such a way that $c \leqq 1, d \leqq 1$, $\gamma \geqq 0$, $\delta \geqq 0$, by mapping the intersections of opposite *nonparallel* sides (if any) onto the *positive* x-axis and y-axis.

Unfortunately, the inverse functions $r(x, y)$ and $s(x, y)$ associated with the mapping (A.4) are *not rational*; they involve square roots. However, the Jacobian of (A.4) is a linear function

$$J = x_r y_s - x_s y_r = 1 - \gamma s - \delta r. \tag{A.5}$$

Hence Newton's method is easily applied to solve for $r(x, y)$ and $s(x, y)$, taking $r_0 = x$ and $s_0 = y$ as starting values. And numerical *cubature* with respect to area in the (x, y)-plane is easily carried out parametrically in the (r, s)-plane using the formula

$$\iint \varphi(x, y)\, dx\, dy = \iint (1 - \gamma s - \delta r)\psi(r, s)\, dr\, ds, \tag{A.6}$$

where

$$\psi(r, s) \equiv \varphi(x, (r, s), y(r, s)).$$

Hence it is a straightforward matter to apply Rayleigh–Ritz–Galerkin methods (see [17]) to each 2-cell or "element." And in each cell, the arguments used in [5] apply parametrically to show that *approximation by interpolation* has $O(h^4)$ accuracy is one uses combination of univariate cubic Hermite or spline interpolation along mesh lines and interpolation to edge values in 2-cells.

REFERENCES

1. J. H. Ahlberg, E. N. Nilsen, and J. L. Walsh, "The Theory of Splines and their Applications," Academic Press, New York (1967).

1a. A. K. Aziz, ed., "Mathematical Foundations of the Finite Element Method with Applications to Partial Differential Equations," Academic Press, New York, 1972.

2. R. E. Barnhill, G. Birkhoff, and W. J. Gordon, Smooth interpolation in triangles, *J. Approx. Theory* **8** (1973), 114–128.

3. G. Birkhoff, The role of algebra in computing, *SIAM–AMS Proc. IV* (1971), 1–48.

4. G. Birkhoff, Interpolation to boundary data in triangles, *J. Math. Anal. Appl.* **42** (1973), 474–484.

5. G. Birkhoff, J. C. Cavendish, and W. J. Gordon, Multivariate approximation by univariate interpolants, *Proc. Nat. Acad. Sci. U.S.A.* **71** (1974), 3423–3425.

5a. G. Birkhoff and S. Gulati, Optimal few-point discretizations of linear source problems, *SIAM J. Numer. Anal.* **11** (1974), 700–728.

5b. G. Birkhoff and W. J. Gordon, *J. Approx. Theory* **1** (1968), 199–208.

5c. G. Birkhoff and H. L. Garabedian, *J. Math. Phys.* **39** (1960), 258–268.

6. G. Birkhoff and L. Mansfield, Compatible triangular finite elements, *J. Math. Anal. Appl.* **47** (1974), 31–53.

7. G. D. Birkhoff, General mean-value and remainder theorems, *Trans. Amer. Math. Soc.* **7** (1906), 107–136.

8. L. E. J. Brouwer, Beweis der Invarianz der Dimensionenzahl, *Math. Ann.* **70** (1911), 161–165; "Collected Works," Vol. 2, pp. 430–434.

8a. R. Courant, Variational methods for the solution of problems of equilibrium and vibrations, *Bull. Amer. Math. Soc.* **49** (1943), 1–23.

9. P. J. DAVIS, "Interpolation and Approximation," Ginn (Blaisdell), Boston, Massachusetts, 1963.
9a. C. de Boor, "A Practical Guide to Splines," Springer-Verlag, Berlin and New York, 1978.
10. C. A. FELIPPA AND R. W. CLOUGH, Finite element stiffness methods by different variational principles in elasticity, *SIAM-AMS Proc. II* (1970), 210–252.
11. H. L. GARABEDIAN, ed., "Approximation of Functions," Elsevier, Amsterdam, 1965.
11a. C. M. HALL AND W. W. MEYER, *J. Approx. Theory* **16** (1976), 105–112.
12. L. V. KANTOROVICH AND V. I. KRYLOV, "Approximate Methods of Higher Analysis," Interscience–Noordholf, New York, 1958.
12a. R. E. LYNCH AND J. R. RICE, *Proc. Nat. Acad. Sci. U.S.A.* **75** (1978), 2541–2544.
12b. D. MANGERON, *Rend. Accad. Sci. Fis .Mat. Napoli* **2** (1932), 28–40.
13. L. MANSFIELD, Interpolation to boundary data in tetrahedra, *J. Math. Anal. Appl.* **56** (1976), 137–164.
14. H. POINCARÉ, Analysis situs, *J. Ec. Polyt.* **2** (1895), 1–121; *Oeuvres*, Vol. 6, 193–289.
14a. G. PÓLYA, *Z. Angew. Math. Mech.* **11** (1931), 445–449.
14b. G. PÓLYA, "Studies in Mathematics and Mechanics presented to Richard von Mises," Academic Press, New York, 1954.
15. W. RITZ, Uber eine neue Methode zur Lösung gewisser Variationsprobleme, *J. Reine Angew. Math.* **135** (1908), 1–61.
16. I. J. SCHOENBERG, ed., "Approximation with Special Emphasis on Spline Functions," Academic Press, New York, 1969.
16a. M. H. SCHULTZ, "Spline Analysis," Prentice–Hall, Englewood Cliffs, New Jersey, 1973.
17. G. STRANG AND G. J. FIX, "An Analysis of the Finite Element Method," Prentice–Hall, Englewood Cliffs, New Jersey, 1973.
17a. J. L. SYNGE, "The Hypercircle Method in Mathematical Physics," Cambridge Univ. Press, London and New York, 1957.
17b. J. L. SYNGE AND W. PRAGER, *Quart. Appl. Math.* **5** (1947), 241–269.
18. E. L. WACHSPRESS, "A Rational Finite Element Basis," Academic Press, New York, 1975.
18a. J. L. WALSH, "Interpolation and Approximation by Rational Functions in the Complex Domain," Colloquium Publication, Vol. XX, 5th ed., Amer. Math. Soc., Providence, Rhode Island, 1969.
18b. A. ZENISEK, *J. Approx. Theory* **7** (1933), 334–351.
19. O. C. ZIENKIEWICZ, "The Finite Element Method in Engineering Science," McGraw–Hill, New York, 1967, 1971.
20. M. ZLÁMAL, On the finite element method, *Numer. Math.* **12** (1968), 394–409.

The author wishes to thank Professors Carl de Boor and Lois Mansfield for helpful criticisms. The research reported was partly supported by the U.S. Office of Naval Research.

DEPARTMENT OF MATHEMATICS
HARVARD UNIVERSITY
CAMBRIDGE, MASSACHUSETTS

On the Optimal Shape of a Flexible Cable

BERNARD D. COLEMAN

0. INTRODUCTION

Richard Duffin, Greg Knowles, and I have been working together on problems arising in the optimal design of load bearing bodies. In our first study [1], we considered optimization problems with constraints which can be expressed as linear integral inequalities. In a forthcoming article [2], we shall discuss applications of a general theory of monotonic transformations to problems which involve nonlinear inequality constraints. In the present paper I describe some elementary problems which give rise to an integral inequality which is not linear but can be cast, by an appropriate substitution, into the form of a differential inequality solvable by quadrature.

A thin, flexible, inextensible cable of given length and given uniform mass density, which has one end held fixed and is subject to the force of gravity and an end load whose horizontal and vertical components are specified, is here called *optimal* if its cross-sectional area varies with arc length s in such a way that the mass of the cable is rendered as small as possible subject to the condition that the magnitude S of the tensile stress (per unit of area) acting across each (normal) transverse section not exceed a prescribed safe stress. The problems that are discussed here are equivalent to that of finding the dependence of ζ on s for an optimal cable. It will be shown that such problems can be solved by a method which requires only mathematics available to the early cultivators of rational mechanics.†

In accord with an occasional practice of the early masters of our science, I here take the liberty of observing that the "key" to the present problems

† I refer here to the calculus, as employed by Leibniz, John Bernoulli, and James Bernoulli in their research on the catenary and its generalizations. For a readable and thorough account of the history of the problem of determining the curve assumed by flexible cable in equilibrium, see Truesdell [3].

ISBN 0-12-178150-X

is the fact that *for an optimal cable, S is independent of s*; this fact permits one to show that the curve assumed by an optimal cable in equilibrium is characterized by a linear relation between the x-coordinate and the tangent angle θ at a point on the curve. Indeed, the solution of the problem of finding the shape of an optimal cable is contained in the phrase: *Ut abscissa sic tangentis angulus.*

1. THE DESIGN OF AN OPTIMAL CABLE

Consider a flexible inextensible cable of uniform density ρ and variable cross-sectional area $\zeta(s)$, with s the distance along the cable. Suppose the cable is thin enough to be regarded as a flexible line. In regions where there are no concentrated forces and the only applied load is the action of gravity, the equations of equilibrium take the form

$$\begin{aligned} \frac{d}{ds}(T\cos\theta) &= -F_x = 0, \\ \frac{d}{ds}(T\sin\theta) &= -F_y = \rho\zeta; \end{aligned} \tag{1.1}$$

here F_x and F_y are the x and y components of the applied force per unit length; $T = T(s)$ is the magnitude of the total tensile force $\mathbf{T}(s)$ acting across the transverse cross section at s, and $\theta = \theta(s)$ is the angle which the tangent to the flexible line makes with the horizontal direction. Suppose further that, for each s, the tensile force $\mathbf{T}(s)$ is uniformly distributed over the transverse section at s; the magnitude of the tensile stress per unit of area is then

$$S(s) = T(s)/\zeta(s). \tag{1.2}$$

Let s be measured from a point with Cartesian coordinates $(0, y_0)$ which is the left-hand end of a portion $\mathscr{P}$ of the cable of total length L; thus, at the left end of $\mathscr{P}$

$$x = 0, \qquad y = y_o, \qquad s = 0. \tag{1.3}$$

Let the coordinates of the right-hand end of $\mathscr{P}$ be $(X, Y + y_o)$, with $X > 0$, so that, at the right end,

$$x = X > 0, \qquad y = y_o + Y, \qquad s = L. \tag{1.4}$$

The horizontal and vertical components of $\mathbf{T}$ are

$$T_x = T\cos\theta, \qquad T_y = T\sin\theta, \tag{1.5}$$

and (1.1) yields, for each s in $(0, L)$,

$$T_x(s) = H = \text{const}, \tag{1.6a}$$

$$T_y(s) = V + \rho \int_0^s \zeta(\sigma)\, d\sigma, \tag{1.6b}$$

where

$$V = \lim_{s \to 0+} T_y(s) \tag{1.7}$$

is the total vertical force exerted by $\mathscr{P}$ on whatever is attached to the left end of $\mathscr{P}$. Because

$$T = \sqrt{T_x^2 + T_y^2}, \tag{1.8}$$

(1.2) and (1.6) yield

$$S(s) = \frac{1}{\zeta(s)} \sqrt{H^2 + \left(V + \rho \int_0^s \zeta(\sigma)\, d\sigma\right)^2}. \tag{1.9}$$

Let

$$\theta_o = \lim_{s \to 0+} \theta(s), \qquad \theta_L = \lim_{s \to L-} \theta(s), \tag{1.10}$$

$$T_o = \lim_{s \to 0+} T(s), \qquad T_L = \lim_{s \to L-} T(s). \tag{1.11}$$

By (1.5), (1.6), and (1.7), the "initial inclination" θ_o of $\mathscr{P}$ is given by

$$\theta_o = \tan^{-1}(V/H), \tag{1.12}$$

and, in view of (1.8),

$$T_o = \sqrt{H^2 + V^2}. \tag{1.13}$$

The numbers H and V may be called the "horizontal and vertical components of the applied load." (More precisely, H and V are the x and y components of the force which $\mathscr{P}$ exerts on the object attached to its left end, and thus $-H$ and $-V$ are the x and y components of the force which that object exerts on $\mathscr{P}$.) It follows from (1.5)–(1.8) that

$$\theta_L = \tan^{-1}\left(\frac{M + V}{H}\right), \qquad T_L = \sqrt{H^2 + (M + V)^2}, \tag{1.14}$$

where

$$M = \rho \int_0^L \zeta(s)\, ds \tag{1.15}$$

is the mass of $\mathscr{P}$.

For future use, let

$$B := \sinh^{-1}(V/H). \tag{1.16}$$

By (1.13) and the formula $\cosh^2 z - \sinh^2 z = 1$,

$$B = \cosh^{-1}(T_o/H). \tag{1.17}$$

Suppose that ρ and L are prescribed in advance, and for each $M > 0$ let C_M^+ be the set of all positive continuous functions on $[0, L]$ which obey (1.15).

Experience indicates that the smaller the maximum value attained by the tensile stress S in $\mathscr{P}$, the longer $\mathscr{P}$ will last without breaking. One is therefore led naturally to the following problem:

(I) Given $M > 0$, $V > 0$, and $H > 0$, find the infimum, as ζ varies over C_M^+, of $\sup_{0<s<L} S(s)$, or, equivalently, the infimum of the numbers $\bar{S}$ for which

$$\bar{S} \geqq \sup_{0<s<L} S(s), \tag{1.18}$$

and if there is a function ζ in C_M^+ for which this infimum is attained, find it.

A closely related problem is

(II) Given $V > 0$, $H > 0$, and a specified "safe stress" $\bar{S} > 0$, find the infimum of the numbers M for which there is ζ in C_M^+ such that (1.18) holds; if this infimum is attained, find the function ζ which realizes it.

The solutions to these problems are given by the following proposition.

Proposition 1. *If $\bar{S} > 0$, $M > 0$, $H > 0$, $V > 0$, and ζ in C_M^+ are such that*

$$\bar{S} \geqq S(s) \qquad \text{for all} \quad s \in (0, L), \tag{1.19}$$

then

$$\bar{S} > \frac{\rho L}{\sinh^{-1}((M + V)/H) - B}, \tag{1.20a}$$

i.e.,

$$M > H \sinh\left(\frac{\rho L}{\bar{S}} + B\right) - V, \tag{1.20b}$$

unless

$$\bar{S} = S(s) \qquad \text{for all} \quad s \in (0, L), \tag{1.21}$$

in which case

$$\bar{S} = \frac{\rho L}{\sinh^{-1}((M + V)/H) - B}, \tag{1.22a}$$

i.e.,

$$M = H \sinh\left(\frac{\rho L}{\bar{S}} + B\right) - V, \tag{1.22b}$$

and

$$\bar{\zeta}(s) = \frac{H}{\bar{S}} \cosh\left(\frac{s\rho}{\bar{S}} + B\right). \tag{1.23}$$

Moreover, if (1.22) *holds there is precisely one function* $\bar{\zeta}$ *in* C_M^+ *for which* (1.19) *holds; this function is given by* (1.23) *and for it* (1.19) *reduces to* (1.21).

Proof. In view of (1.9), the relation (1.19) can be written

$$\bar{\zeta}(s)\bar{S} \geqq \sqrt{H^2 + \left(V + \rho \int_0^s \bar{\zeta}(\sigma)\, d\sigma\right)^2}, \tag{1.24}$$

or

$$\varphi'(s)\bar{S}/\rho \geqq \sqrt{H^2 + [V + \varphi(s)]^2}, \tag{1.25}$$

where

$$\varphi(s) := \rho \int_0^s \bar{\zeta}(\sigma)\, d\sigma. \tag{1.26}$$

Clearly, when $\bar{\zeta}$ is in C_M^+,

$$\varphi(L) = M. \tag{1.27}$$

Since $\varphi(0) = 0$, (1.25) yields, for each Z in $[0, L]$,

$$\frac{\bar{S}}{\rho} \int_0^{\varphi(Z)} \frac{d\varphi}{\sqrt{H^2 + [V + \varphi]^2}} \geqq \int_0^Z ds, \tag{1.28}$$

i.e.,

$$\frac{\bar{S}}{\rho}\left[\sinh^{-1}\left(\frac{\varphi(Z) + V}{H}\right) - \sinh^{-1}\left(\frac{V}{H}\right)\right] \geqq Z. \tag{1.29}$$

Moreover, by the continuity of $\bar{\zeta}$ and of the integrands in (1.28), when (1.19) holds the inequality in (1.29) is strict if and only if there is an s in $(0, Z)$ at which the inequality (1.25) is strict, i.e., at which $S(s) < \bar{S}$. On putting $Z = L$

in (1.29) and recalling (1.27), one obtains (1.20), unless (1.21) holds, in which case (1.29) reduces to the equation

$$\frac{\bar{S}}{\rho}\left[\sinh^{-1}\left(\frac{\varphi(Z)+V}{H}\right)-B\right]=Z, \tag{1.30}$$

which yields (1.22) when $Z = L$. On putting $Z = s$ in (1.30) and solving for φ, one finds that

$$\varphi(s)=H\sinh\left(\frac{s\rho}{\bar{S}}+B\right)-V, \tag{1.31}$$

and differentiation of this formula yields (1.23). Now, (1.29) is equivalent to (1.19), and, as has been shown, if equality holds in (1.29) for the largest value of Z [i.e., if (1.22) holds] then (1.19) holds if and only if (1.31) holds for all s [i.e., if and only if ζ is given by (1.23)] in which case (1.19) reduces to (1.21), and this completes the proof. ■

In the case of the problem (I), the proposition tells us that when the mass M of $\mathscr{P}$ and the "applied loads" V and H are specified, the smallest possible value of the upper bound $\bar{S}$ for the tensile stress S in $\mathscr{P}$ is that shown (1.22a); this smallest value of $\bar{S}$ can be achieved in only one way: by letting the cross-sectional area of P vary with s as shown in (1.23), and when this is done, the tensile stress S across each transverse section of $\mathscr{P}$ will equal $\bar{S}$.

The formula (1.23) for the cross-sectional area gives also the solution to the problem (II). For when (1.23) holds, M is given by (1.22b) and this is, by (1.20b), the smallest possible value of M compatible with the prescribed "safe stress" $\bar{S}$ and "applied loads," V and H.

Because the function $\bar{\zeta}$ of (1.23) gives the solutions to both of the problems (I) and (II), a cable (or a portion $\mathscr{P}$ of a cable), for which the cross-sectional area varies with s as shown in (1.23) with B as in (1.17) and $\bar{S}$ as in (1.22a), will here be said to be *optimal for its applied loads.*

Employing the relations (1.13) and (1.14) one can write as follows the formulas (1.21) and (1.22a) for the tensile stress in an optimal cable:

$$S(s)\equiv\bar{S}\equiv\rho L\left[\ln\frac{M+V+T_L}{V+T_o}\right]^{-1}\qquad\text{for all}\quad s\in(0,L). \tag{1.32}$$

2. THE CURVE ASSUMED BY AN OPTIMAL CABLE IN EQUILIBRIUM

Let $\mathscr{P}$ be in equilibrium and subject (at its left-hand end) to an applied load with horizontal and vertical components H and V, let the spatial coordinates of the endpoints of $\mathscr{P}$ be as in (1.3) and (1.4), and suppose that

$\mathscr{P}$ is optimal for H and V. The relation (1.21) tells us that the tensile stress $S(s)$ in $\mathscr{P}$ is equal to the constant $\bar{S}$ of (1.22a). The equilibrium shape of $\mathscr{P}$ is a curve $\mathscr{C}$ which may be described by giving the y-coordinate as a function of the x-coordinate. It is convenient, however, to first calculate x as a function of θ, i.e., $x = x(\theta)$, along $\mathscr{C}$.

In view of (1.2), (1.5), and (1.6), the relation (1.21) can be written

$$\bar{S} = [H \sec \theta(s)]/\bar{\zeta}(s); \tag{2.1}$$

moreover, by (1.5), $T_y = T_x \tan \theta$, and hence (1.6) yields

$$H \tan \theta = V + \rho \int_0^s \bar{\zeta}(\sigma)\, d\sigma; \tag{2.2}$$

therefore,

$$H[\sec^2 \theta(s)] \frac{d}{ds} \theta(s) = \rho\bar{\zeta}(s). \tag{2.3}$$

Clearly, (2.1) and (2.3) imply that

$$\frac{d}{d\theta} s(\theta) = \frac{1}{\rho} \bar{S} \sec \theta, \tag{2.4}$$

and since

$$\frac{dx}{d\theta} = \frac{dx}{ds}\frac{ds}{d\theta} = (\cos \theta) \frac{d}{d\theta} s(\theta), \tag{2.5}$$

we have the following remarkably simple relation for the dependence of x on θ:

$$\frac{d}{d\theta} x(\theta) = \frac{\bar{S}}{\rho}, \tag{2.6}$$

or, in the notation of (1.10) and (1.12),

$$x = x(\theta) = (\theta - \theta_o)\frac{\bar{S}}{\rho} = \frac{\bar{S}}{\rho}\left[\theta - \tan^{-1}\left(\frac{V}{H}\right)\right]. \tag{2.7}$$

To obtain y as a function of x, one need now merely note that, because $dy/dx = \tan \theta$, (2.7) yields

$$\frac{d}{dx} y(x) = \tan\left(\theta_o + \frac{\rho}{\bar{S}} x\right). \tag{2.8}$$

If the location of the origin is chosen so that in (1.3)

$$y_o = \frac{\bar{S}}{\rho} \ln \sec \theta_o, \tag{2.9}$$

or, equivalently, [by (1.5) and (1.11)],

$$y_o = \frac{\bar{S}}{\rho} \ln \frac{H}{T_o}, \tag{2.10}$$

with T_o as in (1.13), then (2.8) implies that

$$y = \frac{\bar{S}}{\rho} \ln \sec\left(\theta_o + \frac{\rho}{\bar{S}} x\right). \tag{2.11}$$

This proves

Proposition 2. *A cable $\mathscr{P}$ which is optimal for its applied loads assumes in equilibrium a curve $\mathscr{C}$ for which the angle θ of the tangent is a linear function of the abscissa x:*

$$\theta(x) = \frac{\rho}{\bar{S}} x + \tan^{-1}\left(\frac{V}{H}\right); \tag{2.12}$$

here ρ is the density, V is the vertical component and H the horizontal component of the load applied to the left-hand end $(0, y_o)$ of $\mathscr{P}$, and $\bar{S}$ is the uniform tensile stress in $\mathscr{P}$. In terms of the Cartesian coordinates $\tilde{x}$, $\tilde{y}$ defined by $\tilde{x} = \rho x/\bar{S}$, $\tilde{y} = \rho y/\bar{S}$, the curve $\mathscr{C}$ has the following description:

$$\tilde{y} = \ln \sec(\tilde{x} + \tan^{-1}(H/V)). \tag{2.13}$$

Note. Because

$$\frac{dy}{d\theta} = \frac{dy}{ds}\frac{ds}{d\theta} = (\sin\theta)\frac{d}{d\theta} s(\theta), \tag{2.14}$$

(2.4) yields not only (2.6) but also the following simple relation:

$$\frac{d}{d\theta} y(\theta) = \frac{\bar{S}}{\rho} \tan\theta. \tag{2.15}$$

It follows from (2.15) or from (2.7) and (2.11) that when the location of the origin is chosen so that (2.9) holds,

$$y = y(\theta) = \frac{\bar{S}}{\rho} \ln \sec\theta. \tag{2.16}$$

The equations (2.7) and (2.16) give a parametric representation $\theta \mapsto (x, y)$ of the curve $\mathscr{C}$ which is valid and complete, even if one ignores the geometric meaning of θ.

From (2.16), (1.5), and (1.6a) one may read off the following interesting relation between the magnitude T of the tension at a point in $\mathscr{P}$ and the elevation y of that point:

$$T = T(y) = He^{\rho y/\bar{S}}. \tag{2.17}$$

The equation (2.16) yields also the following formulas for Y in (1.4):

$$Y = \frac{\bar{S}}{\rho} \ln \frac{\sec \theta_L}{\sec \theta_o} = \frac{\bar{S}}{\rho} \ln \frac{T_L}{T_o}. \tag{2.18}$$

By integrating (2.4) one may easily obtain the following formulas for the arc length s from the left end of $\mathscr{P}$ to the place where the tangent angle is θ:

$$s = \frac{\bar{S}}{\rho} \ln \frac{\tan \frac{1}{2}(\theta + \frac{1}{2}\pi)}{\tan \frac{1}{2}(\theta_o + \frac{1}{2}\pi)}. \tag{2.19}$$

In particular,

$$L = \frac{\bar{S}}{\rho} \ln \frac{\tan \frac{1}{2}(\theta_L + \frac{1}{2}\pi)}{\tan \frac{1}{2}(\theta_o + \frac{1}{2}\pi)}. \tag{2.20}$$

It is clear that when L is finite and $-\frac{1}{2}\pi < \theta_o < \frac{1}{2}\pi$, the function $\theta = \theta(s)$ cannot achieve the value $\frac{1}{2}\pi$. Thus, in view of (1.12), (1.14), and (2.18), Proposition 2 has the following corollary:

Proposition 3. *The span X of a cable $\mathscr{P}$ which is optimal for its applied load obeys the formula*

$$X = \frac{\bar{S}}{\rho}(\theta_L - \theta_o) = \frac{\bar{S}}{\rho}\left[\tan^{-1}\left(\frac{M + V}{H}\right) - \tan^{-1}\left(\frac{V}{H}\right)\right] \tag{2.21}$$

and is always less than $\pi\bar{S}/\rho$. The ratio T_L/T_o of the magnitudes of the forces $\mathbf{T}_L$ and $\mathbf{T}_o$ applied to the ends of $\mathscr{P}$ is determined as follows by the mass density ρ of $\mathscr{P}$, the uniform tensile stress $\bar{S}$ in $\mathscr{P}$, and the difference Y in the elevation of the ends of $\mathscr{P}$:

$$T_L/T_o = e^{\rho Y/S}. \tag{2.22}$$

REFERENCES

1. B. D. Coleman, R. J. Duffin, and G. Knowles, A class of optimal design problems with linear inequality constraints, *Istit. Lombardo Accad. Sci. Lett. Rend. A* **112** (1978).
2. B. D. Coleman, R. J. Duffin, and G. Knowles, Theory of monotonic transformations applied to optimal design problems, *Arch. Rational Mech. Anal.* (1979), in press.
3. C. Truesdell, "The Rational Mechanics of Flexible or Elastic Bodies 1638–1788. Introduction to Leonhardi Euleri Opera Omnia, Seriei Secundae." Societatis Scientiarium Naturalium Helveticae, Vol. X et XI, Orell Füssli, Zürich, 1960.

This research was supported by the U.S. National Science Foundation.

DEPARTMENT OF MATHEMATICS
CARNEGIE–MELLON UNIVERSITY
PITTSBURGH, PENNSYLVANIA

Theory and Applications of Mixed Finite Element Methods

G. J. FIX

M. D. GUNZBURGER

R. A. NICOLAIDES

1. INTRODUCTION

Mixed finite element schemes are finite element approximations based on stationary variational principles as contrasted with those which yield strict maxima or minima. There are two different contexts where stationary schemes have been found useful in practice. The first is typified by the Poisson equation where one uses a stationary principle such as Kelvin's principle [14] in order to make Dirichlet boundary conditions natural or to get higher accuracy in the gradients of the solution [2, 3]. In this context the choice of a stationary principle is based purely on practical considerations, for there also exist principles yielding minima as solutions. The Dirichlet principle [14] is such an example for the Poisson equation. In the second class of problems, on the other hand, all of the physically natural principles are intrinsically stationary. The principal examples are the Stokes equations and Maxwell's equations.

The fact that a particular variational method is stationary has serious implications for finite element approximations. For example, it is a classical result that finite element approximations based on the Dirichlet principle will be unconditionally stable in a suitable sense and convergence depends only on the ability to approximate in the finite element spaces [15]. This is not true for the Kelvin principle and to obtain convergence and stability certain conditions must be satisfied which restrict the types of grids that can be used [2, 3]. The goal of this paper is to develop a general theory which contains

ISBN 0-12-178150-X

both necessary and sufficient conditions for convergence and stability of the mixed methods.

A great deal has been written about mixed methods including some fundamentally important papers by Brezzi [1] and others in the French school. Nevertheless, the existing theories have not fully explained the actual performance of these methods especially as regards rates of convergence. A primary goal of the theory in this paper is to close that gap.

2. EXAMPLES

Mixed methods can be viewed as abstract Galerkin methods applied to operator equations of the following form:

$$\begin{pmatrix} A & D^* \\ D & 0 \end{pmatrix} \begin{pmatrix} u \\ \varphi \end{pmatrix} = \begin{pmatrix} f \\ g \end{pmatrix}. \tag{1}$$

The Galerkin method is defined by selecting spaces $\mathscr{V}_0$, $\mathscr{S}_0$ with inner products $(\cdot, \cdot)$, $\langle \cdot, \cdot \rangle$. The meaning of (1) is that $f \in \mathscr{V}_0$, $g \in \mathscr{S}_0$ are given and we seek $u \in \mathscr{V}_0$, $\varphi \in \mathscr{S}_0$ for which

$$(Au, v) + \langle \varphi, Dv \rangle = (f, v), \tag{2}$$

$$\langle Du, \psi \rangle = \langle g, \psi \rangle \tag{3}$$

for all $v \in \mathscr{V}_0$ and $\psi \in \mathscr{S}_0$. The mixed method consists of selecting finite dimensional finite element spaces $\mathscr{V}_h \subseteq \mathscr{V}_0$, $\mathscr{S}_h \subseteq \mathscr{S}_0$ and seeking $u_h \in \mathscr{V}_h$, $\varphi_h \in \mathscr{S}_h$ such that (2)–(3) holds (with φ_h, u_h replacing φ, u) for all $\psi \in \mathscr{S}_h$, $v \in \mathscr{V}_h$. Once a basis is chosen for the latter, (2)–(3) reduces to a system of algebraic equations.

Example 1. Consider the Poisson equation where

$$\Delta\varphi = g \quad \text{in } \Omega, \tag{4}$$

$$\varphi = 0 \quad \text{on } \Gamma_D, \tag{5}$$

$$\operatorname{grad} \varphi \cdot \mathbf{v} = 0 \quad \text{on } \Gamma_N. \tag{6}$$

Ω is a bounded open region in $\mathbb{R}^n$ having $\Gamma_D \cup \Gamma_N$ as its boundary and $\mathbf{v}$ as its outer normal. The function g is assumed known and in $L^2(\Omega)$.† The Dirichlet principle [14] characterizes the solution as the strict minimum of a convex functional, and this has been the basis for most finite element approximations [15]. The mixed principle, however, is based on Kelvin's

† Throughout this paper $H^r(\Omega)$ stands for the rth Sobolev space with norm $\|\cdot\|_r$ (cf. [11], but we shall put $L^2(\Omega) = H^0(\Omega)$ [9, 11]).

principle [14] which in some sense is the dual of the Dirichlet principle [2, 3]. It works with the gradient $\mathbf{u} = \operatorname{grad} \varphi$ and the spaces

$$\mathscr{V} \equiv \mathbf{H}^1(\Omega), \qquad \mathscr{V}_0 \equiv \{\mathbf{v} \in \mathscr{V} : \mathbf{v} \cdot \mathbf{v} = 0 \quad \text{on} \quad \Gamma_N\}. \tag{7}$$

It asserts that $\mathbf{u} \in \mathscr{V}_0$ minimizes

$$\int_\Omega \mathbf{v} \cdot \mathbf{v}$$

for all $v \in \mathscr{V}_0$ for which $\operatorname{div} \mathbf{v} = g$. The scalar φ enters as a Lagrange multiplier for the constraint $\operatorname{div} \mathbf{v} = g$. In particular, letting

$$\mathscr{S}_0 \equiv \operatorname{div}(\mathscr{V}_0), \tag{8}$$

an equivalent statement of Kelvin's principle is that $\mathbf{u} \in \mathscr{V}_0$, $\varphi \in \mathscr{S}_0$ is uniquely determined by requiring

$$\int_\Omega \{\mathbf{u} \cdot \mathbf{v} + \varphi \operatorname{div} \mathbf{v}\} = 0, \tag{9}$$

$$\int_\Omega \psi \operatorname{div} \mathbf{u} = \int_\Omega g\psi \tag{10}$$

to hold for all $v \in \mathscr{V}_0$ and $\psi \in \mathscr{S}_0$. Letting $(\cdot, \cdot)$ and $\langle \cdot, \cdot \rangle$ be L^2 inner products (9)–(10) reduces to (2)–(3) (with $f = 0$). The operator A in this case is the identity and

$$D\mathbf{v} = \operatorname{div} \mathbf{v}. \tag{11}$$

Example 2. The second example is provided by the Stokes equations representing an incompressible flow with velocity $\mathbf{u}$ and pressure p linearized about a flow with velocity $\mathbf{U}$ with $\operatorname{div}(\mathbf{U}) = 0$. The equations of motion [16] are

$$-\Delta \mathbf{u} + (\mathbf{U} \cdot \operatorname{grad})\mathbf{u} + \operatorname{grad} p = f \quad \text{in} \quad \Omega, \tag{12}$$

$$\operatorname{div} \mathbf{u} = 0 \quad \text{in} \quad \Omega, \tag{13}$$

$$\mathbf{u} = \mathbf{0} \quad \text{on} \quad \Gamma. \tag{14}$$

To define the mixed variational formulation we introduce the space

$$\mathscr{V}_0 \equiv \{\mathbf{v} \in \mathscr{V} : \mathbf{v} = \mathbf{0} \quad \text{on} \quad \Gamma\} = \mathbf{H}_0^1(\Omega) \tag{15}$$

[with $\mathscr{V}$ defined by (7)]. We view $\operatorname{div} \mathbf{u} = 0$ as a constraint on the momentum equation, and hence view the pressure as a Lagrange multiplier [16]. This leads to the choice (8) for $\mathscr{S}_0$ where $\mathscr{V}_0$ is now given by (15). The variational principle is obtained by dotting (12) with a generic $\mathbf{v} \in \mathscr{V}_0$ and integrating by parts. This gives

$$\int_\Omega \left\{ \sum_j \operatorname{grad} u_j \cdot \operatorname{grad} v_j + (\mathbf{U} \cdot \operatorname{grad})\mathbf{u} \cdot \mathbf{v} - p \operatorname{div} \mathbf{v} \right\} = \int_\Omega \mathbf{f} \cdot \mathbf{v}. \quad (16a)$$

Similarly

$$\int_\Omega \operatorname{div} \mathbf{u}\, \psi = 0, \quad (16b)$$

and (16a)–(16b) is to hold for all $\mathbf{v} \in \mathscr{V}_0$ and $\psi \in \mathscr{S}_0$. Observe that this has the standard form (2)–(3) if we let $(\cdot, \cdot)$ and $\langle \cdot, \cdot \rangle$ be L^2 inner products with

$$(A\mathbf{v}, \mathbf{w}) = \int_\Omega \left\{ \sum_j \operatorname{grad} v_j \operatorname{grad} w_j + (\mathbf{U} \cdot \operatorname{grad})\mathbf{v} \cdot \mathbf{w} \right\}. \quad (17)$$

Again $D\mathbf{v} = \operatorname{div} \mathbf{v}$.

It should be observed that in the partial differential equation (PDE) (12) the pressure is determined only up to an additive constant. In the variational formulation, however, the use of the spaces $\mathscr{V}_0$, $\mathscr{S}_0$ gives the implicit normalization.

$$\int_\Omega q = 0 \quad \text{if} \quad q \in \mathscr{S}_0, \quad (18)$$

since any such q is equal to $\operatorname{div} \mathbf{v}$ for some $\mathbf{v}$ vanishing on Γ.

Example 3. We consider the eddy current problem arising from Maxwell's equations [10, 12]. Here we have a current $\mathbf{u}$ satisfying

$$\operatorname{div} \mathbf{u} = 0 \quad \text{in } \Omega, \quad (19)$$

and a potential φ for which

$$A(\mathbf{u}) + \operatorname{grad} \varphi = 0 \quad \text{in } \Omega, \quad (20)$$

where

$$A[\mathbf{u}](\mathbf{x}) = \frac{1}{\sigma} \mathbf{u}(\mathbf{x}) + \frac{i\omega\mu}{C^2} \int_\Omega \frac{\mathbf{u}(\mathbf{y})\, dy}{|\mathbf{y} - \mathbf{x}|} \quad (21)$$

and σ, ω, μ, C are positive constants.

The boundary conditions take the form

$$\mathbf{u} \cdot \mathbf{v} = u_N \quad \text{on} \quad \Gamma_N, \quad (22)$$

$$\varphi = \varphi_D \quad \text{on} \quad \Gamma_D, \quad (23)$$

where $\Gamma = \Gamma_D \cup \Gamma_N$ is the boundary of Ω. The variational formulation of this problem is similar to the Stokes equations except for the inhomogeneous boundary data. In particular, let

$$\mathscr{V}_0 = \{\mathbf{v} \in \mathscr{V} : \mathbf{v} \cdot \mathbf{v} = 0 \quad \text{on} \quad \Gamma_N\}, \qquad \mathscr{S}_0 = \operatorname{div}(\mathscr{V}_0),$$

and let $\mathbf{u}_N$ be some function in $\mathscr{V}$ for which $\mathbf{u}_N \cdot \mathbf{v} = u_N$ on Γ_N.

We seek a pair $\mathbf{u}$, φ with $\mathbf{u} - \mathbf{u}_N \in \mathscr{V}_0$ and $\varphi \in \mathscr{S}_0$ for which

$$\int_\Omega \{A(\mathbf{u}) \cdot \mathbf{v} - \varphi \operatorname{div} \mathbf{v}\} = -\int_{\Gamma_D} \varphi_D \mathbf{v} \cdot \mathbf{v} \tag{24}$$

and

$$\int_\Omega \psi \operatorname{div} \mathbf{u} = 0 \tag{25}$$

for all $\mathbf{v} \in \mathscr{V}_0$ and $\psi \in \mathscr{S}_0$. Letting $(\cdot, \cdot)$, $\langle \cdot, \cdot \rangle$ be the L^2 inner products and $D = \operatorname{div}$ we have the standard form (2)–(3).

Example 4. Our final example is drawn from acoustics where for simplicity we assume a homogeneous uniform medium with a periodic disturbance of frequency ω [4]. Letting p denote the pressure, $\hat{\mathbf{u}}$ the velocity, and $\mathbf{u} = i\omega\hat{\mathbf{u}}$, the equations of motion are

$$\mathbf{u} + \operatorname{grad} p = \mathbf{f} \quad \text{in} \quad \Omega, \tag{26}$$

$$\operatorname{div} \mathbf{u} - \omega^2 p = g \quad \text{in} \quad \Omega \tag{27}$$

with

$$p = 0 \quad \text{on} \quad \Gamma_D, \tag{28}$$

$$\mathbf{u} \cdot \mathbf{v} = 0 \quad \text{on} \quad \Gamma_N. \tag{29}$$

Here the spaces are

$$\mathscr{V}_0 = \{\mathbf{v} \in \mathscr{V} : \mathbf{u} \cdot \mathbf{v} = 0 \quad \text{on} \quad \Gamma_N\}, \qquad \mathscr{S}_0 = \operatorname{div}(\mathscr{V}_0), \tag{30}$$

and the variational formulation is

$$\int_\Omega \{\mathbf{u} \cdot \mathbf{v} - p \operatorname{div} \mathbf{v}\} = \int_\Omega f \cdot \mathbf{v}, \tag{31}$$

$$\int_\Omega \{\psi \operatorname{div} \mathbf{u} - \omega^2 p\psi\} = \int_\Omega g\psi. \tag{32}$$

Observe that this problem has the slightly different form

$$\begin{bmatrix} I & D^* \\ D & C \end{bmatrix} \begin{bmatrix} \mathbf{u} \\ p \end{bmatrix} = \begin{bmatrix} \mathbf{f} \\ g \end{bmatrix} \tag{33}$$

in that the (2, 2) position is nonzero. This, surprisingly, introduces nontrivial complications into the general theory and we shall treat (33) separately in Section 4.

3. BASIC THEORY

Let $\mathscr{V}_0 \subseteq \mathscr{V}$, $\mathscr{S}_0$ be linear spaces, and let $\|\cdot\|$, $|\cdot|$ be norms on $\mathscr{V}_0$, $\mathscr{S}_0$ with $(\cdot, \cdot)$, $\langle \cdot, \cdot \rangle$ being inner products. We do not assume these spaces are complete, and in fact they are not in some of the examples of the previous section.

Let $A: \mathscr{V}_0 \to \mathscr{V}_0$, $D: \mathscr{V}_0 \to \mathscr{S}_0$ be linear operators defined on all of $\mathscr{V}_0$ with

$$\mathscr{S}_0 \equiv D[\mathscr{V}_0]. \tag{1}$$

Our basic problem can be stated as follows. Given $f \in \mathscr{V}_0$, $g \in \mathscr{S}_0$ find $u \in \mathscr{V}_0$, $\varphi \in \mathscr{S}_0$ for which

$$(Au, v) + \langle \varphi, Dv \rangle = (f, v), \tag{2}$$

$$\langle Du, \psi \rangle = \langle g, \psi \rangle \tag{3}$$

hold for all $v \in \mathscr{V}_0$ and $\psi \in \mathscr{S}_0$.

Our theory will also require that we work with a "transpose" of D which in our examples reduces to the gradient operator. In particular, we take $\mathscr{S} \subseteq \mathscr{S}_0$ and

$$G: \mathscr{S} \to \mathscr{V} \tag{4}$$

with

$$(G\psi, v) = \langle \psi, Dv \rangle \qquad \text{all} \quad \psi \in \mathscr{S}, \quad v \in \mathscr{V}_0. \tag{5}$$

For Examples 1, 3, and 4 the space $\mathscr{S}$ is given by

$$\mathscr{S} = \{\psi \in H^1(\Omega) : \psi = 0 \qquad \text{on} \quad \Gamma_D\} \cap \mathscr{S}_0,$$

while in the Stokes equations we can use $\mathscr{S} = H^2(\Omega) \cap \mathscr{S}_0$ without boundary conditions. In all cases $G\psi = -\operatorname{grad} \psi$.

A final ingredient of our theory is a "dual norm" which reduces to the norm on $H^{-1}(\Omega)$ in the examples. In particular, we put for $f \in \mathscr{S}_0$

$$|f|_* = \sup_{0 \neq \theta \in \mathscr{S}_0} \frac{\langle f, \theta \rangle}{\|G\theta\|}. \tag{6}$$

That (6) is finite and bounded by $|f|$ will follow from the basic assumptions which we now introduce.

Assumptions. (A1) *There is a number* α, $1 < \alpha < \infty$, *and a norm* $\|\cdot\|_a$ *on* $\mathscr{V}_0$ *such that for each* v *in* $\mathscr{V}_0$

$$\|v\|_a \geqq \|v\|$$

and

$$\|v\|_a^2 \leqq \alpha(Av, v), \qquad |(Av, w)| \leqq \alpha\|v\|_a\|w\|_a$$

for all v, w *in* $\mathscr{V}_0$.

(A2) *The operator* G *in* (4) *satisfies*

$$\|G\psi\| \geqq |\psi|$$

for all ψ *in* $\mathscr{S}$.

(A3) *For each* y *in* $\mathscr{V}_0$ *there are* w, z *in* $\mathscr{V}_0$ *for which*

$$y = w + z$$

and

$$Dz = 0, \qquad \|w\|_a \leqq |Dy|_*.$$

Under suitable conditions these assumptions are valid for each of the examples of Section 2. In particular, (A1) is certainly valid in Example 1 with $\|\cdot\|_a$ and $\|\cdot\|$ being the L^2 norm since A in that case is the identity. The same is true in Example 3 since the integral operator makes no contribution to (Av, v). In the Stokes equations we again let $\|\cdot\|$ denote the L^2 norm but now

$$\|\mathbf{v}\|_a^2 = \int_\Omega \left\{\sum_j \operatorname{grad} v_j \cdot \operatorname{grad} v_j + \mathbf{v}\cdot\mathbf{v}\right\}. \tag{7}$$

Since $\operatorname{div} \mathbf{U} = 0$, the second term of (17) in Section 1 does not make a contribution to (Av, v) and hence (A1) is valid.

In the first and third examples (A2) follows from Friedrichs' inequality [11] provided Γ_D has positive measure in Γ. In the Stokes equations it follows from Poincaré's inequality [11] once we note that each $\psi \in \mathscr{S} \subseteq \mathscr{S}_0$ has zero mean value.

Finally in each of these cases (A3) reduces to the fundamental decomposition [8] of a vector field $\mathbf{v}$ into the gradient of a scalar $w = G\xi$ plus a divergence free part z.

To approximate we introduce the finite dimensional space

$$\mathscr{V}_h \subseteq \mathscr{V}_0$$

and put

$$\mathscr{S}_h \equiv \operatorname{div}(\mathscr{V}_h). \tag{8}$$

It is important to note that the choice (8) significantly distinguishes the present theory from the work of Brezzi [1], Fortin [5], and Raviart and Thomas [13]; the work by Girault [6] on the Stokes equations; and the work by Nedelec [12] on the eddy current problem. It is a natural relation in this setting once one observes that the solution u is a critical point of the function

$$\Phi(v) = \tfrac{1}{2}(Av, v) - (f, v)$$

subject to div $v = g$ and $v \in \mathscr{V}_0$. The scalar φ is thus a Lagrange multiplier for the constraint div $v = g$; and (2)–(3) is an expression for the vanishing of the first variation.

The approximation

$$u_h \in \mathscr{V}_h, \qquad \varphi_h \in \mathscr{S}_h$$

is characterized by requiring

$$(Au_h, v_h) + \langle \varphi_h, Dv_h \rangle = (f, v_h), \tag{9}$$

$$\langle Du_h, \psi_h \rangle = \langle g, \psi_h \rangle \tag{10}$$

to hold for all $v_h \in \mathscr{V}_h$ and $\psi_h \in \mathscr{S}_h$. Once a basis has been selected for $\mathscr{S}_h$ and $\mathscr{V}_h$, (9)–(10) reduces to a system of algebraic equations. The goal of this paper is to study the stability of this discrete system and to bound the errors $u - u_h$, $\varphi - \varphi_h$ in terms of the best approximations to u, φ in $\mathscr{V}_h$, $\mathscr{S}_h$.

To this end let

$$e_a^h(v) \equiv \inf_{v_h \in \mathscr{V}_h} \|v - v_h\|_a, \tag{11a}$$

$$e^h(v) \equiv \inf_{v_h \in \mathscr{V}_h} \|v - v_h\|, \tag{11b}$$

and

$$\varepsilon^h(\psi) \equiv \inf_{\psi_h \in \mathscr{S}_h} |\psi - \psi_h| \tag{12}$$

denote the errors in the best approximation of $v \in \mathscr{V}_0$ and $\psi \in \mathscr{S}_0$. In addition,

$$\delta_h \equiv \sup_{\|G\psi\| \leqq 1} \inf_{\psi_h \in \mathscr{S}_h} |\psi - \psi_h| \tag{13}$$

will enter our theory.

To understand (11)–(13) consider first the Poisson equation or the eddy current problem where $\|\cdot\|_a$, $\|\cdot\|$, and $|\cdot|$ are L_2 norms. Introducing a decomposition of Ω into triangles with maximum diameter $h > 0$ (see Fig. 1), we let $\mathscr{V}_h$ denote the space of continuous vector fields in $\mathscr{V}_0$ which are

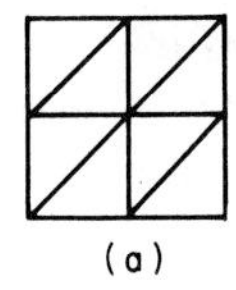

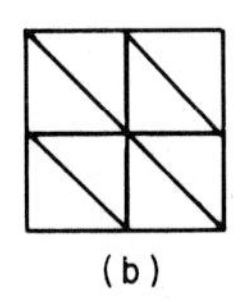

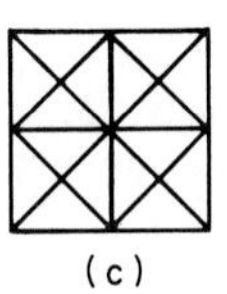

Fig. 1. (a) Right directional grid. (b) Left directional grid. (c) Criss-cross grid.

linear in each subdivision. A classical result from finite element theory is that

$$e^h(v) = e_a^h(v) \leqq Ch^2\|v\|_2,$$

where $\|\cdot\|_2$ is the norm on $H^2(\Omega)$ and C is an absolute constant [15]. The space $\mathscr{S}_h$, being the divergence of vector fields in $\mathscr{V}_h$, consists of piecewise constant functions. For the right and left directional grid in Fig. 1 $\mathscr{S}_h$ is in fact all piecewise constant functions on the respective grids [2, 3], but with the criss-cross grid it is a somewhat smaller space. In particular it is the linear hull of the union of the spaces $\mathscr{S}_h$ associated with the directional grids [2, 3]. In either case,

$$\varepsilon^h(\psi) \leqq Ch\|\psi\| \qquad \text{and} \qquad \delta_h \leqq Ch$$

provided $\psi \in H^1$ and $\|\cdot\|$ is the norm on $H^1(\Omega)$.

For the Stokes problem $\|\cdot\|_a$ denotes the norm (7). Thus we have

$$e_a^h(v) \leqq Ch\|v\|_2$$

with $e^h(v)$, $\varepsilon^h(\psi)$, and δ_h as above.

Fundamental to our theory is the discrete analog of the decomposition in (A3) which we call the discrete decomposition property (DDP).

Definition 1. The space $\mathscr{V}_h$ satisfies the DDP with constant C_G, $0 < C_G < \infty$ if and only if the following holds. Given $y_h \in \mathscr{V}_h$ there is a $w_h \in \mathscr{V}_h$ and $z_h \in \mathscr{V}_h$ for which

$$y_h = w_h + z_h$$

and

$$Dz_h = 0, \qquad \|w_h\|_a \leqq C_G|Dy_h|_*.$$

Our next result shows that DDP is both necessary and sufficient for solubility and stability of (9)–(10).

Theorem 1. *Let the DDP hold. Then for any A satisfying* (A1), (9)–(10) *has a unique solution $u_h \in \mathscr{V}_h$ and $\varphi_h \in \mathscr{S}_h$. Moreover,*

$$\|u_h\|_a \leqq C_G(\alpha^2 + 1)|g|_* + \alpha\|f\|, \tag{14}$$

$$|\varphi_h| \leqq \alpha C_G\|u_h\|_a + C_G|f|. \tag{15}$$

The converse is also true.

Proof. Existence can be established using arguments similar to Brezzi [1] and thus is omitted here.

To prove (14)–(15) let

$$u_h = w_h + z_h. \tag{16}$$

As in Definition 1 (with $y_h = u_h$) note that (10) implies

$$|Du_h|_* \leqq |g|_*. \tag{17}$$

Thus

$$\|w\|_a \leqq C_G |Du_h|_* \leqq C_G |g|_*. \tag{18}$$

To estimate z_h we let $v_h = z_h$ in (9) to get

$$(Az_h, z_h) + (Aw_h, z_h) = (f, z_h). \tag{19}$$

The inequality (14) follows from (A1), (16), (18), and (19).

To prove the second inequality we note that since $\varphi_h \in \mathscr{S}_h = D[\mathscr{V}_h]$ we have

$$\varphi_h = Dy_h \tag{20}$$

for some y_h in $\mathscr{V}_h$. Decomposing as in Definition 1 and letting $v_h = w_h$ in (9) we obtain

$$\langle \varphi_h, \varphi_h \rangle = (f, w_h) - (Au_h, w_h). \tag{21}$$

But

$$\|w_h\|_a \leqq C_G |Dy_h|_* = C_G |\varphi_h|_*. \tag{22}$$

Moreover, since $\|G\theta\| \geqq |\theta|$ by (A2) we have from (6)

$$|\varphi_h|_* \leqq |\varphi_h|. \tag{23}$$

The combination of these results with (A1) gives (15).

To prove the converse suppose $y_h \in \mathscr{V}_h$ is given. Then $g_h = Dy_h \in \mathscr{S}_h$. We solve (9)–(10) in the special case where $f = 0$ and $A = I$; i.e., we find $w_h \in \mathscr{V}_h$, $r_h \in \mathscr{S}_h$ for which

$$(w_h, v_h) + \langle r_h, Dv_h \rangle = 0, \tag{24}$$

$$\langle Dw_h, \psi_h \rangle = \langle g_h, \psi_h \rangle, \tag{25}$$

all $v_h \in \mathscr{V}_h$, $\psi_h \in \mathscr{S}_h$. The inequality (14) gives with $f = 0$

$$\|w_h\|_a \leqq C|g|_* = C|Dy_h|_*. \tag{26}$$

Putting $z_h = y_h - w_h$ we have $Dz_h = 0$. ∎

Theorem 1 shows that DDP is both necessary and sufficient for stability. It is of fundamental importance that not all grids satisfy DDP. In particular, it has been shown [2, 3] that the right and left directional grids for linear elements on triangles do not have this property while the criss-cross grid does. The instabilities that are present in the former grids are illustrated in numerical experiments reported in [2, 3].

We conclude this section with an error analysis of the approximations u_h, φ_h to u, φ. The first part does not use DDP and shows among other things that u_h is the best approximation to u in the seminorm $|D(\cdot)|$. This of course does not rule out instabilities since $|Dw|$ may be small and yet w may have a large "divergence free" component. It is at this stage where the DDP plays a crucial role.

Lemma 1. *For all v_h in $\mathscr{V}_h$*

$$\|D(u - u_h)\| \leqq \|D(u - v_h)\|. \tag{27}$$

Moreover,

$$|D(u - u_h)|_* \leqq \delta_h \|D(u - u_h)\|. \tag{28}$$

Proof. Combining (2)–(3) and (9)–(10) gives

$$(A(u - u_h), v_h) + \langle \varphi - \varphi_h, Dv_h \rangle = 0, \tag{29}$$

$$\langle D(u - u_h), \psi_h \rangle = 0, \tag{30}$$

for all $v_h \in \mathscr{V}_h$ and $\psi_h \in \mathscr{S}_h$. The inequality (27) follows immediately from (30) and (10).

To prove (28) note that for any ψ in $\mathscr{S}$ and ψ_h in $\mathscr{S}_h$

$$\begin{aligned} |\langle D(u - u_h), \psi \rangle| &= |\langle D(u - u_h), \psi - \psi_h \rangle| \\ &\leqq |D(u - u_h)||\psi - \psi_h|. \end{aligned}$$

Taking inf over $\psi_h \in \mathscr{S}_h$ and sup over $\psi \in \mathscr{S}$ with $\|G\psi\| \leqq 1$ gives (28). ■

Lemma 1 is typical of mixed methods. One has a best approximation in some nonstandard norm and this leads to an estimate in a dual norm. Neither results eliminates instabilities, and to do this we need DDP.

Theorem 2. *Let DDP hold. Then*

$$\|u - u_h\|_a \leqq C_1 e_a^h(u) + C_2 \delta_h \varepsilon^h(Du), \tag{31}$$

$$|\varphi - \varphi_h| \leqq 2\varepsilon^h(\varphi) + \alpha C_G \|u - u_h\|_a, \tag{32}$$

where

$$C_1 = (1 + \alpha^2)(1 + C_G), \qquad C_2 = C_G(1 + \alpha^2). \tag{33}$$

Proof. The analysis is virtually identical to the stability analysis in Theorem 1. In particular, let $\hat{\varphi}_h$, $\hat{u}_h$ be the best approximation to φ, u in $\mathscr{S}_h$, $\mathscr{V}_h$; i.e.,

$$\|u - \hat{u}_h\|_a = e_a^h(u), \qquad |\varphi - \hat{\varphi}_h| = \varepsilon^h(\varphi).$$

We use DDP to decompose

$$u_h - \hat{u}_h = w_h + z_h.$$

Observe that

$$C_G^{-1}\|w_h\|_a \leqq |D(u_h - \hat{u}_h)|_* \leqq |D(u - u_h)|_* + |D(u - \hat{u}_h)|_*. \tag{34}$$

The first term is covered in Lemma 1 and to estimate the second we note

$$|D(u - \hat{u}_h)|_* \leqq \| u - \hat{u}_h\|_a = e_a^h(u). \tag{35}$$

The estimate for $\|z_h\|_a$ follows from (29) which we rewrite as

$$(A(u_h - \hat{u}_h), v_h) + \langle \varphi_h - \hat{\varphi}_h, Dv_h\rangle = (A(u - \hat{u}_h), v_h) + \langle \varphi - \hat{\varphi}_h, Dv_h\rangle. \tag{36}$$

The estimate is completed by letting $v_h = z_h$ and using (A1) as in Theorem 1.

To prove (32) we write

$$\varphi_h - \hat{\varphi}_h = Dy_h \tag{37}$$

for some $y_h \in \mathscr{V}_h$. Following the pattern in Theorem 1 we decompose

$$y_h = w_h + z_h. \tag{38}$$

Putting $v_h = y_h$ in (36) gives

$$\langle \varphi_h - \hat{\varphi}_h, Dw_h\rangle = (A(u - u_h), w_h) + \langle \varphi - \hat{\varphi}_h, Dw_h\rangle. \tag{39}$$

Note that the second term in (39) is zero since $D\omega_h \in \mathscr{S}_h$ and $\hat{\varphi}_h$ is the best approximation to φ. Thus

$$|\varphi_h - \hat{\varphi}_h|^2 \leqq \alpha C_G\|u - u_h\|_a|\varphi_h - \hat{\varphi}_h|_*. \tag{40}$$

from which (32) is a direct consequence. ■

Remark. For the Poisson equation with linear elements on the criss-cross grid, Theorem 2 asserts that the L^2 error in $u - u_h$ is of order $O(h^2)$. This sharpens the $O(h)$ estimate in [1, 5, 6, 13]. The L^2 error in the scalar φ is $O(h)$, the same as predicted in [1, 5, 6, 13]. However, observe that (40) implies that the difference $\varphi_h - \hat{\varphi}_h$ in L^2 is of order $O(h^2)$. Since $\hat{\varphi}_h$ is the best L^2 approximation to φ, this means that the mean value of φ over a given triangle is actually approximated to $O(h^2)$. This phenomenon is illustrated in the numerical experiments in [2, 3].

Remark. For the Stokes problem Theorem 2 asserts that the H^1 error $\|\mathbf{u} - \mathbf{u}_h\|_a$ in the velocities is $O(h)$ as is the L^2 norm in the pressure if linear elements on a criss-cross grid are used for $\mathscr{V}_h$. It is also true that the L^2 errors in the velocities are of order $O(h^2)$. The proof of this uses the now standard duality argument [15] and is omitted here.

4. LEAST SQUARES TECHNIQUES

We turn now to the acoustic problem in Example 4 of Section 2, which has the canonical form

$$\begin{bmatrix} I & G \\ D & C \end{bmatrix} \begin{bmatrix} u \\ \varphi \end{bmatrix} = \begin{bmatrix} f \\ g \end{bmatrix}. \tag{1}$$

Numerical work reported in [2, 3, 4] indicates that least squares techniques are particularly adapted to this class of problem. The basic idea is a rather simple one; namely, one minimizes the residual error

$$\|v + G\psi - f\|^2 + \|Dv + C\psi - g\|^2 \tag{2}$$

over appropriate spaces. In the context of Section 2 this reduces to the minimization of

$$\int_\Omega |\mathbf{v} + \operatorname{grad}\psi - f|^2 + \int_\Omega |\operatorname{div}\mathbf{v} - \omega^2\psi - g|^2. \tag{3}$$

The vector $\mathbf{v}$ lies in

$$\mathscr{V}_0 \equiv \{\mathbf{v} \in \mathbf{H}^1(\Omega) : \mathbf{v}\cdot\mathbf{v} = 0 \quad \text{on} \quad \Gamma_N\}, \tag{4}$$

while the scalar ψ lies in

$$\mathscr{S}_1 \equiv \{\psi \in H^1(\Omega) : \psi = 0 \quad \text{on} \quad \Gamma_D\}. \tag{5}$$

It is possible to introduce boundary integrals that render the boundary conditions [(28)–(29), Section 2] natural, but here we incorporate them into the spaces $\mathscr{V}_0$, $\mathscr{S}_1$ for simplicity. The first theoretical work on this method was by Jesperson [7] who showed that the error in the scalar, $|\varphi - \varphi_h|$, converged at optimal rates. Latter numerical experiments [2, 3] indicated that serious instabilities in u_h, on the other hand, can arise, e.g., on directional grids just as they do for the methods in Section 3. In this section we show that the situation is almost identical to the last section in that optimal convergence to u is obtained when DDP is satisfied on $\mathscr{V}_h$ with a constant C_G independent of h.

To define the abstract context we let $H \supseteq \mathscr{V}_0$, $\mathscr{S}_1$ linear spaces with norms $\|\cdot\|$, $|\cdot|$ and inner products $(\cdot,\cdot)$, $\langle\cdot,\cdot\rangle$. For the model problem in Example 4, H is $\mathbf{H}^0(\Omega)$ and $\|\cdot\|$, $|\cdot|$ are L^2 norms.

Assumptions. (B1) *$C: \mathscr{S}_1 \to \mathscr{S}_1$ is linear with*

$$|C\psi| \leqq \gamma|\psi|, \qquad \text{all} \quad \psi \in \mathscr{S},$$

for some $0 < \gamma < \infty$.

(B2) *$G: \mathscr{S}_1 \to H$ satisfies $\|G\psi\| \geqq |\psi|$, all $\psi \in \mathscr{S}_1$.*

(B3) *$D: \mathscr{V}_0 \to \mathscr{S}_0 \equiv D[\mathscr{V}_0]$ with $(G\psi, v) = \langle \psi, Dv \rangle$, all $\psi \in \mathscr{S}_1$, $v \in \mathscr{V}_0$.*†

(B4) *The decomposition property* [(A3), Section 3] *is valid.*

(B5) *For each $f \in \mathscr{S}_0$ there is a $\psi \in \mathscr{S}_1$ such that $G\xi \in \mathscr{V}_0$ and*

$$(DG - C)\xi = f.$$

Moreover, $C\xi \in \mathscr{S}_1$ and

$$\|G(I + C)\xi\| \leqq C_R |f|, \qquad |DG\xi| \leqq C_R |f|.$$

Finally, if $f \in \mathscr{S}_1$, then $DG\xi \in \mathscr{S}_1$ and

$$\|GDG\xi\| \leqq C_R \|Gf\|.$$

We also assume that solvability and the first set of inequalities hold if C is replaced with its adjoint C^* with respect to $\langle \cdot, \cdot \rangle$.

The first four assumptions are easily verified for the model problem. The last assumption involves the solvability and regularity of

$$(\Delta + \omega^2)\xi = f \quad \text{in} \quad \Omega, \tag{6}$$

$$\xi = 0 \quad \text{on } \Gamma_D, \qquad \text{grad } \xi \cdot \mathbf{v} = 0 \quad \text{on} \quad \Gamma_N. \tag{7}$$

This is equivalent to the solvability and regularity of the original physical problem once [(26)–(27), Section 3] have been combined. Recall that if ω^2 is not an eigenvalue of Δ with boundary conditions, then

$$(DG - C)\xi = (\Delta + \omega^2)\xi = f \tag{8}$$

certainly has a solution ξ satisfying (7); i.e., $\xi \in \mathscr{S}_1$ and $G\xi \in \mathscr{V}_0$ [9]. Moreover, the first of the inequalities in (B5), which can be written

$$\|\text{grad } \xi + \omega^2 \xi\|_0 \leqq C_R \|f\|_0, \qquad \|\Delta \xi\|_0 \leqq C_R \|f\|_0, \tag{9}$$

are valid. The last inequality follows if

$$\|\xi\|_3 \leqq C_R \|f\| \tag{10}$$

is valid. The latter will hold if the boundary Γ of Ω is smooth, but it is not valid for regions with reentrant corners, e.g., the L-shaped region.

To define the approximation we introduce finite dimensional subspaces

$$\mathscr{S}_h \subseteq \mathscr{S}_1, \qquad \mathscr{V}_h \subseteq \mathscr{V}_0. \tag{11}$$

(Here the relation [(8), Section 3] is not assumed.) We determine $u_h \in \mathscr{V}_h$, $\varphi_h \in \mathscr{S}_h$ by minimizing

$$\|v^h + G\psi^h - f\|^2 + |Dv^h + C\psi^h - g|^2 \tag{12}$$

† In particular, $|\cdot|$, $\langle \cdot, \cdot \rangle$ are defined on $\mathscr{S}_0$, $\mathscr{S}_0 \times \mathscr{S}_0$.

over v^h in $\mathscr{V}_h$ and ψ^h in $\mathscr{S}_h$. Defining the bilinear form on $\mathscr{S}_1 \times \mathscr{V}_0$

$$a((\psi, v), (\xi, w)) = (v + G\psi, w + G\xi) + \langle Dv + C\psi, Dw + C\xi \rangle, \quad (13)$$

this is equivalent to requiring

$$a((u_h, \varphi_h), (\psi^h, v^h)) = (f, v^h + G\psi^h) + \langle g, Dv^h + C\psi^h \rangle \quad (14)$$

for all v^h in $\mathscr{V}_h$ and ψ^h in $\mathscr{S}_h$.

Let $|||\cdot|||$ denote the norm associated with (13); i.e.,

$$|||(\psi, v)||| = a((\psi, v), (\psi, v))^{1/2}. \quad (15)$$

Observe there is a constant C_a for which

$$|||(\psi, v)||| \leqq C_a\{\|G\psi\| + \|v\| + |Dv|\} \quad (16)$$

holds for all $\psi \in \mathscr{S}_1$ and $v \in \mathscr{V}_0$.

We retain the notation of Section 3 and let

$$e_h(v) \equiv \inf_{v^h \in \mathscr{V}_h} \|v - v^h\|, \qquad \varepsilon_h(\psi) \equiv \inf_{\psi^h \in \mathscr{S}_h} |\psi - \psi^h| \quad (17)$$

denote the errors in the best approximations. A common choice of spaces is to use piecewise linear functions for both $\mathscr{V}_h$ and $\mathscr{S}_h$. For this choice

$$e_h(v) \leqq Ch^2\|v\|_2, \qquad \varepsilon_h(\psi) \leqq Ch^2\|\psi\|_2. \quad (18)$$

In place of the $\mathscr{S}_h$ defined by [(13), Section 3] we have two slightly more complicated error terms arising in the theory. The first is

$$\sigma_h \equiv \sup_{|DG\eta| \leqq 1} \inf_{\eta_h \in \mathscr{S}_h} \|G(\eta - \eta_h)\|. \quad (19)$$

If $\mathscr{S}_h$ consists of piecewise linear elements, then

$$\sigma_h \leqq (\text{const})h \quad (20)$$

follows directly from the result

$$\inf\|\text{grad}(\eta - \eta_h)\|_0 \leqq Ch\|\eta\|_2 \quad (21)$$

of approximation theory [15]. The second error term is

$$\gamma_h \equiv \sup_{\|GDG\eta\| \leqq 1} \inf_{(\eta^h, v^h) \in \mathscr{S} \times \mathscr{V}_h} \{\|G(\eta - \eta^h)\| + |D(G\eta - v^h)| + \|G\eta - v^h\|\}. \quad (22)$$

We assert that

$$\gamma_h \leqq (\text{const})h \quad (23)$$

if linear elements are used in both $\mathscr{S}_h$ and $\mathscr{V}_h$. Indeed, the first term follows from (17) while the second and third terms follow from

$$\inf\|\text{div}(\text{grad}\,\eta - v^h)\|_0 \leqq Ch\|\eta\|_3, \quad (24)$$

$$\inf\|\text{grad}\,\eta - v^h\| \leqq Ch\|\eta\|_2. \quad (25)$$

As in Section 3 our theory splits into two parts the first independent of DDP, and in the second it plays a crucial role. The first part is analogous to the analysis in Section 3; namely, we start by showing that the approximates $\{\varphi_h, u_h\}$ are best in some nonstandard norm. This leads to an estimate in the dual norm

$$|f|_* \equiv \sup_{\|G\theta\| \leqq 1} |\langle f, \theta \rangle|. \tag{26}$$

DDP enters the theory at this point to convert these into L_2 estimates.

Lemma 1. *Let*

$$\varepsilon = \varphi - \varphi_h, \qquad e = u - u_h. \tag{27}$$

Then

$$|||(\varepsilon, e)||| \leqq |||(\varphi - \psi^h, u - v^h)||| \tag{28}$$

for all ψ^h in $\mathscr{S}_h$ and v^h in $\mathscr{V}_h$. Moreover,

$$|De + C\varepsilon|_* \leqq C_a C_R \gamma_h |||(\varepsilon, e)|||. \tag{29}$$

Proof. The inequality (28) is an immediate consequence of the orthogonality

$$a((\varepsilon, e), (\psi^h, v^h)) = 0 \qquad \text{all} \quad \psi^h \quad \text{in} \quad \mathscr{S}_h, \quad v^h \quad \text{in} \quad \mathscr{V}_h. \tag{30}$$

To prove (29) let $\theta \in \mathscr{S}$, be given. We use (B5) to solve

$$(DG - C)\xi = \theta \tag{31}$$

for $\xi \in \mathscr{S}_1$. Note that

$$a((\varepsilon, e), (\xi, -G\xi)) = \langle De + C\varepsilon, C\xi - DG\xi \rangle = \langle De + C\varepsilon, \theta \rangle$$

But by regularity in (B5)

$$\|GDG\xi\| \leqq C_R \|G\theta\|.$$

Thus using orthogonality to write

$$a((\varepsilon, e), (\xi, -G\xi)) = a((\varepsilon, e), (\xi - \xi_h, -G\xi + v_h)),$$

we have

$$\frac{a((\varepsilon, e), (\xi - \xi_h, -G\xi + v_h))}{\|GDG\xi\|} \geqq \frac{1}{C_R} \frac{\langle De + Ce, \theta \rangle}{\|G\theta\|}.$$

Taking the sup over $\theta \in \mathscr{S}_1$ and using (12) gives (29). ∎

We now can establish the result first proved by Jesperson [7] for the Poisson equation, namely, that the rate of convergence in L_2 of the scalar φ_h is optimal (with or without DDP).

Theorem 1. *There are constants $0 < C_i < \infty$ depending only on C_R and C_a such that*

$$|\varepsilon| \leqq (C_1\gamma_h + C_2\sigma_h)|||(\varepsilon, e)|||. \tag{32}$$

The result is an immediate consequence of the following.

Lemma 2.

$$|\varepsilon| \leqq C_R\{|De + C\varepsilon|_* + \sigma_h C_a|||(\varepsilon, e)|||\}. \tag{33}$$

Proof. We solve

$$(DG - C^*)\xi = \varepsilon$$

for $\xi \in \mathscr{S}_1$. Note that

$$\begin{aligned} a((\varepsilon, e), (\xi - \xi_h, 0) &= a((\varepsilon, e), (\xi, 0)) \\ &= \langle De + C\varepsilon, \xi + C\xi\rangle + \langle \varepsilon, \varepsilon\rangle. \end{aligned}$$

Thus

$$|\varepsilon|^2 \leqq |De + C\varepsilon|_* \|G(\xi + C\xi)\| + |||(\varepsilon, e)|||\,|||(\xi - \xi_h, 0)|||.$$

Regularity gives

$$\|G(I + C)\xi\| \leqq C_R|\varepsilon|.$$

Moreover,

$$|||(\xi - \xi_h, 0)||| \leqq C_a\|G(\xi - \xi_h)\|.$$

Taking inf over ξ_h in $\mathscr{S}_h$ gives

$$\|(\xi - \xi_h, 0)\| \leqq \sigma_h C_a\|DG\xi\| \leqq \sigma_h C_a C_R|\varepsilon|.$$

Combining these results we obtain (33). ■

We now turn to the last part of the analysis where DDP comes into play.

Lemma 3. *Let DDP hold and let $\hat{u}_h$ be a best approximation in the sense that*

$$\|u - \hat{u}_h\| = e_h(u). \tag{34}$$

Then

$$\|u_h - \hat{u}\| \leqq C_G|D(u_h - \hat{u}_h)|_* + e_h(u). \tag{35}$$

Proof. We use DDP to write

$$u_h - \hat{u}_h = \omega_h + z_h$$

as in [Definition 1, Section 3]. Note that

$$\|\omega_h\| \leqq |D(u_h - \hat{u}_h)|_*.$$

Also note that from (30) we have

$$a((\varphi, u - \hat{u}_h), (0, v^h)) = a((\varphi_h, u_h - \hat{u}_h), (0, v^h))$$

for all $v^h \in \mathscr{V}_h$. Letting $v^h = z_h$ and recalling $Dz_h = 0$, $(z_h, \omega_h) = 0$, this reduces to

$$(u, z_h) = (u_h, z_h).$$

Thus

$$\|z_h\| \leqq \|u - \hat{u}_h\| = e_h(u). \quad \blacksquare$$

Theorem 2. *Let DDP hold. Then there are constants* $0 < C_i < \infty$ *depending only on* C_R, C_a, γ *such that*

$$\|u - u_h\| \leqq (C_1\sigma_h + C_2\gamma_h)|||(\varepsilon, e)||| + C_3 C_G e_h(u). \tag{36}$$

Proof. Note that

$$|D(u_h - \hat{u}_h)|_* \leqq |D(u - u_h)|_* + |D(u - \hat{u}_h)|_*.$$

As in Section 2

$$|D(u - \hat{u}_h)|_* = \sup \frac{\langle D(u - \hat{u}_h), \theta\rangle}{\|G\theta\|}$$

$$= \sup \frac{(u - \hat{u}_h, G\theta)}{\|G\theta\|} \leqq \|u - \hat{u}_h\|.$$

Also using Lemma 1

$$|D(u - u_h)|_* \leqq C_a C_R \gamma_h |||(\varepsilon, e)||| + \gamma|\varepsilon|_*.$$

Combining these results with $|\varepsilon|_* \leqq |\varepsilon|$ we obtain (36).

Remark. If linear elements are used for both $\mathscr{S}_h$ and $\mathscr{V}_h$ and if the criss-cross grid is used in the latter, then $O(h^2)$ approximations are obtained in both the scalar and vector. If directional grids are used in $\mathscr{V}_h$, then the error in the scalar remains at $O(h^2)$, however instabilities occur in the vector. This is illustrated in the numerical results in [2, 3].

REFERENCES

1. F. Brezzi, On the existence, uniqueness and approximation of saddle point problems arising from Lagrangian Multipliers. *RAIRO* **8** (1974) 129–150.
2. G. J. Fix, M. D. Gunzburger, and R. A. Nicolaides, On mixed finite element methods for first order elliptic systems, *Numer. Math.* (to appear).
3. G. J. Fix, M. D. Gunzburger, and R. A. Nicolaides, On finite element methods of the least squares type, *Comput. Math. Appl.* **5** (1979).

4. G. J. FIX AND M. D. GUNZBURGER, On numerical methods for acoustics problems, ICASE Report 78-15, *Comput. Math. Appl.* (to appear).
5. M. FORTIN, Resolution numerique des equations de Navier–Stokes par des elements finis de type mixed, *2nd Internat. Symp. Finite Elem. Flow Problems, Liqure*, 1976.
6. V. GIRAULT, A mixed finite element method for the stationary Stokes Equations, *SIAM J. Numer. Anal.* **15**, (1978), 3.
7. D. C. JESPERSON, On least squares decomposition of elliptic boundary value problems, *Math. Comp.* **31** (1977).
8. O. A. LADYZHENSKAYA, "The Mathematical Theory of Viscous Incompressible Flow," Gordon & Breach, New York, 1963.
9. J. L. LIONS AND E. MAGENES, "Nonhomogeneous Boundary Value Problems," Springer-Verlag, Berlin and New York, 1973.
10. R. C. MACCAMY AND J. H. MCWHIRTER, Eddy Currents in Thin Sheets, Tokamak Fusion Test Reactor Project, TCTR-227, Westinghouse Fusion Power Systems Dep., Pittsburgh, Pennsylvania.
11. J. NECAS, "Les Methodes Directes en Theorie des Equations Elliptiques," Academia, Prague.
12. J. C. NEDELEC, Computation of eddy currents on a surface in R^3 by finite element methods, *SIAM J. Numer. Anal.* **15** (1978), 3.
13. P. A. RAVIART AND J. M. THOMAS, A mixed finite element method for second order elliptic problems, *in* "Mathematical Aspects of Finite Element Methods," Lecture Notes in Mathematics, Springer-Verlag, Berlin and New York, 1975.
14. J. SERRIN, Mathematical principles of classical fluid mechanics, *in* "Encyclopedia of Physics," Vol. 8, Springer-Verlag, Berlin and New York, 1959.
15. G. STRANG AND G. J. FIX, "An Analysis of the Finite Element Method," Prentice–Hall, Englewood Cliffs, New Jersey, 1973.
16. R. TEMAM, "Navier–Stokes Equations, Theory and Numerical Analysis," North–Holland Publ. Amsterdam, 1977.

G. J. Fix

and

R. A. Nicolaides
DEPARTMENT OF MATHEMATICS
CARNEGIE–MELLON UNIVERSITY
PITTSBURGH, PENNSYLVANIA

M. D. Gunzburger
DEPARTMENT OF MATHEMATICS
UNIVERSITY OF TENNESSEE
KNOXVILLE, TENNESSEE

Equipartition of Energy for Symmetric Hyperbolic Systems

JEROME A. GOLDSTEIN

JAMES T. SANDEFUR, JR.

1. INTRODUCTION

Consider the one dimensional wave equation

$$\frac{\partial^2 v}{\partial t^2} = \frac{\partial^2 v}{\partial x^2} \qquad (t, x \in \mathbb{R} = (-\infty, \infty)).$$

The general solution has the form

$$v(t, x) = \varphi(x - t) + \psi(x + t),$$

where φ and ψ are functions defined on $\mathbb{R}$. The functions φ and ψ are determined by the initial data f_1, f_2, where $v(0, x) = f_1(x)$, $(\partial v/\partial t)(0, x) = f_2(x)$. The energy

$$\mathscr{E} = \int_{-\infty}^{\infty} \left| \frac{\partial v}{\partial t}(t, x) \right|^2 dx + \int_{-\infty}^{\infty} \left| \frac{\partial v}{\partial x}(t, x) \right|^2 dx$$

is independent of t; assume it is finite. Denote the first term

$$\int_{-\infty}^{\infty} \left| \frac{\partial v}{\partial t}(t, x) \right|^2 dx$$

by $K(t)$ and call it the kinetic energy of v at time t. Similarly call the second term the potential energy $P(t)$ of v at time t. If the initial data f_1, f_2 have compact support, then an easy, straightforward calculation yields the existence of a constant $T > 0$ such that $K(t) = P(t) = \mathscr{E}/2$ for all t with

ISBN 0-12-178150-X

$|t| > T$. A simple density argument then yields: for all solutions having finite energy,

$$\lim_{t \to \pm\infty} K(t) = \lim_{t \to \pm\infty} P(t) = \mathscr{E}/2.$$

For the n-dimensional wave equation

$$\frac{\partial^2 v}{\partial t^2} = \Delta v \qquad (t \in \mathbb{R}, \quad x \in \mathbb{R}^n)$$

set

$$\mathscr{E} \equiv \int_{-\infty}^{\infty} \left| \frac{\partial v}{\partial t} \right|^2 dx + \int_{-\infty}^{\infty} |\nabla v|^2 \, dx \equiv K(t) + P(t).$$

It was first shown in 1966 by Brodsky [3] that

$$\lim_{t \to \pm\infty} K(t) = \lim_{t \to \pm\infty} P(t) = \mathscr{E}/2 \tag{1.1}$$

holds for all initial data. Shortly thereafter, Lax and Phillips [13, p. 106] and independently Duffin [5] showed that when the initial data is compactly supported one gets

$$K(t) = P(t) = \mathscr{E}/2 \tag{1.2}$$

for all t with $|t|$ sufficiently large when n is odd, whereas this equipartition of energy from a finite time on fails in even dimensions. Duffin used the Paley–Wiener theorem for the Fourier transform in his approach, while Lax and Phillips used the Radon transform.

In this paper we shall establish equipartition of energy results for a large class of symmetric hyperbolic systems. We shall work in a Hilbert space framework, using a generalized version of the notions introduced in our recent paper [11]. Our equipartition of energy results will be in the limiting sense of (1.1). There are no known definitive abstract results for equipartition of energy from a finite time on such as (1.2), although a few special cases have been treated by Goldstein [9] and Bobisud and Calvert [2]. We shall indicate how equipartition of energy from a finite time on does hold for a special class of symmetric hyperbolic systems.

2. THE FRAMEWORK

Let H be a self-adjoint operator on a complex infinite dimensional Hilbert space $\mathscr{H}$. Associated with H are the abstract Schrödinger initial value problem

$$i\frac{du}{dt} = Hu, \qquad u(0) = f \tag{2.1}$$

and the unitary group $\{e^{-itH} : t \in \mathbb{R} = (-\infty, \infty)\}$. The unique solution of (2.1) is given by

$$u(t) = e^{-itH}f.$$

This is a continuously differentiable solution for f in $\mathscr{D}(H)$, the domain of H; it is generalized solution otherwise. We shall call $\mathscr{E}(g) = \|g\|^2$ the *energy* of a vector $g(\in \mathscr{H})$. Since e^{-itH} is unitary, the energy of the solution of (2.1) is conserved:

$$\mathscr{E}(u(t)) = \mathscr{E}(f) \qquad \text{for all} \quad t \in \mathbb{R}. \tag{2.2}$$

For our study of equipartition of energy, it is convenient to introduce the notion of a division of energy. We begin with a preliminary notion.

Definition 2.1. A *finite projection system* is a set of self-adjoint projections $P_1, P_2, \ldots, P_N$, $2 \leqq N < \infty$, such that $P_j P_k = 0$ for $j \neq k$ and $\sum_{j=1}^N P_j = I =$ this identity operator. A *proper finite projection system* is a finite projection system in which each P_j has norm one (and therefore is nontrivial).

Definition 2.2. A self-adjoint operator H on $\mathscr{H}$ admits a *division of energy* if there is a finite projection system $P_1, \ldots, P_N$ such that for $\mathscr{E}^j(g) \equiv \mathscr{E}(P_j g) = \|P_j g\|^2$ for $g \in \mathscr{H}$, we have

$$\mathscr{E}^j(u(t)) = \mathscr{E}^j(f) \qquad \text{for all} \quad t \in \mathbb{R},$$

where u is the solution of (2.1) and $f \in \mathscr{H}$ is arbitrary. [Necessarily, $\mathscr{E}(f) = \sum_{j=1}^N \mathscr{E}^j(f)$ for all $f \in \mathscr{H}$.] $\mathscr{E}^j(f)$ will be called the *jth partial energy of f*.

Here are two examples of division of energy. First, let

$$H = \begin{bmatrix} H_1 & 0 \\ 0 & H_2 \end{bmatrix} \qquad \text{on} \quad \mathscr{H} = \mathscr{K} \oplus \mathscr{K},$$

where $\mathscr{H}_1$, $\mathscr{H}_2$ are self-adjoint on K. Let $P_1\left[\begin{smallmatrix} a \\ b \end{smallmatrix}\right] = \left[\begin{smallmatrix} a \\ 0 \end{smallmatrix}\right]$, $P_2\left[\begin{smallmatrix} a \\ b \end{smallmatrix}\right] = \left[\begin{smallmatrix} 0 \\ b \end{smallmatrix}\right]$ for all $\left[\begin{smallmatrix} a \\ b \end{smallmatrix}\right] \in \mathscr{H}$. Since

$$e^{itH} = \begin{bmatrix} e^{itH_1} & 0 \\ 0 & e^{itH_2} \end{bmatrix},$$

it follows easily that $\{P_1, P_2, \mathscr{E}^j(f) = \|P_j f\|^2\}$ defines a division of energy for H on $\mathscr{H}$.

Next, let $\mathscr{N}(H)$ be the null space of H, a self-adjoint operator on $\mathscr{H}$. Let P_1 be the orthogonal projection onto $\mathscr{N}(H)$, and let $P_2 = I - P_1$ be the orthogonal projection onto $\overline{\mathscr{R}(H)}$, the closure of the range of H. Then $\{P_1, P_2, \mathscr{E}^j(f) = \|P^j f\|^2\}$ defines a division of energy for H. The finite projection system P_1, P_2 is proper if and only if $\{0\} \neq \mathscr{N}(H) \neq \mathscr{H}$.

The following results illustrate the notion of equipartition of energy. Let A be a self-adjoint operator on a complex Hilbert space $\mathscr{K}$. The problem

$$v''(t) + A^2 v(t) = 0 \qquad (t \in \mathbb{R}), \tag{2.3}$$

$$v(0) = \tilde{f}_1, \qquad v'(0) = f_2 \in \mathscr{D}(A) \tag{2.4}$$

is well posed, assuming $f_1 = A\tilde{f}_1 \in \mathscr{D}(A)$; here prime denotes differentiation with respect to t. Define the *energy* of the solution to be

$$\mathscr{E}(f) = \|v'(t)\|^2 + \|Av(t)\|^2; \tag{2.5}$$

it depends on the data

$$f = \begin{bmatrix} f_1 \\ f_2 \end{bmatrix}$$

but not on t. One can rewrite (2.3), (2.4) as a system (2.1) with

$$u = \begin{bmatrix} Av \\ v' \end{bmatrix}, \qquad H = i\begin{bmatrix} 0 & A \\ -A & 0 \end{bmatrix}, \qquad \text{and} \qquad f = \begin{bmatrix} f_1 \\ f_2 \end{bmatrix}$$

in $\mathscr{H} = \mathscr{K}_1 \oplus \mathscr{K}$, where $\mathscr{K}_1$ is the completion of $\mathscr{R}(A)/\mathscr{N}(A)$. When A is injective (i.e., one-to-one), $\mathscr{K}_1 = \mathscr{K}$. Moreover, (2.5) becomes (2.2). Let

$$\mathscr{E}_K(t) = \|v'(t)\|^2, \qquad \mathscr{E}_P(t) = \|Av(t)\|^2$$

denote the *kinetic* and *potential energies* at time t. Two results of Goldstein [8, 9] are as follows.

Theorem 2.1. *The energy is equipartitioned, i.e.,*

$$\lim_{t \to \pm\infty} \mathscr{E}_K(t) = \lim_{t \to \pm\infty} \mathscr{E}_P(t) = \frac{\mathscr{E}(f)}{2}$$

for all initial data f if and only if e^{itA} converges to zero in the weak operator topology as $t \to \pm\infty$.

Theorem 2.2. *The energy is equipartitioned in the Césaro sense, i.e.,*

$$\lim_{T \to t\infty} \frac{1}{T} \int_0^T \mathscr{E}_K(t)\, dt = \lim_{T \to \pm\infty} \frac{1}{T} \int_0^T \mathscr{E}_P(t)\, dt = \frac{\mathscr{E}(f)}{2}$$

for all initial data f if and only if A is injective.

Definition 2.3. Let the self-adjoint operator H on $\mathscr{H}$ admit a division of energy; let $P_1, \ldots, P_N$ be the associated projections.

Fix k, $1 \leq k \leq N$. The kth partial energy is said to admit *equipartition of energy* if there is a proper finite projection system $Q_1, \ldots, Q_M$ and positive constants $c_1, \ldots, c_M$ such that

$$\|Q_j(P_k u(t))\|^2 \to c_j \|P_k f\|^2$$

for $1 \leqq j \leqq M$ as $t \to \pm\infty$ for all $f \in \mathscr{H}$, where u is the solution of (2.1).

Necessarily we have $\sum_{j=1}^{M} c_j = 1$. Also, as noted in [11], each Q_j has infinite dimensional range and does not commute with H. If $P_1 = I$, $P_j = 0$ for $j \geqq 2$ and $k = 1$ in the above definition, we get the definition of equipartition of energy given in [11]. Now we want to reformulate Theorems 2.1 and 2.2 to indicate the convenience of the current setup. But first we introduce a division of energy.

Let A be a self-adjoint operator on $\mathscr{K}$. We use the notation of the paragraph containing (2.3). Let P_1 [resp. Q] denote the orthogonal projection of $\mathscr{K} \oplus \mathscr{K}$ [resp. $\mathscr{K}$] onto $\mathscr{N}(A) \oplus \mathscr{N}(A)$ [resp. $\mathscr{N}(A)^\perp$]. Let $P_2 = I - P_1$. Then $\mathscr{E}^j = \|P_j u(t)\|^2$ does not depend on t for $j = 1, 2$.

Theorem 2.1′. *$\mathscr{E}^2$ is equipartitioned, i.e.,*

$$\mathscr{E}_K^2(t) = \|Qv'(t)\|^2 \qquad \text{and} \qquad \mathscr{E}_P^2(t) = \|Av(t)\|^2 = \|QAv(t)\|^2$$

converge to $\frac{1}{2}\mathscr{E}^2(f)$ as $t \to \pm\infty$ for all data f if and only if weak $\lim_{t\to\pm\infty} e^{itA} g = 0$ *for all $g \in \mathscr{N}(A)^\perp = QK$.*

Theorem 2.2′. *$\mathscr{E}^2$ is always equipartitioned in the Césaro sense.*

Theorems 2.1′ and 2.2′ coincide with Theorems 2.1 and 2.2 when $\mathscr{N}(A) = \{0\}$, but Theorems 2.1′ and 2.2′ give useful information when $\mathscr{N}(A) \neq \{0\}$. Theorems 2.1′ and 2.2′ are proved (but not explicitly stated) in [8, 9].

We need one final bit of terminology.

Definition 2.4. A differential equation will be said to admit a *division* (*and equipartition*) *of energy* if it can be written in the form (2.1) in such a way that the operator H admits a division (and equipartition) of energy.

Then we can say that Eq. (2.3) admits a division and equipartition of energy when the hypotheses of Theorem 2.1′ hold.

3. THE SYMMETRIC HYPERBOLIC SYSTEMS

Let $A_1, \ldots, A_n$ be real symmetric $m \times m$ matrices. For $\xi = (\xi_1, \ldots, \xi_n) \in \mathbb{R}^n$ let

$$A(\xi) = \sum_{j=1}^{n} A_j \xi_j$$

be the symbol of the matrix differential operator

$$A(D) = \sum_{j=1}^{n} A_j D_j,$$

where $D_j = -i\,\partial/\partial x_j$. The operator $A(D)$ is self-adjoint on the Hilbert space $\mathcal{H} = L^2(\mathbb{R}^n; \mathbb{C}^m)$; its domain is

$$\mathcal{D}(A(D)) = \left\{u \in \mathcal{H} : \int_{\mathbb{R}^n} |A(\xi)\hat{u}(\xi)|^2\, d\xi < \infty\right\},$$

where $\hat{u}$ [or $\Phi(u)$] denotes Fourier transform of u, i.e.,

$$\hat{u}(\xi) = (\Phi u)(\xi) = (2\pi)^{-n/2} \int_{\mathbb{R}^n} e^{-i\xi\cdot x}\, u(x)\, dx.$$

For a very special class of symmetric hyperbolic systems of the form

$$i\,\partial u/\partial t = A(D)u, \tag{3.1}$$

Costa [4] used the Radon transform in odd dimensional space (more precisely, he took n odd and $n \geqq 3$) to establish that the energy was asymptotically partitioned into different types of energy as $t \to \pm\infty$. But no equipartition of energy was obtained. More precisely, he showed that $\mathcal{E} = \|u(t)\|^2 = \sum_{j=1}^{N} \mathcal{E}^j(t)$ where $\alpha_j^{\pm} = \lim_{t\to\pm\infty} \mathcal{E}^j(t)$ exists, but there was no guarantee that $\alpha_1^+, \ldots, \alpha_N^+$ are all nonzero; conceivably $\alpha_1^+ = \|u(0)\|^2$ and $\alpha_j^+ = 0$ for $j \geqq 2$, which is not an interesting case.

We shall show that division of energy can be combined with equipartition of energy to explain the energy theoretic aspects of the asymptotic behavior of a large, important case of symmetric hyperbolic systems. An abstract version of this is contained in Theorem 3.1. A refinement is given in Theorem 4.1.

Let $u \in \mathcal{H}$. Think of u as a column vector with entries $u_1, \ldots, u_m$. Let v be the first k entries of u and w the last $m-k$ entries of u, where $1 \leqq k \leqq m-1$. Thus

$$u = \begin{bmatrix} u_1 \\ \vdots \\ u_m \end{bmatrix} \in L^2(\mathbb{R}^n; \mathbb{C}^m), \qquad v = \begin{bmatrix} u_1 \\ \vdots \\ u_k \end{bmatrix} \in L^2(\mathbb{R}^n; \mathbb{C}^k),$$

$$w = \begin{bmatrix} u_{k+1} \\ \vdots \\ u_m \end{bmatrix} \in L^2(\mathbb{R}^n; \mathbb{C}^{m-k}), \qquad u = v \oplus w = \begin{bmatrix} v \\ w \end{bmatrix},$$

$$L^2(\mathbb{R}^n; \mathbb{C}^m) = L^2(\mathbb{R}^n; \mathbb{C}^k) \oplus L^2(\mathbb{R}^n; \mathbb{C}^{m-k}).$$

Then Eq. (3.1) can be rewritten as

$$\begin{bmatrix} v \\ w \end{bmatrix}_t = \begin{bmatrix} \Gamma_1(D) & \Lambda(D) \\ -\Lambda^*(D) & \Gamma_2(D) \end{bmatrix} \begin{bmatrix} v \\ w \end{bmatrix},$$

where $\Gamma_1(D)$ and $\Gamma_2(D)$ are skew-adjoint. We *assume* $\Gamma_1(D) = \Gamma_2(D) = 0$. The resulting equation

$$\begin{bmatrix} v \\ w \end{bmatrix}_t = \begin{bmatrix} 0 & \Lambda(D) \\ -\Lambda^*(D) & 0 \end{bmatrix} \begin{bmatrix} v \\ w \end{bmatrix} \tag{3.2}$$

includes many examples of interest, such as the (linearized) equations of acoustics, elasticity, magnetohydrodynamics, crystal optics, electricity and magnetism, and others. Before deriving an equipartition of energy result for such systems we recall a needed concept.

Let H be a self-adjoint operator on $\mathscr{H}$. Write $H = \int_{-\infty}^{\infty} \lambda \, dE_\lambda$ by the spectral theorem. The *subspace of absolute continuity* of H is

$$\mathscr{H}_{\mathrm{ac}}(H) = \{u \in \mathscr{H} : \lambda \to \|E_\lambda u\|^2 \text{ is absolutely continuous on } \mathbb{R}\}$$

(see [12, p. 516]).

Definition 3.1. A self-adjoint operator H on $\mathscr{H}$ is said to be *essentially absolutely continuous* [resp. *absolutely continuous*] if

$$\mathscr{H} = \mathscr{N}(H) \oplus \mathscr{H}_{\mathrm{ac}}(H).$$

[resp. $\mathscr{H} = \mathscr{H}_{\mathrm{ac}}(H)$].

Thus, informally, an essentially absolutely continuous operator is absolutely continuous off its null space.

Theorem 3.1. *Let $\mathscr{K}_1, \mathscr{K}_2$ be Hilbert spaces and let $F: \mathscr{K}_1 \to \mathscr{K}_2$ be a closed densely defined linear operator. Suppose that the self-adjoint operators F^*F on $\mathscr{K}_1$ and FF^* on $\mathscr{K}_2$ are essentially absolutely continuous. Then the equation*

$$i\frac{d}{dt}\begin{bmatrix} v \\ w \end{bmatrix} = \begin{bmatrix} 0 & F^* \\ F & 0 \end{bmatrix} \begin{bmatrix} v \\ w \end{bmatrix} \tag{3.3}$$

on $\mathscr{K}_1 \oplus \mathscr{K}_2$ admits a division of energy and has a partial energy not associated with a null space which admits equipartition of energy.

Proof. Consider (3.3) with initial data

$$\begin{bmatrix} v(0) \\ w(0) \end{bmatrix} = \begin{bmatrix} x_0 \\ y_0 \end{bmatrix} \in \mathscr{D}\left[\begin{bmatrix} 0 & F^* \\ F & 0 \end{bmatrix}^2\right] = \mathscr{D}\left[\begin{bmatrix} F^*F & 0 \\ 0 & FF^* \end{bmatrix}\right]$$
$$= \mathscr{D}(F^*F) \oplus \mathscr{D}(FF^*). \tag{3.4}$$

Then (3.3) can be rewritten as the system

$$iv' = F^*w, \qquad iw' = Fv, \tag{3.5}$$

which can be differentiated with respect to t to yield

$$i\begin{bmatrix} v' \\ w' \end{bmatrix}' = \begin{bmatrix} 0 & F^* \\ F & 0 \end{bmatrix}\begin{bmatrix} v' \\ w' \end{bmatrix}$$

with initial data

$$v'(0) = -iF^*y_0, \qquad w'(0) = -iFx_0.$$

Let P_0 [resp. Q_0] denote the orthogonal projection of $\mathscr{K}_1$ [resp. $\mathscr{K}_2$] onto the closed subspace $\mathscr{N}(F)$[resp. $\mathscr{N}(F^*)$]. We have

$$v'(0) = -iF^*(I - Q_0)y_0, \qquad w'(0) = -iF(I - P_0)x_0. \tag{3.6}$$

Moreover, $\{v(t), w(t) : t \in \mathbb{R}\}$ determines and is determined by

$$\{v'(t), w'(t) : t \in \mathbb{R}\}$$

together with $Q_0 y_0$ and $P_0 x_0$. This is so because of (3.6) and (3.4), since F^* [resp. F] is injective on $(I - Q_0)(\mathscr{D}(F^*))$ [resp. $[I - P_0)(\mathscr{D}(F))$]. Therefore we can and do identify the system (3.3) with the system

$$i\begin{bmatrix} v' \\ w' \\ P_0 x_0 \\ Q_0 y_0 \end{bmatrix}' = \begin{bmatrix} 0 & F^* & 0 & 0 \\ F & 0 & 0 & 0 \\ 0 & 0 & 0 & 0 \\ 0 & 0 & 0 & 0 \end{bmatrix}\begin{bmatrix} v' \\ w' \\ P_0 x_0 \\ Q_0 y_0 \end{bmatrix} \tag{3.7}$$

which is of the form (2.1) (see Definition 2.4).

Now let P_1 [resp. P_2] denote the orthogonal projection of

$$\mathscr{H} = \mathscr{K}_1 \oplus \mathscr{K}_2 \oplus \mathscr{N}(F) \oplus \mathscr{N}(F^*) \tag{3.8}$$

onto the first two [resp. the last two] components. Then $\{P_1, P_2\}$ defines a division of energy for (3.7), which we can think of as a division of energy for (3.3). To prove the theorem we must show that the partial energy associated with P_1 admits equipartition of energy.

Differentiating (3.5) with respect to t yields

$$\tilde{v}'' + F^*F\tilde{v} = 0, \tag{3.9}$$

$$\tilde{w}'' + FF^*\tilde{w} = 0, \tag{3.10}$$

where $\tilde{v}(t) = v(t) - P_0 x_0$, $\tilde{w}(t) = w(t) - Q_0 y_0$. Letting $\|\cdot\|_j$ be the norm in $\mathscr{K}_j$ for $j = 1, 2$ we define, for $t \in \mathbb{R}$,

$$\mathscr{E}_K(t; 1, 1) = \|\tilde{v}'(t)\|_1^2, \qquad \mathscr{E}_P(t; 1, 1) = \|F\tilde{v}(t)\|_2^2,$$

$$\mathscr{E}_K(t; 1, 2) = \|\tilde{w}'(t)\|_2^2, \qquad \mathscr{E}_P(t; 1, 2) = \|F^*\tilde{w}(t)\|_1^2.$$

These are the kinetic and potential energies associated with (3.9) and (3.10). We denote the associated total energies by $\mathscr{E}(1, j)$, $j = 1, 2$. The next to last entry 1 in the argument of each of these energies refers to the subscript 1 of P_1. $\mathscr{E}(1, 1)$ [or equivalently $\mathscr{E}(1, 2)$, as we shall soon see] is the partial energy which will be shown to be equipartitioned.

Note that

$$\begin{aligned}\mathscr{E}_P(t; 1, 1) &= \langle F\tilde{v}(t), F(\tilde{v})\rangle_2 = \langle F^*F\tilde{v}(t), \tilde{v}(t)\rangle_1 \\ &= \|(F^*F)^{1/2}\tilde{v}(t)\|_1^2,\end{aligned}$$

and similarly

$$\mathscr{E}_P(t; 1, 2) = \|(FF^*)^{1/2}\tilde{w}(t)\|_2^2,$$

where $\langle\cdot, \cdot\rangle_j$ is the inner product in $\mathscr{K}_j$. By hypothesis, both $(F^*F)^\alpha$ and $(FF^*)^\alpha$ are essentially absolutely continuous for $\alpha = 1$; and it follows easily that the same assertion holds for $\alpha = \frac{1}{2}$. It follows from this and Theorem 2.1′ that the energies for (3.9), (3.10) are equipartitioned as long as the initial data is orthogonal to a certain null space. However, the initial data does have this orthogonality property because of the way $\tilde{v}$ and $\tilde{w}$ were constructed. Consequently

$$\mathscr{E}_K(t; 1, j), \mathscr{E}_P(t; 1, j) \to \tfrac{1}{2}\mathscr{E}(1, j) \tag{3.11}$$

for $j = 1, 2$ as $t \to \pm\infty$. On the other hand,

$$\mathscr{E}_P(t; 1, 1) = \|F\tilde{v}(t)\|_2^2 = \|\tilde{w}'(t)\|_2^2 = \mathscr{E}_K(t; 1, 2) \tag{3.12}$$

by (3.5), and similarly

$$\mathscr{E}_K(t; 1, 1) = \mathscr{E}_P(t; 1, 2). \tag{3.13}$$

Let Q_1, Q_2 denote, respectively, the orthogonal projections of $\mathscr{H}$ onto its first and onto its last three components [see (3.8)]. Then Q_1, Q_2 is a proper finite projection system with respect to which the partial energy $\mathscr{E}^1$ associated with P_1 admits equipartition of energy with constants $c_1 = c_2 = \frac{1}{2}$, since we know that, letting u denote the column vector on the right-hand side of (3.7),

$$\begin{aligned}\|Q_1P_1u(t)\|_{\mathscr{H}}^2 &= \|v'(t)\|_1^2 = \|\tilde{v}'(t)\|_1^2 = \mathscr{E}_K(t; 1, 1), \\ \|Q_2P_1u(t)\|_{\mathscr{H}}^2 &= \|w'(t)\|_2^2 = \|\tilde{w}'(t)\|_2^2 = \mathscr{E}_P(t; 1, 1),\end{aligned}$$

and our assertion follows from (3.11). ■

Consider now the equalities (3.12) and (3.13) in the special case of Maxwell's equations. We treat the case where the current and charge densities are zero, and units are chosen so that the speed of light is one. Then $n = 3$,

$m = 6$, $v = E =$ the electric vector (field at time t), $w = B =$ the magnetic vector,

$$F = i\,\text{curl} = \begin{bmatrix} 0 & -D_3 & D_2 \\ D_3 & 0 & -D_1 \\ -D_2 & D_1 & 0 \end{bmatrix} = F^*,$$

and $\mathscr{K}_1 = \mathscr{K}_2 = L^2(\mathbb{R}^3; \mathbb{C}^3)$. In this case F is essentially absolutely continuous. (This follows, for example, from a result of Avila [1].) From the calculus we know that

$$\mathscr{N}(F) = \{h \in L^2(\mathbb{R}^3; \mathbb{C}^3) : h = \nabla\varphi \text{ for some } \varphi \in W^{1,2}(\mathbb{R}^3)\},$$

which is the closure in $\mathscr{K}_1$ of $\{\nabla\varphi : \varphi \in \mathbb{C}^\infty(\mathbb{R}^3),\ \varphi \text{ has compact support}\}$; here $W^{1,2}(\mathbb{R}^3)$ is the usual Sobolev space $\{\psi \in L^2(\mathbb{R}^3) : D_j\psi \in L^2(\mathbb{R}^3) \text{ for } j = 1, 2, 3\}$. Equation (3.12) [resp. (3.13)] says that the potential [resp. kinetic] energy of the electric vector equals the kinetic [resp. potential] energy of the magnetic vector. Thus there are only two kinds of energy, not four, found in the equipartitioning.

The use of the terms kinetic and potential is thus a matter of convenience, not necessity. To further illustrate this, consider the case of the n-dimensional wave equation

$$u_{tt} = \Delta u \qquad (x \in \mathbb{R}^n, \quad t \in \mathbb{R}).$$

If u represents the transversal displacement of an n-dimensional "string," then $\int_{\mathbb{R}^n} |u_t(t, x)|^2\, dx$ is the kinetic energy of the string. On the other hand, if u represents a velocity potential, then $\int_{\mathbb{R}^2} |u_t(t, x)|^2\, dx$ is interpreted as potential energy. Thus one should be cautious in the use of the terms kinetic and potential energy. "kth partial energy" is neutral but bland; terms like kinetic and potential energy are suggestive, but they should not be taken too seriously.

4. MORE ON SYMMETRIC HYPERBOLIC SYSTEMS

We begin by summarizing some aspects of the "potential decompositions" recently obtained by Gilliam and Schulenberger [6, 7].

Let $A(\xi) = \sum_{j=1}^{n} A_j \xi_j$ be as in the first paragraph of Section 3. We suppose that

$$A(D) = \begin{bmatrix} 0 & -i\Lambda(D) \\ i\Lambda^*(D) & 0 \end{bmatrix} \tag{4.1}$$

as in (3.2). Here, as before, $\Lambda(\xi)$ is a $k \times (m-k)$ matrix, $\Lambda^*(\xi)$ is an $(m-k) \times k$ matrix, etc. The symbol $A(\xi)$ is for $\xi \in \mathbb{R}^n \setminus \{0\}$ a Hermitian matrix having eigenvalues $\lambda_j(\xi)$ which we write in ascending order:

$$\lambda_{-N}(\xi) \leqq \cdots \leqq \lambda_{-1}(\xi) \leqq \lambda_0(\xi) = 0 \leqq \lambda_1(\xi) \leqq \cdots \leqq \lambda_N(\xi).$$

For $j \neq 0$, $\xi \to \lambda_j(\xi)$ is homogeneous of degree one and continuous on $\mathbb{R}^n \setminus \{0\}$; moreover, its multiplicity ν_j is constant off a closed set in $\mathbb{R}^n$ of Lebesgue measure zero. Zero may not be an eigenvalue of $A(\xi)$ [i.e., $A(D)$ may be elliptic], but in the applications it usually is; and in all cases our convention will be to include it in the listing of the eigenvalues. [For example, when the classical d'Alembert wave equation is written as a symmetric hyperbolic system in the usual way, the resulting symbol $A(\xi)$ has a nontrivial null space.] Note that $\lambda_j(\xi) = -\lambda_{-j}(\xi)$, and $\nu_j \geqq 1$ for each $j \neq 0$, although $\nu_0 = 0$ is possible.

Let $\hat{P}_j(\xi)$ be the orthogonal projection onto the eigenspace in $\mathbb{C}^m$ corresponding to the eigenvalue $\lambda_j(\xi)$. Then the projections $\hat{P}_0(\xi)$, $\hat{P}_j(\xi) + \hat{P}_{-j}(\xi)$, $j = 1, \ldots, N$ define the resolution of the identity appearing in the spectral decomposition of the matrix $A(\xi)$. In terms of the Fourier transformed projections $P_j = \Phi^{-1}\hat{P}_j\Phi$ on $\mathscr{H} = L^2(\mathbb{R}^n; \mathbb{C}^m)$, set $\mathscr{H}_j = (P_j + P_{-j})(\mathscr{H})$, so that $\mathscr{H} = \bigoplus_{j=0}^N \mathscr{H}_j$. Let H be the self-adjoint operator $A(D)$ and let H_j be the restriction of H to $\mathscr{H}_j$. For $j \geqq 1$, Gilliam and Schulenberger [7] explicitly constructed (with the aid of the Fourier transform) a unitary mapping V_j from $\mathscr{H}_j$ to an auxiliary Hilbert space $\mathscr{K}_j$ such that $H_j = V_j^{-1}K_jV_j$ where K_j is the self-adjoint pseudodifferential operator

$$K_j = \begin{bmatrix} 0 & I_j \\ -\lambda_j(D)I_j & 0 \end{bmatrix};$$

here I_j is the $\nu_j \times \nu_j$ identity matrix times the identity operator on $L^2(\mathbb{R}^n)$. The space $\mathscr{K}_j$ is the completion of the pairs of $\mathbb{C}^{\nu_j}$-valued smooth compactly supported functions on $\mathbb{R}^n$ under the norm

$$\left\| \begin{bmatrix} f_1 \\ f_2 \end{bmatrix} \right\|_{\mathscr{K}_j} = \left\{ \int_{\mathbb{R}^n} [\lambda_j(\xi)^2 |\hat{f}_1(\xi)|^2 + |\hat{f}_2(\xi)|^2] \, d\xi \right\}^{1/2},$$

and K_j has domain

$$\mathscr{D}(K_j) = \left\{ \begin{bmatrix} f_1 \\ f_2 \end{bmatrix} \in \mathscr{K}_j : \int_{\mathbb{R}^n} [\lambda_j(\xi)^4 |\hat{f}_1(\xi)|^2 + \lambda_j(\xi)^2 |\hat{f}_2(\xi)|^2] \, d\xi < \infty \right\}.$$

Let Q_j be the orthogonal projection of $\mathscr{H}$ onto $\mathscr{H}_j$. It is easy to see that the projections $Q_0, Q_1, \ldots, Q_N$ define a division of energy for H. The associated partial energies will be denoted by $\mathscr{E}^j$; thus $\mathscr{E}^j(f) = \|Q_j f\|^2$ for $f \in \mathscr{H}$.

For $j \geqq 1$, the equation $i\, du/dt = H_j u$ in $\mathscr{H}_j$ is (unitarily) equivalent to the equation

$$\frac{d^2 v}{dt^2} + \lambda_j(D)^2 v = 0$$

in the Hilbert space

$$\left\{ g \in L^2(\mathbb{R}^n; \mathbb{C}^{\nu_j}) : \begin{bmatrix} 0 \\ g \end{bmatrix} \in \mathscr{K}_j \right\}.$$

Moreover, it follows from a theorem of Avila [1] that for $j \geqq 1$, $\lambda_j(D)$ is absolutely continuous. Consequently by Theorem 2.1, H_j admits equipartition of energy.

We have therefore proved the following result.

Theorem 4.1. *Let $A(D)$ have the form* (4.1), *and let N be the number of distinct positive eigenvalues of $A(\xi)$; N is well defined since this number is a constant function of ξ a.e. Then there is a finite projection system $Q_0, Q_1, \ldots, Q_N$ on $\mathscr{H}$ which defines a division of energy for $H = A(D)$. Moreover Q_0 is the orthogonal projection onto the null space of H; and the* j*th partial energy $\mathscr{E}^j(f) = \|Q_j e^{-itH} f\|^2$ admits equipartition into its kinetic and potential parts, so that, using obvious notation,*

$$\mathscr{E}_K^j(t),\ \mathscr{E}_P^j(t) \to \tfrac{1}{2}\mathscr{E}^j(f)$$

as $t \to \pm\infty$ for all initial data $f \in \mathscr{H}$ and $j = 1, \ldots, N$.

Thus there are $2N + 1$ kinds of energy associated with (4.1), including one for the zero eigenvalue.

Example 4.1. Consider the equation for elastic waves in a two-dimensional homogeneous medium:

$$\frac{\partial^2 z}{\partial t^2} = \mu_0\, \Delta z + (\lambda_0 + \mu_0)\, \nabla(\nabla \cdot z) \tag{4.2}$$

where $z = \left[\begin{smallmatrix} p \\ q \end{smallmatrix}\right]$ is the displacement vector and $\lambda_0, \mu_0 > 0$ are the Lamé parameters of the medium. In component form this becomes

$$\frac{\partial^2 p}{\partial t^2} = \mu_0\, \Delta p + (\lambda_0 + \mu_0)\left(\frac{\partial^2 p}{\partial x^2} + \frac{\partial^2 q}{\partial x \partial y}\right),$$

$$\frac{\partial^2 q}{\partial t^2} = \mu_0\, \Delta q + (\lambda_0 + \mu_0)\left(\frac{\partial^2 p}{\partial x \partial y} + \frac{\partial^2 q}{\partial y^2}\right).$$

Let

$$u(t, \xi) \qquad \text{for} \quad t \in \mathbb{R} \quad \text{and} \quad \xi = \begin{bmatrix} \xi_1 \\ \xi_2 \end{bmatrix} \in \mathbb{R}^2$$

be the vector in $\mathbb{C}^5$ having as its components (in order):

$$i(\lambda_0+\mu_0)^{1/2}(\xi_1\hat{p}(t,\xi)+\xi_2\hat{q}(t,\xi)), \qquad \mu_0^{1/2}i(\xi_1\hat{p}(t,\xi)-\xi_2\hat{q}(t,\xi)),$$
$$\mu_0^{1/2}i(\xi_1\hat{p}(t,\xi)+\xi_2\hat{q}(t,\xi)), \qquad \partial p(t,\xi)/\partial t,\ \partial q(t,\xi)/\partial t,$$

and let

$$\Lambda(\xi)=\begin{bmatrix}(\lambda_0+\mu_0)^{1/2}\xi_1 & (\lambda_0+\mu_0)^{1/2}\xi_2\\ \mu_0^{1/2}\xi_1 & -\mu_0^{1/2}\xi_2\\ \mu_0^{1/2}\xi_2 & \mu_0^{1/2}\xi_1\end{bmatrix}.$$

Then (4.2) becomes equivalent to $i\,\partial u/\partial t = A(D)u$ where $A(D)$ is defined by (4.1), and the norm on $\mathscr{H} = L^2(\mathbb{R}^2;\mathbb{C}^5)$ is defined by

$$\|u\|_{\mathscr{H}}^2 = \int_{\mathbb{R}^n}\left\{\sum_{j=0}^{2}|u_{2j+1}(x)|^2+\mu_0|\nabla u_2(x)|^2+(\lambda_0+2\mu_0)|\nabla u_4(x)|^2\right\}dx.$$

Define the unitary operator $\sigma(D)$ from $L^2(\mathbb{R}^2,\mathbb{C}^5)$ (with its usual norm) to H by

$$\sigma(\xi)=\begin{bmatrix}\mu_0^{1/2}\alpha & \alpha\beta\gamma|\xi| & -2\alpha\gamma\delta|\xi| & 0 & 0\\ 0 & 2\mu_0^{1/2}\delta & \beta\mu_0^{-1/2} & 0 & 0\\ 0 & 0 & 0 & i\xi_2|\xi|^{-1} & -i\xi_1|\xi|^{-1}\\ \gamma\alpha^2|\xi|^{-1} & \mu_0^{-1/2}\beta\alpha^2 & 2\mu_0^{1/2}\delta\alpha^2 & 0 & 0\\ 0 & 0 & 0 & i\xi_1|\xi|^{-1} & i\xi_2|\xi|^{-1}\end{bmatrix},$$

where $\alpha=(\lambda_0+2\mu_0)^{-1/2}$, $\beta=(\xi_1^2-\xi_2^2)/|\xi|^3$, $\gamma=(\lambda_0+\mu_0)^{1/2}$, and $\delta=\xi_1\xi_2/|\xi|^3$.

If $v=\sigma(D)u$ where $i\,\partial u/\partial t = A(D)u$, then $-i|\xi|v(t,\xi)$ has components $0, \xi_2\hat{p}(t,\xi)-\xi_1\hat{q}(t,\xi), \xi_2\hat{p}_t(t,\xi)-\xi_1\hat{q}_t(t,\xi), \xi_1\hat{p}(t,\xi)+\xi_2\hat{q}(t,\xi), \xi_1\hat{p}_t(t,\xi)+\xi_2\hat{q}_t(t,\xi)$, where $\hat{p}_t=(\partial p/\partial t)^\wedge$, etc., and v satisfies $i\,\partial v/\partial t = H(D)v$ where

$$H(\xi)=\begin{bmatrix}0 & 0 & 0 & 0 & 0\\ 0 & 0 & 1 & 0 & 0\\ 0 & -\mu_0|\xi|^2 & 0 & 0 & 0\\ 0 & 0 & 0 & 0 & 1\\ 0 & 0 & 0 & -(\lambda_0+2\mu_0)|\xi|^2 & 0\end{bmatrix}.$$

Let Q_j be the orthogonal projection onto the jth component in $L^2(\mathbb{R}^2;\mathbb{C}^5)$. Let $Q_S=Q_2+Q_3$, $Q_P=Q_4+Q_5$. Then $\mathscr{E}_S=\|Q_S v(t)\|^2$ and $\mathscr{E}_P=\|Q_P v(t)\|^2$ are conserved quantities; they are the energies of the shear and pressure waves. Also, Q_0 [which projects onto $\mathscr{N}(H(D))$], Q_S, Q_P define a division of energy for $H(D)$. Moreover, $\|Q_0 \sigma u(0)\| = 0$ for all initial data for (4.2); and

$$w_1 = \begin{bmatrix} v_2 \\ v_3 \end{bmatrix} = Q_S v \qquad \text{and} \qquad w_2 = \begin{bmatrix} v_4 \\ v_5 \end{bmatrix} = Q_P v$$

satisfy

$$i \frac{dw_j}{dt} = \begin{bmatrix} 0 & I \\ -c_j \Delta & 0 \end{bmatrix} w_j$$

where $c_1 = \mu_0$ and $c_2 = \lambda_0 + 2\mu_0$. These are classical wave equations in $\mathbb{R}^2$. It follows that $\mathscr{E}_S$ and $\mathscr{E}_P$ each splits into its kinetic and potential parts, and each of these becomes equipartitioned as $t \to \pm\infty$. In this example there are $2N = 4$ kinds of energy, since no energy is associated with Q_0.

A similar analysis can be given for three dimensional elastic waves.

The above example is adapted from [7], which contains a number of other nice examples. The next example, also taken from [7], is of a symmetric hyperbolic system in which $A(D)$ is not of the form (4.1) and yet for which we can deduce equipartition of energy.

Example 4.2. Gilliam and Schulenberger [7] have shown that the operator

$$A(D) = \begin{bmatrix} D_3 & D_1 - iD_2 \\ D_1 + iD_2 & -D_3 \end{bmatrix} \tag{4.3}$$

associated with the neutrino equation $i\,\partial u/\partial t = A(D)u$ on $L^2(\mathbb{R}^3; \mathbb{C}^2)$ is unitarily equivalent to the operator $\left[\begin{smallmatrix} 0 & I \\ \Delta & 0 \end{smallmatrix}\right]$ which governs the classical wave equation $\partial^2 v/\partial t^2 = \Delta v$ for $x \in \mathbb{R}^3$. It follows that the neutrino operator (4.3) admits equipartition of energy.

While Theorem 4.1 gives us $2N + 1$ energies (or $2N$ if zero is not an eigenvalue), there may actually be more. That is, there may be more than two energies associated with a λ_j^2, as the following example illustrates.

Example 4.3. Recall that in Section 3 we considered Maxwell's equations

$$i \begin{bmatrix} v \\ w \end{bmatrix}' = \begin{bmatrix} 0 & F \\ F & 0 \end{bmatrix} \begin{bmatrix} v \\ w \end{bmatrix}$$

where $F = i$ curl and $v(=E)$, $w(=B) \in L^2(\mathbb{R}^3; \mathbb{C}^3)$. The symbol $A(\xi)$ has eigenvalues $\lambda_{\pm 1} = \pm|\xi|, \lambda_0 = 0$, each of multiplicity two. Using ideas similar to those of Example 4.1, one can show that Maxwell's equations are similar to the system $iu' = Hu$ where

$$H = i\begin{bmatrix} 0_2 & 0_2 & 0_1 \\ 0_2 & 0_2 & I_2 \\ 0_2 & -|\xi|^2 I_2 & 0_2 \end{bmatrix}, \qquad 0_2 = \begin{bmatrix} 0 & 0 \\ 0 & 0 \end{bmatrix}, \qquad I_2 = \begin{bmatrix} 1 & 0 \\ 0 & 1 \end{bmatrix}.$$

The energy associated with the null space of $A(D)$ is $\|\nabla \cdot v\|^2 + \|\nabla \cdot w\|^2$, and each summand is in fact a conserved quantity. Also, we have an equipartition of energy associated with the projections onto the second and third pairs of components of the solution u.

Now we can look at this differently by considering $Q_1 = P_3 + P_5$ and $Q_2 = P_4 + P_6$, where P_j is the projection onto the jth component $(1 \leqq j \leqq 6)$. Then $\tilde{u}_j = Q_j u$ satisfies

$$i\tilde{u}_j' = \begin{bmatrix} 0 & I \\ -|\xi|^2 I & 0 \end{bmatrix} \tilde{u}_j, \qquad j = 1, 2,$$

and both these equations are governed by unitary groups. Consequently we have a division of energy associated with $\{P_1, P_2, Q_1, Q_2\}$; and we have an equipartition of energy for both Q_1 and Q_2 (with kinetic and potential energies for each).

In a similar manner it can be shown that if $A(D)$ has the form (4.1), then there is a division and equipartition of energy, and the number of energies is m, the number of components of the solution $u(t) = \exp(-itA(D))f$.

We conclude with a few remarks.

Remark 4.1. Let $i\,\partial u/\partial t = A(D)u$ be a symmetric hyperbolic system satisfying the hypotheses of Theorem 4.1 and those of Costa's theorem [4]. By Theorem 4.1, for $j \geqq 1$, the jth partial energy is asymptotically equipartitioned into its kinetic and potential parts as $t \to \pm\infty$. By Costa's proof, in which the Radon transform gives an explicit translation representation for $e^{itA(D)}$, the energy is equipartitioned from a finite time on [in the sense of (1.2)] whenever the initial data is compactly supported.

Concerning elastic waves in $\mathbb{R}^n$, everything reduces to wave equation operators of the form $\left[\begin{smallmatrix} 0 & I \\ c\Delta & 0 \end{smallmatrix}\right]$ where $c > 0$. Thus, by the results of Lax and Phillips [13] and Duffin [5], we get equipartition of energy from a finite time on when $n = 3$ but not when $n = 2$.

Remark 4.2. (Cf.[4].) Let H_0, H_1 be self-adjoint operators on $\mathscr{H}_0, \mathscr{H}_1$ respectively, and let $J: \mathscr{H}_0 \to \mathscr{H}_1$ be an "identification" operator (e.g., $\mathscr{H}_0 = \mathscr{H}_1$ and $J = I$). Scattering theory (cf. [12, 13]) tells us that in many cases, for the solution u of $i\, du/dt = H_1 u$, $u(0) = f \in \mathscr{H}_1$ there exist $f_\pm \in \mathscr{H}_0$ such that

$$\|u(t) - Je^{-itH_0} f_\pm\|_{\mathscr{H}_1} \to 0$$

as $t \to \pm\infty$. Moreover, as f ranges over $\mathscr{H}_{1,\mathrm{ac}}(H_1)$, both $f_\pm$ range over $\mathscr{H}_{0,\mathrm{ac}}(H_0)$. (This is just the existence and completeness of the wave operators.) Thus for J isometric, the asymptotic behavior of e^{-itH_1} on $\mathscr{H}_{1,\mathrm{ac}}(H_1)$ is determined by the asymptotic behavior of e^{-itH_0} on $\mathscr{H}_{0,\mathrm{ac}}(H_0)$.

Consequently we can extend our results on equipartition of energy to symmetric hyperbolic systems with variable coefficients and/or in exterior domains for which scattering theory has been done. A useful reference for this is [14].

REFERENCES

1. G. S. S. Avila, Spectral resolution of differential operators associated with symmetric hyperbolic systems, *Applicable Anal.* **1** (1972), 283–299.
2. L. Bobisud and J. Calvert, Energy bounds and virial theorems for abstract wave equations, *Pacific J. Math.* **47** (1973), 27–37.
3. A. R. Brodsky, On the asymptotic behavior of solutions of the wave equation, *Proc. Amer. Math. Soc.* **18** (1967), 207–208.
4. D. G. Costa, On equipartition of energy for uniformly propagative systems, *J. Math. Anal. Appl.* **58** (1977), 56–62.
5. R. J. Duffin, Equipartition of energy in wave motion, *J. Math. Anal. Appl.* **32** (1970), 386–391.
6. D. S. Gilliam, Vector potentials for symmetric hyperbolic systems (to appear).
7. D. S. Gilliam and J. R. Schulenberger, A class of symmetric hyperbolic systems with special properties, *Comm. Partial Differential Equations* **4** (1979), 509–536.
8. J. A. Goldstein, An asymptotic property of solutions of wave equations, *Proc. Amer. Math. Soc.* **23** (1969), 359–363.
9. J. A. Goldstein, An asymptotic property of solutions of wave equations, II, *J. Math. Anal. Appl.* **32** (1970), 392–399.
10. J. A. Goldstein and J. T. Sandefur, Jr., Asymptotic equipartition of energy for differential equations in Hilbert Space, *Trans. Amer. Math. Soc.* **219** (1976), 379–406.
11. J. A. Goldstein and J. T. Sandefur, Jr., Abstract equipartition of energy theorems, *J. Math. Anal. Appl.* **67** (1979), 58–74.
12. T. Kato, "Perturbation Theory for Linear Operators," Springer-Verlag, Berlin and New York, 1966.
13. P. D. Lax and R. S. Phillips, "Scattering Theory," Academic Press, New York, 1967.
14. J. R. Schulenberger and C. H. Wilcox, Completeness of the wave operators for perturbations of uniformly propagative systems, *J. Funct. Anal.* **7** (1971), 447–474.

15. H. A. LEVINE, An equipartition of energy theorem for weak solutions of evolutionary equations in Hilbert space: The Lagrange identity method, *J. Diff. Eqns.* **24** (1977), 197–210.

This work was partially supported by N.S.F. Grant MCS78-01251.

Jerome A. Goldstein
DEPARTMENT OF MATHEMATICS
TULANE UNIVERSITY
NEW ORLEANS, LOUISIANA

James T. Sandefur, Jr.
DEPARTMENT OF MATHEMATICS
GEORGETOWN UNIVERSITY
WASHINGTON, D.C.

Geometric Methods in the Existence and Construction of Fixed Points of Nonexpansive Mappings

L. A. KARLOVITZ

1. INTRODUCTION

A mapping $T: C \to X$ defined on a subset C of a Banach space X, with norm $|\cdot|$, is said to be *nonexpansive* if $|Tx - Ty| \leqq |x - y|$ for all $x, y \in C$. This class of mappings has played a central role in the developments of nonlinear functional analysis in recent years. The theory of such mappings is particularly well developed in the case that X is a Hilbert space.

If C is assumed to be convex and weakly compact and if $T: C \to C$ then one of the main open questions is whether T has a *fixed-point* in C, i.e., whether there exists $x \in C$ so that $Tx = x$. If X is a Hilbert space then the answer is affirmative; moreover, fixed points can be constructed as limits, in the norm, of approximate fixed points. If X has a uniformly convex unit ball or, more generally, if X has normal structure in the sense of Krein–Milman then the existence question has an affirmative answer. These basic existence results are due to Browder [2], Göhde [3], and Kirk [9]. The existence theory has only recently been extended beyond the normal structure case in Karlovitz [6]. It has also been extended for somewhat more restricted classes of convex sets in Karlovitz [7] and Odell and Sternfeld [11].

Our purpose here is to extend both the existence and constructive aspects of the theory. This is achieved in two separate developments which, however, have the common theme of making use of Hilbert space ideas and their generalizations. The first development extends the constructive theorem for Hilbert spaces to a large class of Banach spaces which are characterized by the fact that the Birkhoff–James orthogonality relation satisfies a property of approximate symmetry and by the dual space having a uniformly convex

ISBN 0-12-178150-X

unit ball. This includes, in particular, the classical sequence spaces l^p, $1 < p < \infty$. The second development explores the existence question for Banach spaces which are isomorphic to some Hilbert space. It is shown that if the Banach–Mazur distance to some Hilbert space is not too large then the answer is affirmative. It is also shown that the answer remains affirmative for an increased distance for a certain special class of Banach spaces.

2. APPROXIMATE FIXED POINTS

Much of the theory of nonexpansive mappings relies on the use of approximate fixed points, which can be constructed in various ways. The most common is the following. Let $T: C \to C$ be nonexpansive and let y_0 be a given point of C. Define T_s, $0 < s < 1$, by $T_s x = (1 - s)y_0 + sTx$. Then $T_s: C \to C$, and it is a strict contraction with $|T_s x - T_s y| \leqq s|x - y|$. Hence by the Banach contraction principle it has a unique fixed point denoted by x_s, i.e.,

$$T_s x_s = x_s = (1 - s)y_0 + sTx_s. \tag{1}$$

Thus $|Tx_s - x_s| \leqq (1 - s/s)|x_s - y_0| \to 0$ as $s \to 1$. A standard argument also shows that $x_s: (0, 1) \to X$ is *continuous in the norm*, in fact,

$$|x_s - x_t| \leqq (|s - t|/|1 - s|)(|y_0| + |Tx_t|).$$

The following general lemma for nonexpansive mappings, developed in Karlovitz [6, 8], is one of our main tools.

Lemma. *Let X be a Banach space. Let C_0 be a weakly compact convex subset of X, and let $T: C_0 \to C_0$ be nonexpansive. Suppose that C_0 is minimal in the sense that it contains no proper closed convex subsets which are invariant under T. Let $\{x_n\}$ be a sequence of approximate fixed points in C_0, $|Tx_n - x_n| \to 0$. Then for each $x \in C_0$,*

$$\lim_n |x - x_n| = \operatorname{diam} C_0 = \sup\{|u - v| : u, v \in C_0\}. \tag{2}$$

A standard Zorn's lemma argument can be used to establish the existence of a subset C_0 of C which is minimal in the sense of the lemma. If C_0 is a singleton then, by the invariance, it is a fixed point of T. If not, then we know from the discussion above that there exists a sequence of approximate fixed points in C_0. It then follows directly from (2) that X fails to have normal structure. For the convex set C_0 has the property that each of its points is diametral, i.e., the supremum of this distance to other points of the set equals the diameter of the set; while normal structure requires that every bounded convex set contains at least one point which is not diametral. Hence we have given an alternate proof to the classical existence theorems. We have used this lemma in [6, 8] to derive existence results in the case that the space fails normal structure. This is also done in Section 4 below. In each of the latter applications of the lemma, more explicit use of the geometry of the space is made.

3. SPACES WHEREIN THE BIRKHOFF–JAMES ORTHOGONALITY RELATION IS APPROXIMATELY SYMMETRIC

In a general Banach space X we say (see Birkhoff [1] and James [4]) that w is orthogonal to v, $w \perp v$, if $|w| \leqq |w + \lambda v|$ for all scalars λ. In general, the relation $\perp$ is not symmetric. (Indeed, symmetry characterizes Hilbert spaces for dimension strictly greater than two.) We say that the relation $\perp$ is *uniformly approximately symmetric* if for each $x \in X$, $x \neq 0$, and $\varepsilon > 0$ there exists a closed linear subspace $U = U(x, \varepsilon)$ so that

$$U \text{ has finite codimension, and } \delta + |u| \leqq |u + \lambda x|, \text{ for some } \delta > 0, \text{ for each } u \in U, |u| = 1, \text{ and for each } \lambda, |\lambda| \geqq \varepsilon. \tag{3}$$

Examples and Remarks. The geometric interpretation of the property is clear. If X is a Hilbert space we can choose $U(x, \varepsilon) = \{u : x \perp u$, i.e., x and u orthogonal in the inner product$\}$, independently of ε. For U has codimension 1 and, by virtue of $x \perp u$, $|u + \lambda x|^2 = |u|^2 + \lambda^2|x|^2$, from which (3) follows. In the classical sequence spaces l^p, $1 < p < \infty$, the Birkhoff–James relation $\perp$ is also uniformly approximately symmetric. Given $x = (v_i)$ and $\varepsilon > 0$, it is readily shown that we can choose $N = N(x, \varepsilon)$ so that $U = \{u = (\omega_i) : \omega_i = 0, i = 1, \ldots, N\}$ satisfies (3). In the space l^1 as well as in the newly defined space J_0 (see James [5] and Lindenstrauss and Stegall [10]) the relation $\perp$ satisfies (3) with the additional property that U can be chosen to be weak star closed. This is discussed and applied in [7]. On the other hand, in the function spaces $L^p[0, 1]$, $p \neq 2$, the relation $\perp$ fails to be uniformly approximately symmetric; in fact, (3) cannot even be satisfied with $\delta = 0$. The connection of the uniform approximate symmetry of $\perp$ to the so-called Opial condition of convergence is discussed in [7]. The connection to the weak continuity of the duality mapping will be discussed in a forthcoming paper.

Theorem 1. (Existence) *Let X be a Banach space in which the Birkhoff–James relation $\perp$ is uniformly approximately symmetric. Let C be a weakly compact convex subset and $T: C \to C$ nonexpansive. Then T has a fixed point in C; in fact, if $\{x_n\}$ is a sequence of approximate fixed points in C, $|Tx_n - x_n| \to 0$, and $x_n \to z$, in the weak topology, then $Tz = z$.*

(Construction). *Assume, in addition, that the unit ball of the dual space X^* is uniformly convex. Let y_0 be a given point in C, and let x_s, $0 < s < 1$, be approximate fixed points given by (1). Then $|x_s - z| \to 0$ as $s \to 1$, for some $z \in C$ and $Tz = z$.*

A somewhat more general version of Part 1 (existence) is given in Karlovitz [7]. The proof of the new construction of Part 2 (construction) given herein makes use of Part 1, hence it is included. In order to fully see the constructive nature of Part 2, we remark that the approximate fixed points x_s can be found by a standard iterative procedure. A specialization to the Hilbert space case of the following proofs provides a particularly simple approach to the known results; moreover, it provides some of the motivation for the new results.

Proof of Theorem 1. Part 1: By the discussion of Section 2 we can choose a sequence $\{x_n\}$ in C so that $|Tx_n - x_n| \to 0$ as $n \to \infty$. Without loss of generality we may assume that x_n converges in the weak topology to some $z \in C$ and $\lim|x_n - z| = r$. We let $x = Tz - z$. If either $r = 0$ or $x = 0$ then $Tz = z$ and we are done. Assume the converse. By hypothesis there exists a closed linear subspace $U = U(x, 1/2r)$ so that (3) is satisfied. Thus for each n

$$x_n - z = \lambda_n x + u_n + v_n,$$

where $u_n \in U$ and v_n is in the finite dimensional complement of span$\{x, U\}$. By the weak convergence, $\lambda_n, |v_n| \to 0$ and $|u_n| \to r$. So for sufficiently large n, $|1 - \lambda_n|/|u_n| \geqq 1/2r$, and, by virtue of (3),

$$\begin{aligned} |x_n - Tz| &= |x_n - z + z - Tz| = |(\lambda_n - 1)x + u_n + v_n| \\ &\geqq |u_n|(1 + \delta) - |v_n|. \end{aligned}$$

By the nonexpansiveness of T,

$$|x_n - Tz| \leqq |Tx_n - Tz| + |x_n - Tx_n| \leqq |x_n - z| + |x_n - Tx_n|.$$

A combination of the inequalities contradicts: $|v_n|, |x_n - Tx_n| \to 0$ and $|u_n|, |x_n - z| \to r$. This concludes the proof of Part 1.

Part 2: Without loss of generality we may assume that $y_0 = 0$. Let x_s be given by (1), i.e., $Tx_s = (1/s)x_s$. Let $x_n = x_{s_n}$ be a subsequence such that $x_n \to z$, in the weak topology, and $|x_n - z| \to r$ as $s_n \to 1$. By Part 1, $Tz = z$. Hence, by the nonexpansiveness of T, $z \neq 0$. We next show that $r \neq 0$ leads to a contradiction. Given $\varepsilon > 0$, there exists a linear subspace $U = U(z, \varepsilon)$ so that (3) is satisfied. As in Part 1 we have,

$$x_n - z = \delta_n z + u_n + v_n; \quad |\delta_n|, |v_n| \to 0, \quad \text{and} \quad |u_n| \to r, \quad \text{as} \quad n \to \infty, \tag{4}$$

where $u_n \in U$ and v_n is in the finite dimensional complement of span$\{z, U\}$. Choose $f_n^* \in X^*$, $|f_n^*| = 1$ and $\langle f_n^*, x_n - z\rangle = |x_n - z|$. By nonexpansiveness

$$\begin{aligned} \langle f_n^*, (1/s_n)x_n - z\rangle &\leqq |(1/s_n)x_n - z| \\ &= |Tx_n - Tz| \leqq |x_n - z| = \langle f_n^*, x_n - z\rangle. \end{aligned} \tag{5}$$

Hence

$$\langle f_n^*, x_n \rangle \leqq 0 \qquad \text{and} \qquad \langle f_n^*, -z \rangle \geqq |x_n - z|.$$

By (3), $|u_n + \lambda z| \geqq |u_n|(1 + \delta)$ provided that $|\lambda| \geqq \varepsilon |u_n|$. Hence by virtue of the convexity of the function $\psi(\varphi) = |u_n + \varphi z|$ its minimum is achieved at some φ_n which satisfies $|\varphi_n| \leqq \varepsilon |u_n|$. Then $(u_n + \varphi_n z) \perp z$, and we can choose $g_n^* \in X^*$, $|g_n^*| = 1$ and $\langle g_n^*, u_n + \varphi_n z \rangle = |u_n + \varphi_n z|$ and $g_n^*(z) = 0$.

Now, on the one hand,

$$|g_n^* - f_n^*| \geqq |g_n^*(z) - f_n^*(z)| = |f_n^*(z)| \geqq |x_n - z|, \tag{6}$$

and, on the other hand,

$$|g_n^* + f_n^*|/2 \geqq |\langle g_n^* + f_n^*, w_n \rangle|/2 \geqq 1 - |\langle g_n^*, w_n - w_n' \rangle|/2, \tag{7}$$

where $w_n = (x_n - z)/|x_n - z|$ and $w_n' = u_n + \varphi_n z/|u_n + \varphi_n z|$. It follows from (4) and the choice of φ_n that $|w_n - w_n'|$ can be made arbitrarily small by the choice of n and ε. Hence (6) and (7) together contradict the uniform convexity of the unit ball of X^*. Thus $r \to 0$ and $|x_n - z| \to 0$.

We finish the proof by showing that $|x_s - z| \to 0$ as $s \to 1$. Suppose, to the contrary, that $x_n = x_{s_n} \to z_1$ and $y_n = x_{t_n} \to z_2$, in the norm, as s_n, $t_n \to 1$ and $z_1 \neq z_2$. Choose h_n^*, $k_n^* \in X^*$, $|h_n^*| = |k_n^*| = 1$, and $\langle h_n^*, x_n - z_2 \rangle = |x_n - z_2|$, $\langle k_n^*, y_n - z_1 \rangle = |y_n - z_1|$. By nonexpansiveness, as in (5) above, we derive readily that

$$\langle h_n^*, x_n \rangle \leqq 0 \qquad \text{and} \qquad \langle h_n^*, -z_2 \rangle \geqq |x_n - z_2|.$$

Similarly,

$$\langle k_n^*, y_n \rangle \leqq 0 \qquad \text{and} \qquad \langle k_n^*, -z_1 \rangle \geqq |y_n - z_1|.$$

We can choose subsequences, again denoted by h_n^* and k_n^* so that $h_n^* \to h$ and $k_n^* \to k$, in the weak topology. It is apparent that $|h|, |k| \leqq 1$ and, from the convergence of x_n and y_n,

$$\langle h, z_1 - z_2 \rangle = \langle -k, z_1 - z_2 \rangle = |z_1 - z_2|,$$

and

$$\langle h, z_1 \rangle \leqq 0 \qquad \text{and} \qquad \langle -k, z_1 \rangle \geqq |z_1 - z_2|.$$

The latter shows that $h \neq -k$, and the former therefore contradicts the uniform convexity of the unit ball of X^*, in fact, it contradicts even its smoothness. This finishes the proof of Theorem 1. ■

4. SPACES ISOMORPHIC TO SOME HILBERT SPACE

Here we consider the main existence problem for fixed points of nonexpansive mappings in spaces isomorphic to some Hilbert space. The tools and ideas of Section 2 are particularly useful in this context, wherein they can be directly related to Hilbert space geometry. We classify the spaces more precisley by the Banach–Mazur distance $d(X, H)$ of X to some Hilbert space H, i.e.,

$$d(X, H) = \inf |T||T^{-1}|,$$

where the infimum is taken over all continuous invertible linear transformations of X onto H.

Theorem 2. *Let X be a Banach space having Banach–Mazur distance strictly less than $\sqrt{2}$ to some Hilbert space. Let C be a weakly compact convex subset of X, and $T: C \to C$ nonexpansive. Then T has a fixed point in C.*

Moreover, let $X = l^2$ be renormed according to

$$|x| = \max\{|x|_2/A, |x|_\infty\},$$

for some $A \geqq 1$, where $|\cdot|_2$ and $|\cdot|_\infty$ denote, respectively, the l^2 and l^∞ norms on X. Then if $A < \sqrt{3}$ the same conclusion holds.

Remarks. If $1 \leqq A < \sqrt{2}$ then, clearly, the second part of the theorem is a special case of the first.

If the Banach–Mazur distance to some Hilbert space is less than or equal to $\sqrt{3}/\sqrt{2}$ then it is readily shown that the space has normal structure. Hence the assertion of the theorem follows from the classical results discussed above.

The renorming of l^2 given here is due to R. C. James, who gave it as an example of a space isomorphic to some Hilbert space which fails (for $A \geqq \sqrt{2}$) to have normal structure. The existence of a fixed point for $A = \sqrt{2}$ was given in Karlovitz [6]. The method of proof given here is, at once, more general and more straightforward, relying more directly on Hilbert space ideas.

Proof of Theorem 2. Part 1. By hypothesis, the space can be renormed with a Hilbert space norm $|\cdot|_{\mathrm{H}}$, with inner product $\langle \cdot, \cdot \rangle_{\mathrm{H}}$, so that

$$|x|_{\mathrm{H}} \leqq |x| \leqq K|x|_{\mathrm{H}},$$

for each $x \in X$ and some K, $1 \leqq K < \sqrt{2}$. Let C_0 be a subset of C which satisfies the hypotheses of the lemma. The existence of such a subset can easily be shown by a standard Zorn's lemma argument. If C_0 is a singleton, then it is a fixed point of T and we are finished. If not, we choose a sequence of approximate fixed points $\{x_n\}$ in C_0, $|Tx_n - x_n| \to 0$, such that $x_n \to w$, in the weak topology. We let diameter $C_0 = r > 0$. According to the lemma, given $\varepsilon > 0$, we can choose N so large that $|x_n - w| \geqq r - \varepsilon$ for $n \geqq N$. Consequently, $|x_n - w|_{\mathrm{H}} \geqq (r - \varepsilon)/K$. By the weak convergence,

$$\begin{aligned} |x_n - x_N|_{\mathrm{H}}^2 &= |x_n - w|_{\mathrm{H}}^2 + 2|\langle x_n - w, x_N - w\rangle|_{\mathrm{H}} + |x_N - w|_{\mathrm{H}}^2 \\ &\geqq 2(r - \varepsilon)^2/K^2 - \varepsilon, \end{aligned}$$

for n sufficiently large. Since $|x_n - x_N|_{\mathrm{H}} \leqq \operatorname{diam} C_0 = r$, we then have

$$\sqrt{2(r-\varepsilon)^2/K^2} - \varepsilon \leqq r,$$

which is not compatible with $K < \sqrt{2}$ and the arbitrariness of $\varepsilon > 0$. Thus C_0 is a singleton, and we are finished.

Part 2: We define $|x|_{\mathrm{H}} = |x|_2/A$. Then

$$|x|_{\mathrm{H}} \leqq |x| \leqq A|x|_{\mathrm{H}}.$$

We proceed as in Part 1, by showing that a minimal (in the sense of the lemma) set C_0 cannot contain more than one point. Suppose to the contrary that $\operatorname{diam} C_0 = r > 0$. Without loss of generality we may assume that $0 \in C_0$. We shall make use of the following consequence of the lemma. *If* $\{x_n\}$ *is a sequence of approximate fixed points in* C_0 *then* $\lim |x - x_n|_\infty = r$, *as* $n \to \infty$, *for each* $x \in C_0$. We prove this by contradiction. Suppose that for some $x \neq 0 \in C_0 |x - x_n|_\infty \leqq r - \delta$, for some $\delta > 0$ and for all $n \geqq n_0$. Since $|0 - x_n|_\infty \leqq r$, it follows that $|x/2 - x_n|_\infty \leqq r - \delta/2$, for $n \geqq n_0$. By the uniform convexity of $|\cdot|_{\mathrm{H}}$ it follows from $|x - x_n|_{\mathrm{H}} \leqq r$ and $|0 - x_n|_{\mathrm{H}} \leqq r$ that $|x/2 - x_n|_{\mathrm{H}} \leqq r - \tau$ for some $\tau > 0$. Hence $|x/2 - x_n| \leqq r - \min\{\tau, \delta/2\}$, for $n \geqq n_0$, which contradicts assertion (3) of the lemma.

Continuing with the proof of Part 2, we let y_0 be some point of C_0. We let x_s, $0 < s < 1$, be the approximate fixed points constructed according to (2). Let $\{x_{s_n}\}$ be some subsequence such that $x_{s_n} \to w$, in the weak topology, and $s_n \to 1$. Without loss of generality we may assume that $w = 0$. Given $\varepsilon > 0$, according to the remark of the previous paragraph, we can choose s^* so that

$$|x_s|_\infty = |x_s - 0|_\infty \geqq r - \varepsilon \qquad \text{for} \quad s \geqq s^*. \tag{8}$$

Hence there exists a positive integer i so that

$$|x_{s^*,i}| \geqq r - \varepsilon,$$

where the second subscript denotes the ith component of x_{s^*}. Since $x_{s_n} \to 0$, in the weak topology, there exists $s' > s^*$ so that $|x_{s',i}| \leqq \varepsilon$. We define t by

$$t = \sup\{s : s^* < s < s', |x_{s,i}| \geqq r - \varepsilon\}.$$

By the abovementioned norm continuity of the function $x_s : (0, 1) \to X$ it follows that

$$|x_{t,i}| \geqq r - \varepsilon \qquad \text{and} \qquad r - 2\varepsilon \leqq |x_{t',i}| < r - \varepsilon,$$

for some t', with $t < t'$. Hence, by virtue of the inequality (8), there exists a positive integer $j \neq i$ so that

$$|x_{t',j}| \geqq r - \varepsilon.$$

Consequently, $|x_{t'}|_H^2 > ((r - \varepsilon)^2 + (r - 2\varepsilon)^2)/A^2$. By the weak convergence of $x_{s_n} \to 0$ we can choose n' so that $s_{n'} > s^*$ and

$$|\langle x_{t'}, x_{s_{n'}} \rangle_H| \leqq \varepsilon.$$

By (8), $|x_{s_{n'}}|_H^2 \geqq (r - \varepsilon)^2/A^2$. It follows that

$$\begin{aligned} |x_{t'} - x_{s_{n'}}|_H^2 &\geqq |x_{t'}|_H^2 - 2\varepsilon + |x_{s_{n'}}|_H^2 \\ &\geqq 2(r - \varepsilon)^2/A^2 + (r - 2\varepsilon)^2/A^2 - 2\varepsilon. \end{aligned}$$

Since $|x_{t'} - x_{s_{n'}}| \leqq r$ and $A^2 < 3$ we see that the last inequality contradicts the arbitrariness of $\varepsilon > 0$. This finishes the proof of Theorem 2. ■

REFERENCES

1. G. BIRKHOFF, Orthogonality in linear metric spaces, *Duke Math. J.* **1** (1935), 169–172.
2. F. E. BROWDER, Nonexpansive nonlinear operators in a Banach space, *Proc. Nat. Acad. Sci. U.S.A.* **54** (1965), 1041–1044.
3. D. GÖHDE, Über Fixpunkle bei stetigen Selbstabbildungen mit kompaktenIterierten, *Math. Nachr.* **28** (1964), 45–55.
4. R. C. JAMES, Orthogonality and linear functionals in normed linear spaces, *Trans. Amer. Math. Soc.* **61** (1947), 265–292.
5. R. C. JAMES, A separable somewhat reflexive Banach space with nonseparable dual, *Bull. Amer. Math. Soc.* **80** (1974), 738–743.
6. L. A. KARLOVITZ, Existence of fixed points of nonexpansive mappings in a space without normal structure, *Pacific J. Math.* **66** (1976), 153–159.
7. L. A. KARLOVITZ, On nonexpansive mappings, *Proc. Amer. Math. Soc.* **55** (1976), 321–325.
8. L. A. KARLOVITZ, Fixed point theorems for nonexpansive mappings and related geometrical consideration, *in* "Fixed Point Theory and Applications" (S. Swaminathan, ed.), Academic Press, New York, 1976, pp. 91–105.
9. W. A. KIRK, A Fixed point theorem for mappings which do not increase distance, *Amer. Math. Monthly* **76** (1965), 1004–1006.
10. J. LINDENSTRAUSS AND C. STEGALL, Examples of spaces which do not contain l_1 and whose duals are not separable, *Studia Math.* **54** (1975), 81–105.
11. E. ODELL AND Y. STERNFELD, A fixed point theorem in c_0 (to appear).

Research was supported in part by the National Science Foundation.

GEORGIA INSTITUTE OF TECHNOLOGY
SCHOOL OF MATHEMATICS
ATLANTA, GEORGIA

Bounds for Incomplete Polynomials Vanishing at Both Endpoints of an Interval

M. LACHANCE

E. B. SAFF

R. S. VARGA

1. INTRODUCTION

At the *December 1976 Tampa Conference on Rational Approximation with Emphasis on Applications of Padé Approximants*, Lorentz [3] presented results and open problems concerning *incomplete polynomials of type* θ, i.e., for a fixed θ with $0 < \theta \leqq 1$, the set of all real or complex polynomials of the form

$$\sum_{k=s}^{n} \alpha_k x^k, \qquad \text{where} \quad s \geqq \theta \cdot n, \quad n \text{ an arbitrary nonnegative integer.} \tag{1.1}$$

These incomplete polynomials of type θ have been further studied by Saff and Varga [7–9], by Kemperman and Lorentz [1], and by Lorentz [3]. Note that any incomplete polynomial of type θ has, from (1.1), a zero at $x = 0$ of order at least $[n \cdot \theta]$.

In this paper, we consider the more general problem of polynomials constrained to have zeros at *both* endpoints of a finite interval. Without loss of generality, we take this interval to be $[-1, +1]$. By analogy with (1.1), for any $\theta_1 > 0$ and $\theta_2 > 0$ fixed with $\theta_1 + \theta_2 \leqq 1$, we consider here the set of all real or complex polynomials of the form

$$(x-1)^{s_1}(x+1)^{s_2} \sum_{k=0}^{m} \beta_k x^k \qquad \text{where} \quad s_1 \geqq \theta_1 \cdot (s_1 + s_2 + m),$$

$$s_2 \geqq \theta_2 \cdot (s_1 + s_2 + m). \tag{1.2}$$

ISBN 0-12-178150-X

In Section 2, upper bounds for the growth of polynomials of the form (1.2) are determined. In Section 3, an analog of the classical Chebyshev polynomials for constrained polynomials of the form (1.2) is given, and it is shown that the upper bounds of Section 2 are, in a certain limiting sense, *best possible.*

2. GROWTH ESTIMATES FOR CONSTRAINED POLYNOMIALS

In the spirit of two lemmas of Walsh [12, p. 250], we prove the following result of Lemma 2.1 on bounding above the moduli of constrained polynomials. For a related result, see Kemperman and Lorentz [1].

Lemma 2.1. *Let $\mathscr{E}$ be a closed bounded point set, not a single point, whose complement K with respect to the extended complex plane is simply connected. Let $w = \varphi(z)$ denote a function which maps K onto $|w| > 1$, so that the points at infinity correspond to each other. Let the (not necessarily distinct) points α_k, $k = 1, \ldots, m$, lie exterior to $\mathscr{E}$, and let $P(z)$ be a polynomial of degree n ($n \geqq m$) which vanishes at each of the points α_k (with each α_k listed according to its multiplicity). If, on the boundary of $\mathscr{E}$,*

$$|P(z)| \leqq L, \tag{2.1}$$

then, for all z in K,

$$|P(z)| \leqq L|\varphi(z)|^n \prod_{k=1}^{m} \left| \frac{\varphi(z) - \varphi(\alpha_k)}{\overline{\varphi(\alpha_k)}\varphi(z) - 1} \right|. \tag{2.2}$$

Proof. Define $Q(z)$ for z in K by means of

$$Q(z) := \frac{P(z)}{[\varphi(z)]^n} \prod_{k=1}^{m} \left(\frac{\overline{\varphi(\alpha_k)}\varphi(z) - 1}{\varphi(z) - \varphi(\alpha_k)} \right),$$

so that $Q(z)$ is analytic in K, even at infinity. Its modulus is continuous in the exterior of $\mathscr{E}$, and, on the boundary of $\mathscr{E}$, is bounded above by L. Hence, by the maximum principle,

$$|Q(z)| \leqq L, \qquad \text{for all } z \text{ in } K,$$

which gives the desired inequality (2.2). ■

Here, we are interested in polynomials constrained to have certain order zeros at $x = -1$ and at $x = 1$, and we thus introduce the following notation. As usual, for each nonnegative integer n, π_n denotes the set of complex polynomials of degree at most n. For every ordered triple of nonnegative integers (s_1, s_2, m), the set $\pi(s_1, s_2, m)$ is defined by

$$\pi(s_1, s_2, m) := \{(x-1)^{s_1}(x+1)^{s_2}q(x) : q \in \pi_m\}. \tag{2.3}$$

Finally, for each continuous g defined on a compact set B, we set

$$\|g\|_B := \max\{|g(z)| : z \in B\}. \tag{2.4}$$

To obtain growth estimates for constrained polynomials from Lemma 2.1, take the set $\mathscr{E}$ now to be some real interval $[a, b]$, with $-1 < a < b < +1$. Then,

$$z = \psi(w) := \frac{b+a}{2} + \frac{b-a}{2}\left(\frac{w + w^{-1}}{2}\right), \qquad |w| > 1, \tag{2.5}$$

maps the exterior of the unit circle in the w-plane onto $\mathbb{C}^*\backslash[a, b]$ in the z-plane (where $\mathbb{C}^*$ denotes the extended complex plane). The function $\varphi(z)$ required by Lemma 2.1 is then the inverse map of $\psi(w)$, i.e.,

$$w = \varphi(z) = \frac{\sqrt{z-a} + \sqrt{z-b}}{\sqrt{z-a} - \sqrt{z-b}}, \qquad z \in \mathbb{C}^*\backslash[a, b], \tag{2.6}$$

for some suitable branch of the square root function. For this choice of $\mathscr{E}$ and for the choice $\alpha_k = 1$, $1 \leqq k \leqq s_1$, and $\alpha_k = -1$, $s_1 + 1 \leqq k \leqq s_1 + s_2$, we have, as an immediate consequence of Lemma 2.1,

Corollary 2.2. *Let $p \in \pi(s_1, s_2, m)$, and set $n := s_1 + s_2 + m$. Then, for all $z \in \mathbb{C}^*\backslash[a, b]$,*

$$|p(z)| \leqq \|p\|_{[a,b]} |\varphi(z)|^n \left|\frac{\varphi(z) - \varphi(1)}{\varphi(1)\varphi(z) - 1}\right|^{s_1} \left|\frac{\varphi(z) - \varphi(-1)}{\varphi(-1)\varphi(z) - 1}\right|^{s_2}, \tag{2.7}$$

where $\varphi(z)$ is given by (2.6).

In a typical application, it may be known that a polynomial from $\pi(s_1, s_2, m)$ is bounded above in modulus on $[-1, 1]$ by some constant L. We then wish to choose an interval $[a, b]$, strictly contained in $[-1, 1]$, so as to optimize the upper bound of (2.7). In what follows, we examine the behavior of polynomials of large degree from $\pi(s_1, s_2, m)$ where $s_1/n \geqq \varphi_1$ and $s_2/n \geqq \theta_2$, for θ_1 and θ_2 fixed. We then make use of certain limiting results to determine the best choice for $[a, b]$, depending only on the relative orders of contact at $x = -1$ and at $x = 1$. As Jacobi polynomials play a significant role in the determination of such an optimal interval $[a, b]$ and in the results of Section 3, we briefly summarize some of their basic properties.

For the real parameters $\alpha > -1$ and $\beta > -1$, $P_n^{(\alpha,\beta)}(x)$ denotes the classical Jacobi polynomial of degree n. It is well known (cf. Szegö [11, p. 68]) that if

$$h_n^{(\alpha,\beta)} := \int_{-1}^{1} (1-x)^\alpha (1+x)^\beta (P_n^{(\alpha,\beta)}(x))^2\, dx, \tag{2.8}$$

then the sequence $\{P_n^{(\alpha,\beta)}(x)/\sqrt{h_n^{(\alpha,\beta)}}\}_{n=0}^{\infty}$ is orthonormal with respect to the weight function $(1-x)^\alpha(1+x)^\beta$ in the interval $[-1, 1]$, i.e.,

$$\int_{-1}^{1}(1-x)^\alpha(1+x)^\beta\left(\frac{P_n^{(\alpha,\beta)}(x)}{\sqrt{h_n^{(\alpha,\beta)}}}\right)\left(\frac{P_m^{(\alpha,\beta)}(x)}{\sqrt{h_m^{(\alpha,\beta)}}}\right)dx=\delta_{n,m}.$$

The following representation for $h_n^{(\alpha,\beta)}$ will be useful (cf. Szegö [11, p. 68]):

$$h_n^{(\alpha,\beta)}=\frac{2^{\alpha+\beta+1}}{(2n+\alpha+\beta+1)}\frac{\Gamma(n+\alpha+1)\Gamma(n+\beta+1)}{\Gamma(n+1)\Gamma(n+\alpha+\beta+1)}. \tag{2.9}$$

As a well-known consequence of the theory of orthogonal polynomials, (cf. Szegö [11, p. 63]), the unique monic polynomial $q(x)$ in π_n which minimizes the integral

$$\int_{-1}^{+1}(1-x)^\alpha(1+x)^\beta(q(x))^2\,dx$$

is given explicitly by the monic Jacobi polynomial

$$2^n\binom{2n+\alpha+\beta}{n}^{-1}P_n^{(\alpha,\beta)}(x). \tag{2.10}$$

It is further well known that the zeros of $P_n^{(\alpha,\beta)}(x)$, for any $n \geqq 1$ and for any choices of $\alpha > -1$ and $\beta > -1$, all lie in $(-1, +1)$. Concerning the asymptotic location of zeros of particular sequences of Jacobi polynomials, we state

Lemma 2.3. (Moak *et al.* [6]) *Let a_n and b_n denote, respectively, the smallest and largest zeros of $P_n^{(\alpha_n,\beta_n)}(x)$, where $\alpha_n > -1$ and $\beta_n > -1$. Assume that*

$$\lim_{n\to\infty}\frac{\alpha_n}{2n+\alpha_n+\beta_n}=\theta_1 \quad \textit{and} \quad \lim_{n\to\infty}\frac{\beta_n}{2n+\alpha_n+\beta_n}=\theta_2. \tag{2.11}$$

For $\mu := \theta_2+\theta_1$ and $v := \theta_2-\theta_1$, set

$$\begin{aligned} a = a(\theta_1,\theta_2) &:= \mu v - \sqrt{(1-\mu^2)(1-v^2)},\\ b = b(\theta_1,\theta_2) &:= \mu v + \sqrt{(1-\mu^2)(1-v^2)}. \end{aligned} \tag{2.12}$$

Then,

$$\lim_{n\to\infty} a_n = a \quad \textit{and} \quad \lim_{n\to\infty} b_n = b. \tag{2.13}$$

Moreover, the zeros of the sequence $\{P_n^{(\alpha_n,\beta_n)}(x)\}_{n=0}^{\infty}$ are dense in the interval $[a, b]$.

We make use of this last result in selecting the optimal interval $[a, b]$ in Corollary 2.2. To every ordered pair (θ_1, θ_2) from the set

$$\Omega := \{(\theta_1, \theta_2) : \theta_1 > 0, \theta_2 > 0, \theta_1 + \theta_2 < 1\}, \tag{2.14}$$

there corresponds a unique interval $[a, b]$ with $-1 < a < b < 1$ from (2.12). Consequently, for each (θ_1, θ_2) in Ω, there exists a unique mapping function

$$\phi(z) = \phi(z; \theta_1, \theta_2) := \frac{\sqrt{z-a} + \sqrt{z-b}}{\sqrt{z-a} - \sqrt{z-b}}, \tag{2.15}$$

mapping $\mathbb{C}^* \backslash [a, b]$ onto the exterior of the unit circle. For $(\theta_1, \theta_2) \in \Omega$ and $z \in \mathbb{C}^* \backslash [a, b]$, we now define the function $G(z; \theta_1, \theta_2)$ by

$$G(z) = G(z; \theta_1, \theta_2) := |\varphi(z)| \left| \frac{\varphi(z) - \varphi(1)}{\varphi(1)\varphi(z) - 1} \right|^{\theta_1} \cdot \left| \frac{\varphi(z) - \varphi(-1)}{\varphi(-1)\varphi(z) - 1} \right|^{\theta_2}. \tag{2.16}$$

We extend $G(z; \theta_1, \theta_2)$ continuously to the interval $[a, b]$ by defining $G(z; \theta_1, \theta_2) = 1$ for $z \in [a, b]$. We remark that $G(z; \theta_1, \theta_2)$ is continuous in the variables θ_1 and θ_2 and we extend its definition continuously to the closure $\overline{\Omega}$ of Ω. For example, when $\theta_1 + \theta_2 = 1$, we have

$$G(z) = G(z; \theta_1, \theta_2) = (2\theta_1)^{-\theta_1}(2\theta_2)^{-\theta_2}|1 - z|^{\theta_1}|1 + z|^{\theta_2}. \tag{2.17}$$

In the special case when $\theta_1 = 0$, the G function agrees with the corresponding G function defined by Saff and Varga in [7, 8]. To facilitate the statement of the main result of this section, we first mention some simple properties of the function $G(z; \theta_1, \theta_2)$.

Note that, as the points at infinity correspond to each other under the mapping $\varphi(z)$, for (θ_1, θ_2) fixed in Ω, we have

$$G(z; \theta_1, \theta_2) \to \infty \qquad \text{as} \quad |z| \to \infty. \tag{2.18}$$

Moreover, it is evident from (2.16) that, for $(\theta_1, \theta_2) \in \Omega$,

$$G(-1; \theta_1, \theta_2) = 0 = G(1; \theta_1, \theta_2). \tag{2.19}$$

Next, we also have the following result of Lemma 2.4. Its proof, being similar to that of [7, Lemma 4.2], is omitted.

Lemma 2.4. *For (θ_1, θ_2) fixed in Ω, $G(x; \theta_1, \theta_2)$, considered as a function of the real variable x, is strictly decreasing on $(-\infty, -1)$ and on $(b, 1)$ and is strictly increasing on $(-1, a)$ and on $(1, +\infty)$.*

As a consequence of Lemma 2.4 and the properties of (2.18) and (2.19), for each $(\theta_1, \theta_2) \in \Omega$ there exist two unique real points

$$\rho = \rho(\theta_1, \theta_2) < -1 \qquad \text{and} \qquad \sigma = \sigma(\theta_1, \theta_2) > 1 \tag{2.20}$$

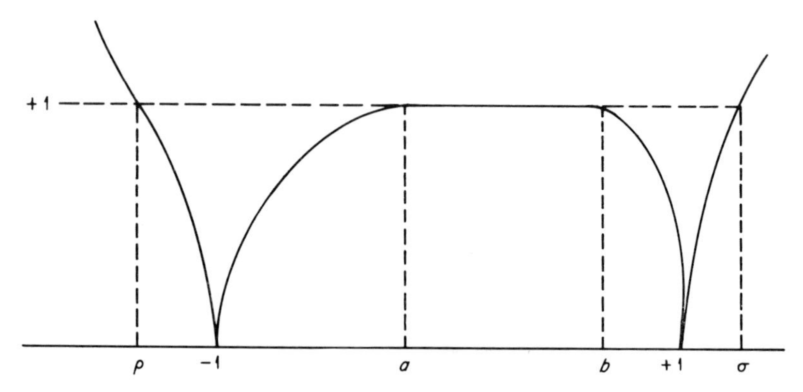

Fig. 2.1. $G(x; 3/9, 5/9)$, x real.

for which we have

$$G(\rho(\theta_1, \theta_2); \theta_1, \theta_2) = 1 = G(\sigma(\theta_1, \theta_2); \theta_1, \theta_2). \tag{2.21}$$

By way of illustration, Fig. 2.1 gives the graph of $G(x; 3/9, 5/9)$ for real values of x.

We comment that, although we do not have an explicit representation for ρ or σ in (2.20), numerical estimates for ρ and σ are easily determined.

Since, for each $(\theta_1, \theta_2) \in \Omega$, the continuous function $G(z; \theta_1, \theta_2)$ vanishes at $z = -1$ and $z = 1$, we have the existence of neighborhoods about $z = -1$ and $z = +1$ for which $G(z; \theta_1, \theta_2) < 1$. Actually, the level curve $G(z; \theta_1, \theta_2) = 1$, enclosing these neighborhoods, traces out a "barbell"-type curve. We illustrate this "barbell" configuration in Fig. 2.2 by graphing the level curve $G(z; 3/9, 5/9) = 1$.

Next, for each pair (θ_1, θ_2) in $\overline{\Omega}$, we define the open set

$$\Lambda(\theta_1, \theta_2) := \{z : G(z; \theta_1, \theta_2) < 1\}. \tag{2.22}$$

We now state the main result of this section.

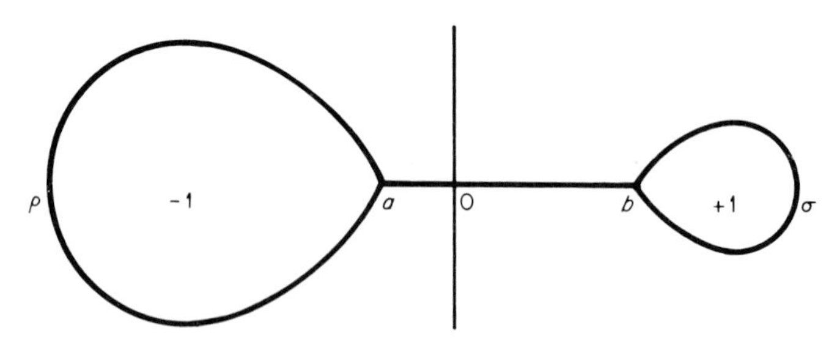

Fig. 2.2. $G(z; 3/9, 5/9) = 1$.

Theorem 2.5. *Let $p \in \pi(s_1, s_2, m)$ with p not identically constant, and set $n := s_1 + s_2 + m$. Then, for all z,*

$$|p(z)| \leqq \|p\|_{[-1,1]}(G(z;\ s_1/n,\ s_2/n))^n. \tag{2.23}$$

Consequently, for $z \in \Lambda(s_1/n, s_2/n)$,

$$|p(z)| \leqq \|p\|_{[-1,1]}(G(z; s_1/n, s_2/n))^n < \|p\|_{[-1,1]}. \tag{2.24}$$

In particular, if $\xi \in [-1, 1]$ is such that $|p(\xi)| = \|p\|_{[-1,1]}$, then

$$a(s_1/n, s_2/n) \leqq \xi \leqq b(s_1/n, s_2/n). \tag{2.25}$$

The statement (2.23) is actually a restatement of Corollary 2.2 in terms of the function $G(z; \theta_1, \theta_2)$. Furthermore, we will show in Section 3 that the inequality (2.23) is *sharp* in a certain limiting sense.

In the next two corollaries, we state convergence and interpolation results for sequences of constrained polynomials.

Corollary 2.6. *Let (θ_1, θ_2) be fixed in $\overline{\Omega}$ and let*

$$\left\{p_i(z) := (z-1)^{s_{1,i}}(z+1)^{s_{2,i}} \sum_{k=0}^{m_i} a_{k,i} z^k\right\}_{i=1}^{\infty}$$

by any infinite sequence of complex polynomials satisfying

$$\lim_{i\to\infty} n_i = \infty \qquad (n_i := s_{1,i} + s_{2,i} + m_i,\ i \geqq 1) \tag{2.26}$$

and

$$\lim_{i\to\infty} s_{1,i}/n_i = \theta_1 \qquad \textit{and} \qquad \lim_{i\to\infty} s_{2,i}/n_i = \theta_2. \tag{2.27}$$

If

$$\limsup_{i\to\infty}(\|p_i\|_{[-1,1]})^{1/n_i} \leqq 1, \tag{2.28}$$

then

$$\lim_{i\to\infty} p_i(z) = 0 \qquad \textit{for all} \quad z \in \Lambda(\theta_1, \theta_2). \tag{2.29}$$

Moreover, for any closed subset B of $\Lambda(\theta_1, \theta_2)$,

$$\limsup_{i\to\infty}(\|p_i\|_B)^{1/n_i} \leqq \|G(\cdot\,; \theta_1, \theta_2)\|_B < 1. \tag{2.30}$$

The proof of Corollary 2.6 follows from applying inequality (2.23) to each polynomial $p_i(z)$ and from the continuity of $G(z; \theta_1, \theta_2)$ in the variables θ_1 and θ_2. We will show in Section 3 that Corollary 2.6 gives the *largest possible* open set of convergence to zero for the class of such polynomials $p_i(z)$.

Our final result of this section concerns polynomials interpolating a function $f(z)$ at the points $z = -1$ and $z = 1$.

Corollary 2.7. *Let $f(z)$ be analytic at $z = -1$ and at $z = 1$. Suppose there exists a sequence of polynomials $\{p_i(z)\}_{i=1}^{\infty}$ with $p_i \in \pi_{n_i}$ $\forall i$, $n_1 < n_2 < \cdots$, where $n_{i+1}/n_i \to 1$ as $i \to \infty$, and such that*

$$\begin{aligned} p_i^{(k)}(1) &= f^{(k)}(1), & k &= 0, 1, \ldots, s_{1,n_i}, \\ p_i^{(k)}(-1) &= f^{(k)}(-1), & k &= 0, 1, \ldots, s_{2,n_i}. \end{aligned} \tag{2.31}$$

Further, suppose that

$$\lim_{i\to\infty} \frac{s_{1,n_i}}{n_i} = \theta_1 \qquad \text{and} \qquad \lim_{i\to\infty} \frac{s_{2,n_i}}{n_i} = \theta_2 \tag{2.32}$$

and that

$$\limsup_{i\to\infty} (\|p_i\|_{[-1,1]})^{1/n_i} \leqq 1. \tag{2.33}$$

Then, $f(z)$ is analytic at each point of $\Lambda(\theta_1, \theta_2)$ and $p_i(z) \to f(z)$ for $z \in \Lambda(\theta_1, \theta_2)$, the convergence being uniform on any closed subset of $\Lambda(\theta_1, \theta_2)$. On a closed subset B of $\Lambda(\theta_1, \theta_2)$ we have the following convergence rate:

$$\limsup_{i\to\infty} \|f - p_i\|_B^{1/n_i} \leqq \|G(\cdot;\ \theta_1,\ \theta_2)\|_B < 1. \tag{2.34}$$

To prove this last result, we consider consecutive differences from the sequence $\{f(z) - p_i(z)\}_{i=1}^{\infty}$ which in turn form a sequence of constrained polynomials $\{p_{i+1}(z) - p_i(z)\}_{i=1}^{\infty}$. This sequence satisfies the hypotheses of Corollary 2.6 and hence tends to zero, geometrically in $\Lambda(\theta_1, \theta_2)$.

As an application of Corollary 2.7, we mention an interpolation problem suggested by Meinardus [5], related to the design of filters. Let g be the real step function defined for all real t by

$$g(t) = \begin{cases} 1 & \text{for } \ t < \frac{1}{2} \\ \frac{1}{2} & \text{for } \ t = \frac{1}{2} \\ 0 & \text{for } \ t > \frac{1}{2} \end{cases} \tag{2.35}$$

and consider the sequence of polynomials $\{q_n(x)\}_{n=0}^{\infty}$ with $q_n \in \pi_{2n+1}$ for all $n \geqq 0$, satisfying the conditions

$$\begin{aligned} q_n^{(k)}(0) &= g^{(k)}(0), & k &= 0, 1, \ldots, n, \\ q_n^{(k)}(1) &= g^{(k)}(1), & k &= 0, 1, \ldots, n. \end{aligned} \tag{2.36}$$

The polynomials q_n are uniquely determined, and, in fact, are explicitly given by

$$q_n(x) = B_{2n+1}(x; g), \qquad n \geqq 0,$$

where

$$B_n(x; f) := \sum_{k=0}^{n} f\left(\frac{k}{n}\right)\binom{n}{k} x^k (1-x)^{n-k}$$

is the *Bernstein polynomial* of degree n for a given real-valued function on $[0, 1]$ (cf. Lorentz [2, p. 4]). It is easily seen that $\{q_n(x)\}_{n=0}^{\infty}$ so defined are uniformly bounded by unity on $[0, 1]$. Setting $p_n(x) := q_n((x+1)/2)$, we find that the sequence $\{p_n(x)\}_{n=1}^{\infty}$ satisfies the hypotheses of Corollary 2.7 with $\theta_1 = \theta_2 = 1/2$. In this case, one readily verifies [cf. (2.20)] that $\rho(1/2, 1/2) = -\sqrt{2}$ and $\sigma(1/2, 1/2) = \sqrt{2}$, so that, by Corollary 2.7, we have for real x that the sequence $\{p_n(x)\}_{n=1}^{\infty}$ converges to 1 for $x \in (-\sqrt{2}, 0)$ and converges to 0 for $x \in (0, \sqrt{2})$, the convergence being *geometric* on any closed subinterval of $(-\sqrt{2}, 0) \cup (0, \sqrt{2})$. Consequently, as $n \to \infty$, $q_n(t) \to g(t)$, geometrically on any closed subinterval of

$$\left(\frac{1-\sqrt{2}}{2}, \frac{1}{2}\right) \cup \left(\frac{1}{2}, \frac{1+\sqrt{2}}{2}\right).$$

More generally, for complex t, Corollary 2.7 implies that the sequence $q_n(t)$ converges geometrically on each closed subset of

$$H = \{t \in \mathbb{C} : (2t-1) \in \Lambda(1/2, 1/2)\}. \tag{2.37}$$

Meinardus [5] has been able to show moreover that the sequence $q_n(t)$ *diverges* for $t < (1-\sqrt{2})/2$, and for $t > (1+\sqrt{2})/2$. We remark that the "overconvergence" properties of Bernstein polynomials have been extensively studied (see Lorentz [2]). In the cases when $\theta_1 + \theta_2 = 1$, the convergence region given, via Corollary 2.7, by

$$\{t \in \mathbb{C} : (2t-1) \in \Lambda(\theta_1, \theta_2)\}, \tag{2.38}$$

and by that of Bernstein polynomials, *coincide*.

3. CONSTRAINED CHEBYSHEV POLYNOMIALS

In [8], the analog of the classical Chebyshev polynomials for polynomials constrained at one endpoint of an interval by a certain order zero were studied. In this section, we extend the definition of these constrained polynomials to the two endpoint case. As a consequence, we shall prove that Theorem 2.5 is best possible in a certain limiting sense.

Proposition 3.1. *For each ordered triple of nonnegative integers* (s_1, s_2, m), *there exists a unique monic polynomial* $Q_{s_1, s_2, m}(x)$ *in* $\pi(s_1, s_2, m)$ [cf. (2.3)] *of precise degree* $n := s_1 + s_2 + m$ *satisfying*

$$\begin{aligned} \|Q_{s_1, s_2, m}\|_{[-1,1]} &= \inf\{\|(x-1)^{s_1}(x+1)^{s_2}x^m \\ &\qquad - g(x)\|_{[-1,1]} : g \in \pi(s_1, s_2, m-1)\} \\ &=: E_{s_1, s_2, m} \end{aligned} \tag{3.1}$$

(*where* $\pi(s_1, s_2, m-1) := \{0\}$ if $m = 0$). *Furthermore, for* $n > 0$, $Q_{s_1, s_2, m}(x)$ *has an alternation set of precisely* $m+1$ *distinct points* $\xi_j^{(s_1, s_2, m)}$, $j = 0, 1, \ldots, m$, *with* [cf. (2.12)]

$$a(s_1/n, s_2/n) \leqq \xi_0^{(s_1, s_2, m)} < \xi_1^{(s_1, s_2, m)} < \cdots < \xi_m^{(s_1, s_2, m)} \leqq b(s_1/n, s_2/n), \quad (3.2)$$

for which

$$Q_{s_1, s_2, m}(\xi_j^{(s_1, s_2, m)}) = (-1)^{s_1 + m - j} E_{s_1, s_2, m}, \qquad j = 0, 1, \ldots, m. \quad (3.3)$$

Proof. As the special cases $s_1 + s_2 = 0$ and $m = 0$ of Proposition 3.1 are clearly true and will in fact be explicitly covered in Proposition 3.2, assume that $s_1 + s_2 > 0$ and that $m > 0$. From general linear approximation theory, it follows that there exists a monic polynomial, say p, in $\pi(s_1, s_2, m)$ with

$$\|p\|_{[-1,1]} = E_{s_1, s_2, m}.$$

As a consequence of Theorem 2.5, the polynomial p also satisfies the extremal problem

$$\|p\|_{[a,b]} = \inf\{\|(x-1)^{s_1}(x+1)^{s_2}x^m - g(x)\|_{[a,b]} : g \in \pi(s_1, s_2, m-1)\},$$

where $a = a(s_1/n, s_2/n)$ and $b = b(s_1/n, s_2/n)$ are defined in (2.12). On this subinterval, the linear space $\pi(s_1, s_2, m-1)$, which has dimension m, satisfies the Haar condition, which guarantees (cf. Meinardus [4, p. 20]) the uniqueness of p. From the same result, we have the existence of an alternation set of at least $m+1$ distinct points in $[a, b]$. If there were $m+k$ alternation points with $k > 1$, the derivative of p, a polynomial of degree $n-1$, would have at least n zeros on $[-1, +1]$, and would, consequently, be identically zero, contradicting the fact that p is monic. Thus, there are precisely $m+1$ distinct alternation points satisfying (3.3). ∎

With the existence of $Q_{s_1, s_2, m}(x)$ for each triple of nonnegative integers, we define

$$T_{s_1, s_2, m}(x) := \frac{Q_{s_1, s_2, m}(x)}{\|Q_{s_1, s_2, m}\|_{[-1,1]}} = \frac{Q_{s_1, s_2, m}(x)}{E_{s_1, s_2, m}} \quad (3.4)$$

to be the (normalized) *constrained Chebyshev polynomial of degree* $s_1 + s_2 + m$ *associated with the set* $\pi(s_1, s_2, m)$ *on the interval* $[-1, 1]$. Let $T_n(x)$ denote as usual the classical Chebyshev polynomial (of the first kind) of degree n, given by

$$T_n(x) = \cos n\theta, \qquad x = \cos\theta.$$

We now summarize some special cases and properties for certain triples (s_1, s_2, m) in

Proposition 3.2. *Set* $x_m := \cos(\pi/2m)$ *for* $m > 0$. *Then,*

(i) $Q_{s_2, s_1, m}(x) = (-1)^{s_1+s_2+m} Q_{s_1, s_2, m}(-x);$

(ii)
$$Q_{0,0,m}(x) = 2^{-m+1} T_m(x),$$
$$E_{0,0,m} = 2^{-m+1},$$
$$\xi_j^{(0,0,m)} = \cos\left(\frac{(m-j)\pi}{m}\right), \qquad j = 0, 1, \ldots, m;$$

(iii)
$$Q_{0,1,m}(x) = \frac{2}{(1+x_{m+1})^{m+1}} T_{m+1}\left[\left(\frac{1-x_{m+1}}{2}\right) + \left(\frac{1+x_{m+1}}{2}\right)x\right],$$
$$E_{0,1,m} = \frac{2}{(1+x_{m+1})^{m+1}},$$
$$\xi_j^{(0,1,m)} = \frac{1}{(1+x_{m+1})}\left\{2\cos\left[\frac{(m-j)\pi}{m+1}\right] + x_{m+1} - 1\right\},$$
$$j = 0, 1, \ldots, m;$$

(iv)
$$Q_{1,1,m}(x) = \frac{2}{(2x_{m+2})^{m+2}} T_{m+2}(x_{m+2} \cdot x),$$
$$E_{1,1,m} = \frac{2}{(2x_{m+2})^{m+2}},$$
$$\xi_j^{(1,1,m)} = \frac{1}{x_{m+2}} \cos\left[\frac{(m+1-j)\pi}{m+2}\right], \qquad j = 0, 1, \ldots, m;$$

(v)
$$Q_{s_1, s_2, 0}(x) = (x-1)^{s_1}(x+1)^{s_2}, \; s_1 + s_2 > 0,$$
$$E_{s_1, s_2, 0} = \left(\frac{2s_1}{s_1+s_2}\right)^{s_1}\left(\frac{2s_2}{s_1+s_2}\right)^{s_2},$$
$$\xi_0^{(s_1, s_2, 0)} = \frac{s_2 - s_1}{s_2 + s_1};$$

(vi)
$$Q_{s,s,1}(x) = x(x^2-1)^s,$$
$$E_{s,s,1} = \left(\frac{2s}{2s+1}\right)^s \left(\frac{1}{2s+1}\right)^{1/2},$$
$$\xi_j^{(s,s,1)} = (-1)^{j+1}\left(\frac{1}{2s+1}\right)^{1/2}, \qquad j = 0, 1;$$

(vii) $Q_{s,s,2m}(x) = (-2)^{-s-m} Q_{0,s,m}(1-2x^2).$

The proofs of the above statements follow easily from the fact (Proposition 3.1) that $Q_{s_1, s_2, m}(x)$ is a monic polynomial and from the existence of the required alternation set.

We now deduce a domination property for $T_{s_1, s_2, m}(x)$ (cf. [8, Proposition 4]).

Theorem 3.3. *Let $p \in \pi(s_1, s_2, m)$, and let*

$$M \geqq \max\{|p(\xi_j^{(s_1, s_2, m)})| : j = 0, 1, \ldots, m\}.$$

If $x \leqq \xi_0^{(s_1, s_2, m)}$ or if $x \geqq \xi_m^{(s_1, s_2, m)}$, then

$$|p(x)| \leqq M |T_{s_1, s_2, m}(x)|. \tag{3.5}$$

Moreover, for each positive integer k and x real,

$$|p^{(k)}(x)| \leqq M |T^{(k)}_{s_1, s_2, m}(x)|, \qquad \text{for} \quad |x| \geqq 1. \tag{3.6}$$

Proof. First, define the polynomial $h(x)$ by means of

$$h(x) := (x - 1)^{s_1}(x + 1)^{s_2} \prod_{j=0}^{m} (x - \xi_j^{(s_1, s_2, m)}).$$

Using the Lagrange interpolation formula, it follows from the definition of $h(x)$ that

$$p(x) = \sum_{j=0}^{m} \frac{h(x)}{(x - \xi_j)} \frac{p(\xi_j)}{h'(\xi_j)}, \tag{3.7}$$

and

$$T_{s_1, s_2, m}(x) = \sum_{j=0}^{m} \frac{h(x)}{(x - \xi_j)} \frac{(-1)^{s_1 + m - j}}{h'(\xi_j)}, \tag{3.8}$$

where $\xi_j = \xi_j^{(s_1, s_2, m)}$, $j = 0, 1, \ldots, m$. Thus, with the hypothesis for M, (3.7) implies that

$$|p(x)| \leqq M \sum_{j=0}^{m} \frac{|h(x)|}{|x - \xi_j| \cdot |h'(\xi_j)|}. \tag{3.9}$$

Next, noting that [8, Proposition 4] establishes the case $s_1 = 0$ of this result, we may assume $s_1 > 0$, which implies $\xi_m < 1$. For $x \in (\xi_m, 1)$, we have, by definition, that $|h(x)| = (-1)^{s_1} h(x)$ and that

$$|h'(\xi_j)| = (-1)^{s_1 + m - j} h'(\xi_j).$$

Consequently, the right-hand side of (3.9) is equal to $(-1)^{s_1} \cdot M \cdot T_{s_1, s_2, m}(x)$. Hence, from (3.9) we have

$$|p(x)| \leqq M |T_{s_1, s_2, m}(x)| \qquad \text{for} \quad x \in (\xi_m, 1).$$

Arguing similarly for $x \in (-\infty, \xi_0)$ and $x \in (1, \infty)$, the first part of this theorem is proved. The second portion can be obtained for real x, $|x| \geqq 1$, by differentiating the formulas (3.7) and (3.8) k times. ■

We may now improve the inequality given in (2.25). Let $p \in \pi(s_1, s_2, m)$ with $m > 0$ and with $p \not\equiv 0$, and suppose that $|p(\xi)| = \|p\|_{[-1,1]}$ where $\xi \in [-1, 1]$. Then, as a consequence of Theorem 3.3,

$$\xi_0^{(s_1, s_2, m)} \leqq \xi \leqq \xi_m^{(s_1, s_2, m)}. \tag{3.10}$$

As an application, consider any $p \in \pi_n$ with $p \not\equiv 0$ and $p(-1) = p(1) = 0$, so that $p \in \pi(1, 1, n-2)$. Then, from (3.10) and Proposition 3.2 (iv), the points ξ in $[-1, +1]$, with $|p(\xi)| = \|p\|_{[-1,+1]}$ satisfy

$$|\xi| \leqq \frac{\cos(\pi/n)}{\cos(\pi/2n)}$$

(cf. Schur [10, Section 5]).

Concerning the behavior of $E_{s_1, s_2, m}$, we have the following generalization of [8, Proposition 8].

Theorem 3.4. *Let (θ_1, θ_2) be fixed in $\overline{\Omega}$ [cf. (2.14)], and let*

$$\{(s_{1,i}, s_{2,i}, m_i)\}_{i=1}^{\infty}$$

be any infinite sequence of ordered triples of nonnegative integers for which

$$\lim_{i\to\infty} n_i = \infty \qquad (n_i := s_{1,i} + s_{2,i} + m_i,\ i \geqq 1), \tag{3.11}$$

and for which

$$\lim_{i\to\infty} \frac{s_{1,i}}{n_i} = \theta_1 \qquad \text{and} \qquad \lim_{i\to\infty} \frac{s_{2,i}}{n_i} = \theta_2. \tag{3.12}$$

Then,

$$\Delta = \Delta(\theta_1, \theta_2) := \lim_{i\to\infty} (E_{s_{1,i}, s_{2,i}, m_i})^{1/n_i}$$
$$= \tfrac{1}{2}\sqrt{(1+\mu)^{1+\mu}(1-\mu)^{1-\mu}(1+\nu)^{1+\nu}(1-\nu)^{1-\nu}}, \tag{3.13}$$

where $\mu := \theta_2 + \theta_1$ and $\nu := \theta_2 - \theta_1$.

Proof. We first introduce the sequence of modified Jacobi polynomials $\{J_i(x)\}_{i=1}^{\infty}$ defined by

$$J_i(x) := (x-1)^{s_{1,i}}(x+1)^{s_{2,i}} P_{m_i}^{(2s_{1,i}, 2s_{2,i})}(x) \Big/ \left[2^{-m_i}\binom{2n_i}{m_i}\right], \qquad i \geqq 1. \tag{3.14}$$

We note that these polynomials are, from (2.10), monic for all $i \geqq 1$. Moreover, from the discussion in Section 2 of their properties the following inequalities are valid for each $i \geqq 1$:

$$\int_{-1}^{1} (J_i(x))^2\, dx \leqq \int_{-1}^{1} (Q_{s_{1,i}, s_{2,i}, m_i}(x))^2\, dx \leqq 2(E_{s_{1,i}, s_{2,i}, m_i})^2. \tag{3.15}$$

Next, on expanding the polynomial $J_i(x)$ in its Legendre polynomial expansion, it follows that (cf. Szegö [11, p. 182])

$$\|J_i\|^2_{[-1,+1]} \leqq \frac{(n_i+1)^2}{2} \int_{-1}^{+1} (J_i(x))^2\, dx, \qquad i \geqq 1,$$

whence, from (3.1),

$$(E_{s_{1,i}, s_{2,i}, m_i})^2 \leqq \|J_i\|^2_{[-1,+1]} \leqq \frac{(n_i+1)^2}{2} \int_{-1}^{+1} (J_i(x))^2\, dx, \qquad i \geqq 1. \tag{3.16}$$

Thus, from (3.15) and (3.16),

$$\Delta = \lim_{i\to\infty} (E_{s_{1,i}, s_{2,i}, m_i})^{1/n_i} = \lim_{i\to\infty} \left(\int_{-1}^{+1} (J_i(x))^2\, dx \right)^{1/2n_i}.$$

From Eq. (2.8) and the definition of $J_i(x)$ in (3.14), we therefore have

$$\Delta = \lim_{i\to\infty} \left[2^{2m_i} \binom{2n_i}{m_i}^{-2} h_{m_i}^{(2s_{1,i}, 2s_{2,i})} \right]^{1/2n_i}.$$

With the definition of $h_m^{(\alpha,\beta)}$ in (2.9) and an application of Stirling's formula, we obtain the value of Δ stated in (3.13). ■

As a consequence of the inequalities (3.15) and (3.16), upper and lower bounds for $E_{s_1, s_2, m}$ are

$$2^{-1/2}L \leqq E_{s_1, s_2, m} \leqq (n+1)L/\sqrt{2}, \tag{3.17}$$

where

$$L = L(s_1, s_2, m) := 2^m \binom{2n}{m}^{-1} (h_m^{(2s_1, 2s_2)})^{1/2},$$

$n := s_1 + s_2 + m$, and where $h_m^{(2s_1, 2s_2)}$ is given in (2.9).

Theorem 3.5. *Let* (θ_1, θ_2) *and* $\{(s_{1,i}, s_{2,i}, m_i)\}_{i=1}^{\infty}$ *be as in Theorem 3.4. Then, for* $z \in \mathbb{C}^* \backslash [a, b]$,

$$\lim_{i\to\infty} |Q_{s_{1,i}, s_{2,i}, m_i}(z)|^{1/n_i} = \Delta(\theta_1, \theta_2) \cdot G(z; \theta_1, \theta_2), \tag{3.18}$$

where $\Delta(\theta_1, \theta_2)$ *is defined in* (3.13) *and where* $G(z; \theta_1, \theta_2)$ *is defined in* (2.16) *and* (2.17). *Furthermore, this limit holds uniformly for any compact subset not containing the interval* $[a, b]$.

Proof. As in Saff and Varga [8], we use a normal families argument. First, for each $i \geqq 1$, set

$$n_i := s_{1,i} + s_{2,i} + m_i, \qquad \theta_{1,i} := s_{1,i}/n_i, \qquad \theta_{2,i} := s_{2,i}/n_i.$$

Setting $\mu_i := \theta_{2,i} + \theta_{1,i}$ and $\nu_i := \theta_{2,i} - \theta_{1,i}$, define a_i and b_i from (2.12), and let $\varphi_i(z)$ be defined from (2.15) for each $i \geqq 1$. Furthermore, set

$$\begin{aligned}\Delta_i &:= \Delta(\theta_{1,i}, \theta_{2,i})\\ &= \tfrac{1}{2}[(1+\mu_i)^{1+\mu_i}(1-\mu_i)^{1-\mu_i}(1+\nu_i)^{1+\nu_i}(1-\nu_i)^{1-\nu_i}]^{1/2}, \qquad i \geqq 1,\end{aligned}$$

and

$$K := \mathbb{C}^*\backslash[a, b] \qquad \text{and} \qquad K_i := \mathbb{C}^*\backslash[a_i, b_i] \qquad \text{for} \quad i \geqq 1,$$

and, for each $i \geqq 1$, set

$$\begin{aligned}u_i(z) &:= \frac{1}{n_i}\ln|Q_{s_{1,i}, s_{2,i}, m_i}(z)|,\\ v_i(z) &:= \ln \Delta_i + \ln|\varphi_i(z)| + \theta_{1,i}\ln\left|\frac{\varphi_i(z) - \varphi_i(1)}{\varphi_i(1)\varphi_i(z) - 1}\right|\\ &\quad + \theta_{2,i}\ln\left|\frac{\varphi_i(z) - \varphi_i(-1)}{\varphi_i(-1)\varphi_i(z) - 1}\right|.\end{aligned}$$

We remark that both the functions $u_i(z)$ and $v_i(z)$ for i fixed are harmonic in K_i with the exception of the points $z = -1$, $z = +1$, and $z = \infty$. In a neighborhood of $z = 1$, we can write

$$u_i(z) = \theta_{1,i}\ln|z - 1| + h_{1,i}(z)$$

and

$$v_i(z) = \theta_{1,i}\ln|z - 1| + \hat{h}_{1,i}(z),$$

where it can be verified that the functions $h_{1,i}(z)$ and $\hat{h}_{1,i}(z)$ are both harmonic at $z = 1$ for all $i \geqq 1$. Furthermore, an analogous representation is true in a neighborhood of $z = -1$. Near, $z = \infty$, we have

$$u_i(z) = \ln|z| + g_i(z), \tag{3.19}$$

and

$$v_i(z) = \ln|z| + \hat{g}_i(z). \tag{3.20}$$

In (3.19) and (3.20), both $g_i(z)$ and $\hat{g}_i(z)$ are harmonic at $z = \infty$ and

$$g_i(\infty) = 0 = \hat{g}_i(\infty).$$

Next, for each $i \geqq 1$, set

$$d_i(z) := u_i(z) - v_i(z).$$

Note that $d_i(z)$ is harmonic in K_i, even for $z = -1$, $z = 1$, and $z = \infty$. As z tends to $[a_i, b_i]$ in K_i, we note that $v_i(z)$ tends to $\ln \Delta_i$, whence

$$\limsup_{\substack{z \to [a_i, b_i] \\ z \in K_i}} d_i(z) \leqq \frac{1}{n_i} \ln E_{s_{1,i}, s_{2,i}, m_i} - \ln \Delta_i. \tag{3.21}$$

From (3.13) and (3.21) it follows that on any closed subset of K, the harmonic functions $d_i(z)$ are, for i sufficiently large, uniformly bounded from above. Hence, the $d_i(z)$ form a normal family on K. If $d(z)$ denotes a limit function of this family, then from (3.12) and (3.21) we have $d(z) \leqq 0$ for all z in K. However, since $d_i(\infty) = g_i(\infty) - \hat{g}_i(\infty) = 0$, $i \geqq 1$, we conclude that $d(z) \equiv 0$ in K. Since $\lim_{i \to \infty} v_i(z) := v(z)$ uniformly on any compact subset of $K \setminus \{-1, 1, \infty\}$, we have $\lim_{i \to \infty} u_i(z) = v(z)$ uniformly on a compact set of $K \setminus \{-1, 1, \infty\}$. Hence,

$$|Q_{s_{1,i}, s_{2,i}, m_i}(z)|^{1/n_i} = e^{u_i(z)} \to e^{v(z)} = \Delta(\theta_1, \theta_2) \cdot G(z; \theta_1, \theta_2),$$

as i tends to infinity, uniformly on any compact set omitting the interval $[a, b]$. ■

In closing this section, we establish the sharpness of Theorem 2.5 in a certain limiting sense. Let $(\theta_1, \theta_2) \in \overline{\Omega}$ and let $\{(s_{1,i}, s_{2,i}, m_i)\}_{i=1}^{\infty}$ be any infinite sequence of ordered triples of nonnegative integers satisfying

$$\lim_{i \to \infty} n_i = \infty \qquad (n_i = s_{1,i} + s_{2,i} + m_i, \quad i \geqq 1),$$

$$\lim_{i \to \infty} \frac{s_{1,i}}{n_i} = \theta_1 \qquad \text{and} \qquad \lim_{i \to \infty} \frac{s_{2,i}}{n_i} = \theta_2.$$

For the normalized Chebyshev polynomial

$$\mathcal{T}_i(z) := T_{s_{1,i}, s_{2,i}, m_i}(z) \tag{3.22}$$

associated with the set $\pi(s_{1,i}, s_{2,i}, m_i)$ defined in (3.4), we apply inequality (2.23) of Theorem 2.5 to obtain

$$|\mathcal{T}_i(z)|^{1/n_i} \leqq \|\mathcal{T}_i\|_{[-1,1]}^{1/n_i} G(z; s_{1,i}/n_i, s_{2,i}/n_i) = G(z; s_{1,i}/n_i, s_{2,i}/n_i).$$

Letting i tend to infinity yields

$$\limsup_{i \to \infty} |\mathcal{T}_i(z)|^{1/n_i} \leqq G(z; \theta_1, \theta_2).$$

But, recalling the definition of Δ in Theorem 3.4, the result of Theorem 3.5 applied to the *normalized* Chebyshev polynomials gives

$$\lim_{i\to\infty} |\mathscr{T}_i(z)|^{1/n_i} = G(z; \theta_1, \theta_2), \tag{3.23}$$

for all $z \notin [a, b]$, so that the inequality (2.23) of Theorem 2.5 is sharp in this limiting sense. We also remark that the sequence $\mathscr{T}_i(z)$ satisfies the hypothesis of Corollary 2.6, and, by (3.23), $\mathscr{T}_i(z)$ diverges for z exterior to the level curve $G(z; \theta_1, \theta_2) = 1$. Thus, Corollary 2.6 gives the largest possible open set of convergence to zero.

REFERENCES

1. J. H. B. Kemperman and G. G. Lorentz, Bounds for polynomials with applications, *Nederl Akad. Wetensch. Proc. Ser. A.* **82** (1979), 13–26.
2. G. G. Lorentz, "Bernstein Polynomials," Univ. of Toronto Press, Toronto, 1953.
3. G. G. Lorentz, Approximation by incomplete polynomials (problems and results), *in* "Padé and Rational Approximation: Theory and Applications" (E. B. Saff and R. S. Varga, eds.), Academic Press, New York, 1977, pp. 289–302.
4. G. Meinardus, "Approximation of Functions: Theory and Numerical Analysis," Springer-Verlag, Berlin and New York, 1967.
5. G. Meinardus, private communication.
6. D. S. Moak, E. B. Saff, and R. S. Varga, On the zeros of Jacobi polynomials $P_n^{(\alpha_n, \beta_n)}(x)$, *Trans. Amer. Math. Soc.* **249** (1979), 159–162.
7. E. B. Saff and R. S. Varga, The sharpness of Lorentz's theorem on incomplete polynomials, *Trans. Amer. Math. Soc.* **249**, (1979) 163–186.
8. E. B. Saff and R. S. Varga, On incomplete polynomials, "Numerische Methoden der Approximationstheorie," Band 4 (L. Collatz, G. Meinardus, and H. Werner, eds.), ISNM 42 Birkhäuser Verlag, Basel and Stuttgart, 1978, pp. 281–298.
9. E. B. Saff and R. S. Varga, Uniform approximation by incomplete polynomials, *Internat. J. Math. Math. Sci.* **1** (1978), 407–420.
10. I. Schur, Über das Maximum des Absoluten Betrages eines Polynoms in einem gegebenen Interval, *Math. Z.* **4** (1919), 271–287.
11. G. Szegö, "Orthogonal Polynomials," Colloquium Publication, Vol. XXIII, 4th Ed., Amer. Math. Soc., Providence, Rhode Island, 1975.
12. J. L. Walsh, "Interpolation and Approximation by Rational Functions in the Complex Domain," Colloquium Publication, Vol. XX, 5th Ed., Amer. Math. Soc., Providence, Rhode Island, 1969.

Research supported in part by the Air Force Office of Scientific Research under Grant AFOSR-74-2729, and by the Department of Energy under Grant EY-76-S-02-2075. The research of E. B. Saff was conducted as a Guggenheim Fellow, visiting at the Oxford University Computing Laboratory, Oxford, England.

M. Lachance

and

E. B. Saff
DEPARTMENT OF MATHEMATICS
UNIVERSITY OF SOUTH FLORIDA
TAMPA, FLORIDA

R. S. Varga
DEPARTMENT OF MATHEMATICS
KENT STATE UNIVERSITY
KENT, OHIO

Several Applications of the Shorted Operator

T. D. MORLEY

In this paper we give a brief introduction to the shorted operator construction of Anderson and Krein, with emphasis on the interplay between the algebraic formulation and the variational formulation. Applications to statistics and electrical networks are given.

1. INTRODUCTION

Consider the simplest of quadratic programs

$$\text{(QP)} \qquad \inf_{Ax=b} (Zx, x).$$

Here A and Z are given linear operators and the vector b is fixed. We seek the infimum of all the values of (Zx, x) (where $(\cdot, \cdot)$ denotes inner product) over all those x's such that $Ax = b$.

Assume Z is positive semidefinite, i.e., $(Zx, x) \geqq 0$ for all x and $Z = Z^*$. (The latter is redundant if Z is defined on a complex inner product space.) Then program **QP** is a convex program. The Lagrange multiplier formulation becomes

$$\text{(K)} \qquad Ax \qquad = b,$$

$$\text{(K}'\text{)} \qquad Zx - A^*v = 0.$$

Modulo a change of variables we may rewrite problem **QP** as

$$\text{(QP}'\text{)} \qquad \inf_{y \in \ker(A)} (Z(c + y), c + y)$$

where $\ker(A)$ is the null space of A and c is any vector such that $Ac = b$. Clearly we may restrict our attention to the case where $b \in \text{range}(A)$. Problem **QP**$'$ depends only on c and $\mathscr{S} = \ker(A)^\perp$. Define the function f: range$(A) \to \mathbb{R}$ by the formula of program **QP**$'$

$$\text{(S)} \qquad f(c) = \inf_{y \in \mathscr{S}^\perp} (Z(c + y), c + y).$$

ISBN 0-12-178150-X

Then if Z is a positive semidefinite operator on a real or complex, finite or infinite dimensional Hilbert space and $\mathscr{S}$ is a closed subspace, then $f(c)$ turns out to be a quadratic form in c,

$$f(c) = (\mathscr{S}(Z)c, c),$$

where the linear operator $\mathscr{S}(Z)$ is called the *shorted operator* of Z with respect to the subspace $\mathscr{S}$. In the case Z is an invertible operator on a finite dimensional space $\mathscr{S}(Z)$ is known as the Schur complement of Z.

Depending on the application, the problem of primary interest may be either **QP** or $\mathbf{K} - \mathbf{K}'$. In the general model of a time invariant electromechanical system considered by Duffin and Morley [10], **K** was the generalization of Kirchhoff's current law, and $\mathbf{K}'$ was the generalization of Kirchhoff's voltage law. The operator Z corresponded to the branch resistances. The operator A corresponds with the node-arc incidence matrix of the underlying graph. The fact that **QP** will give a solution to $\mathbf{K} - \mathbf{K}'$ in the case of ordinary resistive networks was noticed by Maxwell [15] (as pointed out in a footnote by J. J. Thompson). Thus voltage is the Lagrange multiplier of current!

The interplay between program **QP** and system $\mathbf{K} - \mathbf{K}'$ is very useful in the analysis of resistive networks, constructed either of ordinary (scalar) resistors or resistive n-ports, for in these cases the operator Z which arises will be positive semidefinite. (See [7] for the electrical motivation and terminology.)

The connection between program **QP** and system $\mathbf{K} - \mathbf{K}'$ becomes even more important when the quadratic form $x \mapsto (Zx, x)$ is replaced by a nonlinear function, with some assumed monotonicity properties (see [5]).

The Gauss–Markov theorem of statistics leads to a quadratic program of the same form as **QP**. In this case the operator Z will be a convariance and so will be a priori positive semidefinite. The function of $\mathbf{K} - \mathbf{K}'$ in this case is a tool to solve **QP**.

For other applications of the shorted operator see [16] (coding theory), [4] (statistics), [12] (structural mechanics), and [13] (Hilbert space).

Closely related to program **QP** is the following seemingly special case. Given positive semidefinite linear operators R and S and a vector c we wish to determine the value

$$\textbf{(PS)} \qquad g(b) = \inf_{x+y=b} (Rx, x) + (Sy, y).$$

Again there will be a unique positive semidefinite operator denoted $R:S$ such that

$$(R:Sb, b) = \inf_{x+y=b} (Rx, x) + (Sy, y).$$

This was shown in finite dimensions by Anderson and Duffin [3] and later in infinite dimensions by Fillmore and Williams [13] and Anderson and Trapp [6]. The operator $R:S$ is called the *parallel sum.*

That **PS** is a special case of (**KQ**) can be seen by setting

$$A = [I, I], \qquad Z = \begin{bmatrix} R & 0 \\ 0 & S \end{bmatrix},$$

where I is the identity operator. As pointed out by Anderson and Duffin [3], problem **QP** can be viewed as a limit problem of type **PS**.

All linear operators considered in this paper will be bounded operators between Hilbert spaces.

2. THE SHORTED OPERATOR

As before let Z be a positive semidefinite order and let $\mathscr{S}$ be a closed subspace of X. Given an $c \in \mathscr{S}$ we seek an $Z_0 \in \mathscr{S}^\perp$ such that

$$\textbf{(S)} \qquad (Z(c + z_0), c + z_0) = \inf_{z \in \mathscr{S}^\perp} (Z(c + z), c + z).$$

By use of Lagrange multipliers or by considering a variation $\delta z \in \mathscr{S}^\perp$ problem **S** reduces to the problem

$$\textbf{(S}'\textbf{)} \qquad \textit{Find} \quad z_0 \in \mathscr{S}^\perp \quad \textit{such that} \quad Z(c + z_0) \in \mathscr{S}.$$

With respect to a suitable orthonormal basis we may partition Z as

$$Z = \begin{bmatrix} Z_{11} & Z_{12} \\ Z_{21} & Z_{22} \end{bmatrix}$$

with $Z_{11}: \mathscr{S} \to \mathscr{S}$, $Z_{12}: \mathscr{S}^\perp \to \mathscr{S}$, $Z_{21} = Z_{21}^*: \mathscr{S}^\perp \to \mathscr{S}$, $Z_{22}: \mathscr{S}^\perp \to \mathscr{S}^\perp$. With respect to the same orthonormal basis as above we may write any vector u as

$$u = \begin{bmatrix} v \\ w \end{bmatrix}$$

with $v \in \mathscr{S}$, and $w \in \mathscr{S}^\perp$. This notation problem **S**′ can be rewritten as

Given c find an x_0 and w such that

$$\begin{bmatrix} Z_{11} & Z_{12} \\ Z_{21} & Z_{22} \end{bmatrix} \begin{bmatrix} c \\ z_0 \end{bmatrix} = \begin{bmatrix} w \\ 0 \end{bmatrix}.$$

If Z is invertible we may solve for z_0,

$$Z_{21}c + Z_{22}Z_0 = 0, \qquad z_0 = -Z_{22}^{-1}Z_{21}c,$$

and plugging into **S** we conclude

$$((Z_{11} - Z_{12}Z_{22}^{-1}Z_{21})c, c) = \inf_{z \in \mathscr{S}^\perp} (Z(c + z), c + z).$$

If Z is not invertible set $Z^\varepsilon = Z + \varepsilon I$. A simple "$\varepsilon$–$\delta$" proof gives

$$\inf_{z \in \mathscr{S}^\perp} (Z(c + z), c + z) = \lim_{\varepsilon \downarrow 0} \inf_{z \in \mathscr{S}^\perp} (Z^\varepsilon(c + z), c + z).$$

Now a simple argument gives

$$(\mathscr{S}(Z)c, c) = \inf_{z \in \mathscr{S}^\perp} (Z(c + z), c + z),$$

where

$$\mathscr{S}(Z) = \lim_{\varepsilon \downarrow 0}(Z_{11}^\varepsilon - Z_{12}(Z_{22}^\varepsilon)^{-1}Z_{21}).$$

The analysis of this section could have been done in terms of **QP** instead of **S** with no essential change.

3. A GAUSS–MARKOV THEOREM

Consider the linear statistical model

$$\mathbf{z} = Bx + \mathbf{v},$$

where B is a linear operator, x is a vector, and $\mathbf{v}$ is a vector-valued random variable with mean 0 and covariance $V^2 = \operatorname{cov}(\mathbf{v}, \mathbf{v})$. We note that V^2 is positive semidefinite. Given a linear function of x, we wish to estimate it by a linear function of $\mathbf{z}$. Of the many such estimators, one such is the *best linear unbiased estimator* (or BLUE). In finite dimensions with nonsingular covariance the BLUE is given by a classical formula, known as the Gauss–Markov theorem. Albert in [1] extended this formula to the case of a possibly singular covariance V^2. Beutler and Root [9] extended the Gauss–Markov theorem to Hilbert space under the condition that the covariance V^2 is nonsingular. Suppose we are given a linear function $l(x) = (c, x)$.

We wish to find a linear function of $\mathbf{z}$ that estimates this linear function. Thus we wish to find a g such that, in some sense,

$$(g, \mathbf{z}) \approx (c, x).$$

Letting $\mathscr{E}(\cdot)$ denote the expectation of a random variable, let us call g an *unbiased estimator* if

$$\mathscr{E}(g, \mathbf{z}) = (c, x).$$

If g is an unbiased estimator,

$$\mathscr{E}(g, \mathbf{z}) = \mathscr{E}(g, Bx + \mathbf{v}) = \mathscr{E}(g, Bx) + \mathscr{E}(g, \mathbf{v}) = \mathscr{E}(g, Bx) = (g, Bx).$$

Thus g is an unbiased estimator if and only if $B^*g = c$, where B^* denotes the adjoint of B.

Supposing that $c \in \text{range}(B^*)$, the best linear unbiased estimator (BLUE) is the unbiased estimator that minimizes the variance of $(g, \mathbf{z})$ among all unbiased estimators g. But variance $(g, \mathbf{z}) = (V^2 g, g)$ so the problem of finding a BLUE for c becomes the minimization problem

$$\min_{B^*y=c} (V^2 y, y).$$

It should be noted that the expectations and covariance of Hilbert space random variables are defined in terms of the Bochner integral. (See [9] for analytical details of the present problem.)

If the above minimization is replaced by an infimum then the shorted operator construction will give the answer.

Theorem 1 [18]. *Suppose the range of B is closed: then*

$$(B^\dagger \mathcal{S}(K) B^{\dagger *} g, g) \geqq \text{cov}(g, z)$$

where K is the covariance of z, and z is any unbiased estimator of c. Here $\mathcal{S} = \ker(B^*)^\perp$. *Moreover, if*

$$(I - BB^\dagger) K BB^\dagger B^{*\dagger} c \in \text{range}((I - BB^\dagger) K (I - BB^\dagger)),$$

then a best linear unbiased estimator is given by

$$g = (I - ((I - BB^\dagger)((I - BB^\dagger) K (I - BB^\dagger))^\dagger (I - BB^\dagger) K BB^\dagger)) B^{*\dagger} c.$$

In the preceding theorem the notation $A^\dagger$ means the unique linear operator defined on $\text{range}(A) \oplus \text{range}(A)^\perp$ such that $A^\dagger c$ is the $x \in \ker(A)^\perp$ such that

$$A^\dagger c = \begin{cases} Ax = c & \text{if} \quad c \in \text{range}(A) \\ 0 & \text{if} \quad c \in \text{range}(A)^\perp. \end{cases}$$

4. NETWORKS

Suppose we connect two resorters with resistance values R and S in parallel and connect a battery with voltage v and current c. The Kirchhoff and Ohm laws give

(P1) $$x + y = c,$$

(P2) $$Rx = Sy = v.$$

Here x and y are the currents theory of the two resistors. The resistance values R and S are of course strictly positive real numbers. High school algebra then suffices to show

$$v = \frac{1}{(1/R) + (1/S)} c.$$

Now let us consider **P1** and **P2** with R and S positive semidefinite linear operators and c, x, y, v vectors. It is readily seen that **P1**–**P2** are a special case of **K**–**K**′. The corresponding variational formulation is

$$\inf_{x+y=c} (Rx, x) + (Sy, y),$$

which is a special case of **QP**. Thus, the shorted operator gives the existence and uniqueness of a linear operator $A:B$ such that

$$(R{:}Sc, c) = \inf_{x+y=c} (Rx, x) + (Sy, y).$$

This generalization of the connection of ordinary resistors is not electrically specious (see [3] for the physical motivation). This operation was defined (in an equivalent definition) by Anderson and Duffin [3] with later work by others [6, 8, 13]. Fillmore and Williams in [13] gave the first definition (again different from the above) of $A:B$ in infinite dimensional Hilbert space. The preceding definition is due to Ando.

Theorem 2. *Let R and S be positive semidefinite linear operaLrs on a complex Hilbert space U. Then*

(a) *$R{:}S$, the parallel sum of R and S is also positive semidefinite.*

(b) *$R{:}S = S{:}R$.*

(c) *If T is also a positive semidefinite linear operator on U, then $(R{:}S):T = R{:}(S{:}T)$.*

(d) *(Series–parallel inequality.) Given R, S, T, U all positive semidefinite, then $(R:T) + (S:U) \leqq (R+S):(T+U)$*

(e) *(Transformer inequality.) If $P\colon W \to U$ is a linear operator, then $P^*(R{:}S)P \leqq (P^*RP){:}(P^*SP)$.*

(f) *(Parallel–inner product inequality.) $(R{:}Sc, c) \leqq (Rc, c){:}(Sc, c)$.*

(g) *$\|R{:}S\| \leqq \|R\|:\|S\|$.*

Proof. We shall prove (e), for proofs of the others see [3, 6, 13] or [17], for other proofs of (e) see [6] or [13]:

$$\begin{aligned}
(P^*(R{:}S)Pc, c) &= ((R{:}S)Pc, Pc) \\
&= \inf_{x+y=Pc} (Rx, x) + (Sy, y) \\
&\leqq \inf_{\substack{x+y=Pc\\ x=Pw\\ y=Pv\\ v+w=c}} (Rx, x) + (Sy, y) \\
&= \inf_{v+w=c} (RPx, Px) + (SPy, Py) \\
&= \inf_{v+w=c} (P^*RPv, v) + (P^*SPu, u) \\
&= ((P^*RP){:}(P^*SP)c, c). \quad \blacksquare
\end{aligned}$$

Let P and Q be orthogonal projections onto range(P) and range(Q) respectively. Halmos asked for a formula for $P \wedge Q$, the infimum of the two projections. This is defined as the orthogonal projection onto $\mathscr{S} =$ range$(P) \cap$ range(Q). The remarkable formula of the following theorem is due to Anderson and Duffin [3] in the finite dimensional case. Filmore and Williams later extended this result to infinite dimensions (see [13]).

Theorem 3. *Let P and Q denote orthogonal projections onto range(P) and range (Q), respectively; then*

$$P \wedge Q = 2P{:}Q,$$

where $P \wedge Q$ is the orthogonal projection onto range(P) $\cap$ range(Q).

Proof. Let $c \in \mathscr{S} =$ range$(P) \cap$ range(Q); then

$$(P{:}Qc, c) = \inf_{x+y=c} (Rx, x) + (Sy, y) = \left(R\frac{c}{2}, \frac{c}{2}\right) + \left(S\frac{c}{2}, \frac{c}{2}\right) = \frac{1}{2}\|c\|^2.$$

The equality follows from "setting the derivative equal to zero."

On the other hand if $c \in \mathscr{S}^{\perp} =$ range$(P) \cap$ range$(Q)^{\perp} =$ range$(P)^{\perp} \oplus$ range$(Q)^{\perp} = \ker(P) \oplus \ker(Q)$ [where $\ker(A)$ denotes the null space of A], then we can find $x_0 \in \ker(P)$, $y_0 \in \ker(Q)$ such that $x_0 + y_0 = c$. Thus we have

$$(P{:}Qc, c) = \inf_{x+y=c} (Px, x) + (Qy, y) = (Px_0, x_0) + (Qy_0, y_0) = 0.$$

But now we are done. With respect to a suitable orthonormal basis we may write any x as

$$\begin{bmatrix} x_1 \\ x_2 \end{bmatrix}$$

with $x_1 \in \mathscr{S}$, and $x_2 \in \mathscr{S}^{\perp}$ with respect to this basis we have shown

$$P{:}Q = \begin{bmatrix} \frac{1}{2}I & * \\ * & 0 \end{bmatrix}.$$

But since $P{:}Q$ is positive semidefinite, an elementary lemma on positive semidefinite operators will show that

$$P{:}Q = \begin{bmatrix} \frac{1}{2}I & 0 \\ 0 & 0 \end{bmatrix},$$

and we are done. ∎

The above proof is taken from [17].

REFERENCES

1. A. ALBERT, The Gauss–Markov theorem for regression models with possibly singular covariances, *SIAM J. Appl. Math.* **24** (1973), 182–187.
2. W. N. ANDERSON, Shorted operators, *SIAM J. Appl. Math.* **20** (1971), 520–525.
3. W. N. ANDERSON AND R. J. DUFFIN, Series and parallel addition of matrices, *J. Math. Anal. Appl.* **26** (1969), 576–594.
4. W. N. ANDERSON, G. D. KLEINDORFER, P. R. KLEINDORFER, AND M. B. WOODROOFE, Consistent estimates of the parameters of a linear system, *Ann. Amer. Statist.* **40** (1969), 2064–2075.
5. W. N. ANDERSON, T. D. MORLEY, AND G. E. TRAPP, Fenchel duality on non-linear networks, *IEEE Trans. Circuits and Systems* Special Issue Math. Found. Systems **25** (1978), 762–765.
6. W. N. ANDERSON AND G. E. TRAPP, Shorted operators II, *SIAM J. Appl. Math.* **28** (1975), 61–71.
7. W. N. ANDERSON AND G. E. TRAPP, A class of Monotone operator functions related to electrical network theory, *Linear Algebra and Appl.* **15** (1976), 53–67.
8. T. ANDO AND K. NISHIO, Characterizations of operations derived from network connections, *J. Math. Anal. Appl.* **53** (1976), 539–549.
9. F. J. BEUTLER AND W. L. ROOT, The operater pseudo-inverse in control and systems identification, *in* "Generalized Inverses and Applications" (M. Z. Nashed, ed.), Academic Press, New York, 1976.
10. R. J. DUFFIN AND T. D. MORLEY, Almost definite operators and electro-mechanical systems, *SIAM J. Appl. Math.* **35** (1978), 21–30.
11. R. J. DUFFIN AND T. D. MORLEY, Inequalities induced by electrical connections, *in* "Recent Applications of Generalized Inverses" (M. Z. Nashed, ed.), Pittman, London, 1978.
12. S. J. FENVES, Structural analysis by networks, matrices, and computers, *J. Structural Div. Proc. ASCE* **92** (1966), 199–221.
13. P. A. FILLMORE AND J. P. WILLIAMS, On operator ranges, *Adv. in Math.* **7** (1971), 254–281.
14. M. A. KREIN, The theory of self-adjoint extensions of semi-bounded Hermitian transformations and applications, *Math. Moscow* **20** (1947), 431–495; **21** (1948), 365–404. (In Russ.)
15. J. C. MAXWELL, "Treatise on Electricity and Magnetism," 3rd Ed., reprinted Dover, New York, 1953.
16. R. J. MCELIESE, E. R. RODEMICH, AND H. C. RUMSEY, JR., The Lovasz bound and some generalizations. *J. Comb. Inform. System Sci.* **3** (1978), 134–152.
17. T. D. MORLEY, Parallel summation, Maxwell's principle, and the infimum of projections, *J. Math. Anal. Appl.* (to appear).
18. T. D. MORLEY, A Gauss–Markov theorem for infinite dimensional regression models with possible singular covariance, *SIAM J. Appl. Math.* (to appear).

Research was partially supported by a grant from the National Science Foundation.

DEPARTMENT OF MATHEMATICS
UNIVERSITY OF ILLINOIS
URBANA, ILLINOIS

Newton's Method for Polynomials with Real Roots and Their Divided Differences

R. N. PEDERSON

1. INTRODUCTION

Let

$$f(x) = \prod_{k=1}^{n} (x - \alpha_k), \qquad \alpha_1 > \alpha_2 > \cdots > \alpha_n, \tag{1.1}$$

be a polynomial with only real roots. Since the roots of the derivatives of f lie between the roots of f, α_1 can be approximated to any degree of accuracy by applying Newton's method with initial guess greater than α_1. With perfect computing one could then compute all of the roots of f by successively dividing out factors corresponding to computed roots. Since, in general, one can only approximate roots of f, it seems worthwhile to study the question of the precision required to compute all of the roots to a preassigned degree of accuracy. In a recent paper Hager and Pederson [2] presented an algorithm for approximating the eigenvalues of a Hermitian matrix by using a recursion formula for the divided differences of the characteristic polynomial with respect to approximations of the roots. Experiments on a variety of matrices produced very good results. In this paper we obtain precise results on the accuracy required in approximating α_1 by $\tilde{\alpha}_1$ in order that the largest root of the divided difference

$$f_1(\tilde{\alpha}_1, x) = \frac{f(x) - f(\tilde{\alpha}_1)}{x - \tilde{\alpha}_1} \tag{1.2}$$

lie within the region of certain convergence of Newton's method for f. We also obtain a partial result toward determining the accuracy with which one must approximate $\alpha_1, \ldots, \alpha_j$ by $\tilde{\alpha}_1, \ldots, \tilde{\alpha}_j$ in order that the divided difference with respect to the latter should have its largest root in the region of certain convergence of Newton's method to α_j.

ISBN 0-12-178150-X

2. CONVERGENCE OF NEWTON'S METHOD FOR DIVIDED DIFFERENCES

The guaranteed convergence of Newton's method for f (1.1) with initial guess greater than α_1 is inherited by the divided difference $f_1(\tilde{\alpha}_1, x)$ as long as $\tilde{\alpha}_1$ is so good an approximation of α_1 that $f(\tilde{\alpha}_1, x)$ has $n-1$ real roots. In this section we obtain conditions under which Newton's method converges even though $f_1(\tilde{\alpha}_1, x)$ has lost some of its real roots.

We consider first the case of approximations $\tilde{\alpha}_1 > \alpha_1$.

Theorem 2.1. *If $\tilde{\alpha}_1 > \alpha_1$ and $x > \alpha_2$, then $f_1(\tilde{\alpha}_1, x)$ and all of its derivatives are positive.*

Proof. Since $f(x) = (x - \alpha_1)f_1(\alpha_1, x)$, it follows easily that

$$f_1(\tilde{\alpha}_1, x) = f_1(\alpha_1, x) + (\tilde{\alpha}_1 - \alpha_1)f_2(\alpha_1, \tilde{\alpha}_1, x), \tag{2.1}$$

where f_2 is the divided difference of $f_1(\tilde{\alpha}_1, x)$ with respect to α_1. The proof easily follows by induction on the degree of the polynomial. ∎

Since there is a neighborhood of α_2 in which Newton's method is guaranteed to converge, the preceding theorem implies that if $\tilde{\alpha}_1 > \alpha_1$ and $\tilde{\alpha}_1 - \alpha_1$ is sufficiently small, then Newton's method applied to $f_1(\tilde{\alpha}_1, x)$, with $\tilde{\alpha}_1$ as initial guess, will converge to the largest root $\tilde{\alpha}_2$ of $f_1(\tilde{\alpha}_1, x)$. It is easy to construct examples where $\tilde{\alpha}_2$ is a much worse approximation of α_2 than $\tilde{\alpha}_1$ is of α_1. It therefore seems worthwhile to consider the question of when $\tilde{\alpha}_2$ is in the region of certain convergence to α_2 for the function f. To this end we consider approximations $\tilde{\alpha}_1 < \alpha_1$.

Theorem 2.2. *If $\tilde{\alpha}_1$ is between α_1 and the largest root of f', then for x greater than the largest zero $\tilde{\alpha}_2$ of $f_1(\tilde{\alpha}_1, x)$ all of the derivatives of $f_1(\tilde{\alpha}_1, x)$ are positive.*

Proof. The hypothesis implies that $f_1(\tilde{\alpha}_1, x)$ has a largest real root $\tilde{\alpha}_2$ in the interval (α_2, α_1). Since $f^{(j)}(\tilde{\alpha}_1) \geqq 0$ for all j, the case $x \geqq \tilde{\alpha}_1$ is disposed of by expanding f about $\tilde{\alpha}_1$. Suppose then that $\tilde{\alpha}_2 < x < \tilde{\alpha}_1$. It follows from the Leibniz formula that

$$f_1^{(k)}(\tilde{\alpha}_1, x) = k!\,\frac{f(\tilde{\alpha}_1) - f(x)}{(\tilde{\alpha}_1 - x)} - \sum_{j=1}^{k} \binom{k}{j} \frac{(k-j)!\,f^{(j)}(x)}{(\tilde{\alpha}_1 - x)^{k-j+1}}. \tag{2.2}$$

Suppose that $f^{(k+1)}(x) \leqq 0$. Because the largest zeros of $f^{(j)}$ decrease with j and each derivative can have at most one zero in (α_2, α_1), it follows that $f^{(j)}(x) < 0$ for $j \leqq k$. Hence all of the terms on the right side of (2.2) are positive. If $f^{(k+1)}(x) > 0$ we use Taylor's formula with the remainder to express the right side of (2.2) as $f^{(k+1)}(y)$, $x < y < \tilde{\alpha}_1$. Since $f^{(k+1)}$ can have at most one root in (α_2, α_1), $f^{(k+1)}(y) > 0$. This completes the proof. ∎

3. INEQUALITIES BASED ON THE ZEROS OF THE SECOND DERIVATIVE

In this section we prove some inequalities relating the roots α_j of (1.1) to those of the second derivative

$$f''(x) = \prod_{j=1}^{n-2} (x - \gamma_j), \qquad \gamma_1 > \gamma_2 > \cdots > \gamma_{n-2}. \tag{3.1}$$

We observe that γ_k is always in the interval $[\alpha_{k+1}, \alpha_k)$ or $(\alpha_{k+2}, \alpha_{k+1})$ so that in the interval $[\alpha_{k+1}, \alpha_k)$ there is always a point where f and f'' have opposite signs. The following theorem is therefore not vacuous.

Theorem 3.1. *If x is a point in the interval $[\alpha_{k+1}, \alpha_k)$ where $f(x)f''(x) \leqq 0$, then*

$$\sum_{j=k+2}^{n} \frac{1}{(\gamma_k - \alpha_j)} \leqq 2 \sum_{j=1}^{k} \frac{1}{(\alpha_j - \gamma_k)}. \tag{3.2}$$

Proof. The identity

$$f''(x) = 2f(x) \sum_{i<j} \frac{1}{(x - \alpha_i)(x - \alpha_j)} \tag{3.3}$$

and the condition $f(x)f''(x) \leqq 0$ imply that

$$\sum_{i<j} \frac{1}{(x - \alpha_i)(x - \alpha_j)} \leqq 0. \tag{3.4}$$

The above inequality may be rewritten

$$\sum_{i=1}^{k} \sum_{j=i+1}^{n} \frac{1}{(x - \alpha_i)(x - \alpha_j)} + \sum_{k+1 \leqq i<j} \frac{1}{(x - \alpha_i)(x + \alpha_j)} \leqq 0. \tag{3.5}$$

After omitting the terms in the first sum for which $j \leqq k$ by virtue of their positivity and using the identity

$$2 \sum_{k+1<i<j} \frac{1}{(x - \alpha_i)(x - \alpha_j)} = \left(\sum_{i=k+1}^{n} \frac{1}{(x - \alpha_i)} \right)^2 - \sum_{i=k+1}^{n} \frac{1}{(x - \alpha_i)^2} \tag{3.6}$$

in the second, we obtain

$$2\sum_{i=1}^{k} \frac{1}{(x - \alpha_i)} \sum_{j=k+1}^{n} \frac{1}{(x - \alpha_j)} + \left(\sum_{j=k+1}^{n} \frac{1}{(x - \alpha_j)} \right)^2 \leqq \sum_{j=k+1}^{n} \frac{1}{(x - \alpha_j)^2}. \tag{3.7}$$

The proof is completed by substituting the inequality

$$\sum_{j=k+1}^{n} \frac{1}{(x - \alpha_j)^2} \leqq \frac{1}{(x - \alpha_{k+1})} \sum_{j=k+1}^{n} \frac{1}{(x - \alpha_j)} \tag{3.8}$$

in the term on the right side of (3.7) and by canceling a common factor. ∎

We next prove an inequality in which it is convenient to introduce fictitious roots of multiplicity k defined by

$$\sum_{j=1}^{k} \frac{1}{(\alpha_j - x)} = \frac{k}{(\tilde{\alpha}_k(x) - x)}, \qquad x \in (\alpha_{k+1}, \alpha_k). \tag{3.9}$$

The quantities $\tilde{\alpha}_k(x)$ are somewhat artificial in that they depend on x; nevertheless they are useful.

Theorem 3.2. *If x is a point in $[\alpha_{k+1}, \alpha_k)$ where $f(x)f''(x) \leqq 0$ and*

$$\frac{k}{(\tilde{\alpha}_k(x) - x)} < \frac{1}{(x - \alpha_{k+1})}, \tag{3.10}$$

then

$$\sum_{j=k+2}^{n} \frac{1}{(x - \alpha_j)} \leqq \frac{k}{(\tilde{\alpha}_k(x) + k\alpha_{k+1} - (k+1)x)} \tag{3.11}$$

Proof. By replacing k by $k + 1$ in (3.5), we have

$$\sum_{i=1}^{k+1} \sum_{j=i+1}^{n} \frac{1}{(x - \alpha_i)(x - \alpha_j)} + \sum_{k+2 \leqq i < j} \frac{1}{(x - \alpha_i)(x - \alpha_j)} \leqq 0. \tag{3.12}$$

Since $x \in [\alpha_{k+1}, \alpha_k)$, the terms in the second sum in (3.12) are all positive and may be omitted. Hence, after rewriting the first sum, we have

$$\sum_{i=1}^{k+1} \sum_{j=k+2}^{n} \frac{1}{(x - \alpha_i)(x - \alpha_j)} + \sum_{i=1}^{k} \sum_{j=i+1}^{k+1} \frac{1}{(x - \alpha_i)(x - \alpha_j)} \leqq 0. \tag{3.13}$$

The terms in the second sum in (3.13) for which $j \leqq k$ are positive and may be omitted. Hence

$$\sum_{i=1}^{k+1} \frac{1}{(x - \alpha_i)} \sum_{j=k+2}^{n} \frac{1}{(x - \alpha_j)} \leqq \frac{1}{(x - \alpha_{k+1})} \sum_{i=1}^{k} \frac{1}{(\alpha_i - x)}. \tag{3.14}$$

The proof is completed by substituting the definition (3.9) of $\tilde{\alpha}_k(x)$ into (3.14). ■

The preceding theorem yields the following simply stated condition that the kth root of f be simple.

Theorem 3.3. *If there is a $t \in [0, 1/(2k + 1)]$ such that the kth largest root γ_k of f'' satisfies, with $x_k = \max\{\alpha_{k+1}, \gamma_k\}$, the inequality*

$$\gamma_k \leqq \alpha_{k+1} + t(\tilde{\alpha}_k(x_k) - \alpha_{k+1}), \tag{3.15}$$

then α_{k+1} is a simple root and

$$\alpha_{k+1} - \alpha_{k+2} \geqq \frac{(1 - (2k + 1)t)}{k} (\tilde{\alpha}_k(x_k) - \alpha_{k+1}). \tag{3.16}$$

Proof. By applying Theorem 3.2 and omitting all but the term corresponding to $j = k + 2$ on the left side of (3.11), we obtain

$$(2k + 1)x_k \geqq \tilde{\alpha}_k(x_k) + k\alpha_{k+1} + k\alpha_{k+2}. \tag{3.17}$$

The proof is completed by combining (3.15) and (3.17) ■

When the condition (3.15) is satisfied, it is possible to obtain a generalization of Theorem 3.2 for points x where f and f'' do not necessarily have opposite signs.

Theorem 3.4. *Let* (3.15) *be satisfied and set* $x_k = \max\{\alpha_{k+1}, \gamma_k\}$. *Then for* $x_k \geqq x \geqq \alpha_{k+2}$ *we have*

$$\sum_{j=k+2} \frac{1}{(x - \alpha_j)} \leqq \frac{k(x_k - \alpha_{k+2})}{(x - \alpha_{k+2})(\tilde{\alpha}_k(x_k) + k\alpha_{k+1} - (k + 1)x_k)}. \tag{3.18}$$

Proof. Write

$$\sum_{j=k+2}^{n} \frac{1}{(x - \alpha_j)} = \sum_{j=k+2}^{n} \frac{1}{(x_k - \alpha_j)} + \sum_{j=k+2}^{n} \frac{(x_k - x)}{(x - \alpha_j)(x_k - \alpha_j)} \tag{3.19}$$

use the inequality $(x - \alpha_j) \geqq (x - \alpha_{k+2})$ in the second term on the right in (3.19) and then apply Theorem 3.2. ■

4. MONOTONE AND CONTRACTION CONVERGENCE OF NEWTON'S METHOD FOR THE SECOND LARGEST ROOT

Newton's method, with initial guess in the indicated interval, will converge to α_2 if either

$$f(x), \quad f'(x), \quad \text{and} \quad f''(x) \qquad \text{have the same sign in } (\alpha_2, \alpha_2 + \delta) \tag{4.1}$$

or

$$\left| \frac{d}{dx}\left(x - \frac{f(x)}{f'(x)}\right) \right| < 1 \qquad \text{in} \quad [\alpha_2 - \delta, \alpha_2 + \delta]. \tag{4.2}$$

In this section we obtain a precise condition on $\tilde{\alpha}_1$ in order that the largest root $\tilde{\alpha}_2$ of the divided difference $f_1(\tilde{\alpha}_1, x)$ serve as an admissible initial guess for either (4.1) or (4.2). We first obtain a condition for monotone convergence.

Theorem 4.1. *Let f be given by* (1.1) *and define* $K_n = ne^2/(2 - 4/n)$. *If*

$$\gamma_1 \geqq \alpha_2 + (\alpha_1 - \alpha_2)/K_n \qquad \text{and} \qquad 0 \leqq (\alpha_1 - \tilde{\alpha}_1) \leqq (\alpha_1 - \alpha_2)/K_n, \tag{4.3}$$

then the largest zero $\tilde{\alpha}_2$ of $f_1(\tilde{\alpha}_1, x)$ is in the interval (α_2, γ_1).

Proof. It suffices to prove that $f(\tilde{\alpha}_1)f(\gamma_1) < 1$. Let us write this ratio as

$$\frac{f(\tilde{\alpha}_1)}{f(\gamma_1)} = \frac{(\alpha_1 - \tilde{\alpha}_1)(\tilde{\alpha}_1 - \alpha_2)}{(\alpha_1 - \gamma_1)(\gamma_1 - \alpha_2)} \prod_{j=3}^{n} \left(1 + \frac{(\tilde{\alpha}_1 - \gamma_1)}{(\gamma_1 - \alpha_j)}\right). \tag{4.4}$$

By applying the inequality $(1 + x) \leqq e^x$ to the product in (4.4) and then invoking Theorem 2.1 with $k = 1$, we obtain

$$\frac{f(\tilde{\alpha}_1)}{f(\gamma_1)} \leqq \frac{(\alpha_1 - \tilde{\alpha}_1)(\tilde{\alpha}_1 - \alpha_2)}{(\alpha_1 - \gamma_1)(\gamma_1 - \alpha_2)} e^2. \tag{4.5}$$

Since the roots of the derivatives of a polynomial with only real roots are increasing functions of the roots, the largest possible value of γ_1 is attained by the polynomial

$$(x - \alpha_1)(x - \alpha_2)^{n-1}. \tag{4.6}$$

The proof is completed by substituting the largest root of the second derivative of (4.6) into the right side of (4.5). ■

When the hypothesis of Theorem 4.1 is not satisfied, we shall show that $\alpha_2 - \tilde{\alpha}_2$ is Lipschitz continuous in $\alpha_1 - \tilde{\alpha}_1$ with a reasonably small Lipschitz constant. As a preparation for this result, we derive an a priori estimate which bounds $\tilde{\alpha}_2$ away from α_1.

Theorem 4.2. *If the largest zero γ_1 of f'' satisfies*

$$\gamma_1 \leqq (\alpha_1 + \alpha_2)/2 \tag{4.7}$$

and if

$$0 < \alpha_1 - \tilde{\alpha}_1 < (\alpha_1 - \alpha_2)/4e^2 \tag{4.8}$$

then the largest zero of the divided difference $f_1(\tilde{\alpha}_1, x)$ is in the interval $(\alpha_2, (\alpha_1 + \alpha_2)/2)$.

Proof. The inequality (4.7) implies that f and f'' have opposite signs at the point $(\alpha_1 + \alpha_2)/2$. As a consequence of Theorem 2.1 we may replace γ_1 in (4.5) by $(\alpha_1 + \alpha_2)/2$ to obtain

$$\frac{f(\tilde{\alpha}_1)}{f(\gamma_1)} \leqq 4e^2 \frac{(\alpha_1 - \tilde{\alpha}_1)}{(\alpha_1 - \alpha_2)}. \tag{4.9}$$

The proof follows immediately from (4.8) and (4.9). ■

We now have the tools necessary to obtain a functional inequality between the relative errors

$$X = \frac{(\tilde{\alpha}_2 - \alpha_2)}{(\alpha_1 - \alpha_2)} \quad \text{and} \quad Y = \frac{(\alpha_1 - \tilde{\alpha}_1)}{(\alpha_1 - \alpha_2)} \tag{4.10}$$

where $\tilde{\alpha}_1$ is an approximation to α_1 and $\tilde{\alpha}_2$ is the largest root of $f_1(\tilde{\alpha}_1, x)$.

Theorem 4.3. *If the largest zero γ_1 of f'' satisfies $\gamma_1 \leqq \alpha_2 + \rho(\alpha_1 - \alpha_2)$ for some ρ, $0 \leqq \rho < \frac{1}{3}$, and if $0 \leqq \alpha_1 - \tilde{\alpha}_1 < (\alpha_1 - \alpha_2)/4e^2$, then*

$$X(1 - X) \leqq Y(1 - Y)\exp(1/(1 - 3\rho)). \tag{4.11}$$

Proof. By Theorem 4.2 we have the inequalities $\alpha_1 > \tilde{\alpha}_1 > \tilde{\alpha}_2 > \alpha_2$ so that X and Y are positive. Since $f_1(\tilde{\alpha}_1, \tilde{\alpha}_2) = f(\alpha_1, \alpha_2) = 0$,

$$[f_1(\tilde{\alpha}_1, \tilde{\alpha}_2) - f_1(\alpha_1, \tilde{\alpha}_2)] + [f(\alpha_1, \tilde{\alpha}_2) - f(\alpha_1, \alpha_2)] = 0. \tag{4.12}$$

By using standard identities for divided differences (see Norlund [3]), one obtains from (4.12)

$$X(1 - X) = Y(1 - Y)\prod_{j=3}^{n} \frac{(\tilde{\alpha}_1 - \alpha_j)}{(\tilde{\alpha}_2 - \alpha_j)}. \tag{4.13}$$

An application of the inequality $1 + x \leqq e^x$ yields

$$\prod_{j=3}^{n} \frac{(\tilde{\alpha}_1 - \alpha_j)}{(\tilde{\alpha}_2 - \alpha_j)} \leqq \exp\left\{\sum_{j=3}^{n} \frac{(\tilde{\alpha}_1 - \tilde{\alpha}_2)}{(\tilde{\alpha}_2 - \alpha_j)}\right\}. \tag{4.14}$$

Since $0 \leqq \tilde{\alpha}_1 - \tilde{\alpha}_2 < \alpha_1 - \alpha_2$ and $\tilde{\alpha}_2 - \alpha_j > \alpha_2 - \alpha_j$, we may deduce from Theorem 3.4 with $y = \alpha_2 + \rho(\alpha_1 - \alpha_2)$ and $x = \alpha_2$ that

$$\sum_{j=3}^{n} \frac{(\tilde{\alpha}_1 - \tilde{\alpha}_2)}{(\tilde{\alpha}_2 - \alpha_j)} \leqq \frac{(y - \alpha_3)(\alpha_1 - \alpha_2)}{(\alpha_2 - \alpha_3)(\alpha_1 + \alpha_2 - 2y)}. \tag{4.15}$$

After substituting the definition of y into the right side of (4.15) one obtains an expression which is decreasing in $\alpha_2 - \alpha_3$ so the proof of Theorem 4.3 follows from (3.16). ∎

By combining the previous two theorems we are now able to obtain a bound for the shift in α_2 which results from an error in α_1.

Theorem 4.4. *If $0 < (\alpha_1 - \tilde{\alpha}_1) < (\alpha_1 - \alpha_2)/4e^2$ and the largest zero γ_1 of f'' satisfies*

$$\gamma_1 \leqq \alpha_2 + \rho(\alpha_1 - \alpha_2), \qquad 0 \leqq \rho < \tfrac{1}{6}, \tag{4.16}$$

then the largest zero of the divided difference $f_1(\tilde{\alpha}_1, x)$ satisfies

$$0 \leqq (\tilde{\alpha}_2 - \alpha_2) < 2\exp[1/(1 - 3\rho)](\alpha_1 - \tilde{\alpha}_1). \tag{4.17}$$

Proof. The hypothesis implies that the quadratic equation

$$t(1 - t) - Y(1 - Y)\exp[1/(1 - 3\rho)] = 0 \tag{4.18}$$

has two real roots. Hence X is either less than the smallest root or greater than the largest root of (4.18). The a priori estimate of Theorem 4.2 rules out the latter possibility. The proof is completed by direct estimation of the smallest root of (4.18). ∎

We next obtain a precise condition under which the Newton operator

$$T(x) = x - f(x)/f'(x) \tag{4.19}$$

is a contraction operator.

Theorem 4.5 *If the largest root* γ_1 *of* f'' *satisfies*

$$\gamma_1 \leqq \alpha_2 + \rho(\alpha_1 - \alpha_2), \qquad 0 \leqq \rho \leqq \tfrac{1}{7}, \tag{4.20}$$

then for $|x - \alpha_2| \leqq \rho(\alpha_1 - \alpha_2)$ *we have the estimates*

$$(x - \alpha_2)\frac{f'(x)}{f(x)} \geqq \frac{1 + 2\rho}{1 + \rho} - \frac{\rho}{1 - 4\rho}, \tag{4.21}$$

$$\left|\frac{(x - \alpha_2)f''(x)}{f(x)}\right| \leqq \frac{7}{2(\alpha_1 - \alpha_2)}, \tag{4.22}$$

and that hence

$$|T'(x)| < 6|(x - \alpha_2)|/(\alpha_1 - \alpha_2). \tag{4.23}$$

Proof. It follows from Theorem 3.4 that for $|x - \alpha_2| \leqq \rho(\alpha_1 - \alpha_2)$

$$\sum_{j=3}^{n} \frac{1}{(x - \alpha_j)} \leqq \frac{(\alpha_2 - \alpha_3) + \rho(\alpha_1 - \alpha_2)}{[(\alpha_2 - \alpha_3) - \rho(\alpha_1 - \alpha_2)](1 - 2\rho)}. \tag{4.24}$$

Because the right side of (4.24) increases as $(\alpha_1 - \alpha_3)$ decreases, it is a consequence of (3.16) that

$$\sum_{j=3}^{n} \frac{1}{(x - \alpha_j)} \leqq \frac{1}{(1 - 4\rho)(\alpha_1 - \alpha_2)}. \tag{4.25}$$

The proof of (4.21) now follows from (4.25) and the identity

$$(x - \alpha_2)\frac{f'(x)}{f(x)} = \frac{(\alpha_1 + \alpha_2 - 2x)}{(\alpha_1 - x)} + \sum_{j=3}^{n} \frac{(x - \alpha_2)}{(x - \alpha_j)}. \tag{4.26}$$

In order to prove (4.22), we use the identity

$$\frac{(x - \alpha_2)f''(x)}{f(x)} = \frac{-2}{(\alpha_1 - x)} + \frac{2(\alpha_1 + \alpha_2 - 2x)}{(\alpha_1 - x)} \sum_{j=3}^{n} \frac{1}{(x - \alpha_j)} + (x - \alpha_2)\left\{\left(\sum_{j=3}^{n} \frac{1}{(x - \alpha_j)}\right)^2 - \sum_{j=3}^{n} \frac{1}{(x - \alpha_j)^2}\right\}. \tag{4.27}$$

By virtue of (3.6), the term within the braces is positive. Hence, as a consequence of (4.25), when $x \leqq \alpha_2$ the left side of (4.27) has the upper bound

$$\frac{-2}{(\alpha_1 - x)} + \frac{2(\alpha_1 + \alpha_2 - 2x)}{(\alpha_1 - x)(1 - 4\rho)(\alpha_1 - \alpha_2)} \tag{4.28}$$

and the lower bound

$$\frac{-2}{(\alpha_1 - x)} - \frac{\rho}{(1 - 4\rho)^2(\alpha_1 - \alpha_2)}. \tag{4.29}$$

When $x \geqq \alpha_2$ the left side of (4.27) has the upper bound

$$\frac{-2}{(\alpha_1 - x)} + \frac{2(\alpha_1 + \alpha_2 - 2x)}{(\alpha_1 - x)(1 - 4\rho)(\alpha_1 - \alpha_2)} + \frac{\rho}{(1 - 4\rho)^2(\alpha_1 - \alpha_2)} \tag{4.30}$$

and the lower bound

$$-2/(\alpha_1 - x). \tag{4.31}$$

In any of the cases (4.28)–(4.30) we have the bound (4.22). The proof is completed by computing $T'(x)$ and applying (4.21) and (4.22) to obtain (4.23). ■

5. THE REGION OF MONOTONE CONVERGENCE OF NEWTON'S METHOD FOR THE $(k + 1)$st ROOT

In order to generalize Theorem 4.1 it is necessary to exploit more fully the constraint implied by the vanishing of the second derivative. The method is that of Lagrange multipliers which has been used in a rich variety of applied problems in [1].

Theorem 5.1. *Let K_n be defined as in the statement of Theorem 4.1. If $\alpha_k > \tilde{\alpha}_k > \alpha_k - (\alpha_k - \alpha_{k+1})/K_n$ and the kth largest root γ_k of f'' satisfies $\gamma_k \geqq \alpha_{k+1} + (\alpha_k - \alpha_{k+1})/K_n$, then the divided difference $f_1(\tilde{\alpha}_k, x)$ has a root $\tilde{\alpha}_{k+1}$ in the interval (α_{k+1}, α_k).*

Proof. Let us write

$$\frac{f(\tilde{\alpha}_k)}{f(\gamma_k)} = \frac{(\alpha_k - \tilde{\alpha}_k)(\tilde{\alpha}_k - \alpha_{k+1})}{(\alpha_k - \gamma_k)(\gamma_k - \alpha_{k+1})} \prod_{j \neq k, k+1} \frac{(\tilde{\alpha}_k - \alpha_j)}{(\gamma_k - \alpha_j)}. \tag{5.1}$$

The proof will follow from the facts that the product in (5.1) is bounded from above by e^2 and γ_k is bounded from above by the largest root of the second derivative of $(x - \alpha_k)(x - \alpha_{k+1})^{n-k}$. In order to prove that the product is bounded by e^2, we define

$$\varphi(x) = \prod_{j \neq k, k+1} (1 + x_j), \qquad x_j = \frac{(\tilde{\alpha}_k - \gamma_k)}{(\gamma_k - \alpha_j)}. \tag{5.2}$$

The constraint that γ_k is the kth largest root of f'' is then

$$2 \sum_{i<j} x_i x_j = \left(\sum_{j=1}^{n} x_j \right)^2 - \sum_{j=1}^{n} x_j^2 = 0. \tag{5.3}$$

From the definition (5.2) of the x_j's and the ordering (1.1) of the roots α_j, we obtain the additional constraint

$$-1 \leqq x_k \leqq \cdots \leqq x_1 < 0 \qquad \text{and} \qquad x_{k+1} \geqq \cdots \geqq x_n > 0. \tag{5.4}$$

Our problem is now to show that the maximum of (5.2) subject to the constraints (5.3) and (5.4) is less than e^2. Since the parameters x_k and x_{k+1} do not appear explicitly in the expression for φ we regard them to be fixed. Hence, if in (5.4) we allow the possibilities $x_1 = 0$ or $x_n = 0$, the extremum problem achieves a maximum. In view of the definition of the x_j's the cases $x_1 = 0$ and $x_n = 0$ correspond to polynomials of degree less than n. Theorem 3.1 gives the bound e^2 for third degree polynomials. Hence, by induction we may assume that the maximum is achieved at a point where $x_1 \neq 0$ and $x_n \neq 0$.

In order to ensure that we have an interior maximum, we rule out the possibility of the first k variables being equal. If $x_1 = \cdots = x_k$, then after applying the inequality $1 + x_j \leqq \exp(x_j)$, we have

$$\varphi(x) \leqq (1 + x_k)^{k-1} \exp\left(\sum_{j=k+1}^{n} x_j\right). \tag{5.5}$$

An application of Theorem 3.1 then gives

$$\varphi(x) \leqq (1 + x_k)^{k-1} \exp(-2kx_k). \tag{5.6}$$

The hypothesis implies that $(1 + x_k)\exp(-2x_k) < 1$. It follows that if φ achieves a value greater than e^2, then $x_k < x_1$.

In order that the boundary not be unnecessarily complicated, we now enlarge the domain to include $-1 \leqq x_j \leqq x_k$, $j \leqq k$, and $0 \leqq x_j \leqq x_{k+1}$, $j \geqq k + 1$. If the coefficient of x_n in the first form of the constraint (5.7) is zero, then the second form shows that all of the variables are zero. Hence we may regard x_n as the dependent variable. The method of Lagrange multipliers now gives

$$\frac{\varphi(x)}{(1 + x_i)} - 2\lambda\left[\sum_{j=1}^{n} x_j - x_i\right] = 0, \tag{5.7}$$

valid for each i such that $x_i \neq x_k$ or x_{k+1} and for the dependent variable x_n. After solving (5.7) for $\varphi(x)$, and subtracting the expressions corresponding to $i = 1$ and $i = n$, we obtain

$$(x_1 - x_n)\left[\sum_{j=2}^{n-1} x_j - 1\right] = 0. \tag{5.8}$$

The fact that x_1 and x_n have opposite signs them implies that

$$\sum_{j=2}^{n-1} x_j = 1. \tag{5.9}$$

By using the inequality $(1 + x_j) \leqq \exp(x_j), j \geqq k + 1$ in (5.2) and then applying (5.9), one arrives at

$$\varphi(x) \leqq \left\{\prod_{j=1}^{k-1}(1 + x_j)e^{-x_j}\right\}\exp[x_n - x_{k+1} - x_k + x_1 + 1]. \quad (5.10)$$

The proof now follows immediately from (5.4). ∎

6. CONTRACTION CONVERGENCE OF NEWTON'S METHOD TO THE $(k + 1)$st LARGEST ROOT

By analogy with the alternative of Section 4 for the second largest root, if the kth largest root γ_k of f'' does not satisfy the condition of Theorem 5.1, then in a reasonably large interval about α_{k+1} there is contraction convergence of Newton's method to α_{k+1}.

Theorem 6.1. *If the kth largest root γ_k satisfies the inequality $\gamma_k \leqq \alpha_{k+1} \leqq \rho(\alpha_k - \alpha_{k+1})$, the Newton operator is a contraction operator.*

Proof. It follows from Theorems 3.3 and 3.4 that

$$(x - \alpha_{k+1})f'(x)/f(x) \geqq 1 - \rho/[1 - 3k + 1)\rho]. \quad (6.1)$$

As a consequence of the inequality

$$\sum_{j \neq k+1} \frac{1}{(x - \alpha_j)^2} \leqq \frac{1}{(\alpha_k - x)} \sum_{j=1}^{k} \frac{1}{(\alpha_j - x)} + \frac{1}{(x - \alpha_{k+1})} \sum_{j=k+2}^{n} \frac{1}{(x - \alpha_j)}, \quad (6.2)$$

the identity

$$\frac{(x - \alpha_{k+1})f''(x)}{f(x)} = 2 \sum_{j \neq k+1} \frac{1}{(x - \alpha_j)} + (x - \alpha_{k+1})\left[\left(\sum_{j \neq k+1} \frac{1}{(x - \alpha_j)}\right)^2 - \sum_{j \neq k+1} \frac{1}{(x - \alpha_j)^2}\right], \quad (6.3)$$

together with Theorems 3.3 and 3.4, it follows that

$$\left|\frac{(x - \alpha_{k+1})f''(x)}{f(x)}\right| \leqq \left[\frac{2k}{(1 - (3k + 1)\rho)} + \frac{2\rho k^2}{(1 - (3k + 1)\rho)^2}\right]\frac{1}{(\alpha_k - \alpha_{k+1})}. \quad (6.4)$$

By virtue of (6.1) and (6.4), we now have the inequality

$$\left|\frac{f(x)f''(x)}{[f'(x)]^2}\right| \leqq \frac{2\rho k}{(1 - 5k\rho)^2}. \quad (6.5)$$

The right side of (6.5) is easily shown to be bounded from above by 0.8 when $\rho \leqq \frac{1}{10}k$. This completes the proof. ∎

REFERENCES

1. R. Duffin, E. Peterson, and C. Zener, "Geometric Programming," Wiley, New York, 1967.
2. W. W. Hager and R. N. Pederson, "Newton's Method for Eigenvalues" (to appear).
3. N. E. Norlund, "Differenzenrechnung," Springer-Verlag, Berlin, 1924.

This work is supported by NSF Grant MCS77-09237.

AMS (MOS) 1970 Subject Classifications: 65F15 and 65H05

DEPARTMENT OF MATHEMATICS
CARNEGIE-MELLON UNIVERSITY
PITTSBURGH, PENNSYLVANIA